10th International Conference Numerical Analysis in Engineering (NAE 2017)

IOP Conference Series: Materials Science and Engineering
Volume 308

Banda Aceh, Indonesia
24 – 25 August 2017

ISBN: 978-1-5108-5889-3
ISSN: 1757-8981

Printed from e-media with permission by:

Curran Associates, Inc.
57 Morehouse Lane
Red Hook, NY 12571

proceedings.com

Some format issues inherent in the e-media version may also appear in this print version.

Printed by Curran Associates, Inc. (2018)

For permission requests, please contact the Institute of Physics at the address below.

Institute of Physics
Dirac House, Temple Back
Bristol BS1 6BE UK

Phone: 44 1 17 929 7481
Fax: 44 1 17 920 0979

techtracking@iop.org

Additional copies of this publication are available from:

Curran Associates, Inc.
57 Morehouse Lane
Red Hook, NY 12571 USA
Phone: 845-758-0400
Fax: 845-758-2633
Email: curran@proceedings.com
Web: www.proceedings.com

Table of contents

Volume 308

10th International Conference Numerical Analysis in Engineering
24–25 August 2017, Banda Aceh, Indonesia

Accepted papers received: 24 January 2018
Published online: 19 February 2018

Preface

Papers

IOP Conf. Series: Materials Science and Engineering **308** (2018) 011001 doi:10.1088/1757-899X/308/1/011001

FOREWORD FROM CHAIRMAN

On The 10th International Conference on Numerical Analysis in Engineering (10th NAE)
BANDA ACEH, August 24-25, 2017

Good morning and best wishes for all of us.
Dear Ladies and Gentlemen,
First of all, let us praise and gratitude to the Almighty God for His blessings and permission of His grace so we can be present and come together during this blessed day on the 10th International Conference on Numerical Analysis in Engineering (NAE) 2017. It is our pleasure to welcome all of us to Banda Aceh and to this high-quality conference. I am honored to inform you that organizer and IOP has signed an Agreement to publish the proceeding of this conference in Material Science and Engineering; Conference Series. The proceeding will be indexed in several reputable databases such as Scopus, Thompson, Compendex, etc.
FYI, after call for paper was announced several months ago, almost 100 papers have been submitted to organizer. After the review process, only about 53 papers are accepted.

Ladies and Gentlemen,
I would to inform that our authors and or participants today come from 5 different countries. They are from Malaysia, Japan, Taiwan, Uganda, and Indonesia. This event is expected to become an excellent platform for knowledge exchange among academician, researchers, scientists, engineers, students and all related stakeholders. By gathering here, we expect the conference will offer the participants to network, collaborate, as well as to share research finding and experience. So please build your network.
On behalf of organizer we would like to thank our keynote speakers today. Professor Akiyuki Takahashi from Tokyo University of Science and Prof Kamal Arifin from University Kebangsaan Malaysia. Your contributions to this international conference are highly appreciated. Our thank you also goes to our international board and steering committee. I would also express special thanks to all of the founders; Prof Hiromi Homma, Prof Masanori Kikuchi, Prof Kokuo Kishimoto, and Prof Yasuhiro Kanto.
Finally, I would like to express my gratitude to all distinguished speakers and all respected participants for contributing in this conference. Countless appreciation to the organizing commitee for the hard work in realizing this conference.

I earnestly hope that this conference and the time you spend here will provide you with a fruitful and rewarding experience.

Thank you for all of you.

Dr. Eng Himsar Ambarita
10th NAE Chair

IOP Conf. Series: Materials Science and Engineering **308** (2018) 011001 doi:10.1088/1757-899X/308/1/011001

Conference Activities

IOP Conf. Series: Materials Science and Engineering **308** (2018) 011001 doi:10.1088/1757-899X/308/1/011001

IOP Conf. Series: Materials Science and Engineering **308** (2018) 011001 doi:10.1088/1757-899X/308/1/011001

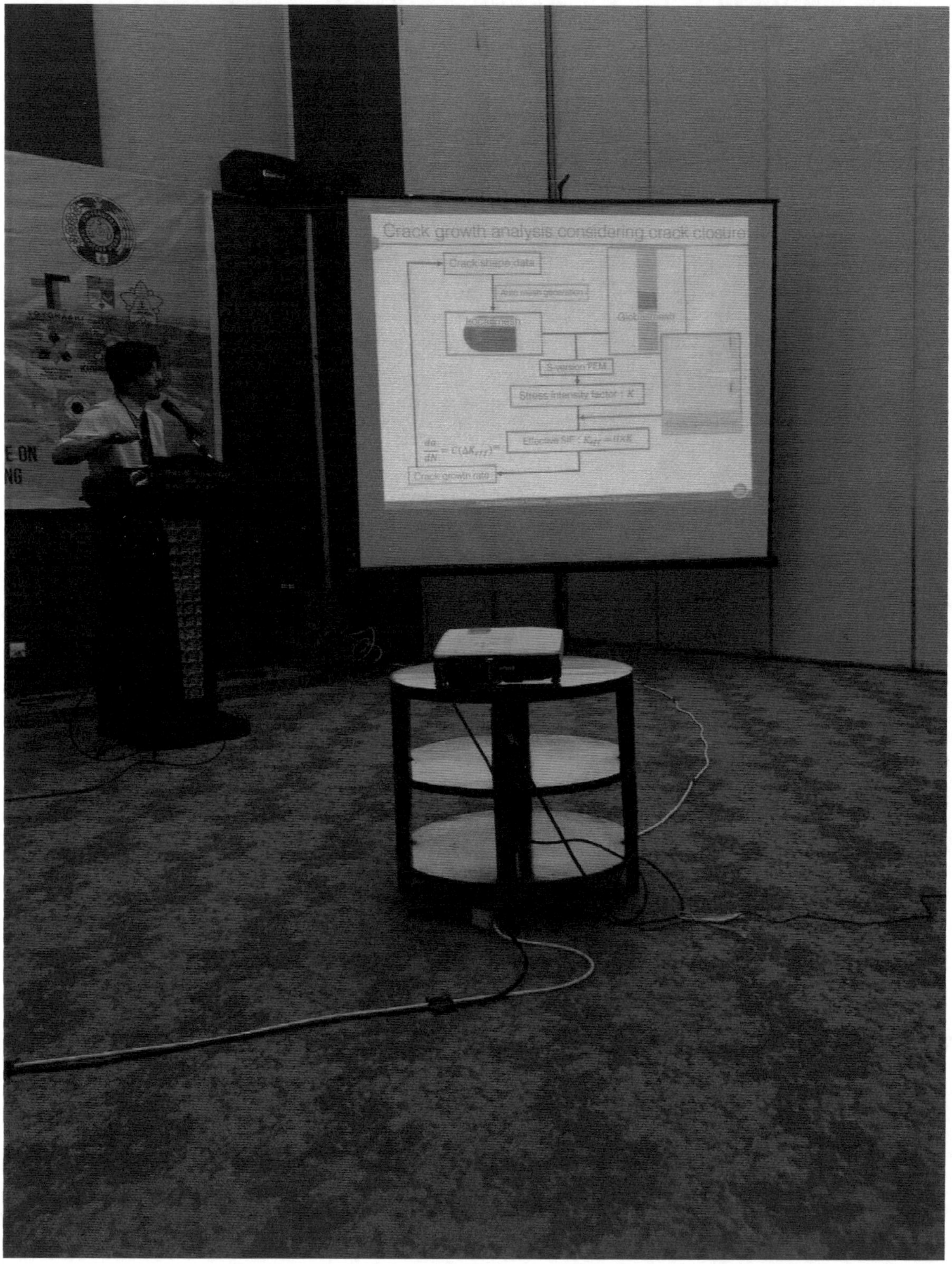

IOP Conf. Series: Materials Science and Engineering **308** (2018) 011002 doi:10.1088/1757-899X/308/1/011002

CONFERENCE INFORMATION

Numerical analysis utilizing the advantage of the development of computer hardware and software has had an immerse impact in many fields of engineering in recent years as it can provide not only fast but also comprehensive solutions in engineering problems.

The 10th International Conference on Numerical Analysis in Engineering (NAE) 2017 offers a place and opportunities for researchers and engineers from academic, industries and other sectors to exchange their scientific and technological information. NAE 2017 is also provided for students to present their research papers.

The conferece is organized by University of Sumatera Utara, in cooperation with Toyohashi University of Technology (TUT), Toyohashi – Japan; JSME – ICTS (International Chapter for Indonesian Section); Universiti Kebangsaan Malaysia (UKM), Bangi, Kuala Lumpur – Malaysia; Badan Kejuruan Mesin - Persatuan Insinyur Indonesia (BKM-PII); Universitas Syiah Kuala (UNSYIAH); IKATM USU (Mechanical Engineering Alumni Association of USU), and other academic institutions in Japan.

Background: Materials, Science, and Engineering

The conference covers the following topics approached by the numerical or the experimental method (but not limited to):

- Stress, strain and deformation analysis (linear, nonlinear, viscoelastic, viscoplastic, foams, seismic,etc)
- Fracture analysis (fracture mechanics, fatigue, creep, etc)
- Damage analysis
- Engineering dynamics (vibration, rotation of shaft, noise, multibody dynamics, etc)
- Mechanical elements (gear, screw, welding, joint, bearing, etc)
- Heat and mass transfer (heat transfer, combustion, refrigeration, chemical reaction, diffusion, etc)
- Fluid dynamics (laminar flow, turbulent flow, fluid machines, drain, dam, irrigation, etc)
- Material forming (metal, plastics, composite, etc)
- Tribology (friction, lubrication, etc)
- New algorithms for numerical analysis (Neural network, Genetic algorithm, Evolutionary algorithm, etc)
- Computational method in engineering and science
- FEM Application in Geotechnical and Structural Engineering
- Artificial Intelligence Application in Engineering, such as Expert System, Pattern Recognition, Neural Network Genetic Algorithm, etc
- Biomaterial and Biomechanics
- Biomedical Engineering
- Nano science and technology

IOP Conf. Series: Materials Science and Engineering **308** (2018) 011002 doi:10.1088/1757-899X/308/1/011002

Keynote Speakers:

1. Prof. Yoshitake WADA (Kindai University)
2. Prof. Ahmad Kamal Ariffin (Universiti Kebangsaan Malaysia)

Honorary Board:

1. Prof. H. Mohamad Nasir, Ph.D., Ak (Ministry of Research, Technology, and Higher Education Republic of Indonesia)
2. Dr. Muhammad Dimyati (Directorate General Empowerment on Research and Development, of Ministry of Reseach, Technology, and Higher Education Republic of Indonesia)
3. Prof. Dr. Runtung, SH, M.Hum (Rector, University of Sumatera Utara (USU)
4. Prof. Dr. Eng Samsul Rizal (Rector, UNSYIAH)
5. Prof. Dr. Satrio Soemantri Brojonegoro (Visiting Professor of Toyohashi University of Technology
6. Prof. Emeritus Masanori Kikuchi (Tokyo University of Science)

International Board:

1. Prof. Emeritus Hiromi Homma (Toyohashi University of Technology)
2. Prof. Emeritus Masanori Kikuchi (Tokyo University of Science)
3. Prof. Masashi Daimaruya (Muroran Institute of Technology)
4. Prof. Kikuo Kishimoto (Tokyo Institute of Technology)
5. Prof. Yasuhiro Kanto (Ibaraki University)

Scientific Committee:

1. Prof. Michihisa Koyama (Kyushu University)
2. Prof. Emeritus Masanori Kikuchi (Tokyo University of Technology)
3. Prof. Hideki Kawai (Muroran Institute of Technology)
4. Prof. Bustami Syam (University of Sumatera Utara)
5. Prof. Ahmad Kamal Ariffin (Universiti Kebangsaan Malaysia)
6. Prof. Yasunori Kikuchi (The University of Tokyo)
7. Prof Dr. T.M Indra Mahlia (Universiti Tenaga Nasional, Malaysia)
8. Assoc. Prof. Ing. Petr Valášek, Ph.D (Czech University of Life Sciences Prague)
9. Assoc. Prof. Dr. Gürkan A. K. GÜRDİL (Ondokuz Mayis University, Turkey)
10. Hasan Ozcan, Ph.D (Karabuk University, Turkey)
11. Dr. Edward Halawa (Charles Darwin University, Australia)

IOP Conf. Series: Materials Science and Engineering **308** (2018) 011002 doi:10.1088/1757-899X/308/1/011002

Organizing Committee:

International Chair : Prof. Bustami Syam
Prof. Masanori Kikuchi
Prof. Ahmad Kamal Ariffin

Local Organizing Committee

Chairman	: Dr. Eng. Himsar Ambarita
Co-chair	: Dr. M. Sabri (Regular Session) Dr. Bode (Student Session) Dr. Ir. Mirza Irwansyah, MBA, MLA
Secretary	: Dr. Eng. Taufik Bin Nur
Member	: Dr. Adi Setiawan Dr. Emerson P. Sinulingga Dr. Perwira M. Tarigan Dr. -Ing. Ikhwansyah Isranuri Dani Gunawan, ST., MT.
Treasury	: Melani
Secretariat	: Yetti Utami
Supporting Staff	: Firdus Wahyu Hardiansyah Anggia Murni Vera Jumiati

Contact Info:

Organizing Committee NAE 2017
Email: nae@usu.ac.id

IOP Conf. Series: Materials Science and Engineering **308** (2018) 011003 doi:10.1088/1757-899X/308/1/011003

Peer review statement

All papers published in this volume of *IOP Conference Series: Materials Science and Engineering* have been peer reviewed through processes administered by the proceedings Editors. Reviews were conducted by expert referees to the professional and scientific standards expected of a proceedings journal published by IOP Publishing.

IOP Conf. Series: Materials Science and Engineering **308** (2018) 012001 doi:10.1088/1757-899X/308/1/012001

Identifying strengths and weaknesses of Quality Management Unit University of Sumatera Utara software using SCAMPI C

D Gunawan[*1], A Amalia[2], R F Rahmat[1], M A Muchtar[1], I Siregar[3]
[1]Department of Information Technology, Universitas Sumatera Utara, Jl. dr. Mansur No. 9 Kampus USU, Medan 20155, Indonesia
[2]Department of Computer Science, Universitas Sumatera Utara, Jl. dr. Mansur No. 9 Kampus USU, Medan 20155, Indonesia
[3]Department of Industrial Engineering, Universitas Sumatera Utara, Jl. dr. Mansur No. 9 Kampus USU, Medan 20155, Indonesia

*E-mail: danigunawan@usu.ac.id

Abstract. Identification of software maturity level is a technique to determine the quality of the software. By identifying the software maturity level, the weaknesses of the software can be observed. As a result, the recommendations might be a reference for future software maintenance and development. This paper discusses the software Capability Level (CL) with case studies on Quality Management Unit (Unit Manajemen Mutu) University of Sumatera Utara (UMM-USU). This research utilized Standard CMMI Appraisal Method for Process Improvement class C (SCAMPI C) model with continuous representation. This model focuses on activities for developing quality products and services. The observation is done in three process areas, such as Project Planning (PP), Project Monitoring and Control (PMC), and Requirements Management (REQM). According to the measurement of software capability level for UMM-USU software, turns out that the capability level for the observed process area is in the range of CL1 and CL2. Planning Project (PP) is the only process area which reaches capability level 2, meanwhile, PMC and REQM are still in CL 1 or in performed level. This research reveals several weaknesses of existing UMM-USU software. Therefore, this study proposes several recommendations for UMM-USU to improve capability level for observed process areas.

1. Introduction

An institution called Quality Management Unit USU (UMM-USU) requires the development of new features to support business process in quality assurance. However, the existing software does not help software scalability because it was not built on top of the mature framework. Current software was developed based on problems that appeared all of a sudden. As a result, there are many independent small modules in UMM. This software development was not started with the plan to integrate among modules. Furthermore, the software lacks available documentation to explain the features of each module. Also, UMM-USU software will continue to evolve from time to time. Therefore, the stakeholders decide to rebuild UMM-USU software to replace the existing software and emphasize the project planning to accommodate the future development. This activity should be scrutinized, understood and fully supported [1].

To avoid the scalability issue, we propose the measurement of existing UMM-USU software capability level to identify its level. The capability software level identification aims to observe the drawbacks in existing software and recommend the improvement of software capability level. This level can be used as the base for the next software development. The measurement can be obtained by implementing a framework model called Capability Maturity Model Integration (CMMI). It contains best practices in development activity and maintenance which cover the whole lifecycle of the products from the beginning to the delivery and maintenance.

This paper is organized as follows. In section 2, we the related works of CMMI. In section 3, we describe the rationale for inclusion or exclusion of process area. Finally, we draw some conclusions and future works in section 4.

2. Related Works

Previous research has analyzed the impact of CMMI implementation on the quality issues on a case study methodology and investigates the real-life situation in a Russian engineering company [2]. It finds that application of CMMI yields significant improvements in engineering project performance regarding cost, time reductions, and quality management. Therefore, CMMI can be considered not only as a tool for process optimization but also as an instrument of quality improvement.

Another research addressed the problem by demystifying measurement concepts and terms in CMMI [3]. It also clarifying the natural evolution of measurement practices that should occur as organizations strive to improve their processes across all the CMM levels.

The implementation of CMMI also had been done in Deutsche Bank which is leading the way in the introduction of CMMI-SVC [4]. It shows the financial services sector can also utilize the model. The successful implementation was also followed by intensive work to implement, complete the identified priority and prepare for the final appraisal (SCAMPI A).

Another research about the implementation of CMMI-DEV had been conducted [5]. This study focuses on CMMI and the maintenance processes during the maintenance phase. The result shows that 11 of 22 process areas explicitly address software maintenance. Three of the 11 process areas such as requirements development, requirements management, and technical solution has more extended guidelines towards software maintenance.

3. Measuring UMM-USU Based on SCAMPI C

The measurement of existing software was done by using Standard CMMI Appraisal Method for Process Improvement class C (SCAMPI C). An appraisal is an activity that helps to identify the strengths and weaknesses of the organization's processes. This action includes three phases such as Plan and Prepare for Appraisal, Conduct Appraisal, and Report Result. This research only measures three process areas such as Project Planning (PP), Project Monitoring and Control (PMC) and Requirements Management (REQM). These process areas have been identified as the central issue of the UMM-USU software development.

3.1 Plan and Prepare Appraisal

Plan and prepare appraisal is a phase to determine the minimum requirements for conducting planning processes. This phase consists of the following processes: analyze appraisal, develop appraisals plan, select and prepare the team, prepare participants and obtain initial objective evidence, and prepare for collection of objective evidence. Capability measurement is done by observing requirement components which have been fulfilled. Requirement components are the CMMI components that are essential to achieving process improvement in a given process area. The required components in CMMI have specific and generic goals. As the basis for deciding process area satisfaction in appraisals, CMMI uses goal satisfaction [6]. Generic and specific goals of the given process areas are shown in table 1.

To achieve the improvement of the UMM-USU software, we observe expected components from the process areas. Expected components are the CMMI components that describe important activities to achieve a required CMMI component. Expected components is a guidance for those who are willing to implement improvements or perform appraisals. The expected components in CMMI are the specific and generic practices which are shown in table 2.

Table 1. Generic and Specific Goals of Each Process Area.

Generic Goal (GG)	Process Area	Specific Goal (SG)
1. Achieve Specific Goals 2. Institutionalize a Managed Process	PP	1. Estimates of project planning parameters are established and maintained. 2. A project plan is established and maintained as the basis for managing the project
3. Institutionalize a Defined Process 3.1 Establish a Defined Process,	PMC	1. Actual project performance and progress are monitored against the project plan. 2. Corrective actions are managed to closure when the project's performance or results deviate significantly from the plan.
3.2 Collect Process Related Experiences.	REQM	1. Requirements are managed, and inconsistencies with plans and work products are identified.

Table 2. Generic and Specific Practices of Each Process Area.

Generic Practice (GP)	Process Area	Specific Practice (SP)
1.1 Perform Specific Practices, 2.1 Establish an Organizational Policy, 2.2 Plan the Process, 2.3 Provide Resources, 2.4 Assign Responsibility, 2.5 Train People, 2.6 Control Work Product, 2.7 Identify and Involve Relevant Stakeholder, 2.8 Monitor and Control the Process, 2.9 Objective and Evaluate Adherence 2.10 Review Status with Higher Level Management.	PP	1.1 Establish a top-level work breakdown structure (WBS) to estimate the scope of the project. 1.2 Establish and maintain estimates of work product and task attributes. 1.3 Define project lifecycle phases on which to scope the planning effort. 1.4 Estimate the project's effort and cost for work products and tasks based on estimation rationale. 2.1 Establish and maintain the project's budget and schedule. 2.2 Identify and analyze project risks. 2.3 Plan for the management of project data. 2.4 Plan for resources to perform the project. 2.5 Plan for knowledge and skills needed to perform the project. 2.6 Plan the involvement of identified stakeholders. 2.7 Establish and maintain the overall project plan. 3.1 Review all plans that affect the project to understand project commitments. 3.2 Adjust the project plan to reconcile available and estimated resources. 3.3 Obtain commitment from relevant stakeholders responsible for performing and supporting plan execution.
3.1 Establish and maintain the description of a defined process. 3.2 Collect process related experiences derived from planning and performing the process to support the future use and improvement of the organization's processes and process assets.	PMC	1.1 Monitor actual values of project planning parameters against the project plan. 1.2 Monitor commitments against those identified in the project plan. 1.3 Monitor risks against those risks identified in the project plan. 1.4 Monitor the management of project data against the project plan. 1.5 Monitor stakeholder involvement against the project plan. 1.6 Periodically review the project's progress, performance, and issues. 1.7 Review the project's accomplishments and results at selected project milestones. 2.1 Collect and analyze issues and determine corrective actions to address them. 2.2 Take corrective action on identified issues. 2.3 Manage corrective actions to closure.
	REQM	1.1 Develop an understanding with the requirements providers on the meaning of the requirements. 1.2 Obtain commitment to requirements from project participants. 1.3 Manage changes to requirements as they evolve during the project. 1.4 Maintain bidirectional traceability among requirements & work products. 1.5 Ensure that project plans and work products remain aligned with the requirements.

3.2 Conduct Appraisal

To conduct appraisal phase, several steps are required to be taken, such as Examine Objective Evidence, Document Objective Evidence, Verify Objective Evidence, Validate Preliminary Appraisal Outputs and

Generate Appraisal Result. To analyze the current condition of existing UMM-USU software, we conducted the interview with IT manager who designed the UMM software. We also interviewed the head of UMM-USU to determine the satisfaction level of current UMM-USU software. The questions are designed according to specific practice from each process area. The head of UMM-USU granted us the privilege to access existing documents. The result of appraisal process area identification for each specific goals and practices are shown as quilt chart in figure 1.

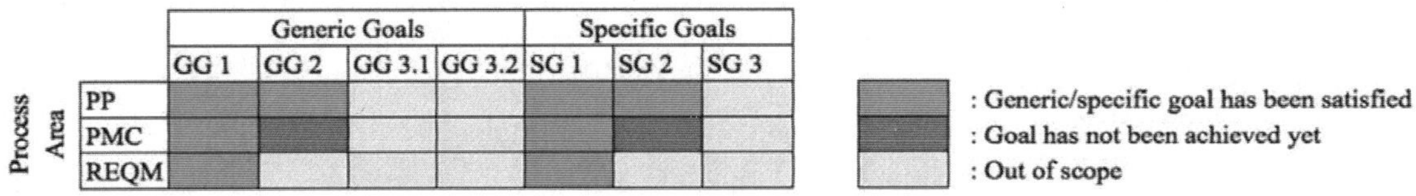

Figure 1. The result of appraisal process area identification.

According to guilt chart in figure 1, the capability level can be determined and illustrated in figure 2.

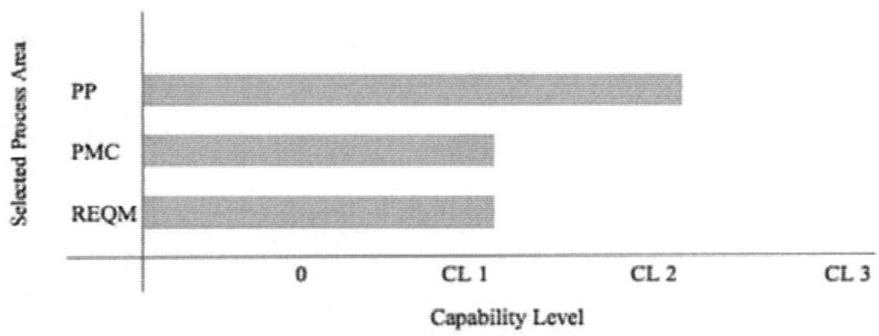

Figure 2. Guilt chart for the capability level.

We observed the specific practice to determine the required improvement. According to figure 3, improvement recommendation is given to the specific practice that has low or medium status.

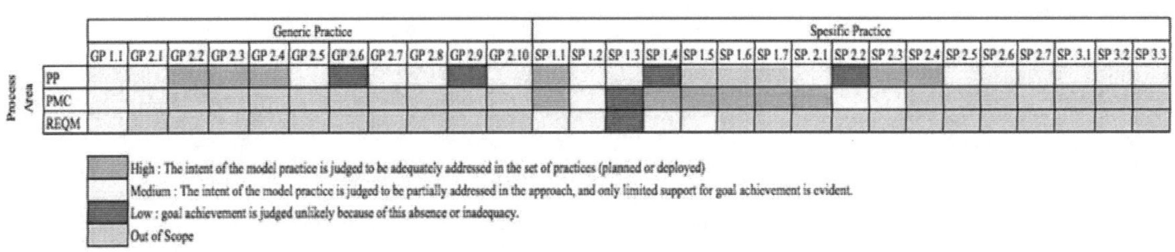

Figure 3. Improvement recommendation.

4. Weaknesses and Recommendation

4.1 Weaknesses and Recommendation for process area Project Planning

The purpose of Project Planning (PP) is to establish and maintain plan that define project activities. After the measurement had been done for PP process area, there are several SP which has high, medium and low goal achievement. For the specific practice PP, there are only three SP (SP 1.1, SP 2.3 and SP 2.4) which have been implemented consistently. There are two SP which are not implemented, they are SP 1.4 and SP 2.2. The following table 3 has weaknesses and recommendation for SP which has medium and low goal achievement.

Table 3. Weaknesses and recommendation for process area Project Planning.

No	Specific Practice	Weakness	Recommendation
1	1.2	Not all established and maintained attribute tasks like software functions are estimated	Estimate all the established and attribute tasks.
2	1.3	Only small scope of project lifecycle is defined and there is no documentation of estimated planning effort.	Estimate all planning effort project lifecycle
3	1.4	There is no estimation of the project's effort and cost for work products and tasks based on estimation rationale.	Estimate all of project's effort and cost.
4	2.1	Only some of the project's budget and schedule are established and maintained because there is no procedure and regulation in UMM-USU to calculate the cost estimation of project's budget.	Establish and maintain all of the project's budget and schedule.
5	2.2	There is no Identification and analyzability of project risk because the lack of human resources who capable to analyze the project risk.	Hire an expert or train a person in UMM-USU to understand and capable of analyzing the project risk.
6	2.5	Only some of attribute tasks are planned to perform the knowledge and skills needed.	Plan all knowledge and skill needed to perform attribute tasks
7	2.6	Only some attribute tasks are planned to involve of identified stakeholders.	Plan the involvement identified stakeholders for all tasks
8	2.7	Not all project plan is established and maintained.	Plan to establish and maintain overall project
9	3.1	Only some plans that affect the project are reviewed.	Review all plans that affect the project.
10	3.2	Only some project plans are adjusted to reconcile available and estimated resources.	Adjust project plans to reconcile available and estimated resources if needed.
11	3.3	Only some commitment from relevant stakeholders is obtained to responsible for performing and supporting plan execution.	Obtain commitment from relevant stakeholders responsible.

4.2 Weaknesses and recommendation for process area Project Monitoring and Control

The purpose of Project Monitoring and Control (PMC) is to provide an understanding of the project's progress. Appropriate correction actions can be taken when the project's performance deviates significantly from the plan. As shown in figure 3, there are four SP that have to be improved. Those are SP 1.2, SP 2.2 and SP 2.3 (medium status) and SP 1.3 (low status). On the other hand, some specific practices have adequately addressed, such as SP 1.1, SP 1.4, SP 1.5, SP 1.6, SP 1.7, and SP 2.1. This research proposes the recommendations for process area Project Monitoring and Control as shown in table 4.

Table 4. Weaknesses and Recommendation for process area Project Monitoring and Control.

No	Specific Practice	Weakness	Recommendation
1	1.2	Not all commitments against those identified in the project plan are monitored.	Monitor all commitments against those identified in the project plan.
2	1.3	There is monitoring risks against those risks identified in the project plan.	Monitor risks against those risks identified in the project plan.
3	2.2	Not all corrective action on identified issues are taken	Take corrective action on identified issues.
4	2.3	Not all corrective actions to closure are managed	Manage corrective actions to closure.

4.3 Weaknesses and recommendation for process area REQM

Requirement Management (REQM) is intended to help the requirements management of a project's products and product components, and also to identify inconsistencies among those requirements and the project's plans and work products. As shown in figure 3, there are five SP that have to be improved. Those are SP 1.1, SP 1.2, SP 1.4 and SP 1.4 (medium status) and SP 1.3 (low status). None of these SP has adequately addressed in the set of practices (planned or deployed). This research proposes the recommendations for process area Requirement Management as shown in table 5.

Table 5. Weaknesses and recommendation for process area Requirement Management

No	Specific Practice	Weakness	Recommendation
1	1.1	Too few discussions are set to develop an understanding with the requirements providers on the meaning of the requirements.	Develop an understanding with the requirements providers on the meaning of the requirements with more discussion
2	1.2	There is no document of the impact caused if the requirements are changed are during the project process. The commitment from project participants is not written in contract.	Obtain commitment to requirements from project participants.
3	1.3	The changes of requirements are not manageable	Manage changes to requirements as they evolve during the project.
4	1.4	Only some requirements of bidirectional traceability are maintained	Maintain bidirectional traceability among requirements and work products.
5	1.5	Only some of project plans and work products remain aligned with the requirements are ensured	Ensure that project plans and work products remain aligned with the requirements.

4.4 Report Result

The report result consists of Deliver Appraisal Report and Package and Archive Appraisal Package. All the weaknesses and recommendations that have been analyzed to improve the quality existing UMM-USU software. These weaknesses and recommendations have been given to UMM-USU management. The proposed recommendations could be a guideline to develop a new software with established project planning.

5. Conclusion

After measuring the capability level of existing UMM-USU software, the capability level for the observed process area is CL 1 and CL 2. Process area Planning Project (PP) has reached capability level 2 (CL 2). On the other hand, Project Monitoring and Control (PMC) and Requirement Management (REQM) are still in the level 1 (CL 1). These result leads to the proposed recommendations for the existing UMM-USU software. The proposed recommendations could be designed to be a guideline to develop the next UMM-USU software.

References

[1] Fuggetta A and Di Nitto E 2014 Software process *Proc. of the on Future of Software Engineering* pp 1–12
[2] Titov S, Bubnov G, Guseva M, Lyalin A, and Brikoshina I 2016 Capability maturity models in engineering companies: case study analysis *6th Seminar on Industrial Control Systems: Analysis, Modeling and Computation* **2** pp 4–7
[3] Weber C and Layman B 2002 Measurement Maturity and the CMM: How Practices Evolve as Processes *Softw. Qual. Prof.* **4** pp 6–20
[4] Reuber J, Brenner M, and Lentz S 2014 CMMI for Services Successful Maturity Level Assessment, using the Example of Deutsche Bank
[5] Jansson A-S 2007 Software maintenance and process improvement by CMMI.
[6] CMMI Product Team 2010 CMMI for Development Version 1.3.

IOP Conf. Series: Materials Science and Engineering **308** (2018) 012002 doi:10.1088/1757-899X/308/1/012002

Software requirements elicitation to support internal monitoring of quality assurance system for higher education in Indonesia

A Amalia*[1], D Gunawan[2], S M Hardi[1], D Rachmawati[1]

[1]Department of Computer Science, Universitas Sumatera Utara, Jl. dr. Mansur No. 9 Kampus USU Medan 20155

[2]Department of Information Technology, Universitas Sumatera Utara, Jl. dr. Mansur No. 9 Kampus USU Medan 20155

*E-mail: amalia@usu.ac.id

Abstract. The Internal Quality Assurance System (in Indonesian: SPMI (*Sistem Penjaminan Mutu Internal*) is a systemic activity of quality assurance of higher education in Indonesia. SPMI should be done by all higher education or universities in Indonesia based on the Regulation of the Minister of Research, Technology and Higher Education of the Republic of Indonesia Number 62 of 2016. Implementation of SPMI must refer to the principle of SPMI that is independent, standardize, accurate, well planned and sustainable, documented and systematic. To assist the SPMI cycle properly, universities need a supporting software to monitor all the activities of SPMI. But in reality, many universities are not optimal in building this SPMI monitoring system. One of the obstacles is the determination of system requirements in support of SPMI principles is difficult to achieve. In this paper, we observe the initial phase of the engineering requirements elicitation. Unlike other methods that collect system requirements from users and stakeholders, we find the system requirements of the SPMI principles from SPMI guideline book. The result of this paper can be used as a choice in determining SPMI software requirements. This paper can also be used by developers and users to understand the scenario of SPMI so that could overcome the problems of understanding between this two parties.

1. Introduction

The Government of Indonesia participates in ensuring the provision of higher education in Indonesia. Therefore, the Indonesian government has released regulations to support the Quality Assurance System for Higher Education (in Indonesian: SPM Dikti (*Sistem Penjaminan Mutu Pendidikan Tinggi)*). Higher Education is a highly qualified education that produces graduates who have a capability in develop their potential in science and/or technology that useful for society and nation [1]. SPM Dikti consists of Internal Quality Assurance System (In Indonesian: SPMI (*Sistem Penjaminan Mutu Internal)* and External Quality Assurance System (In Indonesian: SPME (*Sistem Penjaminan Mutu Eksternal).* SPMI is implemented internally by all universities in Indonesia and SPME is held by the government through higher education accreditation. SPMI is a substantial activity to achieve a well-guaranteed business process that yields qualified and standardized university products. The Government of Indonesia through the Ministry of Research, Technology and Higher Education has released a guideline for the implementation of SPMI which was recently revised in 2016. This instruction shall be followed by all providers of higher education in Indonesia in the mode most appropriate to history, vision, mission,

IOP Conf. Series: Materials Science and Engineering **308** (2018) 012002 doi:10.1088/1757-899X/308/1/012002

mandate, culture of the college organization concerned. These guidelines contain the stages that higher education providers must follow to implement the internal quality assurance system. SPMI is a systemic activity of quality assurance for higher education in Indonesia to control and improve the accomplishment of higher education in well planned and continuously events [1]. The foundation of SPMI is the Regulation of the Minister of Research, Technology and Higher Education of the Republic of Indonesia No. 62 of 2016 on the Quality Assurance System of Higher Education. SPMI through the activity cycle: Determination, Implementation, Evaluation, Control, Enhancement or can abbreviate as DIECE (in Indonesia: PPEPP) as described on figure 1.

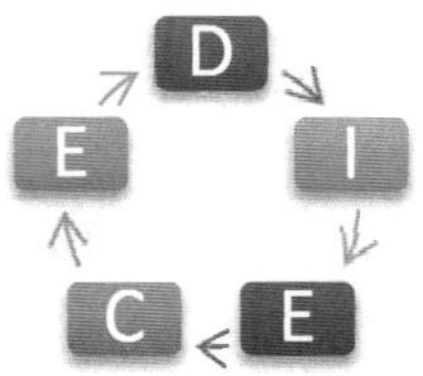

Figure 1. SPMI Cycle

(D)etermination : activities to establish the standard which consists of Higher Education National Standard (In Indonesia :SN Dikti) and Standar Dikti that are defined by the University.
(I) mplementation: activities to implement the SN Dikti and University Standard;
(E)valuation : activities to examine whether the SN Dikti already established by the university or not.
(C)ontrol : activities to analyze the root cause if some SN Dikti is not achieved
(E)nhancement : activities to improve SN Dikti to higher standard

In the implementation of the SPMI cycle, every university should apply the SPMI principle which consists of:

- Autonomous/independent, SPMI is developed and implemented independently by every college. This SPMI should be performed in all level on University like n department (program study), faculty and other units.
- Standardized, SPMI is applied based on SN Dikti established by the government and the university.
- Accurate, SPMI employ accurate data and information from a valid database.
- Well Planned and Sustainable, SPMI is implemented by using 5 (five) quality assurance steps: Determination, Implementation, Evaluation, Control, Enhancement or can abbreviate as DIECE
- Documented and systematic, all levels in the SPMI cycle are systematically recorded.

To assist the SPMI cycle properly, universities need software on monitor all the activities of SPMI. But in reality, many universities are not optimal in building this SPMI monitoring system. One of the obstacles is the determination of system requirements in support of SPMI principles is difficult to achieve. SPMI software should support the implementation of DIECE cycle independently, accurately, standardize, well planned, continuously, documented and systematically. Implementation of SPMI should also involve all stakeholders in the university. Therefore, it requires a process engineering requirements in collecting software requirements based on the principles of SPMI. The requirements engineering is the first phase of software engineering process, in which user requirements are gathered, understood, and specified [2]. The early stage of requirements engineering is requirements elicitation. Requirements elicitation is practice of collecting requirements of a system from user, customers, and stakeholders. This process may appear simple: ask the customer, the users, and others what the objectives of the system is, what is to be accomplished and how the system or software fits into the needs of business. But on SPMI software design it is not that simple. A large number of stakeholders involved in the SPMI activities made the access and authorized user's level to become very wide and complex. Another problem is the implementation of autonomous SPMI's action chooses SPMI application model to become unlimited. Some universities do not have full understanding of the problem domain to deploy SPMI, and it causes problems of perception where the customers/users are not entirely sure of what is needed. For that purpose, we offer a concept to collect SPMI software requirements based on SPMI's principle from SPMI guidelines.

The purpose of this paper is to provide a concept of requirements collection based on the business scenario of SPMI implementation process. This idea can be an option for universities in Indonesia to collect the requirements to build the SPMI software. This paper can also overcome the problems understanding between developers and users in conceive the SPMI Scenario model.

2. Related Works

Related works about quality control on higher education observation had been done by [3-8]. But most of these researchers focus on applying quality control practice in a college. Requirements engineering research has been done by previous researchers [2, 9-12]. The purpose of [2] is to analyze the problems of the requirements engineering from several viewpoints. According to [2] one of the issues of this process is users have incomplete understanding of their needs. The work done by [9] observed the roadmap of requirements engineering. A study by [10] identified and analyzed existing prioritization techniques in the context of the formulated research questions. This study observes various software engineering literature sources that discuss prioritization in the determination of software requirements. Other research is to consider about systematic review of requirement elicitations techniques done by [6][11]. Another study done by [6] concluded that identified requirements must be clear, consistent, modifiable and traceable to produce a quality product. Research [12] describes a computational mechanism for deriving use cases from object system models, simple rules for use case composition, taxonomies of classes of exceptions which give rise to alternative courses in scenarios, and a computational mechanism for generation of multiple scenes from use case specifications. Based on these earlier studies it can be concluded, there are no single techniques which fulfill all the demand of requirement elicitation and information gathering. One of the most important goals of elicitation is to find out what problem needs to be solved and hence identify system boundaries, identification of stakeholder and user classes.

3. Research Methodology

The purpose of this paper is to provide a concept elicitation requirements based on the principle of SPMI. The stages of this research are SPMI Model Determination, SPMI Principles Observation, Scenario of Business Process Determination and Requirements Elicitation. The SPMI principle is a foundation in determining the requirements elicitation. The meaning of requirements itself is a condition or ability that must be met by the software to fulfil the specification standards based on user requests or based on a valid document like SPMI's guideline.

3.1 SPMI Model Determination

Based on the higher education assignment there are two models in implementation of SPMI. The first model is the college establish a Quality Assurance Unit (in Indonesian: UMM (*Unit Manajemen Mutu*). With this model, all the implementation of quality assurance at the university will be arranged by this unit. The second model is the implementation of SPMI integrated into the university's management. Universities can choose from either of these models or can also do a combination of these two models. In this study, we used the first model of SPMI implementation and involved three main components. These components are Quality Assurance Unit at department/program study/level unit (In Bahasa Indonesia: GJM / GKM), internal auditor and quality assurance unit at university level. GJM / GKM provides internal quality monitoring and controlling activities on their department. The GJM/GKM should do quality self-assessment on their unit. Internal auditors are staffs of the university that capable and certified to do auditing in department level. The report of the self-assessment from GJM/GKM and the report from internal auditor will be analyzed by UMM staff, and the result is given to the university leader.

3.2 SPMI Principles Observation

The implementation of the DIECE cycle (in Indonesian: PPEPP (*Perencanaan, Pelaksanaan, Evaluasi Pelaksanaan, Pengendalian, Pelaksanaan, Peningkatan*) for SPMI activities must be implemented by all universities in Indonesia. In the implementation of SPMI must refer to the principle of SPMI that is independent, accurate, national standardize, planned, sustainable, documented and systematic. Autonomous means in SPMI implementation, the universities are not justified in involving external parties. Therefore, all SPMI activities are carried out autonomously including selection of SPMI model and SPMI scenario selection. The implementation of SPMI should be standardized. It means that the university should establish the standards or SN Dikti based on the government regulation and university policy. The application of SPMI must also be accurate based on a valid source like university database. Implementation of SPMI also must be well planned and documented.

3.3 Scenario of Business Process Determination

Selection of SPMI implementation scenario is the independent right of the university. Each university can determine the model of the plot that best suits with the university policy. Each scenario should be able to cover the principle of SPMI. In this research, we offer a scenario of SPMI activity as in figure 2.

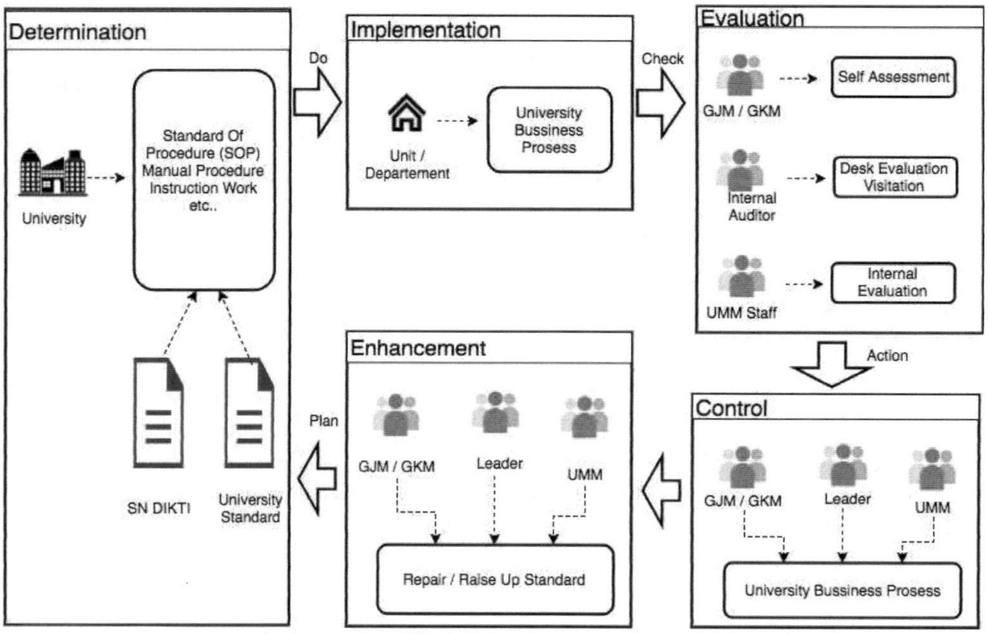

Figure 2. Scenario of SPMI Implementation at University

We offer a continuous DIECE cycle scenario with Deming's philosophy to ensure constant quality control. The scenario cycle processes are Plan, Do, Check and Act process. The *plan* stage is a stage of determining the SN Dikti. The SN Dikti can be described as a group of policies, regulations, standards, and rules that established based on higher education national standards and university's standard. All the level and university's component from university's level should *Do* these SN Dikti as university's business process guideline. In the *evaluation* stage, university's level like unit/department/program study/faculty has to examine and verify whether the SN Dikti that established in *plan* stage under the implementation. The examination is performed by university's unit level through their GJM/GKM, internal auditors, and UMM staffs. GJM/GKM assess the accordance with self-assessment examination. The internal auditors evaluate the compliance through auditing process to all university's level. Furthermore, UMM staffs will analyze and report the comparison result of self-assessment from GJM / GKM with auditing result from internal auditors. After that, based on this report, the university leader, GJM/GKM and UMM can analyze the existing SN Dikti to repair the standards or raise up the standard. These new enhancement standards become new SN Dikti to the next quality control cycle.

IOP Conf. Series: Materials Science and Engineering **308** (2018) 012002 doi:10.1088/1757-899X/308/1/012002

3.4 Requirements Elicitation

Requirements elicitation is the practice of collecting the system requirements. We collect the system requirements from SPMI scenario business process based on SPMI guideline. In this stage, we substitute the scenario business process into software functions. We observe the procedures that can be supported by the software and implement them into Unified Modelling Language which is Use Case and Sequence Diagram. Use case is a diagram to describe a function or procedure that can be performed by an actor system. There are four levels actor in this software, each of these actor levels has its authority in access some functions in the software. The level actors in this software are:

- U-QA is an actor level in charge of managing SPMI software. These actors could be the staffs of Quality Assurance Unit (UMM) in University level.
- D-QA is an actor level that responsible for the implementation of quality assurance at the faculty, department or unit level. There are two kind actors in this level, for faculty level or GJM staff and department/unit level or GKM staff
- The internal auditor is an actor level that capable of auditing the quality control process of department/unit in the university.
- University leader is an actor level for the leader of university like Rector, Dean, department/unit leader. The use case and sequence diagram can be seen in figure 2 and figure 3.

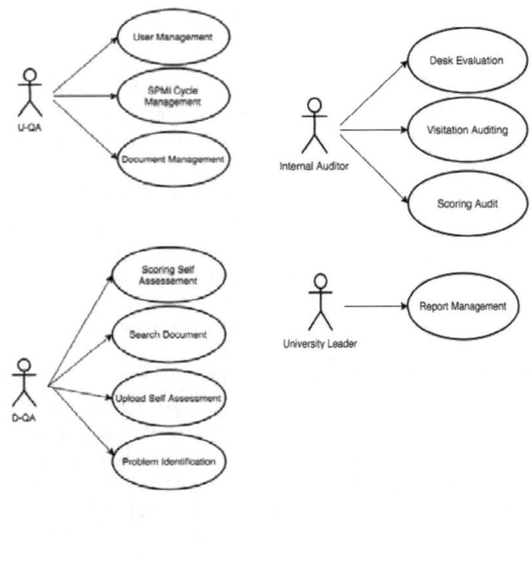

Figure 3. Use Case Diagram

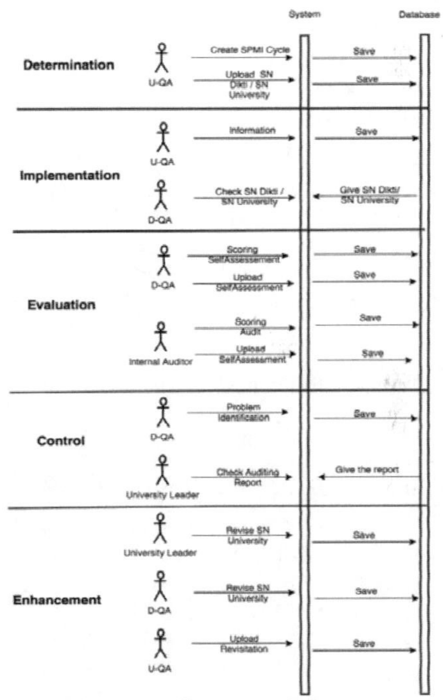

Figure 4. Sequence Diagram

Some functions can only be done by a certain actor level and certain phase only. Based on the scenario and sequence diagrams that have been described in the previous stage, there are functional requirements required for SPMI software design. These functions/procedures list can be seen in Table 1. All these functions are integrated into a whole software that can support SPMI activities.

IOP Conf. Series: Materials Science and Engineering **308** (2018) 012002 doi:10.1088/1757-899X/308/1/012002

Table 1. Functional Requirements

Scope Function	Information	Functional Requirements
User Management	A set of functions to manage the actors/users who involved in the SPMI system	▪ Function for user login settings ▪ Function to divide user permissions based on user level authority ▪ Enables/disables users in certain SPMI cycles.
SPMI Cycle Management	A set of functions to manage SPMI cycle	▪ Function to start the new cycle of SPMI. ▪ Function to view the old cycle documentation. ▪ Search cycle based on keywords.
Document Management	A set of functions to manage documents.	▪ Function to upload documents based on user authority ▪ Function to search for documents ▪ Function to create metadata automatically for documents uploaded by user.
Self Assessment / Scoring	A set of functions to do self-assessment	▪ Function to display a questionnaire assessment. ▪ Function to receive input assessment by the user assessment. ▪ Function to calculate the scoring of the assessment. ▪ Function to display the scoring in graphical form
Internal Auditing	A set of tasks for internal auditing	▪ Function to check unit/department scoring result (desk Evaluation) ▪ Function to input scoring value according to field visit. ▪ Function to calculate the scoring audit. ▪ Function to display comparison result of unit/ department scoring VS auditor result scoring. ▪ Function to send audit results to nit/department and Quality Assurance Unit.
Problem Identification	A set of functions for input problem identification	▪ Function for input identification problem if scoring value is low. ▪ Function to report if we want to revise SN Dikti or SN University.
Quality Assurance Report Management	A set of functions for report management	▪ Function to find report auditing a unit/department. ▪ Function to display reporting comparisons between multiple unit/department ▪ Function to recap the audit report.

4. Conclusion

The result of this paper is functional requirements in the design of software support SPMI. Stages of collecting elicitation requirements are done by observing the SPMI guidelines and principles which are then formulated in a choice of business process scenario. Users or clients can add specific requirements according to the needs of each college. This paper can also be a bridge for developers and users to understand the scenario of SPMI implementation better so that the problems of understanding between the two parties can be overcome.

References

[1] Kementrian Riset, Teknologi dan Pendidikan Tinggi [Ministry of Research, Technology and Higher Education] 2016 *Pedoman Sistem Penjaminan Mutu Pendidikan Tinggi [Guidance of Quality Assurance System*] (Direktorat Penjaminan Mutu)

[2] Wahono R S 2003 Analyzing Requirements Engineering Problems *Proc of IECI* pp 55-58

[3] Hou A Y C 2014 Quality in Cross-Border Higher Education and Challenges for the Internationalization of National Quality Assurance Agencies in the Asia-Pacific region: the Taiwanese experience *J Studies in Higher Education* 39 pp 135-152

[4] Lim D B 1999 Quality Assurance in Higher Education in Developing Countries *J Assessment & Evaluation in Higher Education* **24** pp 379-390

[5] Suprihatin K, Abdul H, Bin R, Yusof M 2015 Development Design Model of Academic Quality Assurance at Private Islamic University Jakarta Indonesia, *J Higher Education Study* **5** pp 6–29

[6] Pandey D, Suman U, Ramani A.K 2010 An Effective Requirement Engineering Process Model for Software Development and Requirements Management, *Proc. Recent Technologies in Communication and Computing* (*ARTCom*), *2010 Int. Conference* on IEEE pp 287-291

[7] Gamar MM, Arifin M, Ulfatin N 2016 Preparation and Implementation of Quality Assurance System Based on Accreditation of Study Programs in Higher Education *Journal of Social Sciences* (*COES&RJ-JSS)* **5** pp 238–251

[8] Abdullah F 2006 The development of HEdPERF: a new measuring instrument of service quality for the higher education sector *International Journal of Consumer Study* **30** pp 569–581

[9] Nuseibeh B and Easterbrook S 2000 Requirements Engineering: A Roadmap, *Proc of the Conference on The Future of Software Engineering* pp 35–46

[10] Achimugu P, Selamat A, Ibrahim R, Mahrin M N R 2014 A Systematic Literature Review of Software Requirements Prioritization Research *Int. Softw. Technol.* **56** pp 568–585

[11] Khan S, Dulloo A B, Verma M 2014 Systematic Review of Requirement Elicitation Techniques *Int. Journal of Inf and Comp Technol* **4**:2 pp 133–138.

[12] Maiden N M 1998 CREWS-SAVRE : Scenarios for Acquiring and Validating, *Domain Modelling for Interactive Systems Design* Springer US pp 1–31

IOP Conf. Series: Materials Science and Engineering **308** (2018) 012003 doi:10.1088/1757-899X/308/1/012003

Three-pass protocol scheme for bitmap image security by using vernam cipher algorithm

D Rachmawati[*], M A Budiman, L Aulya

Departemen Ilmu Komputer, Fakultas Ilmu Komputer dan Teknologi Informasi, Universitas Sumatera Utara, Jl. Universitas No. 9-A, Medan 20155, Indonesia.

[*]E-mail: dian.rachmawati@usu.ac.id

Abstract. Confidentiality, integrity, and efficiency are the crucial aspects of data security. Among the other digital data, image data is too prone to abuse of operation like duplication, modification, etc. There are some data security techniques, one of them is cryptography. The security of Vernam Cipher cryptography algorithm is very dependent on the key exchange process. If the key is leaked, security of this algorithm will collapse. Therefore, a method that minimizes key leakage during the exchange of messages is required. The method which is used, is known as Three-Pass Protocol. This protocol enables message delivery process without the key exchange. Therefore, the sending messages process can reach the receiver safely without fear of key leakage. The system is built by using Java programming language. The materials which are used for system testing are image in size 200x200 pixel, 300x300 pixel, 500x500 pixel, 800x800 pixel and 1000x1000 pixel. The result of experiments showed that Vernam Cipher algorithm in Three-Pass Protocol scheme could restore the original image.

1. Introduction

Nowadays, sending images via internet network is very susceptible to the technological crimes such as interruptions, intercepts, and modifications. Without a security guarantee, intruder can get the message easily.

Image is one of multimedia in the form of visual information. Images are created from pixels. These pixels consist of colors namely: red, green and blue referred to as RGB [3]. Image security is indispensable because there have been many cases related to images modified by intruder. For that reason, the image is very necessary to be secured and kept confidential in order not to be abused by unauthorized persons.

Various security techniques have been developed to protect and maintain the confidentiality of messages, one of them is cryptographic techniques. Cryptography is both the science and arts of maintaining the message's security by encoding it into a form that can no longer be understood.

Many algorithms can be used for cryptographic process, but in this research the author will discuss about Vernam Cipher algorithm. The Vernam Cipher algorithm uses a stream cipher where the cipher is derived from the XOR operation between the plaintext bits and the key bits [1]. The security of the Vernam Cipher algorithm relies heavily on the key exchange process. If the key is leaked, the security of this algorithm will collapse. Therefore, a method that minimizes the key leakage during the exchange of messages is required. The method used is known as Three-Pass Protocol. This protocol enables message delivery without key exchange [4]. So the messages sent can reach the receiver safely without fear of key leakage. By using a fast internet connection, Three-Pass Protocol scheme is very applicable.

In this modern era, smartphones have grown rapidly around the world. The public adaptation of smartphones is very fast, especially with the presence of the first version of the operating system Android which was officially released in September 2008 [5]. Android gets attention from people around the world including Indonesia. A survey in April-May 2013 found that Android is the most popular platform that make it the most widely used smartphone operating system in the world.

2. Method

This research used Vernam Cipher algorithm in Three-Pass Protocol scheme for securing bitmap image.

2.1. Vernam Cipher Algorithm

Vernam Cipher is a cryptographic algorithm invented by Major J. Maugborne and G. Vernam in 1920's [1]. The Vernam Cipher algorithm is adopted from One Time Pad Cipher, in which case characters are replaced by bits (0,1). In other words, Vernam Cipher is another version of One Time Pad Cipher [1].

Vernam Cipher algorithm is a symmetric of the key type cryptography algorithm. The key used to perform encryption and decryption use same key [7]. For binary messages, OTP encryption uses the XOR operation between each bit of the message with bits corresponding to the private key [1]. In this method, plaintext is converted into ASCII code and then subjected to XOR operation against keys that have been converted into the ASCII code.

Using One Time Pad, each key letter is used once for a single message and it is not reused. The key character stream length is equal to the message length. One time pad contains rows of randomly generated key characters. One pad is only used once to encrypt the message, after which the pad that has been used is destroyed (hence called one time). One encryption is only possible if both sender and receiver are in the same ownership of the key. Therefore, both parties must exchange their keys before the sending process begins. This means that safe communication is expected and planned within a certain time frame [6].

In general, the Vernam Cipher algorithm scenario is as follows:

1. Alice and Bob agree to use a key to encrypt and decrypt the message.
2. Then the key is used to encrypt messages from Alice to Bob.
3. Bob decrypts the message from Alice using the key that they had agreed on earlier, so that the original message can be read.

Here is the encryption and decryption formula of Vernam Cipher:

$$C \equiv P \oplus K \tag{1}$$

$$P \equiv C \oplus K \tag{2}$$

where C is ciphertext, P is plaintext, and K is key.

2.2. Three-Pass Protocol

Three-Pass Protocol is a framework that allows the sender to send encrypted messages to the recipient without need to distribute the sender's key to the receiver [4]. This process is referred to as Three-Pass Protocol because the sender and receiver perform three exchange stages to encrypt the message.

The protocol was developed by Adi Shamir in 1980's [2]. The basic concept of Three-Pass Protocol is that each party has a private encryption key and a personal decryption key. Both parties independently use the key to encrypt messages and to decrypt messages [2]. The scheme of the Three-Pass Protocol can be seen in figure 1.

IOP Conf. Series: Materials Science and Engineering **308** (2018) 012003 doi:10.1088/1757-899X/308/1/012003

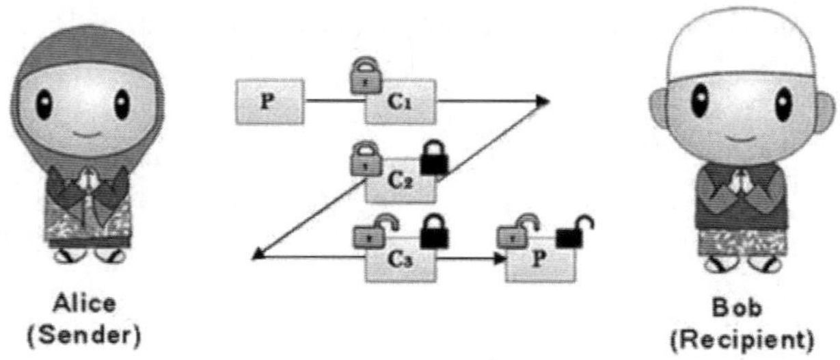

Figure 1. Scheme of Three-Pass Protocol

Three-Pass Protocol has several stages to deliver messages from the sender to the recipient. Here are the steps:

a. Alice (the message sender) has a K_A key. Alice encrypts the message with her own key and sends an encrypted message (C_1) to Bob (the recipient of the message).
b. Bob (recipient of the message) has a K_B key. Bob encrypts C_1 message with his private key and sends the encrypted message again (C_2) to Alice.
c. Alice who has received C_2 decrypts the message by using his private key and sends the decrypted message (C_3) to Bob.
d. Bob who has received C_3 also decrypt the message with his own private key to get the original message (P) sent by Alice.

Example:

- The value of RGB pixel (3.3) is (26,8,95) with sender key (K_A) : 00010101 and recipient key (K_B) : 00011101, the calculation process of encryption and decryption can be seen in figure 2.

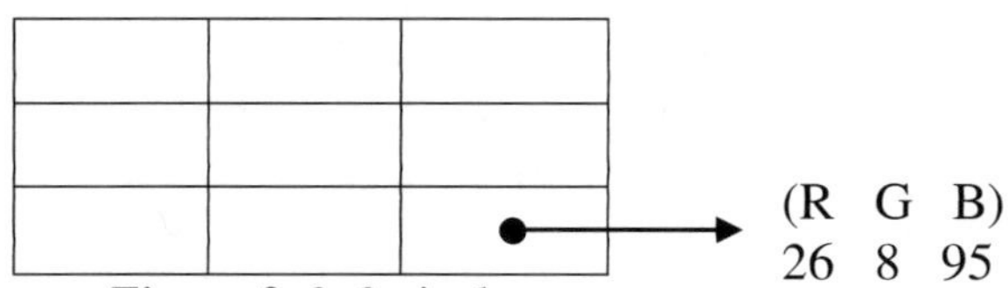

Figure 2. 3x3 pixels

1. First change value of RGB from decimal to binary:

	R	G	B
Decimal	26	8	95
Binary	00011010	00001000	01011111

Then obtained "00011010 00001000 01011111" as plaintext (P).

2. Then encrypt plaintext with sender key (K_A) as follows:

Table 1. First stage pixel encryption calculation process

Plaintext (P)	Key (K_A)	$C_1 = (P \oplus K_A)$	Decimal
00011010	00010101	00001111	15
00001000	00010101	00011101	29
01011111	00010101	01001010	74

3. After C_1 is obtained, encrypt C_1 with recipient key (K_B) as follows:

Table 2. Second stage pixel encryption calculation process

Ciphertext (C_1)	Key (K_B)	$C_2 = (C_1 \oplus K_B)$	Decimal
00001111	00011101	00010010	18
00011101	00011101	00000000	0
01001010	00011101	01010111	87

4. After C_2 is obtained, encrypt C_2 with sender key (K_A) as follows:

Table 3. Third stage pixel encryption calculation process

Ciphertext (C_2)	Key (K_A)	$C_3 = (C_2 \oplus K_A)$	Decimal
00010010	00010101	00000111	7
00000000	00010101	00010101	21
01010111	00010101	01000010	66

5. After C_3 is obtained, encrypt C_3 with recipient key (K_B) as follows:

Table 4. Calculation original pixel decryption

Ciphertext (C_3)	Key (K_B)	$P = (C_3 \oplus K_B)$	Decimal
00000111	00011101	00011010	26
00010101	00011101	00001000	8
01000010	00011101	01011111	95

Table 4 shows that results of the pixel decryption is suitable to the pixels before encryption that is "26,8,95". It means, the use of Vernam Cipher algorithm in Three-Pass Protocol met parameters of data integrity.

3. Results and Discussions

The experiments were conducted on Windows 7 Notebook with Intel inside processor, 32-bit architecture, and 2048MB RAM. The development environment used for coding Java programming language is IDE Eclipse. And to run the execution, we used android on Xiaomi Redmi 4A with Qualcomm Snapdragon 425 quad-core 1.4 GHz processor and 2048MB RAM. The results of the experiments are presented in tables 5 and 6.

Table 5. The results of algorithm testing with process time.

Image 200 x 200 pixels				
Original image	First Ciphertext (C_1)	Second Ciphertext (C_2)	Third Ciphertext (C_3)	Decryption

Time Process (second) 0.285		0.254	0.244	0.249
Average Time (second)				0.258
Image 300 x 300 pixels				
	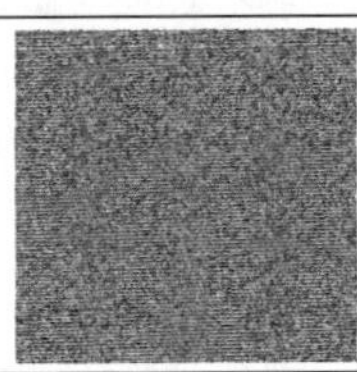	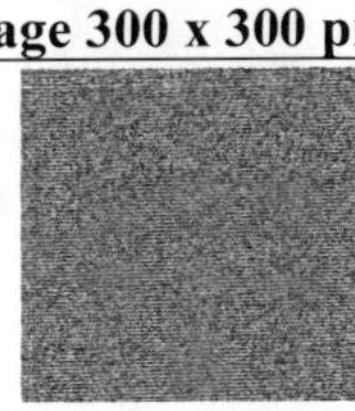	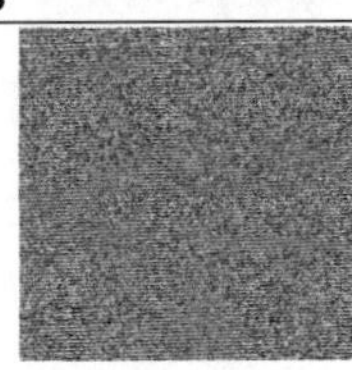	

IOP Conf. Series: Materials Science and Engineering **308** (2018) 012003 doi:10.1088/1757-899X/308/1/012003

Time Process (second) 0.547	0.54	0.513	0.523	
Average Time (second)			0.53075	
Image 500 x 500 pixels				

Time Process (second) 1.455	1.451	1.403	1.415	
Average Time (second)			1.431	
Image 800 x 800 pixels				

Time Process (second) 3.749	3.778	3.595	3.627	
Average Time (second)			3.68725	
Image 1000 x 1000 pixels				

Time Process (second) 5.629	5.745	5.533	5.718	
Average Time (second)			5.66125	

In Table 5, there are five columns. The first column is the original image with a variety of sizes, the second column is the result of first stage image encryption, the third column is the result of second stage image encryption, the fourth column is the result of third stage image encryption, and the fifth column is the result of the original image decryption. From the testing of encryption and decryption process using image size 200x200 pixels, 300x300 pixels, 500x500 pixels, 800x800 pixels, and 1000x1000 pixels, the average time for each execution is 0.258 seconds, 0.53075 seconds, 1.431 seconds, 3.68725 seconds, and 5.66125 seconds. Figure 3 illustrates the acquired time relationship of encryption process to the size of an image.

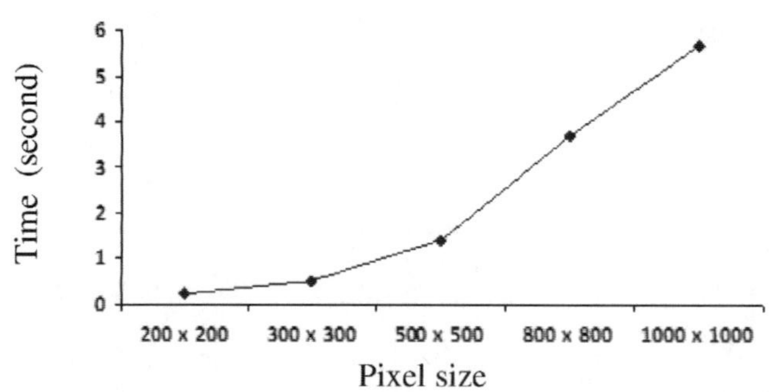

Figure 3. Graph of the relationship between the size of image with the time

Figure 3 shows that the time required to encrypt image 200x200 pixel is 0.3 seconds, 300x300 pixel is 0.5 seconds, 500x500 pixel is 1.4 seconds, 800x800 pixel is 3.7 seconds, and the 1000x1000 pixel is 5.7 seconds. In Figure 3, it could be seen that the relationship between the pixel size of an image is linearly straight to the time of the encryption process. That is, the larger the pixel size, the time spent on program execution will also be longer.

Table 6. Measure of image quality for 5 sample images.

Resolution (pixel)	MSE	PSNR
200 x 200	0	Infinity
300 x 300	0	Infinity
500 x 500	0	Infinity
800 x 800	0	Infinity
1000 x 1000	0	Infinity

In table 6, it could be seen that original image and reconstruction image are exactly the same.

4. Conclusion

In conclusion, encryption and decryption process on image security using Vernam Cipher algorithm in Three-Pass Protocol scheme has fulfilled the integrity of data. The pixel size of the image is linearly straight with time. The larger the image pixel size, the longer it takes for encryption process. Furthermore, the image of the reconstruction is the same as the original image before encryption with the value of MSE and PSNR 0 and infinity.

Acknowledgments

The authors gratefully acknowledge that the present research is supported by Fund Dissemination IPTEKS Research Results for Lecturers / Researchers Universitas Sumatera Utara.

References

[1] Nagaraj N 2012 *Commun Nonlinear* Sci Numer Simulat **17** pp 4029-4036.
[2] Oktaviana B and Siahaan A P U 2016 *IOSR Journal of Computer Engineering (IOSR-JCE)* **18**:4 pp 26–29
[3] Rachmawati D, Amalia, Surya J 2017 *IOP Conf. Series: Journal of Physics: Conf. Series* **801**
[4] Rahim, R and Ikhwan A 2016 *International Journal of Science and Research (IJSR)* **5**:11 pp 102–104.
[5] Saparkhojayev N, Shakhov E, Mailybayev Y 2016 *Journal of Physics:Conference Series* **710**
[6] Sharma S and Gupta V 2013 *Int. Journal of Innovative Research in Science, Engineering and Technology* **2**:6 pp 2248-2250
[7] Zaeniah and Purnama B E 2015 *Int. Journal of Advanced Computer Science and Applications* **6**:9 pp 292–297

IOP Conf. Series: Materials Science and Engineering **308** (2018) 012004 doi:10.1088/1757-899X/308/1/012004

Lean manufacturing analysis to reduce waste on production process of fan products

I Siregar*[1], A A Nasution[2], UAndayani[3], R M Sari[1], K Syahputri[1], Anizar[1]

[1]Department of Industrial Engineering, Universitas Sumatera Utara, Medan – Indonesia, Almamater Rd., Campus USU, Medan – Sumatera Utara, 20155
[2]Department of Accounting, Universitas Sumatera Utara, Medan – Indonesia, Almamater Rd., Campus USU, Medan – Sumatera Utara, 20155
[3]Department of Information Technology, Universitas Sumatera Utara, Medan – Indonesia, University Rd., Campus USU, Medan – Sumatera Utara, 20155

*E-mail: ikhsan.siregar@usu.ac.id

Abstract. This research is based on case study that being on electrical company. One of the products that will be researched is the fan, which when running the production process there is a time that is not value-added, among others, the removal of material which is not efficient in the raw materials and component molding fan. This study aims to reduce waste or non-value added activities and shorten the total lead time by using the tools Value Stream Mapping. Lean manufacturing methods used to analyze and reduce the non-value added activities, namely the value stream mapping analysis tools, process mapping activity with 5W1H, and tools 5 whys. Based on the research note that no value-added activities in the production process of a fan of 647.94 minutes of total lead time of 725.68 minutes. Process cycle efficiency in the production process indicates that the fan is still very low at 11%. While estimates of the repair showed a decrease in total lead time became 340.9 minutes and the process cycle efficiency is greater by 24%, which indicates that the production process has been better.

1. Introduction

Competition nowadays is very rapid in the industrial sector, demanding every company engaged in manufacturing to improve business strategy. This research site is one of the manufacturing companies that produce electronic goods in North Sumatra. The production process that takes place in the company shows the existence of time that is not value added among others the inefficient material transfer in the raw material parts and molding the fan component still using manual way, the length of time waiting for the fan component after the next molding process on the raft at the assembly station. The length of production lead time in the fan production process in the company causes the company can not meet the needs of customers who have been ordered. These problems will adversely affect the company both financially and in partnership and show the lack of productivity of the company in increasing the production capacity of the fan product.

Lean Manufacturing is a set of techniques that when combined and run well will reduce and then eliminate waste [1]. Reducing waste is intended for all activities undertaken on the production floor is an activity that has value added activity, aims to improve competitiveness through the production process effectively and the use of resources efficiently. In this study, the effort made is to define the elements of the manufacturing cycle to reduce waste. The approach used is value stream mapping.

Value stream mapping is a process mapping tool that serves to identify the flow of material and information on the production process from raw material to finished product. Value stream mapping is represented by symbols which representing activity [2]. Activity is grouped into value added and non-value added, so it can be easily known which activity could give added value and those who do not give added value.

The successful application of lean manufacturing by using value stream mapping to reduce total lead time has been proven through scientific research [3]. In his journal that leads on current state maps of 29.43 days can be reduced to the inventory level of the process, which reduces inventory levels to 18.04 days to 15 days. This proves that the use of value stream mapping can reduce total lead time.

2. Research Methods

The research instrument used in this research is digital stopwatch used in data collection to measure the time of fan production process. Value stream mapping is used in data processing to visualize the real state of the production process and determine the classification of value added activity and non-value added activity. Observations made based on The Maytag Company that is with the provision of 10 times observations for activities that take place in the cycle of about 2 minutes or less while 5 times observation for activities that take place in the cycle time greater than 2 minutes.

3. Results and Discussions

3.1. Formation of Current State Map

Current State Map is a description of the ongoing production process within the company covering material flow and information flow. The current state map consists of several steps starting from determining the value stream manager to forming the entire plant flow map.

3.1.1. Determination of Value Stream Manager

Value Stream Manager is someone who understands the whole production process that occurs in detail and has an important role in the production process so as to provide complete information and can assist in providing suggestions for improvement of production processes.

3.1.2. Formation of SIPOC Diagram

SIPOC diagram used in this study to analyse and describe about how the process can provide services to customers. The formation of this diagram helps in explaining a complex project and its scope is unclear.

The elements used in the diagram are:

1. Supplier : Raw material storage warehouse and additional spare parts
2. Input : Plastic pellets, calcium carbo, colour flour, and additional spare parts
3. Process : Process stages that change from raw materials and spare parts into products.
4. Output : Fan type 1651 KP
5. Customer : Consumer

3.1.3. Overall Map Formation (Current State Map)

At this stage each process along the value stream is combined with the flow of material and information into a single stream within the plant. Understanding of the two streams as follows:

1. Material flow.
   Material flow describes the movement of the main material in the process of production process along the value stream.
2. Information flow.
   The flow of information that companies use there are two types, namely:

a. Manual Information Flow.

This information flow occurs between the production managers of each process that takes place on the production floor. The schedule given is the daily activity schedule after getting the adjustment of the amount of ingredients entered.

b. Electronic Information Flow.
Information delivered using electronic devices. This information flow takes place between the production planning department and the marketing department, raw material suppliers, and between the marketing department and the customers.

Current state map that has been equipped with material flow and information flow and lead time bar can be seen in figure 1. In making this current state map, the time is divided into two, ie production lead time indicating the existence of non-value added time and process time Cycle in the production process) which is value added time.

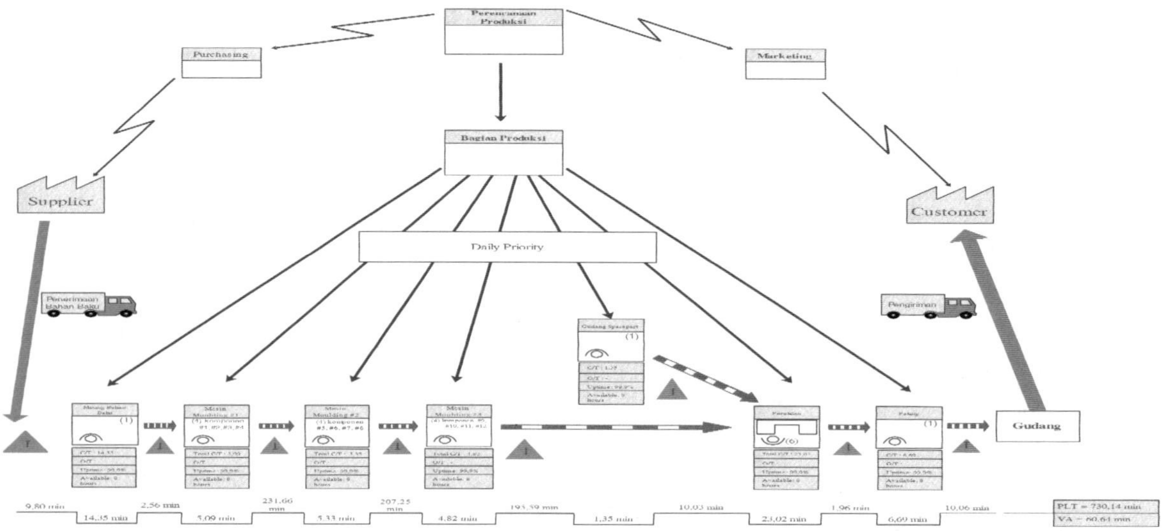

Figure 1. Current State map

3.1.4. Takt Time Calculation

Takt time shows how often a product should be produced to meet consumer demand. Production time available with shift is 8 hours / day on Monday to Saturday. Rest period of 1 hour / day on Monday to Saturday. Within 1 month consists of 4 weeks and within 1 week there are 6 working days [4].

$$\text{Takt time} = \frac{available\ work\ time\ per\ week}{customer\ demand\ per\ week}$$

Customer demand rate per week = (1800 units/month) / (4 weeks/month) = 473 unit/weeks
Available time percentage obtained information that the production of fan type 1651 KP reaches 45% compared with other types of products produced in the company. So that available time of 480 minutes working hours per day, for 1651 KP fan is 216 minutes or 12,960 minutes. Takt time calculation is done starting at each work station, and starting from the last process is on the product gasket until the mixing of raw materials.

Customer demand rate per week on fan packing product will be the basis of customer demand rate calculation in previous process, that is assembling process in accordance with push system principle.

Comparison between takt time and cycle time of production obtained from observation can be seen in Table 1.

Table 1. Comparison of Cycle Time and Takt Time

Activity	Cycle Time	Takt time
Mixing raw materials	21,36	28,51
Molding dop hinge	46,62	27,36
Molding rotary switch	35,11	27,36
Molding dynamo's dwell	51,55	27,36
Molding dynamo's cap	50,15	27,36
Molding plastic nuts	38,42	27,36
Molding fan leaf	49,80	27,36
Molding casing logo	35,73	27,36
Molding hinge fan stand	39,86	27,36
Molding front casing	35,48	27,36
Molding back casing	35,45	27,36
Molding casing button	34,66	27,36
Molding fan tread stand	47,87	27,36
Installation of bottom tread with spare parts	9,20	28,80
Assembly	18,32	28,80
Packing	13,40	28,51

From the table above the time process under takt time indicates the process run faster or can meet the demand. The production process in this state is not very good, but it needs to be analyzed whether there is a surplus of labor that can be reduced to balance the workload of other stations. Process time under takt time is the process of mixing raw materials, molding rotary rotary switch components, molding chasing component logo, front chasing component molding, rear chassis component molding, button chasing component molding, spare part assembly, assembly assembly, and Packing.

Stations with processing time above takt time indicate the process is running slower than it should be. The process is molding component dop hinge, house dynamo component molding, dynamo-housing component molding, molding plastic nuts components, molding fan leaf components, molding fan stand hinges, and molding stand fan footprint components. Improvements that can be made to adjust to takt time are by reducing batch size of production, improving working methods by reducing inefficient activity or increasing labour amount.

3.2. Current State Map Analysis

3.2.1. Waste Identification with PAM

In identifying the existence of non-value added activities for the company, these observations are made directly to the company's production floor. With the observation supported by interviews with employees and supervisors, therefore can be seen whether or not there is waste in the company. In addition, a method that can be used to identify waste is Process Activity Mapping (PAM). Recapitulation of Process Activity Mapping (PAM) results can be seen in table 2.

Table 2. Recapitulation of Number and Time of PAM Result of Fan Production Process Type 1651 KP

Activity	Numbers	Time (Minutes)
Operation (O)	22	77,74
Transportation (T)	19	72,58

IOP Conf. Series: Materials Science and Engineering **308** (2018) 012004 doi:10.1088/1757-899X/308/1/012004

Inspection (I)	5	15,52
Delay (D)	13	581,40

3.2.2. VA dan NVA Analysis

In this stage will be grouped activities that include value added and non-value added. The value of value added time is obtained from the process time contained in the current state map. While non value added time is the lead time of each process which is also found in current state map. Grouping for value added activity can be seen in table 3. Grouping for non value added activities can be seen in table 4.

Table 3. Value Added Activity

No	Value Added Activity	Time (Minutes)
1	Raw materials are weighed	1,95
2	The calcium carbo extract is weighed	1,82
3	The colour flour enhancement is weighed	1,19
4	Mixing Raw Materials	12,18
5	Issued from mixing machine	1,59
6	Molding dop hinge	1,17
7	Molding rotary switch	1,22
8	Molding dynamo's dwell	1,23
9	Molding dynamo's cap	1,3
10	Molding plastic nuts	1,17
11	Molding fan leaf	1,38
12	Molding casing logo	1,19
13	Molding hinge fan stand	1,4
14	Molding front casing	1,2
15	Molding back casing	1,17
16	Molding casing button	1,17
17	Molding fan tread stand	1,11
18	Installation of bottom tread with spare parts	1,28
19	Assembly	26,98
20	Packing	16,04
	Total	77,74

Table 4. Non Value Added Activity

No	Non Value Added Activity	Time (Minutes)
1	The raw material is brought to the mixing machine	1,32
2	Additional calcium carbo ingredients are brought to the mixing machine	1,28
3	The colour flour enhancer is brought to the mixing machine	1,19
4	Moved to a molding machine	2,56

5	Components stacked waiting in transport	572,58
6	Bring it to the assembly line	61,92
7	Spare parts waiting for the raft	8,82
8	Spare parts are transported to the assembly line	1,21
9	Vane testing	1,96
10	Check the assembly results	7,54
11	Transport to a temporary buildup	3,12

Comparison between value added (VA) and non-value added (NVA) can be clearly seen in figure 2.

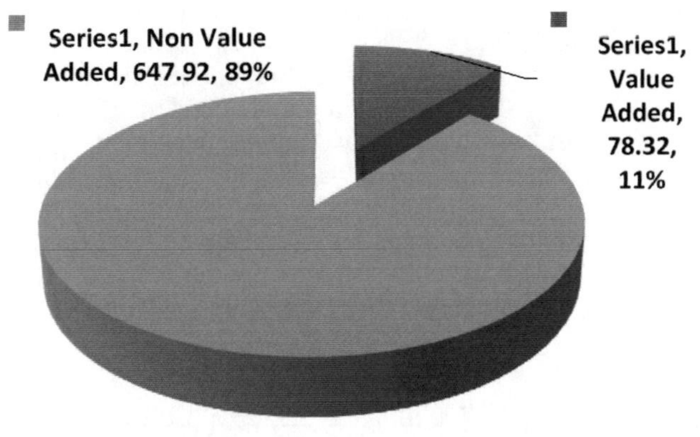

Figure 2. Comparison between Value Added and Non Value Added

3.2.3. Cycle Time Analysis

The cycle time of each process that is a value added time sometimes has non value added elements in it called non value creating time. Overly long cycle times in the processing of a process can also lead to non-value creating time. For example, in parts molding components, long cycle time causes workers on the molding component is often idle so that this activity is classified as waste. Analysis of improvements made to reduce cycle time that will be explained in the proposed further improvement given.

3.2.4. PAM Analysis by Using 5W1H Method

Based on the results of the identification of process activities contained in the production line will be eliminated or reduce activities that are considered not have added value. In addition, it will also consider improvements in activities that can minimize production process time to be shorter. There are some non-value-added and evaluable activities totalling 17, among others, 2, 4, 6, 11, 15, 18, 21, 24, 27, 30, 33, 36, 39, 42, 45, 47 and 59 This is based on direct observation and discussion result due to excessive activity and production process and not added value so it will affect to lead time.

3.2.5. Waste Analysis in Process

Waste in waiting time that often occurs in the production process at the factory, after that sought the cause by using tools 5 why that can be seen in Table 5. From table 5, it can be concluded that the root of the problem of the occurrence of waste of waiting time due to the limited number of material handling and the number of workers for some long production process that takes a long time to workmanship requires more time. In addition to analyse the waste that occurred, also conducted an analysis of the results of calculation take time lower than cycle time that occurs in some processes.

Table 5. Use of the 5 Whys Method in Troubleshooting

No	Problem: There is a long lead time production in the form of waiting time	
1.	*Why*	The flow of the production process is gradual or not flowing
2.	*Why*	The product components accumulate a long time to perform the next stage of the production process.
3.	*Why*	The old product component is transported to the assembly line
4.	*Why*	Material transportation is repeated using manual method
5.	*Why*	Limited number of material handling available and irregular workplace

3.3. Formation of Future State Map

3.3.1. Preparation of Improvement Efforts

After it is known that the cause of the waste occurring that causes the high value of non-value added time, then further effort will be made to reduce the non-value added time based on the root cause analysis of the waste that has been explained previously. Based on the root of the problem described above, it can be made a corrective action plan to minimize waste in the form of waiting time. The root of the problem shows the limited number of material handling and the number of workers and the irregularity of the work area on the production floor so that it needs to be repaired. Alternative possible improvements are made to each production activity to reduce waste in the form of waiting time, irregularity of the workplace, and the number of operators waiting. The alternatives can be seen in table 6.

Table 6. Alternative Improvements for Each Production Activity

No	Roots of Problems	Alternatives Improvement
1.	Raw materials and auxiliary materials in the raw material warehouses await the stages of the process before mixing the raw materials.	Adds the number of workers and job arrangements to the mixing of raw materials
2.	Manual removal resulting in repetitive transfers	Adding semi-automatic material handling to make it easier for employees to do material handling.
3.	The length of waiting time for the product components from the result of molding machine	Improved working procedures in work procedures and operator expertise in using molding machines
4.	The existence of waiting products assembled on assembly lines between assemblers one with other assemblers	Improve working procedures and also train operators to balance during assembly process.
5	Irregularity of the breaking of the components coming from the molding machine to assembly section	Adding the container as a place to lay WIP components so as not to scatter.

3.3.2. Depiction of Future State Map

Improvements made are drawn in the future state map and calculated lead time result proposed improvement. The time change included is a time change that can be observed or estimated from the current state. The picture of the proposed future state map can be seen in Figure 3. The implementation of the 5 whys strategy and the analysis using PAM (process activity mapping) on the fan production process will be able to reduce the product components that accumulate and reduce the waiting time that occurs. So it can reduce waste (waste) in the process of fan production.

4. Conclusions

Based on the theoretical basis, the results of research and analysis have been done then it can be concluded that there is several non-value added activities which in a very significant amount at present state map with total time of 647.94 minutes of total lead time of 725.68 minutes. Through estimation of improvement result obtained Total lead time in future state map is 250,80 minutes. Value stream mapping showed a decrease in total lead time of 340.9 minutes. The saving of working hours will directly increase the fan production capacity without having to increase the input used in the production process resulting in increased productivity.

Acknowledgement

The authors gratefully acknowledge that the present research is supported by Ministry of Research and Technology and Higher Education Republic of Indonesia. The support was under the research grant Research Grant of Applied Product (in Bahasa: Hibah Penelitian Produk Terapan) of Year 2017.

References

[1] Wilson L 2010 *How To Implement Lean Manufacturing* (New York: McGraw-Hill)
[2] Nash M and Polling S 2008 *Mapping The Total Value Stream* (Taylor and Francis Group)
[3] Salunke S S and Hebbar S 2015 Value Stream Mapping: A Continuous Improvement Tool for Reduction in Total Lead Time *International Journal of Current Engineering and Technology* India: Engineering Management, Manipal Institute of Technology
[4] Batra, Kritika Et al. 2015 *Study on Application of Lean Manufacturing Tools for Performance Improvement. India*; Department of Mechanical Engineering, Delhi Technological University

IOP Conf. Series: Materials Science and Engineering **308** (2018) 012005 doi:10.1088/1757-899X/308/1/012005

Design of production process main shaft process with lean manufacturing to improve productivity

I Siregar[*1], A A Nasution[2], U Andayani[3], Anizar[1], K Syahputri[1]

[1]Department of Industrial Engineering, Universitas Sumatera Utara, Medan 20155, Sumatera, Indonesia.
[2]Department of Accounting, Universitas Sumatera Utara, Medan 20155, Sumatera, Indonesia.
[3]Department of Information Technology, Universitas Sumatera Utara, Medan 20155, Sumatera, Indonesia.

[*]E-mail: ikhsan.siregar@usu.ac.id

Abstract. This object research is one of manufacturing companies that produce oil palm machinery parts. In the production process there is delay in the completion of the Main shaft order. Delays in the completion of the order indicate the low productivity of the company in terms of resource utilization. This study aimed to obtain a draft improvement of production processes that can improve productivity by identifying and eliminating activities that do not add value (non-value added activity). One approach that can be used to reduce and eliminate non-value added activity is Lean Manufacturing. This study focuses on the identification of non-value added activity with value stream mapping analysis tools, while the elimination of non-value added activity is done with tools 5 whys and implementation of pull demand system. Based on the research known that non-value added activity on the production process of the main shaft is 9,509.51 minutes of total lead time 10,804.59 minutes. This shows the level of efficiency (Process Cycle Efficiency) in the production process of the main shaft is still very low by 11.89%. Estimation results of improvement showed a decrease in total lead time became 4,355.08 minutes and greater process cycle efficiency that is equal to 29.73%, which indicates that the process was nearing the concept of lean production.

1. Introduction

This research was conducted on a palm oil processing machine spare parts manufacturing company in North Sumatra. The products produced are Main Shaft, Cone, Intermediate Gear, Protect Nut, Pulley, Flexible Couple, Press Cage, Screw Press and Digester. The production system used is make to order where the product will be made if the order has been received from the consumer with the order rules that have arrived first, will be served first (First Come First Serve).

The company experienced problems in carrying out its production. The Company has delayed the settlement of orders that are not in accordance with the agreement when the order is received. Delay in order completion occurs repeatedly with quite large number of delays. Company data shows that the largest number of delays occurred in Main Shaft products with a delay in August of 24 units, September at 12 units and in October 40 units. Delays in product completion will have a chain effect on subsequent product settlements and will inhibit production flow.

Delays in product completion indicate the low productivity of the company in terms of resource utilization. Efforts to increase productivity can be done by increasing the efficiency of resource use. A job will be said to be resolved efficiently if its completion time is shortest [1]. One way to reduce lead time according to Wilson [2] is to eliminate unnecessary activities in the production process. Thus, improvements by reducing and eliminating non value added activity in the production process can improve efficiency. For that the company must know the activities that provide added value (value added activity) and activities that do not provide added value (non-value added activity). One approach that can be used to reduce and eliminate non-value added activity is Lean Manufacturing.

Lean manufacturing is a set of techniques that when combined and run well will reduce and then eliminate waste. Previous research has found that the elimination of non-value added activity by the Lean Manufacturing approach can decrease the production time from 6900 seconds and the time after recommendation of improvement turns to 4645 seconds. This means that the production required process time is reduced 32.68% (2255 seconds) before [3]. Another study also found that Lead Time in paper production amounted to 162 minutes, after the proposed improvement was achieved Lead Time reduction of 72 minutes. So Lead Time obtained for 90 minutes, by reducing the waiting time on arrival of raw material until the production floor process [4]. The research got the same result that is decreasing Lead Time through waste elimination. Elimination of waste is intended for all activities undertaken on the floor of production is an activity that provides added value (value added activity), resulting in increased productivity and competitiveness through increasing the effectiveness of production processes and the efficiency of the use of resources (resources).

2. Research Method

Based on the method used, this research includes action research because this research aims to get a model of company production process design that can increase productivity. Judging from the level of explanation, this research includes descriptive research, because this research describes the systematic, factual and accurate about the facts and the properties of a particular object [5]. The object of this research is the activity done in the main shaft production process itself.

The research instrument used in this research is a Seiko brand digital stopwatch in data collection which used is to measure the time of Main Shaft production process. Present state maps are used in data processing whose function is to visualize the real state of the production process and determine the classification of value added activity and non-value added activity.

Stages in the implementation of research that is:

1. Analysis of delay in order completion.
2. Analysis of non-value added activity with value stream mapping analysis tools and classification of activity. Calculation of Process Cycle Efficiency is done based on value added time and total lead time resulted from activity classification which become indicator in lean manufacturing approach.
3. Elimination of non-value added activity with the 5 Whys tool and application of pull demand system.
4. Preparation of Future State Value Stream Mapping and estimation of improvement results.
5. Conclusion.

3. Results and Discussions

3.1. Order delay analysis

The delay in order completion occurs repeatedly with a small number of delays as shown in table 1.

Table 1. Main Shaft Completion Delay

Spare part	Bulan	Amount (Unit)	Delay (Days)
Main Shaft	August	24	14

	September	12	3
	October	40	5
Cone	August	16	7
	September	16	5
	October	50	2
Cone Guide	August	16	3
	September	16	3
	October	20	2
Hidrolik	August	8	2
	September	12	7
	October	24	4
Intermediate gear	August	16	4
	September	12	4
	October	31	3

Company data show that the largest number of delays occurred in Main Shaft products with a delay in August of 24 units, September at 12 units and in October 40 units.

3.2. Non value added activity analysis

Non value-added activity analysis that occurs in main shaft production process is done with value stream mapping analysis tools and classification of activity.

3.2.1. Value stream mapping

Value stream mapping is an analysis tool in lean manufacturing that represents where waste occurs although it does not directly give effect to the production process. Creating a map of each process category (Door-to-Door Flow) requires production process information to be recapitulated in one process box. For that we need to understand the order of main shaft production process. The order of main shaft production process is:

1. Measurement and cutting process
2. Rolling ring plate process
3. Welding process
4. The process of lathe
5. Milling process
6. Painting process
7. Packing process.

The information collected from each work station is included in the process box. The Present state map illustrates the real condition of the entire process that takes place on the production floor in the process box and the inventory and work in process images that are waiting for each process. Present state map can be seen in figure 1.

3.2.2. Determination of activity classification

Based on the Present state map in Figure 1, the production floor activity will be classified into value added activity and non-value added activity. Activities that are classified into value added activity are activities to convert raw materials into products according to consumer demand. While the activity that is classified non value added activity is activity that does not give change to raw material become product according to consumer demand. Classification of activities can be seen in table 2. The classification indicates that Value added activity based on consumer point of view is 1,295,08 minutes. While non-value added activity amounted to 9,509,51 minutes. Non value added activity that occurs is very significant and is a waste that needs to be reduced and eliminated.

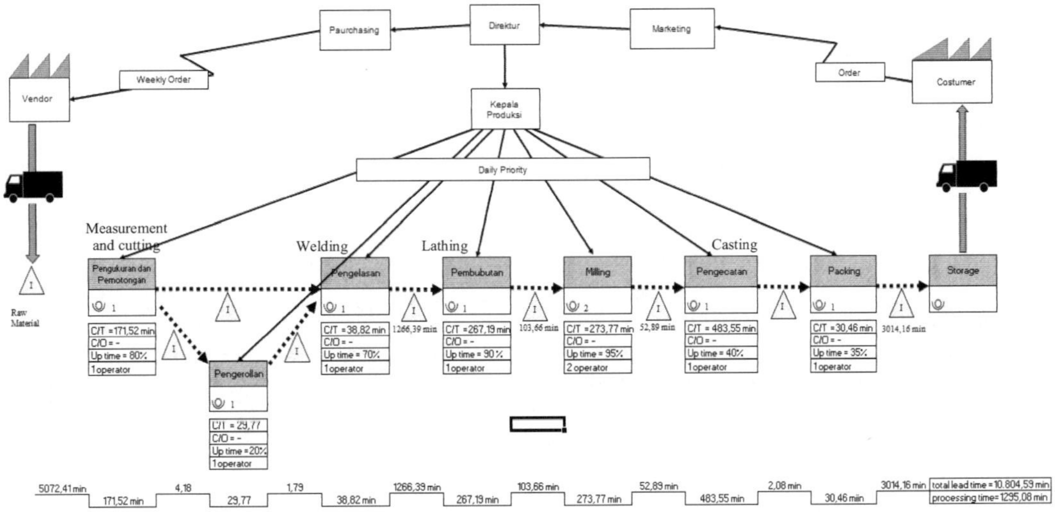

Figure 1. Present state map

Table 2. Value Added and Non-Value Added Time

No.	Activity	Non value Added	Value Added
1	Iron bars are stored in temporary storage	5.072,41	
2	Measurement and cutting		171,52
3	Rolling		29,77
4	Welding		38,82
5	Main shaft is waiting to be turned in Lathe	1.266,39	
6	Lathing		267,19
7	Main shaft is waiting to be taken to the milling station	23,06	
8	Main shaft waiting for milling	80,60	
9	Milling		273,77
10	Main shaft waiting to be taken to painting station	52,89	
11	Painting		483,55
12	Packing		30,46
13	Main shaft waiting to be sent	3.014,16	
	Total	9.509,51	1.295,08

The classification indicates that Value added activity based on consumer point of view is 1,295,08 minutes. While non-value added activity amounted to 9,509,51 minutes. Non value added activity that occurs is very significant and is a waste that needs to be reduced and eliminated.

3.2.3. Calculation of Process Cycle Efficiency

George (2002) says that a company is said to have implemented the concept of Lean Manufacturing if it has a value of process cycle efficiency of 30% which means that the value-added activity reaches 30% of the activity time as a whole. From table 2 above, it can be seen that the large Value added time based on consumer point of view of 1.295.08 minutes. While non-value added time of 9,509,51 minutes. The calculation of process cycle efficiency is as follows:

$$Process\ Cycle\ Efficiency = \frac{Value\ added\ time\ x\ 100\%}{Total\ lead\ time} \quad (1)$$

$$= 11.98\%$$

Where:
Total Lead Time = Sum of the value added time with the non-value added time.

3.3. Elimination of Non Value Added Activity

Lean manufacturing approach used to reduce and eliminate non value added activity is 5 whys tool to find root of problem and arrangement of alternative corrective action and application of pull demand system.

3.3.1. Whys Methods

Classification of activities indicates a major waste of this company is the waste of waiting time. This is supported by the results of discussions with production heads that indicate a waste of waiting time. Waiting time is caused because the main shaft production process runs step by step and does not flow. This problem will look for the root cause with the help of the 5 Whys tool.

Table 3. Alternative Problems Improvement

No.	Roots of Problems	Alternative Improvement
1	*Supplier stacked orders for a week*	Communicate to suppliers to make written cooperation contracts
2	The main shaft line production is out of balance	Modify the chuck on the available lathe in order to be able to do the main shaft
3	The number of cranes is limited to serve the needs of all stations	Performing crane addition as the main material handling tool on the production floor
4	*The main shaft is temporarily stored at the lathe station*	Provide work instruction sheet to create work standardization
5	Avoid to turn the engine on and off repeatedly	Consider the amount of lead time of the main shaft production process compared to the setup of the milling machine
6	Transportation equipment is unable to service delivery of the product	Performing additional product delivery vehicles

Having known the root of the problem that causes the high value of non-value added time, then further will be prepared for improvement efforts to reduce non value added time based on the root cause of waste. Alternative improvements that can be made to minimize and eliminate non value added activity in the form of waiting time can be seen in table 3.

Recall Kanban is a kanban card containing the information, quantity, and type of product to be taken from the previous process. The withdrawal kanban is designed uniformly for the whole process of main shaft production, withdrawal kanban can be seen in figure 2.

IOP Conf. Series: Materials Science and Engineering **308** (2018) 012005 doi:10.1088/1757-899X/308/1/012005

No.		Previous Process
Part Code		
Part Name		
Machine Type		
		Next Process
Amount	Production Line	

Figure 2. Recall Kanban

Recall Kanban is a kanban card containing the information, quantity, and type of product to be taken from the previous process. The withdrawal kanban is designed uniformly for the whole process of main shaft production, withdrawal kanban can be seen in figure 2.

Kanban production orders are kanban cards containing information, quantities, and types of products to be made in the previous process. Production order orders are made in the same format to be used throughout the production process of making the shaft. The format of the kanban card production design commands that will be used in the production process of making the main shaft can be seen in figure 3.

3.3.2. Pull Demand System Implementation

To apply lean concept, production control is cultivated as pull system. The application of pull demand system in the lean concept is done by using Kanban cards. Kanban is a tool to achieve just in time production. With the achievement of just in time production system, the waste of time waiting for main shaft production process can be eliminated.

Kanban to be designed in this research is Kanban withdrawal (recall kanban) and kanban production orders (production kanban).

No.		Process
Part Code		
Part Name		
Machine Type		
Amount		

Figure 3. Kanban Production Order

Application of pull demand system with Kanban tools there are some rules that must be understood and followed by all operators.

3.4. Future state value stream mapping

Preparation of value stream mapping after improvement is drawn in the future state map and calculated the lead time of the proposed improvement. The time change included is a time change that can be observed or estimated from the current state. Value stream mapping in the form of future state maps can be seen in figure 4.

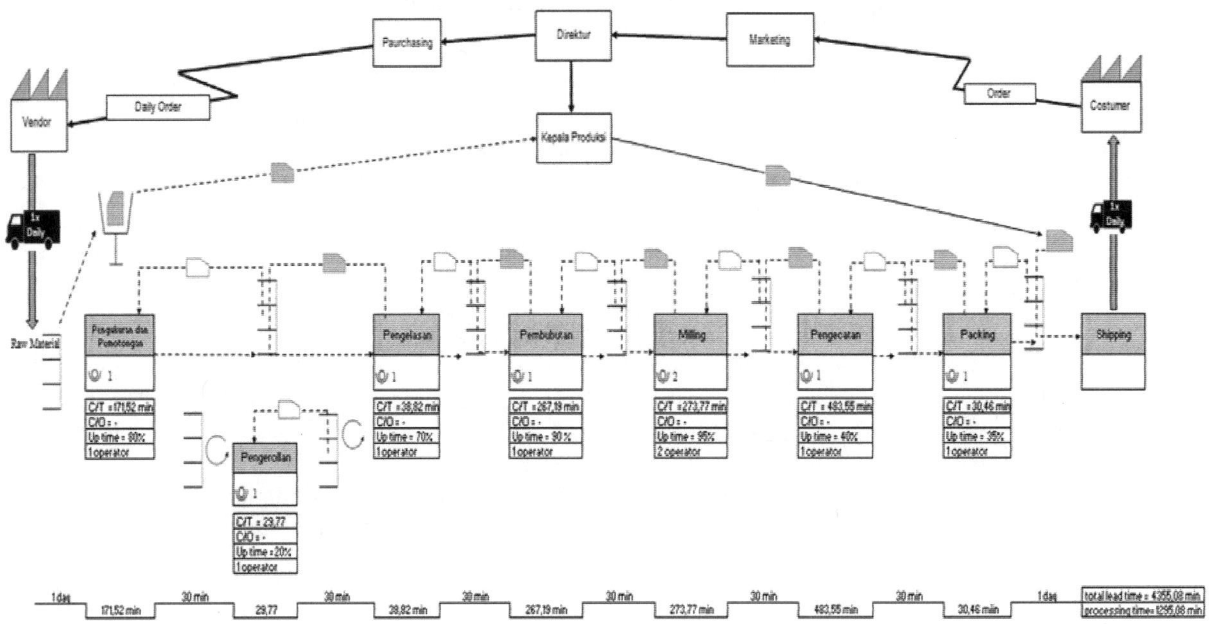

Figure 4 ***Future State Map***

Future state map improvement results showed a decrease in total lead time to 4,355,08 minutes. Total lead time has a direct effect on the process cycle efficiency, to see how much improvement of process cycle efficiency after estimation is done as follows:

$$Process\ Cycle\ Efficiency = \frac{Value\ added\ time\ x\ 100\%}{Total\ lead\ time} \tag{2}$$

$$= 29.73\%$$

The process cycle efficiency after estimation has a higher value compared to process cycle efficiency before the improvement of 29.73%. Process cycle efficiency of main shaft production process increased by 17.84% from 11.89%. Improved process cycle efficiency shows significant savings on employees' working hours. Lead time main shaft production process of 10,804.59 minutes down to 4,355,08 minutes.

4. Conclusion

Based on the theoretical basis, the results of research and analysis have been done then it can be concluded that there are very significant non-value added activity at present state map there are six activities with total time of 9,509.51 minutes of total lead time of 10804.59 minutes. Through estimation of improvement results obtained Total lead time in the future state map of 4.355.08 minutes. Value stream mapping showed a decrease in total lead time of 6,449.51 minutes. The saving of working hours will directly increase the production capacity of the main shaft without having to increase the input used in the production process resulting in increased productivity.

References

[1] Sritomo W 2000 *Ergonomic: study of moving and time I* [in Indonesia: Studi Gerak dan Waktu I] (Surabaya: Guna Widya)

[2] Lonnie W 2010 *How to Implement Lean Manufacturing* (USA: McGraw-Hill Company)

[3] Devis Z 2011 Penerapan Lean Manufacturing guna Meminimasi Waste pada Lantai Produksi di PT Kharisma Esa Ardi Surabaya [implement lean manufacturing for reducing waste on the production floor] (Surabaya: UPN "VETERAN")

[4] Zaenal F, Laksono S M 2011 Implementasi Lean Manufacturing untuk Peningkatan Produktivitas *Nasional seminar proceeding Teknologi Manajemen XIII* (Surabaya: ITS)

[5] Sukaria S 2012 *Research Method* [in Indonesia: *Metode Penelitian*] (Medan: USU Press)

IOP Conf. Series: Materials Science and Engineering **308** (2018) 012006 doi:10.1088/1757-899X/308/1/012006

Analysis of bullwhip effect on supply chain with Q model using Hadley-Within approach

I Siregar*[1], A A Nasution[2], N Matondang[1], M R Persada, K Syahputri[1]

[1]Department of Industrial Engineering, Universitas Sumatera Utara, Medan – Indonesia

[2]Department of Accounting, Universitas Sumatera Utara, Medan – Indonesia

*E-mail: ikhsan.siregar@usu.ac.id

Abstract. This research held on a tapioca flour industry company that uses cassava as raw material to produce tapioca starch product. Problems that occur in this company is inaccurate planning, consequently there is a shortage of variation between the number of requests with the total supply is met, so it is necessary to do research with the formulation of the problem that is how to analyze the Bullwhip Effect on the supply chain using Q model through Hadley-Within approach so as not to disturb the product distribution system at the company. Product distribution system at the company, obtained by the number of requests. The 2015 forecast result is lower than actual demand for distributors and manufactures in 2016 with average percentage difference for Supermarket A distributor, Supermarket B and manufacturing respectively 38.24%, 89.57% and 43.11% . The occurrence of information distortion to the demand of this product can identify the existence of bullwhip effect on the supply chain. The proposed improvement to overcome the bullwhip effect is by doing inventory control policy with Q model using Hadley-Within approach.

1. Introduction

Lean manufacturing concept is widely applied in the world, some of the previous studies that discussed lean manufacturing as in lean improvement research were then simulated and investigated before being implemented, allowing systematic design of slim improvement [1]. Lean manufacturing is also used for maintenance management, which is the implementation of maintenance management and lean manufacturing techniques in maintenance workshops to eliminate losses due to damage and to improve productivity and employee motivation in companies that produce dairy products [2]. In addition, this study also discusses the supply chain, as previous research which states that the merger of the bullwhip effect method with supply chain [3]. And other research mentioned about forecasting supply chain, almost same with the method applied in this research [4] [5]. In lean manufacturing practices such as manufacturing processes and equipment, manufacturing planning and scheduling, visual Information systems, supplier relationships, customer relations, labour and product development & technology [6]. Some even combine kanban method with lean manufacturing [7]. In addition there is also a combination degan fuzzy method [8]. This research combines three methods namely kanban, supply chain and lean manufacturing.

This company is a company that processing sweet potato into tapioca flour as the finished product. The marketing area of tapioca flour products of this company is generally in Sumatera area and more focused in Medan area and developed to the area of Aceh, Padang, Jambi, Pekanbaru, and Palembang.

IOP Conf. Series: Materials Science and Engineering **308** (2018) 012006 doi:10.1088/1757-899X/308/1/012006

The company must be able to plan and control the finished product inventory appropriately to meet the demand and stock of the company's products.

Output product is the most produced products by the company compared to other types of flour products. And the production is make to stock so that it can cause the accumulation of products (over stock) or product shortage (stock out) if the planning is not accurate. Production data of flour products in 2015 can be seen in table 1.

Table 1. Demand and Sales Data Year 2015 (boxes)

Month	Demand (Distributor)	Sales (Manufacture)	Variance Value
January	2602	2551	51
February	2720	2510	210
March	2512	2413	99
April	2548	2443	105
May	2790	2495	295
June	2606	2432	174
July	2667	2475	192
August	2446	2463	17
September	2850	2464	386
Oktober	2669	2543	126
November	2555	2424	131
Desember	2458	2389	69
Total	31423	29602	1821

Table 1 above shows the variance values between the number of requests and the number of sales that resulted in stock outs during the 2015 period of 1821 boxes which could not be met. In August 2015 there was a stockpiling of inventory (over stock). Overstock and stock outs are caused by demand that is likely to change. The change causes a demand distortion. This distortion leads to inaccurate demand, so there is a varied difference between the number of requests and the number of sales. This phenomenon is called the bullwhip effect. In handling probabilistic phenomena taken with the Q model because it is responsive in case of lack of goods compared with other models. In addition, the amount of safety reserves is more accurate because it is set simultaneously with cost optimization. Q model inventory policy with Haldey-Within approach.

2. Research Method

The object of this research is distributor of Lotte Mart and distributor of Berastagi Supermarket with observed data that is product demand data. The selection of distributors due to the area of distributor marketing is very broad, in addition to these distributors also have large retailers so that the number of requests received by the company is also very large.

The methodology used to collect data is as follows:

1. Conduct direct observation of the object of research.
2. Utilizing historical data
3. Library study from various sources

Then do the data processing obtained from the data collection in the company. Data processing using inventory control policy with model Q using Hadley-Within approach, so obtained by economical lot order size (qo) and security reserve (ss) optimal.

3. Result and Discussion

3.1. Bullwhip Effect Analysis

The size of the bullwhip effect in a supply chain echelon is the ratio between the variance coefficient of demand (Cout) and the coefficient of sales variance (Cin).
Systematically can be written:

$$\omega = \frac{Cout}{Cin} \tag{1}$$

$$Cout = \frac{\sigma[Dout\ (t,t+T)]}{\mu[Dout\ (t,t+T)]} \tag{2}$$

$$Cin = \frac{\sigma[Din\ (t,t+T)]}{\mu[Din\ (t,t+T)]} \tag{3}$$

where, σ = demand standard deviation
μ = average demand
t = t time interval
T = horizontal demand t time interval

Dout (t,t+T) and Din (t,t+T) are the request over a time interval (t, t + T) and will be written as Dout and Din. Based on the result of identification of bullwhip effect, bullwhip effect value is less than one on Lotte Mart distributor (0,1982), Berastagi Supermarket (0,2425), and also its manufacture (0,2382). The value of the bullwhip effect is less than one, meaning there is no demand variability in Lotte Mart's distributor supply chain, and distributor Berastagi Supermarket, as well as its manufacturing supply chain. However, the value of the bullwhip effect that is smaller than one indicates an increase in the variability of sales of Gunung Agung products.
Based on the identification of bullwhip effect, it is known that the cause of bullwhip effect is:

1. Demand Forecasting Updating
   The inaccuracy of demand made by the manufacture resulted in the variability of demand in the supply chain.
2. Lot Sizing
   Lot Sizing is necessary because the production process and product delivery will not be economical if done in small size. A relatively stable end-of-day customer request will turn into a weekly or biweekly order from a retailer so that the distribution centre will receive a more volatile order than the demand faced by the retailer.
3. Rationing and Shortage Gaming
   Distributors and retailers often do rationing, when the demand is often not entirely fulfilled in full, the result is often when the inventory is sufficient, the distributor and retailer change or cancel their order it will damage the market information on the supply chain.
4. Price Fluctuation
   Promotional policies in the form of discounting lead to changes in demand. Demand jumps at the time of discounting and decreases when the discount diverges.
   The delay in order completion occurs repeatedly with a small number of delays as shown in table 1.

3.2. Inventory Policy with Q model

The inventory policy with the Q model using the Haldey-Within method has an advantage in the determination of the optimal lot size and the safeguard reserves are easily solved analytically because the solution search is done iteratively. Comparison of inventory calculation includes optimal inventory policy and total inventory cost for Lotte Mart, Berastagi Super Market, and its manufacturing chain can be seen in tables 2 – 4.

Table 2. Calculation of Inventory Control of Lotte Mart Distributor in 2016

Inventory control	Notation	Lotte Mart
Average demand	D	1675
Lack of inventory	N	1
Order lot size	q_o^*	24
Safety Stock	Ss	6
Reorder point	r^*	1
Level of service	η	98 %
Total inventory cost	Ot	Rp 84,856,875

Table 3. Calculation of Inventory Control of Distributor Berastagi Supermarket in 2016

Inventory control	Notation	Brastagi
Average demand	D	1158
Lack of inventory	N	1
Order lot size	q_o^*	20
Safety Stock	Ss	3
Reorder point	r^*	6
Level of service	η	98 %
Total inventory cost	Ot	Rp 58,877,025

Table 4. Calculation of Manufacturing Inventory Control In 2016

Inventory control	Notation	Manufacturing
Average demand	D	2834
Lack of inventory	N	1
Order lot size	q_o^*	31
Safety Stock	Ss	9
Reorder point	r^*	16
Level of service	η	99 %
Total inventory cost	Ot	Rp 143,134,642

Several approaches can be proposed to overcome the bullwhip effect i.e.:

1. Information Sharing
   Sharing demand information to all subjects in the supply chain, including distribution centres, factories, or suppliers of components or raw materials. Forecast errors across supply chain lines can be reduced by better exchange of information. Improved management and forecasting techniques
2. Improved Demand Management / Forecasting can be done by improving the techniques of forecasting, especially in the manufacturing sector to obtain more accurate demand forecasting results.
3. Creating stable price.
   To avoid a forward buying reaction, the frequency and intensity of partial promotional activities like this should be reduced and more directed to continuous price reductions so as to create

programs such as everyday low price. Or even if promotional activities or price reductions are made, all parties to the supply chain must know the program well so that it is not mistaken in estimating the real demand.

4. Reduction of fixed costs
   For procurement sector, lot size of orders can be reduced by eliminating excessive administrative activities by eliminating activities that do not provide added value such as optimization.
   Company data show that the largest number of delays occurred in Main Shaft products with a delay in August of 24 units, September at 12 units and in October 40 units.

4. Conclusion

Based on the analysis of processing and discussion of data, it can be taken some conclusions, namely:

1. Based on forecasting results in 2015 lower than actual demand on distributors and manufactures in 2016 with average percentage of difference for distributors Lotte Mart, Berastagi Supermarket, and manufacturing respectively by 38.24%, 89.57%, and 43.11%.
2. The result of forecasting comparison of bullwhip effect value of demand in 2015 with actual demand for 2016 for distributor of Lotte Mart, Berastagi Supermarket and at its manufacturing chain is 0,1982; 0.2425, and 0.2382.
3. Proposed Solution to overcome the bullwhip effect at PT. Florindo Makmur is by reducing the bullwhip effect by paying attention to the needs and safety supply using the Q model through the Hadley-within approach, then determining the economical lot order size, the safety reserves and the total inventory expectations.

References

[1] Omogbai O and Salonitis K 2016 Manufacturing system lean improvement design using discrete event simulation *Procedia CIRP* **57** pp 195-200

[2] Arslankaya S and Atay H 2015 Maintenance management and lean manufacturing practices in a firm which produces dairy products *Procedia-Social and Behavioral Sciences* **207** pp 214-224

[3] Buchmeister B, Friscic D, Palcic I 2014 Bullwhip effect study in a constrained supply chain *Procedia Engineering* **69** pp 63-71

[4] Ali M M, Babai M Z, Boylan J E, Syntetos A A 2017 Supply chain forecasting when information is not shared *European Journal of Operational Research* ***260***:3 pp 984-994

[5] Lee T and Nam H 2016 An Empirical Study on the Impact of Individual and Organizational Supply Chain Orientation on Supply Chain Management *The Asian Journal of Shipping and Logistics* **32**:4 pp 249-255

[6] Wahab A N A, Mukhtar M, Sulaiman R 2013 A conceptual model of lean manufacturing dimensions *Procedia Technology* **11** pp 1292-1298

[7] Rahman N A A, Sharif S M, Esa M M 2013 Lean manufacturing case study with Kanban system implementation *Procedia Economics and Finance* **7** pp 174-180

[8] Oleghe O and Salonitis K 2016 Variation modeling of lean manufacturing performance using fuzzy logic based quantitative lean index *Procedia CIRP* **41** pp 608-613

IOP Conf. Series: Materials Science and Engineering **308** (2018) 012007 doi:10.1088/1757-899X/308/1/012007

Utilisation of polyurethane composit with 50% composition of roystonea regia fiber as noise reduction panel on car hood

Ikhwansyah*[1], Mulia[1], S Gunawan[2,3], R D W Lubis[1]

[1]Mechanical Engineering, University of Sumatera Utara, Medan 20155, Indonesia
[2]Mechanical Engineering, Sekolah Tinggi Teknologi Sinar Husni, Medan 20373, Indonesia
[3]Mechanical Engineering, University of Medan Area, Medan 20223, Indonesia

*E-mail: isranuri@yahoo.com

Abstract. The objective is to get the characteristics of noise reduction, noise reduction level, variety of measurement spaces, and knowing the process in making acoustic material of natural fiber becomes noise reduction on a car hood. The process of making noise reduction material used casting method and pressed by using molded press. The composition of noise reduction material consist of 50% roystonea regia by 32 mesh and 50% combined by gypsum and polyurethane. The result shows that the average result of noise reduction at X_1- side is 5,7% and X_2- side is 3,9%, X_1+ side is 0,9% and X_2+ side is 6,2%, Z_1- side is 8,9% and Z_2- side is 10,1%, Z_1+ side is 9,7% and Z_2+ side is 10,01%. The main conclusion of the study shows that a noise reduction which made of roystonea regia with 32 mesh mixed by matrix of polyurethane and gypsum is appropriate for noise reduction on car hood.

1. Introduction

The development of automobile production in Indonesia increase annualy. Each car has capability in reducing a noise that resulted from engine running system. The material for noise reduction has different noise reduction level depends on its brand and type. The higher noise reduction level in cabin space, the lower noise comes in the car. Most of the noise resulted from outside of the car as the engine other vehicles, crowded street, raindrop on the roof of a car, even the sounds of an engine running is still very noisy from the inside of cabin. The noise resulted from outside of the car can reduce the feeling of comfortable and damn distracting the driver while the driver is driving a car. The noise points that come to cabin of car resulted from: (1) door trim that reduce resonance on the license body plate, (2) cabin wall to reduce the noisy engine running that enter to cabin, (3) spakbor (the wheels space) to reduce a rumbling of road noise when the car runs, (4) engine hood to reduce noise of a machine and detain heat to prevent the hood paint not faded quickly, (5) the cabin floor to maximize noise reduction of wheel and power train derived from the bottom of the car, (6) the car roof to reduce heat and minimize noise of raindrop [1,2].

Basically, noise reduction has the same theory, namely reduces the ability of sound waves to do penetration in the panels of vehicles. Based on reduction material used, there are solid and liquid material. The solid materials are asphalt, foam, and butyl rubber. But now there are a variation experimented, there are distinction of noise reduction thickness, the aluminum foil layers, and the difference in the contours of noise reduction. The advantages and disadvantages of every material such as: (1) asphalt, the advantages are reachable price, the ability of reduction is very good, many of the variation thickness, easy in installing and good in heat resistant. The disadvantages are very heavy (for one sheet by a size 60x60 cm needs 2 mm thickness and by weighed 2 pounds, while for 1 panel of the door at least needs two sheets). (2) Foam, the advantage is low price, it has many thickness variations, and light. The disadvantages is difficult in installment, the ability of noise reduction is low, thick, and cannot be used for wet. (3) Butyl rubber, the advantage is the ability of noise reduction is good, the mass is lighter until 80% than asphalt at the same thickness, easy in installment but the price is expensive [3].

As long as the expansion of science and technology, it is found that a noise reduction from porous material, fibrous material, resonators and panels. Many of porous and fibrous material can be used as a composite of noise reduction. Recent decades, the use of natural fibers have many demands because the quality of low density, low price, renewable, low costs production, good in mechanical and physical properties and abounding [4]. The use of natural fibers initially by the utilization of waste long-lived plants such as areca and coconut, which can reduce the effects of global warming. One of a natural fiber which become an objective of research is the fibers of roystonea regia or in the world trade known as Areca Catechu L. This plant comes from South Asia and Southeast Asia and some of them grew up in India [5].

Roystonea regia has researched with variations mesh 10, 14, and 32 with contained 30%, 40%, 50%, 60%, and 70% of fibers. The result of the experimental got the best noise reduction coefficient obtained in composition of 50% fiber and 32 mesh [6]. By the experimental, researchers want to canvass the utilization of roystonea regia fibers with composition 50% fiber and 32 mesh to be a noise reduction on Daihatsu Xenia car hood. In this study the noise reduction will be investigated experimentally by influencing variation of measurement distances. The main objective is to explore the decline of noise reduction to get information about noise reduction characteristic of 50% roystonea regia by 32 mesh.

2. Method

In this study a mini bus which is used for human transportation will be used as a tested car. The specifications of the car are presented in Table 1. It is three cylinders four stroke engine fueled by gasoline.

Table 1. Specification of the car

No	Parameter	Value
1	Commercial name/model	Daihatsu Xenia 1.0
2	Number of cylinder/stroke	Three-cylinder/four strokes engine
3	Cooling system	Water cooled
4	Bore × Stroke	72 × 81 mm
5	Maximum output	65 PS/ 5600 rpm
6	Engine weight	995 kg

2.1 Experimental Apparatus

In this study the making of noise reduction material on car hood for xenia used fibers of roystonea regia as a filler, gypsum and milionate MR-200 of polyisocianate and JKR-7631L of polyol as composite of matrix polymeric foam material. The process of making specimens started from making fiber, as shown in figure 1. Where the first process is cutting down the roystonea regia, making it in a part of cutting, placing to an oblique position for drying process, after the roystonea regia totally dry, blend it to be a fiber. After becomes a fiber, it sieves by using screen mesh to get roystonea regia fiber by 32 mesh.

The next step after produced roystonea regia fiber by 32 mesh, furthermore it is added by matrixes (gypsum and polyurethane gradually). Gypsum mixed into roystonea regia fibers and stirred until evenly distributed, then added the milionate MR-200 of polyisocianate and stirred until evenly distributed, then added JKR-7631L of polyol alternately into a mixture until evenly distributed. The combination of filler and matrix is poured to a mold of xenia car hood. The technique of making specimen used hand lay-up method and pressed by lock mold. It is like shown as figure 2.a. The drying process needs about 20 minutes minimum and 30 minutes maximum, then the product detachable from the mold. And the result of molding process produces an acoustic material as shown in figure 2.b.

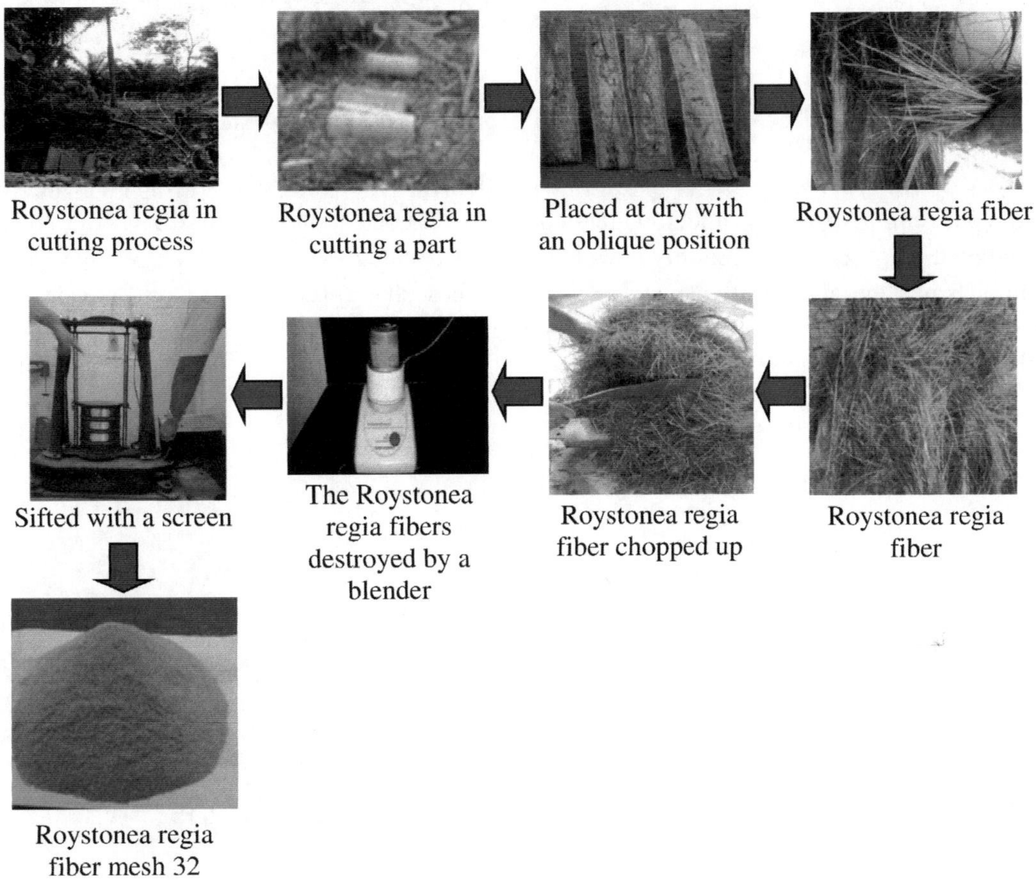

Figure 1. Preparation process in making roystonea regia fiber to be meshed 32

In this study the method of measurement is directly measured by digital sound level meter (SPL) instrument to get the values of noise reduction. In order to perform the study, an experimental apparatus has been designed and developed as shown in figure 3.

In measurement process, the SPL moved by the distance as 10 cm, 20 cm, 30 cm, 40 cm, 50 cm, 60 cm, 70 cm, 80 cm, 90 cm, and 100 cm sequentially for each side. It moved for X- as front car hood side, X+ as behind car hood side, Z- as left car hood side, and Z+ as right car hood side.

IOP Conf. Series: Materials Science and Engineering **308** (2018) 012007 doi:10.1088/1757-899X/308/1/012007

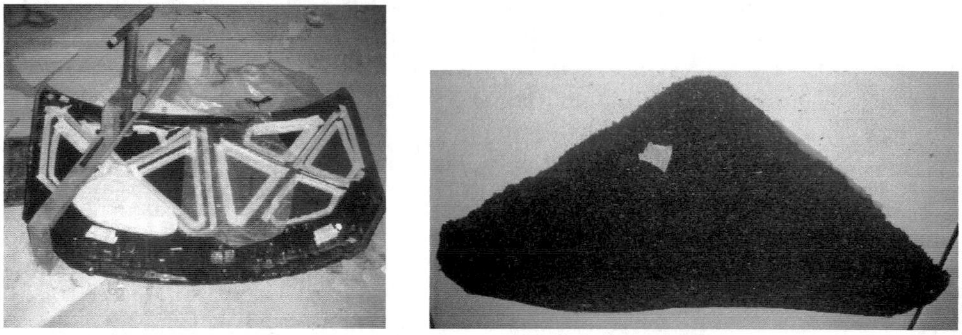

(a) (b)

Figure 2. car hood as lock mold to produce acoustic material

The measurement condition is devided by four kinds of measurement, firstly is without engine load and noise reduction, secondly is without engine load but with noise reduction, thirdly is with engine load and without noise reduction, fourthly is with engine load and noise reduction. The engine load is the condition when the air conditioner and the body electrical is turned on. All the measurement process is in stationary engine run condition. The datas got from the sides that measured will be shown as a grafic to get the conclusion for the experiment.

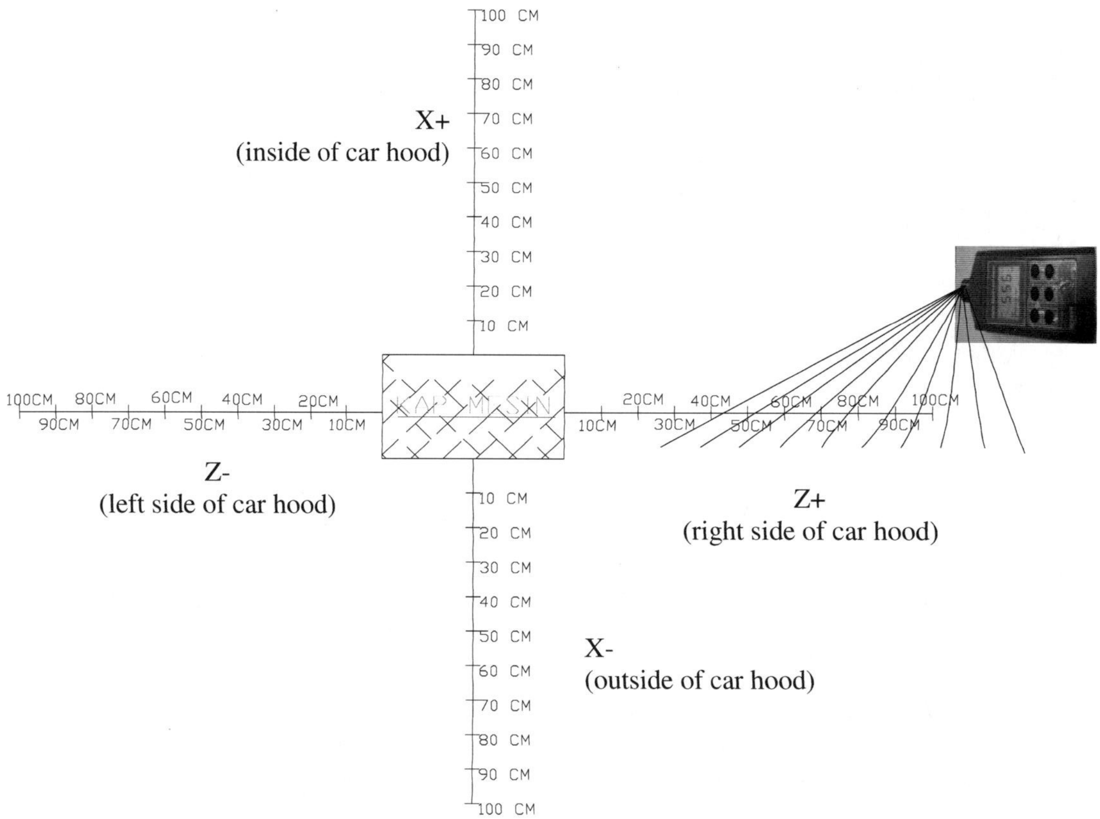

Figure 3. The measurement side of hood

2.2 Noise Reduction Analysis

Sound intensity is the flow of energy carried by sound waves in a region per unit area, the Intensity of Sound is very important to understand the total of radiation from a sound source and the pressure. It can be calculated by using the following equation:

$$I = \frac{p^2}{\rho . c} \tag{1}$$

Where I is Intensity [J/m²s] a function of square pressure (p2), it can develop an equation for sound pressure level (SPL) as follows;

$$PWL = 10 \log \frac{P_1{}^2}{P_0{}^2} \tag{2}$$

Or:

$$SPL = 20 \log \frac{p_1}{p_2} \tag{3}$$

Where:

P_0 = reference pressure as pressure the noise incapable of being heard on a frequency 1000 Hz.
P_1 = working pressure;

3. Result and Discussion

The results will be discussed in two subsections; they are coefficient of 50% fiber of roystonea regia and distance variation of measurement.

3.1 Coefficient of 50% fiber of roystonea regia

From the eksperiment obtained the data of noise reduction coefficients (α) with frequency variations in using impedance tube and the results shown on Table 2.

Table 2. The eksperiment data of noise reduction coefficient with 32 mesh and composition 50% of roystonea regia fiber

Frequency (Hz)	A	B	X1	X2	k	α
125	0.448	0.239	0.275	0.2	2.289	0.716
250	1.232	0.768	0.275	0.2	4.579	0.611
500	0.586	0.383	0.275	0.2	9.159	0.572
1000	0.502	0.308	0.275	0.2	18.318	0.623
1500	0.052	0.038	0.275	0.2	27.477	0.453
2000	0.667	0.403	0.275	0.2	36.637	0.635

The result of Table 2 changed to a chart will shown as Figure 4. Where it apparently increase the coefficients at the frequency of 125 Hz namely 0.7162 and when the frequency of 125 Hz to the frequency of 250 Hz experienced reduction in the value of the coefficients employ the namely 0.6107.The noise reduction coefficients at the frequency of 500 Hz is 0.5719 and deriving from the frequency of 250 Hz, but increasing at the frequency of 1000 Hz namely 0.623.The noise reduction coefficients at the frequency of 1500 Hz derive from frequency 1000 Hz namely 0.4532 and at the frequency of 2000 Hz namely 0.6346.So the noise reduction coeffisient produced of material 50% fibers of roystonea regia with mesh 32 and 50 % gypsum and polyurethane with the impedance tube methods is an average number of experiment by frequency variations of 0,6016.

IOP Conf. Series: Materials Science and Engineering **308** (2018) 012007 doi:10.1088/1757-899X/308/1/012007

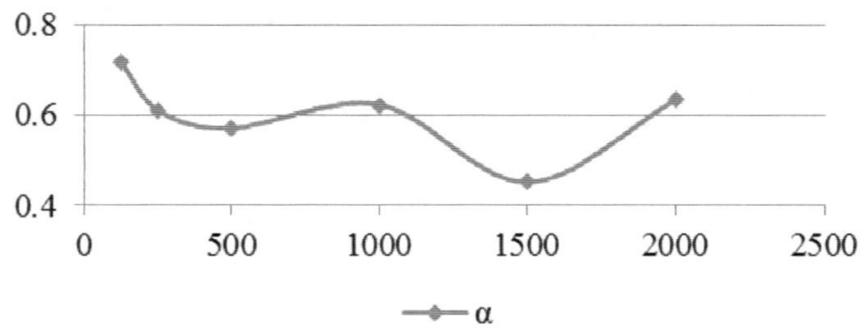

Figure 4. Chart of noise reduction coefficient

3.2 Distance variation of measurement

In this discussion will be noted on car hood sides by the symbols below:

1. Front car hood side in stationer condition without engine load and without noise reduction (X_1-).
2. Front car hood side in stationer condition without engine load and with noise reduction (X_1-‘).
3. Front car hood side in stationer condition with engine load and without noise reduction (X_2-).
4. Front car hood side in stationer condition with engine load and with noise reduction (X_2-‘).
5. Behind car hood side in stationer condition without engine load and without noise reduction (X_1+).
6. Behind car hood side in stationer condition without engine load and with noise reduction (X_1+‘).
7. Behind car hood side in stationer condition with engine load and without noise reduction (X_2+).
8. Behind car hood side in stationer condition with engine load and with noise reduction (X_2+‘).
9. Left car hood side in stationer condition without engine load and without noise reduction (Z_1-).
10. Left car hood side in stationer condition without engine load and with noise reduction (Z_1-‘).
11. Left car hood side in stationer condition with engine load and without noise reduction (Z_2-).
12. Left car hood side in stationer condition with engine load and with noise reduction (Z_2-‘).
13. Right car hood side in stationer condition without engine load and without noise reduction (Z_1+).
14. Right car hood side in stationer condition without engine load and with noise reduction (Z_1+‘).
15. Right car hood side in stationer condition with engine load and without noise reduction (Z_2+).
16. Right car hood side in stationer condition with engine load and with noise reduction (Z_2+‘).

Figure 5 shows that on stationary condition with and without engine load and noise reduction matrial, the waves current of noise flows descently from near of noise source. The best noise reduction percentage is on Z_2 side. It is 10.11% by average noise reduction 6.81 dB. And the lowest noise reduction percentage is on X_1 side. It is 0.89% by average noise reduction 0.5 dB. For comparing the data, it is shown as Table 3 and Table 4 below.

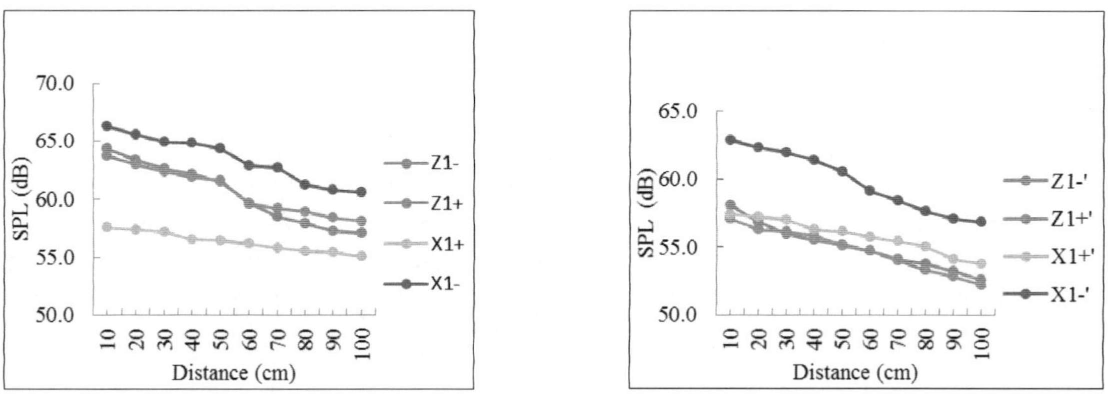

(a) (b)

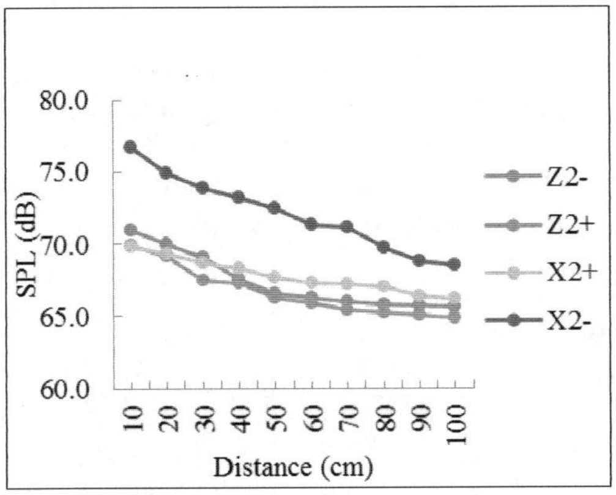

(c)

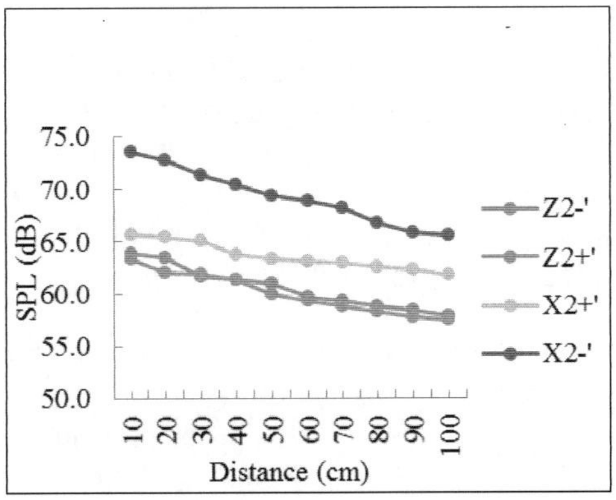

(d)

Figure 5. Result of SPL Measurement

Table 3. Data of SPL measurement at right and left side of car hood

Distance (cm)	Z_1-	Z_1-'	Intensity of Noise Reduction	Pecentage of Noise Reduction	Z_1+	Z_1+'	Intensity of Noise Reduction	Pecentage of Noise Reduction
10	64.3	58.1	6.2	9.642	63.7	57.1	6.6	10.361
20	63.3	56.8	6.5	10.268	63.0	56.3	6.7	10.635
30	62.6	56.0	6.6	10.543	62.3	56.1	6.2	9.952
40	62.1	55.5	6.6	10.628	61.9	55.8	6.1	9.855
50	61.5	55.1	6.4	10.406	61.6	55.2	6.4	10.389
60	59.7	54.7	5	8.375	59.6	54.7	4.9	8.221
70	58.5	54.1	4.4	7.521	59.2	54.0	5.2	8.784
80	57.9	53.8	4.1	7.081	58.9	53.3	5.6	9.507
90	57.3	53.2	4.1	7.155	58.4	52.8	5.6	9.589
100	57.1	52.6	4.5	7.881	58.1	52.2	5.9	10.155
Number of Average			5.44	8.950	Number of Average		5.92	9.745

Distance (cm)	Z_2-	Z_2-'	Intensity of Noise Reduction	Pecentage of Noise Reduction	Z_2+	Z_2+'	Intensity of Noise Reduction	Pecentage of Noise Reduction
10	71.0	63.8	7.2	10.140	69.9	63.3	6.6	9.442
20	70.0	63.5	6.5	9.285	69.2	62.0	7.2	10.404
30	69.1	61.7	7.4	10.709	67.5	61.9	5.6	8.296
40	67.6	61.4	6.2	9.171	67.3	61.4	5.9	8.766
50	66.5	61.0	5.5	8.271	66.3	59.9	6.4	9.653
60	66.3	59.7	6.6	9.955	65.9	59.3	6.6	10.015
70	66.0	59.3	6.7	10.155	65.4	58.8	6.6	10.092
80	65.8	58.8	7	10.638	65.2	58.2	7	10.736
90	65.7	58.4	7.3	11.111	65.0	57.7	7.3	11.231
100	65.6	57.9	7.7	11.737	64.9	57.4	7.5	11.556
Number of Average			6.81	10.117	Number of Average		6.67	10.019

From Table 3 can be concluded that the ratio of the right and the left side of car hood got the highest value of average percantage at the right side with engine load without noise reduction Z_2- is 10.1% (6.81 dB). It is caused by sound that generated by alternator and power steering pump. Furthermore, Table 4 shows the ratio of inside and outside car hood. The highest value of noise reduction average at outside (front side) of car hood by engine load with noise reduction is 6.2% (4.2 dB). It is caused of car hood reduction.

Table 4. Data of SPL measurement at inside (behind) and outside (front side) of car hood

Distance (cm)	X_1+	X_1+'	Intensity of Noise Reduction	Pecentage of Noise Reduction	X_1-	X_1-'	Intensity of Noise Reduction	Pecentage of Noise Reduction
10	57.5	57.4	0.1	0.174	66.3	62.9	3.4	5.128
20	57.4	57.2	0.2	0.348	65.5	62.3	3.2	4.885
30	57.2	57.0	0.2	0.349	64.9	61.9	3	4.622
40	56.5	56.3	0.2	0.354	64.8	61.4	3.4	5.246
50	56.4	56.1	0.3	0.532	64.3	60.5	3.8	5.909
60	56.2	55.7	0.5	0.889	62.9	59.1	3.8	6.041
70	55.8	55.4	0.4	0.716	62.7	58.4	4.3	6.858
80	55.5	55.0	0.5	0.901	61.2	57.6	3.6	5.882
90	55.4	54.1	1.3	2.346	60.8	57.1	3.7	6.085
100	55.1	53.8	1.3	2.359	60.6	56.8	3.8	6.271
Number of Average			0.5	0.897	Number of Average		3.6	5.693
Distance (cm)	X_2+	X_2+'	Intensity of Noise Reduction	Pecentage of Noise Reduction	X_2-	X_2-'	Intensity of Noise Reduction	Pecentage of Noise Reduction
10	69.8	65.7	4.1	5.874	76.7	73.5	3.2	4.172
20	69.4	65.4	4	5.764	74.9	72.7	2.2	2.937
30	68.7	65.0	3.7	5.385	73.9	71.3	2.6	3.518
40	68.3	63.7	4.6	6.735	73.2	70.4	2.8	3.825
50	67.7	63.3	4.4	6.499	72.5	69.4	3.1	4.275
60	67.3	63.1	4.2	6.241	71.3	68.8	2.5	3.506
70	67.2	62.9	4.3	6.398	71.1	68.2	2.9	4.078
80	67.0	62.6	4.4	6.567	69.7	66.8	2.9	4.161
90	66.4	62.3	4.1	6.174	68.8	65.8	3	4.360
100	66.2	61.8	4.4	6.646	68.5	65.6	2.9	4.233
Number of Average			4.22	6.228	Number of Average		2.81	3.907

4. Conclusion

From the experimental data with 50% roystonea regia fibers used impedance tube got the average results of noise reduction coefficient (α) namely 0.6. and the experimental data of noise reduction used SPL such as 6.81 dB for Maximum average on Z_2- and minimum 0.5 dB on X_1+. The decline of noise reduction level by variations of measurement distances used SPL stated that the results of noise coefficients produced noise reduction level linearly. However, The longer measurement distances, the result of measurement will be smaller. The highest noise level was on X_2- and the lowest on X_1+.

5. References

[1] Warman 2016 *Assessment of Sound Absorption Characteristic noise reduction material from Roystonea Regia Fiber by Using Polyurethane and Gypsum as Matrix*, USU, Thesis Chapter 4 pp 61-72
[2] Iqbal 2016 *Performance analysis of Roystonea Regia Fiber To be Speaker Box of Single Reflex Bandpass Type with Matrix Polyurethane and Gypsum* Chapter 4 pp 60-62
[3] Doelle, Leslie 1993 *Environment Acoustic*, Erlangga, Jakarta p1
[4] Decision of Environment Ministry No. 48 year 1996 The Raw Noise Level pp 3-8
[5] Irene Wijaya Elvaswer 2015 *Journal of Noise Reduction Coefficient Testing of Natural Fiber Acoustic Material Waste Tempeh for Noise Controller* **4**:2
[6] Suhada K 2010 *Study of Noise Reduction Coefficient from Composite Materials of palm stem and Gypsum as noise reduction Using Impedance Tube Method*, USU, Thesis Chapter 4 pp 62-68
[7] Asmoro W A 2007 Majalah IPTEK [IPTEK Magazine] **18**:2 p 41
[8] Yusriah L 2014 *Journal of Cleaner Production* **72** pp 174-180
[9] Binoj J S, Raj R E, Sreenivasan V S, Thusnavis G R 2016 *Journal of Bionic Engineering* **13** pp 156-165

IOP Conf. Series: Materials Science and Engineering **308** (2018) 012008 doi:10.1088/1757-899X/308/1/012008

The decline of soil due to the pile of highway project Medan-Kualanamu (STA 35 + 901) with the finite element method

I P Hastuty*, Roesyanto, A B Sihite
Department of Civil Engineering, Universitas Sumatera Utara, Medan, Indonesia

*E-mail: ika.hastuty@usu.ac.id

Abstract. Consolidation is the process of discharge of water from the ground through the pore cavity. Consolidation occurs in soft soil or unstable soil that allows an improvement in order to make the soil more stable. The method of using Prefabricated Vertical Drain (PVD) is one way to improve unstable soils. PVD works like a sand column that can drain water vertically. This study aims to determine the decrease, pore water pressure and soil consolidation rate with Prefabricated Vertical Drain (PVD) and without PVD analytically and using finite element method that affect the duration of soil decline to reach 90% consolidation or in other words soil does not decline anymore. Based on the analytical calculation, the decrease obtained is equal to 0.47 m meanwhile the result of calculation using finite element method is 0.45 m. The consolidation rate obtained from analytical calculation is 19 days with PVD and 115 days without PVD. The consolidation rate obtained from finite element method is 63 days with PVD and 110 days without PVD. And the pore water pressure is 0.92 KN/m^2.

1. Introduction

Geotechnical is an area steeped in civil engineering work relating to the ground. In the soil work process, we often find geotechnical problems such as low soil bearing capacity found on soft ground and a decline due to loading. The load above the soil surface causes the ground underneath to consolidate. Consolidation is the process of discharge of water from the ground through the pore cavity. At the time of consolidation, the ground tension at the beginning of the loading is fully borne by the pore water pressure and will slowly be transferred to the effective stresses of soil, so that the strength and bearing capacity of the soil will increase. The construction of buildings that stand on soft ground needs to consider the magnitude of the decline that will happen to get a good bearing capacity. Thus, the construction process should be carried out after the soil is fully consolidated, so a method of soil improvement is needed to overcome the problem. The selection of soil improvement techniques is influenced by the soil layer, soil characteristics, cost, material availability and experience. One of the soil improvement techniques often used in soft soil problems is preloading and with vertical drainage. Preloading is a method of soil improvement by providing additional load at the construction site until the desired consolidation is reached. The implementation of this method is applied with the use of Prefabricated Vertical Drain (PVD). PVD itself is a geosynthetic material whose concept works like a sand column that can drain water vertically. With the vertical drain, the time required to drain water from the ground is faster and soils consolidate faster too.

1.1. Consolidation

Consolidation is a process of reduction in volume or reduction in pore cavity of a low saturated soil permeability due to loading, wherein the process is affected by the speed of pore water press out of the

IOP Conf. Series: Materials Science and Engineering **308** (2018) 012008 doi:10.1088/1757-899X/308/1/012008

soil cavity. If a layer of saturated soil which has a low permeability is given a load, the pore water pressure in the soil will increase rapidly. Differences in pore water pressure in the soil layer resulted in water flowing into the soil layer with lower pore pressures, followed by settlement in soil. This process takes time due to low soil permeability. There are three different stages which are obtained from the result of consolidation experiment:

Stage I : Initial compression, which is generally caused by preloading.

Stage II : Primary consolidation, the period during which pore pressures are gradually transferred into effective tension, as a result of the discharge of water from the pores of the soil.

Stage III : Secondary consolidation, which occurs after the pore pressure is completely lost.

The compression occurring here is due to the plastic adaptability of the soil grains as shown in figure 1.

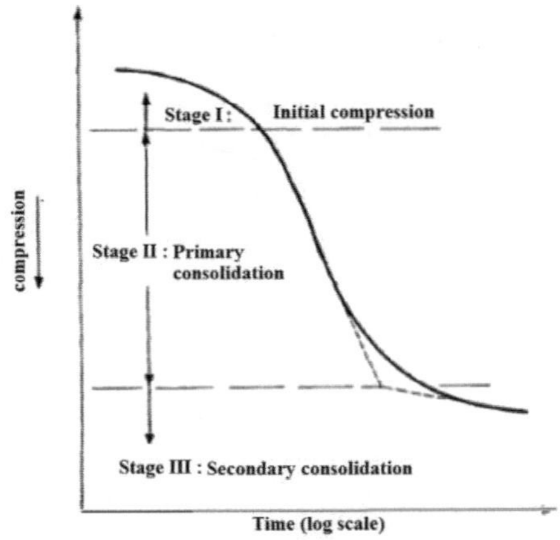

Figure 1. Relation Chart ΔH and log t

The mathematical equation of 1-D consolidated by Terzaghi with parabolic shape:

$$\frac{\partial u}{\partial t} = C_v \frac{\partial^2 u}{\partial y^2} \quad (1)$$

Where in equation one, u is the excess pore pressure (kN/m^2), t is review time (sec), y is review depth (m), $\frac{\partial u}{\partial t}$ is the first derivative of excess pore water pressure against time and $\frac{\partial^2 u}{\partial y^2}$ is the second derivative of excess pore water pressure against depth. The general solution of this equation is:

$$W_{Y,T} = \frac{4}{\pi}\sum_{m=0}^{\infty}\frac{1}{(2m+1)^2} Exp^{-\left[\pi^2(2m+1)^2/4\right]T_v} \sin\frac{\pi}{2}(2m+1)Y \quad (2)$$

With vertical consolidation degrees (Uv):

$$\mathrm{Uv} = \frac{\sqrt{\frac{4\,\mathrm{Tv}}{\pi}}}{[1+(4\mathrm{Tv}/\pi)^{2,8}]^{0,179}} \quad (3)$$

With m = integer number; $T_v = \frac{C_v t}{H^2}$ (time factor)

IOP Conf. Series: Materials Science and Engineering **308** (2018) 012008 doi:10.1088/1757-899X/308/1/012008

The magnitude of primary settlement occurs:

$$S_p = \frac{C_c H}{1+e_0}\log\left(\frac{\sigma'_0+\Delta\sigma'}{\sigma'_o}\right) \tag{4}$$

Where in equation four, σ'_0 is initial effective vertical stress (kN/m^2), $\Delta\sigma'$ is aditional effective vertical stress (kN/m^2), C_c is compression index, H is layer thickness (m) and e_0 is initial void ratio.

Radial consolidation will occur in situations that include drainage of a central source, such as on a vertical drainage used under a pile to accelerate pore water drainage by reducing drainage distance and hence also accelerating consolidation.
The consolidation equation for radial directional drainage is as follows:

$$\frac{\partial u}{\partial t} = C_r\left(\frac{\partial^2 u}{\partial r^2} + \frac{1\partial u}{r\partial r}\right) \tag{5}$$

Where in equation five, *Cr* is radial consolidation coefficient (cm^2/s), *R* is vertical drainage radius (cm).

Assuming a smear zone effect has solved by equal-strain consolidation (Barron, 1948) the solution of the radial consolidation equation is as follows:

$$F = F(n) + F(s)$$

$$U_r = 1 - e^{(-8T_r/F)}$$

$$F(n) = \left(\frac{n^2}{n^2-1}\right)\left(\ln(n) - \frac{3n^2-1}{4n^2}\right)$$

Where:

$$S = \frac{r_s}{r_w}$$

$$n = \frac{d_e}{d_w}$$

$$F(s) = \left[\left(\frac{kh}{ks}\right) - 1\right]\ln\left(\frac{ds}{dw}\right)$$

In general, n> 20 so it can be considered to be l/n = 0 so, the formula F (n) will be a simple formula that is:

$$F(n) = \ln(n) - 3/4 \quad or \tag{6}$$

$$F(n) = \ln(de/dw) - 3/4 \tag{7}$$

Where, *De* is equivalent diameter (after cross-section is converted to circle shape) (m), *Dw* is vertical drainage diameter (m), *Rs* is smear zone radius (m), *Rw* is sand drain radius (m), *Ks* is radial permeability coefficient on smear zone are (1-15) kr (m/sec), *Kr* is radial permeability coefficient are (1-15) kv (m / sec), and C_r is $C_v(k_r/k_v)$ (cm^2/sec).

$$T_r = \frac{C_r t}{d_e^2} \tag{8}$$

The smear zone effect is the reduced coefficient value for the clay soil near the vertical drainage or the vertical drainage diameter used is minimized, this is due to the process of remoulding during the installation of vertical drainage by using a pivot.

1.2. Vertical Drain

Table 1. Prefabricated Vertical Drain Installation Method

Type of Drain	Installation Method	Drain Diameter (m)	Distance between Drain (m)	Maximum length (m)
Sand drain	Pushing into the ground with the vibrating system	0,15 – 0,6	1 - 5	≤ 30
Sand drain	Using a long hollow bar with auger	0,3 – 0,5	2 - 5	≤ 35
Sand drain	Jet grouting method	0,2 – 0,3	2 - 5	≤ 30
prefabricated sand drains (sandwicks)	inserting a tool with vibrating systems using the mandrel tip of the tool, using auger	0.06-0.15	1,2 - 4	≤ 30
Prefabricated band-shaped drains	Inserting the tool into the ground using the mandrel at the end.	0,05 – 0,1 (Equivalent diameter)	1,2 -3,5	≤ 60

Source: *(TenCate Polyfelt® Alidrain Prefabricated Vertical Drain)*

1.3. Finite Element Method

One of the most commonly used computer programs for geotechnical field analysis is the 2D finite element method. There are two types of finite element geometry models commonly used in the 2D finite element method, the Plane-Strain and Axysimetry models. The Plane-Strain model can be used to solve problems in cross-sectional areas with relatively uniform stress and deformation conditions in the direction perpendicular to the cross section. While the Axisymetry model is used for field analysis with radial cross section, with uniform stress and deformation conditions in each radial direction.

Table 2. The Plane-Strain and Axysimetry models

No	Model	Parameter	Remarks
1	Mohr-Coulomb (MC)	E, ν, ϕ, c, dan ψ	Perfect elastic - plastic, is the most widely used models in geotechnical analysis
2	Jointed Rock (JR)	E, ν, G, ϕ, c , ψ	Anisotropic elastic - plastic, used to model the behavior of rock layers that have stratification and certain directions
3	Hardening-Soil (HS)	ϕ, c , ψ, ν, E50, Eur,Eoed, p, Ko, Rf, σ	Isotropic, this model can be used for all types of soil
4	Soft-Soil-Creep (SSC)	ϕ,c, ψ, κ, λ, μ, ν,M, Ko,	Used for soft soil conditions that have time-dependent behavior
5	Soft Soil (SS)	ϕ, c , ψ, κ, λ, ν, M,Ko	The Cam-Clay model, used for the analysis of primary compression and normal consolidated clay

IOP Conf. Series: Materials Science and Engineering **308** (2018) 012008 doi:10.1088/1757-899X/308/1/012008

2. Research Method

The method used in this study is analysis research method:

a. Collect various types of literature in the form of books and scientific papers related to this thesis.
b. Collect data from the result of soil investigation such as standart penetration test (SPT) data , consolidation data, direct shear data, and PVD technical data. The data was obtained from PT. BINA KARYA as Planning Consultant and CHEC - CSCEC - HK - JO as the executing contractor.
c. Analyze the data obtained from the field with reference sources related to the thesis.
d. Calculate the decline of land with or without prefabricated vertical drains (PVD) with the existing methods.
e. Finding the magnitude of decline in soil using finite element method, by modeling the behavior of soil on the Finite Element Method. The soil modeling used is Mohr-Coulomb model.
f. Compare the results between the analytical calculation and the calculation using finite element method, then make conclusions and suggestions.

3. Result and Discussion

3.1 The magnitude of decline

a. Stage 1

$\sigma_{0\text{-}1} = \gamma_{unsat} \times h + (\gamma_{sat} - \gamma_w) \cdot (H_{dr} - 1)$

where, $h = 0{,}7$ m

$\gamma_{unsat} = 9{,}8$ kN/m^3

$\gamma_{sat} = 15{,}2$ kN/m^3

$H_{dr} = 7{,}3/2 = 3{,}65$ m

$\sigma_{0\text{-}1} = (6{,}86 \times 0{,}7) + (15{,}2 - 10) \cdot (3{,}65 - 1)$

$= 20{,}64$ kN/m^2

The magnitude of decline due to layer 1:

$$S_1 = \frac{0{,}942 \cdot 0{,}7}{1+1{,}883} \log\left[\frac{20{,}64 + 6{,}86}{20{,}64}\right]$$

$S_1 = 0{,}23 \cdot 0{,}12$

$S_1 = 0{,}03$ m

b. Stage 2

$\Delta\sigma' = \gamma_{unsatclay} \times H_{clay}$

$= 9{,}42 \times 3{,}3$

$= 31{,}08$ kN/m^2

$\sigma_{0\text{-}2} = \gamma_{unsat} \times h + (\gamma_{sat} - \gamma_w) \cdot (H_{dr} - 1)$

where , $h = 3{,}3$ m

$\sigma_{0\text{-}2} = (9{,}42 \times 3{,}3) + (15{,}15 - 10) \cdot (3{,}65 - 1)$

$= 44{,}73$ kN/m^2

The magnitude of decline due to layer 2:

$$S_2 = \frac{0{,}942 \cdot 3{,}3}{1+1{,}883} \log\left[\frac{44{,}73 + 31{,}08}{44{,}73}\right]$$

$S_2 = 0{,}25$ m

c. Stage 3

$\Delta\sigma' = \gamma_{unsatclay} \times H_{clay}$
$= 10 \times 3{,}3$
$= 33\ kN/m^2$

$\sigma_{0\text{-}3} = \gamma_{unsat} \times h + (\gamma_{sat} - \gamma_w) \cdot (H_{dr} - 1)$

where, h = 3,3 m

$\sigma_{0\text{-}3} = (10 \times 3{,}3) + (15{,}15 - 10) \cdot (3{,}65 - 1)$
$= 47{,}02\ kN/m^2$

The magnitude of decline due to layer 3:

$$S_3 = \frac{0{,}942.\ 3{,}3}{1+1{,}883} \log\left[\frac{47{,}02 + 33}{47{,}02}\right]$$

$S_3 = 0{,}25$ m

So the total primary decline is
$Sp_{total} = S_1+S_2+S_3$
$= 0{,}03$ m + 0,25 m + 0,25 m
$Sp_{total} = 0{,}53$ m
So the magnitude of the soil decline is equal to Sc, where;
Sc = 0,53. U (90%)
= 0.48 m

3.2Rate of Consolidation

From the data obtained
Cv = 0.0114 cm²/sec = 35.95 m²/year
Calculation of declining time rate
Tv 90% = 0.848
Hdr = 7.3 m / 2 = 3.65 m
From equation 2.4 obtained:

$$t = \frac{H^2 dr.T_v}{C_v}$$

$$t = \frac{3{,}65^2.0{,}848}{0{,}0114}$$

$t =$ 115 Hari

Table 3. Calculation Sc

Years	Days	Uv	1 - Uv	Tr	Ur	1 - Ur	U	Sc (m)
0,00	0,00	0,00	1,00	0,00	0,00	1,00	0,00	0,00
0,00	1,00	0,10	0,90	0,04	0,09	0,91	0,18	0,09
0,01	2,00	0,14	0,86	0,08	0,17	0,83	0,28	0,15
0,01	3,00	0,17	0,83	0,13	0,25	0,75	0,37	0,20
0,01	4,00	0,19	0,81	0,17	0,31	0,69	0,45	0,24
0,01	5,00	0,22	0,78	0,21	0,37	0,63	0,51	0,27
0,02	6,00	0,24	0,76	0,25	0,43	0,57	0,56	0,30
0,02	7,00	0,25	0,75	0,29	0,48	0,52	0,61	0,33
0,02	8,00	0,27	0,73	0,34	0,53	0,47	0,66	0,35
0,02	9,00	0,29	0,71	0,38	0,57	0,43	0,69	0,37
0,03	10,00	0,30	0,70	0,42	0,61	0,39	0,73	0,39

0,03	11,00	0,32	0,68	0,46	0,64	0,36	0,76	0,40
0,03	12,00	0,33	0,67	0,50	0,68	0,32	0,78	0,42
0,04	13,00	0,35	0,65	0,54	0,70	0,30	0,81	0,43
0,04	14,00	0,36	0,64	0,59	0,73	0,27	0,83	0,44
0,04	15,00	0,37	0,63	0,63	0,76	0,24	0,85	0,45
0,04	16,00	0,38	0,62	0,67	0,78	0,22	0,86	0,46
0,05	17,00	0,40	0,60	0,71	0,80	0,20	0,88	0,47
0,05	18,00	0,41	0,59	0,75	0,82	0,18	0,89	0,47
0,05	19,00	0,42	0,58	0,80	0,83	0,17	0,90	0,48
0,05	20,00	0,43	0,57	0,84	0,85	0,15	0,91	0,48
0,06	21,00	0,44	0,56	0,88	0,86	0,14	0,92	0,49
0,06	22,00	0,45	0,55	0,92	0,87	0,13	0,93	0,49
0,06	23,00	0,46	0,54	0,96	0,88	0,12	0,94	0,50
0,07	24,00	0,47	0,53	1,01	0,89	0,11	0,94	0,50
0,07	25,00	0,48	0,52	1,05	0,90	0,10	0,95	0,50
0,07	26,00	0,49	0,51	1,09	0,91	0,09	0,96	0,51
0,07	27,00	0,50	0,50	1,13	0,92	0,08	0,96	0,51
0,08	28,00	0,51	0,49	1,17	0,93	0,07	0,96	0,51
0,08	29,00	0,52	0,48	1,21	0,93	0,07	0,97	0,51
0,08	30,00	0,52	0,48	1,26	0,94	0,06	0,97	0,51
0,08	31,00	0,53	0,47	1,30	0,95	0,05	0,97	0,52
0,09	32,00	0,54	0,46	1,34	0,95	0,05	0,98	0,52
0,09	33,00	0,55	0,45	1,38	0,95	0,05	0,98	0,52
0,09	34,00	0,56	0,44	1,42	0,96	0,04	0,98	0,52
0,10	35,00	0,57	0,43	1,47	0,96	0,04	0,98	0,52
0,10	36,00	0,57	0,43	1,51	0,97	0,03	0,99	0,52
0,10	37,00	0,58	0,42	1,55	0,97	0,03	0,99	0,52
0,10	38,00	0,59	0,41	1,59	0,97	0,03	0,99	0,52
0,11	39,00	0,60	0,40	1,63	0,97	0,03	0,99	0,52
0,11	40,00	0,60	0,40	1,68	0,98	0,02	0,99	0,53

3.3 Consolidation time with and without PVD using finite element method

The consolidation time obtained by finite element method until the degree of consolidation reaches 90 is 63 days as shown in figure 2.

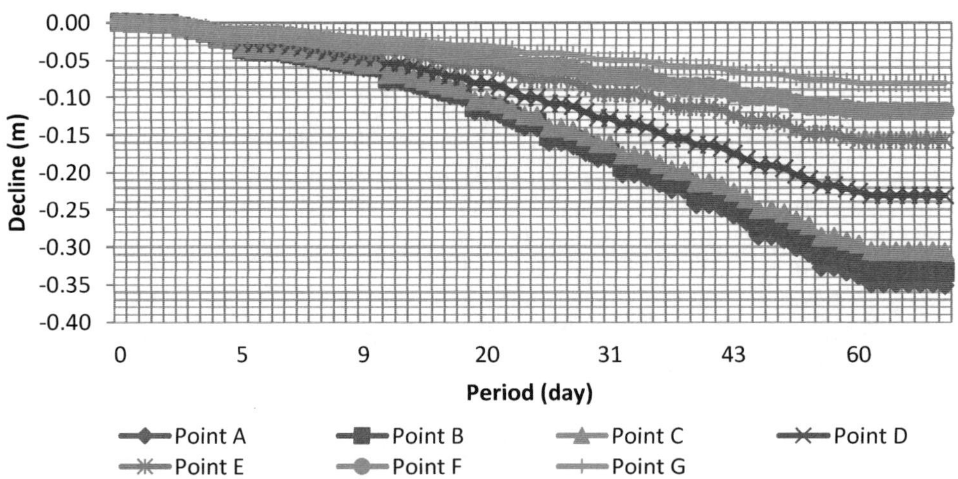

Figure 2. The consolidation time

1. At point A the decline that occurs until the 90% consolidation is 0.35 m
2. At point B the decline that occurs until the 90% consolidation is 0.33 m
3. At point C the decline that occurs until the 90% consolidation is 0.31 m
4. At point D the decline that occurs until the 90% consolidation is 0.23 m
5. At point E the decline that occurs until the 90% consolidation is 0.16 m
6. At point F the decline that occurs until the 90% consolidation is 0.12 m
7. At point G the decline that occurs until the 90% consolidation is 0.01 m

- The result of soil declining occurred due to embankment without the vertical drainage with finite element method. The length of time required to achieve 90% consolidation is 110 days with the magnitude of decline 0.36 m
- The magitude of pore water pressure at 90% of consolidation
  10% x 9,2 (The highest peak of the pore water pressure) = 0,92 KN/m^2

4. Conclusions

1. From the analytical calculation of the soil decline at STA 35 + 901 using prefabricated vertical drain (PVD) to reach 90% consolidation degree obtained a decline of 0.47 m. While the total decline result obtained from Finite Element Method at STA 35 + 901 using prefabricated vertical drain (PVD) to reach 90% consolidation degree is 0.45 m
2. The result of consolidation time calculation with PVD and without PVD:
   - Consolidation time obtained with analytical calculation
     With PVD : 19 days
     Without PVD : 115days
   - Consolidation time obtained with finite element method
     With PVD : 63 days
     Without PVD : 110 days
3. Pore water pressure when it reaches 90% consolidation with PVD occurred on day 9 is 0,92KN/m^2.

Acknowledgements
Financial support for the research was provided by University of Sumatera Utara, Medan

References

[1] Barron R A 1948 Consolidation of Fine-Grained Soils by Drains Wells *Transactions of* ASCE **113**:2346 pp 718 - 754
[2] Bella R A 2011 *Pemodelan Timbunan Pada Tanah Lunak Dengan Menggunakan Program Plaxis* [*Modeling Dosage In Soil By Using Plaxis Program*] (Civil Engineering: Universitas Nusa Cendana)
[3] Bowles, Joseph E 1989 *Physical properties and geotechnical soil 2nd edition* (Erlangga: Jakarta)
[4] Cassagrande, Fadum A R E 1940 *Notes on Soil Testing For Engineering Purposes* (Harvard)
[5] Chai, Miura J C N 1999 Investigation of Factors Affecting Vertical Drain Behaviour *Journal of Geotechnical Engineering ASCE* **125**:3 pp 216 – 226
[6] Chu et al. 2004 Practical Consideration for Using Vertical Drains in Soil Improvement Projects *Geotextiles and Geomembranes* **24**:6 pp 339 - 348
[7] Das, Braja M 1985 *Principle of Geotechnical Engineering I* (Erlangga: Jakarta)
[8] Das, Braja M 1985 *Principle of Geotechnical Engineering II* (Erlangga: Jakarta)
[9] Das, Braja M 2014 *Advanced Soil Mechanics: Fourth Edition* (CRC Press: New York)

IOP Conf. Series: Materials Science and Engineering **308** (2018) 012009 doi:10.1088/1757-899X/308/1/012009

Analysis of the soil reinforcement by using geotextile on the pile of Medan – Kualanamu of highway project (STA 35 + 901) with the finite element method

Ika Puji Hastuty*, Roesyanto, and Agave Manulang
Department of Civil Engineering, Universitas Sumatera Utara, Medan, Indonesia

*E-mail: ika.hastuty@usu.ac.id

Abstract. Consolidation is the process of discharge of water from the soil through pore cavity. Poor subgrade condition which is in the form of plates, is necessary to be repaired so that the subgrade will be able to support the load of construction. One method used as soil improvement is by geotextile. The type of geotextile used on the road construction project (STA 35 + 901) Medan Kualanamu freeway is PP woven polypropleen geotextile. This study aims to determine the magnitude of the settlement, horizontal deformation, tensile strength of geotextile by using finite element method that affect the length of time the land decline to reach 90% consolidation or in other words does not decrease again or is considered zero. The results obtained from the calculation of this study obtained a decrease that occurred using geotextile with finite element method of 0.45 m, the horizontal deformation obtained by using the most extreme elemental method with geotextile was 0.08 m while the horizontal deformation occurring with no geotextile was 0.09 m and the tensile stress obtained by the geotextile tensile stress calculation was 19.51 KN/m^2.

1. Introduction

One of the obstacles on the foundation of road construction is excessive settlement. Foundation settlement process is usually caused because the foundation is not strong enough to withstand the load or the layer of soil under the foundation is compressed due to structural load. Construction has a very varied load, it is consisting of the load of the building itself and the load from the outside (wind, earthquake, etc.).

Generally, the settlement on the ground caused by loading can be divided into two major groups:

- The consolidation settlement, which is the result of changes in the volume of saturated soil as a result of the discharge of water occupying the soil pores
- Immediate settlement, which is the result of elastic deformation of dry, wet, and water-saturated soils without any changes in its water content.

Consolidation is a process of reduction in volume or reduction in pore cavity of a low saturated soil permeability due to loading, wherein the process is influenced by the speed of pore water discharge from the soil cavity. When a layer of saturated soil with a low permeability is given load, then the pore water pressure in the soil will increase immediately. The differences in pore water pressure in the soil layer resulted in water flowing into the soil layer with lower pore pressures, followed by its settlement. This process takes time due to low soil permeability. Consolidation is any process which involves a decrease

in water content of saturated soil without replacement of water by air. In general it is the process in which reduction in volume takes place by expulsion of water under long term static loads.
Geotextile is a thin, flexible, permeable sheet used for stabilization and soil improvement attributed to civil engineering work. Utilization of geotextiles is a modern way of reinforcing soft soil. The function of geotextile besides as reinforcement of soft ground, as the pile ; geotextile is also used as field separator, filter, drainage and as a protective layer.

There are 2 types of geotextile:

1. Woven Geotextile

   Woven geotextile is a type of geotextile made from Polypropilene as the base material. In order to make it easier to apply, Woven Geotextile is made like a rice sack but it is not made of jute material but black from synthetic materials. Woven geotextile has a function as subgrade stabilization material, especially soft soil base so that the soil can be coated and the surface is not easily down, because the tensile of this geotextile is 2 times higher than non-woven geotextile for the same gramation or weight per m^2.

2. Non-woven Geotextile

   Non-woven geotextile is a Fabric Filter shaped like rugs. Generally the base material of non-woven geotextile is made of Polyesther or Polypropylene polymer. Another function of geotextile is as a substitute for gunny sack in the curing process of concrete because it can prevent the occurrence of cracks during the process of draining new concrete.

The advantages of using geotextile:

1. Prevent subbase and base aggregate contamination by soft base soils and distribute an effective traffic load through layers of heap.
2. Eliminate the loss of heap aggregate into soft and basic soil and minimize the costs and additional needs of the 'wasted layer aggregate'.
3. Reduce the thickness of stripping excavation and minimize preparatory work.
4. Increase the aggregate of heat resistance to local collapse on the location of load by strengthening the heap.
5. Reduce uneven settlement and deformation of the finished structure.

1.1. Tensile stress of geotextile

The tensile strength of the geotextile must be strong enough to withstand tears at encounters with collapsing surfaces. In this case geotextile is designed with force in the directions. In the analysis phase consists of two stages, that are:

1. Boundary balance analysis without any reinforcement.
2. Boundary balance analysis with reinforcement

The tensile strength of geotextile can be calculated by this formula:

$$\tau g = (cg + \sigma g \tan\delta) \qquad (1)$$

Where in equation one, τg is geotextile tensile stress (KN/ m^2), Cg is soil adhesion with geotextile (KN/ m^2), (For soft soil, the adhesion between soil and geotextile (cg) can be considered the same as the cohesion (cu) of the soil, so cg = cu), σg is bearing capacity of geotextile (KN/ m^2) and δ is shear angle with geotextile (-).

Bearing capacity of geotextile can be calculated by this formula:

$$\sigma g = \frac{E\epsilon}{a\sqrt{\left(1+\frac{a}{s^2}\right)}} \qquad (2)$$

Where in equation two, σg is geotextile bearing capacity (KN/ m^2), E is modulus of *geotextile*(-), (Efficient value of friction from geotextile to soil (E), for geotextile E = 0.6 - 0.8), ϵ is tension pressure of geotextile (KN/ m^2), a is geotextile length divided by 2 (m) and S is settlement under the load (m)

IOP Conf. Series: Materials Science and Engineering **308** (2018) 012009 doi:10.1088/1757-899X/308/1/012009

1.2. Vertical Drain

Table 1. Prefabricated Vertical Drain Installation Method

Type of Drain	Installation Method	Drain Diameter (m)	Distance between Drain (m)	Maximum length (m)
Sand drain	Pushing into the ground with the vibrating system	0,15 – 0,6	1 - 5	≤ 30
Sand drain	Using a long hollow bar with auger	0,3 – 0,5	2 - 5	≤ 35
Sand drain	Jet grouting method	0,2 – 0,3	2 - 5	≤ 30
prefabricated sand drains (sandwicks)	inserting a tool with vibrating systems using the mandrel tip of the tool, using auger	0.06-0.15	1,2 - 4	≤ 30
Prefabricated band-shaped drains	Inserting the tool into the ground using the mandrel at the end.	0,05 – 0,1 (Equivalent diameter)	1,2 -3,5	≤ 60

Source: *(TenCate Polyfelt® Alidrain Prefabricated Vertical Drain)*

1.3. Finite Element Method

Table 2. The Plane-Strain and Axysimetry models

No	Model	Parameter	Remarks
1	Mohr-Coulomb (MC)	E, ν, ϕ, c, dan ψ	Perfect elastic - plastic, is the most widely used models in geotechnical analysis
2	Jointed Rock (JR)	E, ν, G, ϕ, c , ψ	Anisotropic elastic - plastic, used to model the behavior of rock layers that have stratification and certain directions
3	Hardening-Soil (HS)	ϕ, c , ψ, ν, E50, Eur,Eoed, p, Ko, Rf, σ	Isotropic, this model can be used for all types of soil
4	Soft-Soil-Creep (SSC)	ϕ,c, ψ, κ, λ, μ, ν,M, Ko,	Used for soft soil conditions that have time-dependent behavior
5	Soft Soil (SS)	ϕ, c , ψ, κ, λ, ν, M,Ko	The Cam-Clay model, used for the analysis of primary compression and normal consolidated clay

Source :(Chu et. al. 2004)

2. Research Method

- Collect various types of literatureS and scientific papers related to this thesis.
- Collecting data from the results of soil investigation such as data of standart penetration test (SPT), colsolidation, direct shear, and geotextile technical data. The data was obtained from PT. BINA KARYA as a consultant planner and CHEC - CSCEC - HK - JO as contractor implementer.
- Analyze the data obtained from the field with reference sources related to the thesis.
- Calculate the settlement of soil with or without geotextile with existing methods.

- Calculationg the amount of settlement using finite element method, by modeling the soil behavior on the finite element method. The soil modeling used is Mohr-Coulomb model.
- Comparing the calculation of the soil settlement obtained by analytical method with the calculations performed with finite element method, then make conclusions and suggestions.

3. Result and Discussion

3.1. Magnitude of Settlement

The calculation of settlement obtained by finite element method is 0.45 meters

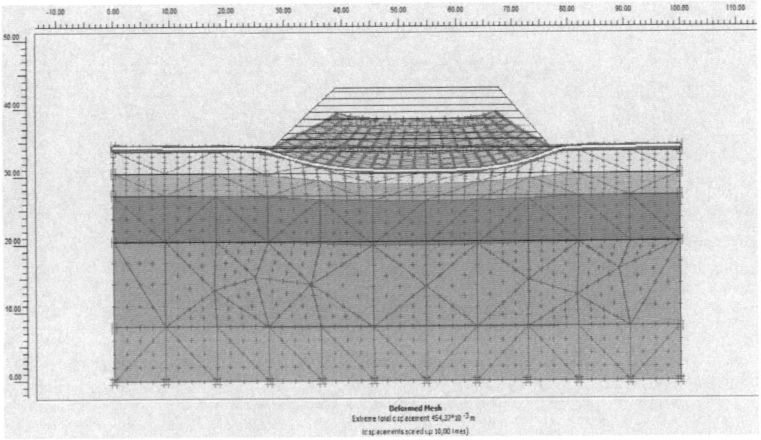

Figure 1. Soil Settlement in Finite Element Method

3.2. Results of Horizontal Deformation Modeling with Finite Element Method

The parameters used in searching the result of horizontal deformation modeling are the same as the settlement parameter as well as the way it works. The result of horizontal deformation calculation obtained by using finite element method and using geotextile is an extreme horizontal deformation which is equal to 0.08 m. While horizontal deformation obtained without using geotextile obtained 0.09 m.

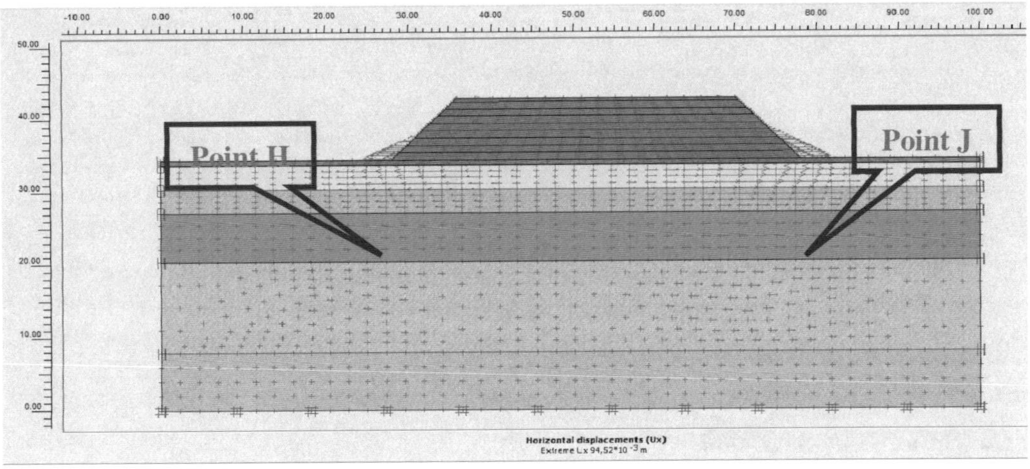

Figure 2. Horizontal Deformation without geotextile

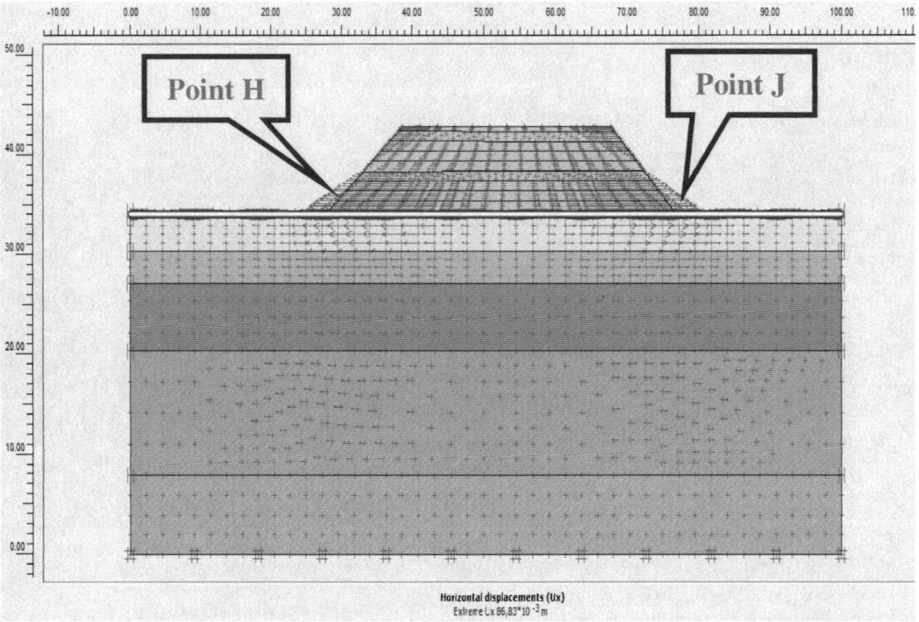

Figure 3. Horizontal Deformation with Geotextile

3.3. Tensile stress on Geotextile

Equation formula 2.6 and 2.7 to calculate the tensile stress of geotextile:

$$\tau g = (cg + \sigma g \tan\delta)$$
$$\tau g = (19{,}5 + \sigma g \tan 7)$$
$$\text{Calculating, } \sigma g = \frac{E\epsilon}{a\,(1+\frac{a}{s^2})}$$
$$\sigma g = \frac{0{,}7\,(60{,}01)}{23\sqrt{(1+\frac{23}{0{,}36^2})}}$$
$$\sigma g = \frac{40{,}007}{307{,}05}$$
$$\sigma g = 0{,}13 \text{ KN}/m^2$$

so,

$$\tau g = (19{,}5 + \sigma g \tan 7)$$
$$\tau g = (19{,}5 + 0{,}13 \tan 7)$$
$$\tau g = 19{,}51 \text{ KN}/m^2$$

The magnitude of Tensile stress on geotextile is 19,51 KN/m^2

4. Conclusions and Sugestions

1. According to the calculation using finite element method, soil settlement on STA 35 + 901 without using Geotextile until it reached 90% of consolidation degree is equal to 0,46 m. While the total

settlement obtained from the finite element method program on STA 35 + 901 using Geotextile until it reached 90% of consolidation degree is 0.45 m.

2. The result of horizontal deformation obtained by using finite element method and located at the point that is observed (the most extreme point H and point J) with geotextile is 0.08 m while horizontal deformation that occurs without geotextile is 0.09 m.
3. The soil improvement technique obtained on the project is that geotextille has the utility of being a separator.
4. According to the calculation, obtained that tensile stress of geotextile is 19,51 Kn/m^2

Acknowledgements
Financial support for the research was provided by University of Sumatera Utara, Medan

References

[1] Bowles, Joseph E 1989 *Physical properties and geotechnical soil 2nd edition* (Erlangga: Jakarta)
[2] Cassagrande, Fadum A R E 1940 *Notes on Soil Testing For Engineering Purposes* (Harvard)
[3] Das, Braja M 1985 *Principle of Geotechnical Engineering I* (Erlangga: Jakarta)
[4] Das, Braja M 1985 *Principle of Geotechnical Engineering II* (Erlangga: Jakarta)
[5] Das, Braja M 2014 *Advanced Soil Mechanics: Fourth Edition* (CRC Press: New York)
[6] Irsyam, Mansyur, *SI-3221 Rekayasa Pondasi* [Foundation Engineering] (ITB: Bandung)

Integration of computer-assisted fracture reduction system and a hybrid 3-DOF-RPS mechanism for assisting the orthopedic surgery

Irwansyah[*1,2], **N P Sinh**[1], **J Y Lai**[1], **T Essomba**[1], **R Asbar**[2], **P Y Lee**[3]

[1]Department of Mechanical Engineering, National Central University, Taoyuan, Taiwan
[2]Department of Mechanical Engineering, Syiah Kuala University, Banda Aceh, Indonesia
[3]Orthopedic Department, Show Chwan Memorial Hospital, Changhua, Taiwan

[*]E-mail: irwansyah@unsyiah.ac.id

Abstract. In this paper, we present study to integrate virtual fracture bone reduction simulation tool with a novel hybrid 3-DOF-RPS external fixator to relocate back bone fragments into their anatomically original position. A 3D model of fractured bone was reconstructed and manipulated using 3D design and modeling software, PhysiGuide. The virtual reduction system was applied to reduce a bilateral femoral shaft fracture type 32-A3. Measurement data from fracture reduction and fixation stages were implemented to manipulate the manipulator pose in patient's clinical case. The experimental result presents that by merging both of those techniques will give more possibilities to reduce virtual bone reduction time, improve facial and shortest healing treatment.

1. Introduction

Utilization the computer-assisted or robotic surgery technology for enhancing orthopedic procedure is still not mature yet. It needs more evident for patients and surgeons to improve the uncertainty and make them sure that those of assisted-technology can provide accurate, predictable and safe treatment. At least two decades research in Computer-Aided Orthopaedic Surgery (CAOS) and Orthopaedic Robotics (CAOR) have happened in design and development level but only a few of them are presently being allowed for clinical trials. In orthopedic clinical practice, complicated bone deformities treatment and unreduced fractures are always problematic issues. In this study, the combination of virtual fracture reduction with a robotic-assisted reduction is proposed to solve those problems. Currently, we have built an integrated preoperative simulation system for orthopedic surgery and have been used as preoperative planning tool for a surgeon to solve clinical cases of bone fracture reduction and implant placement. The system is operated in a PC based environment that integrates the virtual surgery tools in single computer program package, making it easy to implement in clinical applications [1-2]. The aims of our research to study integration of computer-assisted fracture reduction and robotic system. Even though computer-assisted fracture reduction have been developed by several researchers with different level of accuracies, we explore beneficity merging both of cutting-edge-technology. A 3D model of fractured bone is generate directly from a stacks of CT images, segmented in multi region, and virtually reduce and stabilize. A computer-integrated orthopedic system, called FRACAS, was develop for deal with long bone fractures. It replace to use fluoroscopic images to virtual reality display of 3D bone models [3]. A 3D visualization tool was developed to generate a realistic model of the bone-fixator system. The visualization tool has improved the current software to provide a realistic depiction of the treatment procedure [4]. Integrating surgeon instructed, image-guided and robot-assisted applied to reposition long bone fracture. A robotic solution exists to solve the problems of manipulating and reducing long bone fractures [5-6]. Employing 3D imaging and surgical navigation techniques in combination with a robot-assisted reduction procedure to overcome mal-alignment fracture reduction and minimized high x-ray exposures especially for operating team [7]. A new reduction strategy introduced by modification the hexapod computer-assisted fracture reduction system which is detachable, flexible and more precise in coordinate transformations used for closed diaphyseal fracture reduction [8]. A robotic system with 6 d.o.f mobility applied for home positioning of femoral shaft fracture based on Stewart platform [9]. A computer-aided control system might effectively to manipulate external fixators [10]. We outline a system by generating 3D model reconstruction, fracture bone reduction, fragments target coordinate computation, and collecting set of data of rotation and translation fragments displacement to control robot manipulator movement. Then, every single stage of planning reduction simulation and path tracking fracture fragments reduction were presented.

2. Computer-Assisted Fracture Reduction (CAFR)

Reconstruction bone anatomic structure in 3D space is important stages in virtual preoperative planning. Image acquisition, segmentation, virtual fracture reduction and fixation are the virtual surgery simulation features. Focusing on the fracture reduction, the aiming is to position, align the fracture bones and stabilize them into the bone structure. In a simple case, fracture reduction aligns only two bone fragments to their home-position. Meanwhile, for the complex case, fracture reduction needs to align more than two fragments. In some fracture cases, accurate reduction of broken fragments is difficult to achieve due to the desired position of the fragments is not easy to be recognized from available medical images. A 2D standard radiograph or CT images may not be sufficient enough to give a better understanding of complex orthopedic injuries. The virtual preoperative planning attracts surgeon attention to interact with digital medical images platform. It generates 3D model of a fractured bone from a stacked patient's CT images. It also integrates the surgical procedures in a 3D environment to lead advantages of better planning complex fractures. A 3D virtual simulation tool based on a personal computer is employed, enabling surgeons to relocate fracture fragments onto their original position. This laboratory default software, named PhysiGuide, was employed to deal with the reduction of comminuted fracture fragments. It begins with the display and manipulation of CT images on 2D and 3D viewports to allow the user for viewing the fractured bones, as presented in Figure 1.

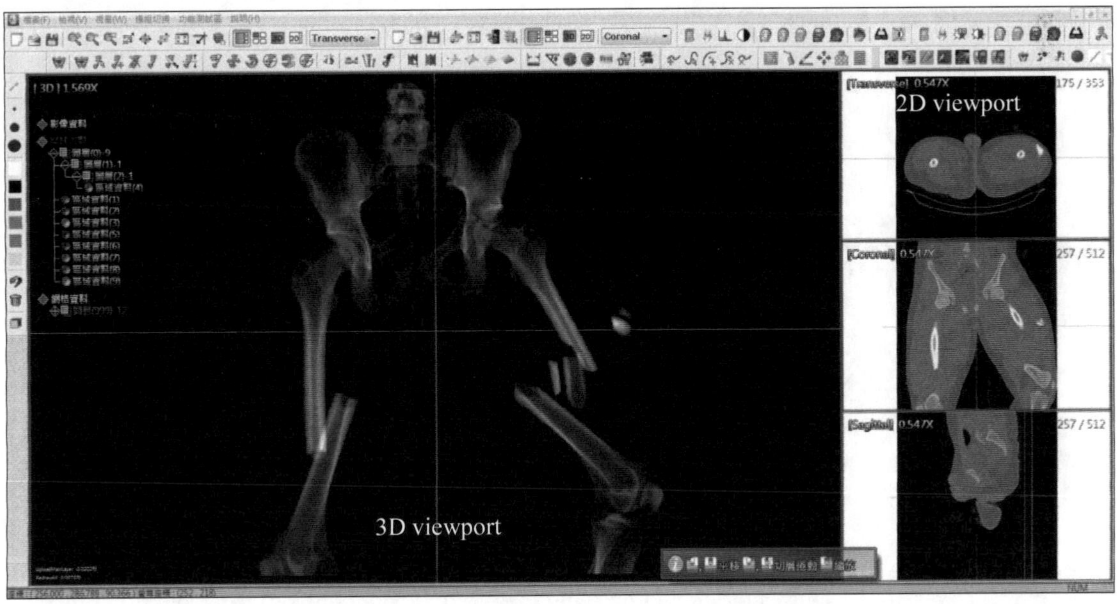

Figure 1. The PhysiGuide software user windows for preoperative planning orthopedic surgery.

Figure 1 presents the comminuted fracture in the hip bone, in left femur, and simple fracture in right femur fragments. In this study, the simple fracture femur is taken as an example due to it frequently encountered injury, the biggest long bone and the one with the most complex post deformity. However, in fact, it is the easiest type of bone to be demonstrated. There are three steps to work into the planning of bone fracture treatment (1) generating anatomical bone and fragments from the scanned 3D images; (2) fixing and stabilizing the fragments to accomplish early recovery and (3) analyzing the virtual fracture reduction under common biomechanical conditions. 3D model of fractured bone is constructed from medical images that represent the bone and fragments involved in a fracture. The 3D models usually represented in a volume data for visualization, in a point cloud for fast interaction and in a triangulation mesh for geometric operation between models purposed. For extracting the meshes from CT stacks, Marching Cubes (MC) algorithm was the most widely used for their simplicity and speed. The size of models generated by these methods depends on the resolution of the medical images and the complexity of the segmented bone structures. Furthermore, the segmentation of medical images is required to obtain the bone regions involved in a fracture. Several studies have been conducted and available on segmentation fractured bone tissue including intensity-based, edge-based, region-based and registration-based. Regardless of the model representing the main purposes of segmentation are to differentiate between bone tissue and other tissues as well as recognition of different bone fragment or labeling [11]. Virtual bone reduction enables to relocate back bone fragments to their original positions and orientations in space. The correspondence between two adjacent bone fragments should be consider to align the complex fracture. In the proposed system, both user interactive and semi-automatic bone reduction method were employed in fragments registration. For user interactive or called also manual bone reduction, each of the bone segments is moved piece by piece through a user interface within three rotations and two translations along the viewing plane [12]. The final result of manual bone reduction must be judged visually. The operating time might lengthen when an excessive number of fragments relocated. In semi-automatic bone reduction, two algorithms are proposed, namely multi-point positioning and mirror positioning. The accuracy of the alignment depends mainly on the accuracy of the point pairs. A registration algorithm [13] is used to align each of the broken fragments with respect to the reference fragment. When all broken fragments are transformed, their might repositioned to original positions in space. For more detailed, a middle shaft fracture femur reduction procedures was presented consists of generating a 3D model from CT-scanned, segmenting and labeling fragments and aligning the two bone fragments into the home position, as depicted in Figure 2.

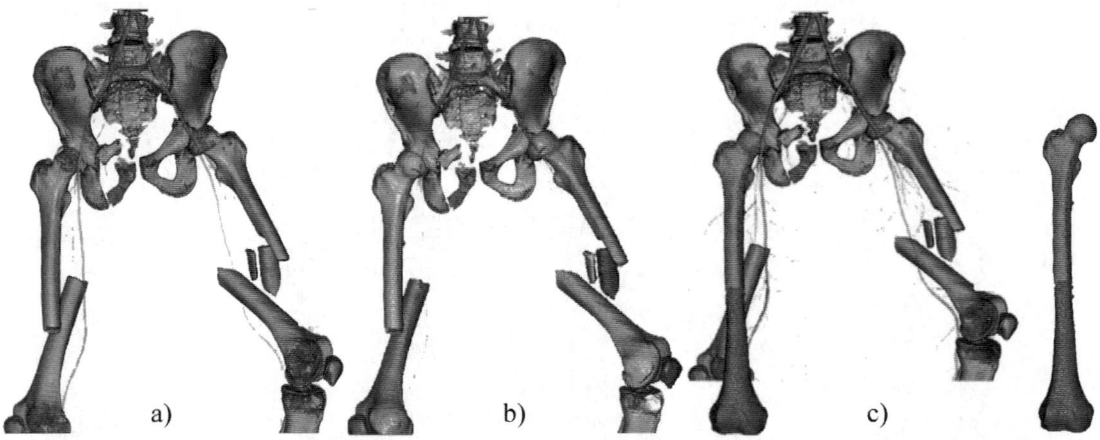

Figure 2. The virtual fracture reduction surgery in the space environment. a) 3D reconstruction from a stacked of CT-images, b) segmentation and labeling the fracture fragments, c) the reduction procedures by paired points registration.

3. Robot-Assisted Fracture Reduction (RAFR)

The robotic manipulator has been widely proposed in orthopedics for assisting fracture fixation, bone lengthening, and deformation correction. The clinician typically drove fractured bone fragments to an anatomically desired position by modifying the rod arms length of the robotic manipulator. This task is accomplished by the clinician with the high requirement in the experience and expertise. As an alternative, the commercial robotic system might hand the same task accompanying with automation mechanism software implementing the mathematical model. The external fixator ring refer to Stewart platform was selected as the main platform. It adopts the analogy of Ilizarov's external fixation device. Stewart platform has several advantageous as a robotic fracture repositioned, in term of high stiffness and precision, repeatability, working space in 6 d.o.f and high load to weight ratio. However, it is not suitable for all application. A 3-RPS (revolute-prismatic-spherical) got a lot of attention, instead of has some ability with Stewart platform, it easy to control and low cost fabrication. According to those reasons, we proposed a novel design of 3 d.o.f star-double triangle spherical parallel manipulator. The basic concept of this manipulator integrating a 3-DOF- Revolute–Prismatic–Sphere and Double Triangular Spherical Planar manipulator into a single device. The main structure comprises a ring base as a fixed platform and equilateral triangular platforms as a moving platform of the robot that fitted into the proximal and distal fragments of the fractured bone using pins or wires. The structure of double-triangle look-like the star composed two combinations of RPS manipulator introduced by K.M. Lee [14] and Triangular Planar-Parallel Robot (PRP) suggested by Daniali [15]. The concept design of this novelty robot shown in Figure 3. This novel manipulator is developed based on three identical RPS manipulator legs. The mechanism consists of the first joint (R-joint) connecting to the base and the last joint (S-joint) attached to the moving platform. The upper side of links are connected to the moving platform by the spherical joint and the other ends of the links are connected to the base platform through the revolute joints. The planar robot is a specific symmetrical closed-loop mechanism, consist of a pair of triangles with straight edges, placed on the top of the frame and another on top of the first triangle sliding along the edge pairs of interconnected points of three fixed edges. The moving platform connected to the base platform by three extensible linear actuators. The lengths of actuators are changed with prismatic joints (P-joints), the platform move on 2 d.o.f rotation and 1 d.o.f in translation. The planar parallel mechanism move in the planar motions.

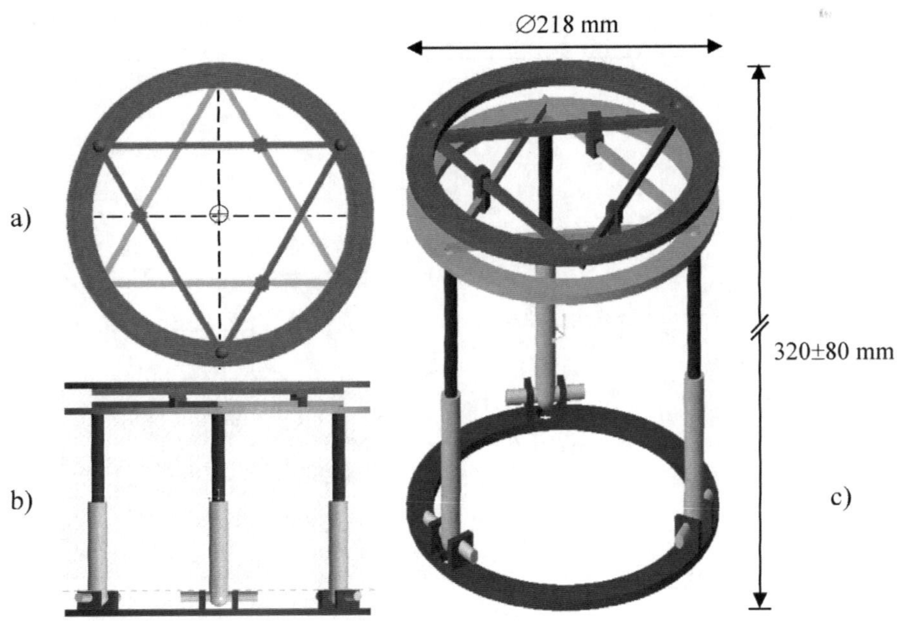

Figure 3. Conceptual design of CAD model RPS manipulator. a) top-view, b) front-view, c) iso-view

IOP Conf. Series: Materials Science and Engineering **308** (2018) 012010 doi:10.1088/1757-899X/308/1/012010

The virtual proposed manipulator able to work in the maximum length 320±80 mm and the diameter of the platform is 218 mm. It also can maneuver in XYZ rotation about ± 15-20° and XY translation ± 50 mm, Z translation movement ±237 ÷ 384 mm. Detailed of the parts dimension and platform configuration of the manipulator is presented in table 1.

Table 1. The virtual proposed manipulator platform configuration parameters.

Parts	Dimensions (mm)		d.o.f	Joint range
Top plate (moving platform)	∅218 x 10		-	-
Base plate (fixing platform)	∅218 x 10		-	-
Linear actuator	∅20 x 320±80		1	80 mm
Ball Joints	∅20		3	± 15°
Revolute Joints	∅5		1	± 45°
Movements	Platform Configuration			
Rotation X- α	± 15°			
Rotation Y- β	± 15°			
Rotation Z- γ	± 20°			
Translation X	± 50 mm			
Translation Y	± 50 mm			
Translation Z	± 237 ÷ 384 mm			

Figure 4a depicts the basic concept and architecture of a hybrid 3-DOF-RPS manipulator used to perform virtual reduction simulation of the deformed and fractured bone. Kinematic structure, the coordinate axes are fixed for different joints of manipulator. The frame O_{xyz} fix on to platform base, whereas O triangle centroid of $A_1A_2A_3$ with Z axis perpendicular to platform base. Frame coordinate of $O'_{x'y'z'}$ and $O''_{x''y''z''}$ laying down on the moving platform $B_1B_2B_3$ and $C_1C_2C_3$ with Z axis pointed upward and normal to platform base as shown in Figure 4b. The spatial 3-RPS parallel manipulator comprises of three spherical joints (S) located at the vertices of the equilateral triangle B_i (i = 1, 2, 3) of the moving platform B. Three revolute joints (R) at the corners of the equilateral triangle A_i (i = 1, 2, 3) of the base platform A. Both of fixed base and moved platform merged by prismatic joints. The planar robot is a unique symmetrical closed-loop mechanism with a pair of straight edges triangle. The upper mechanism (platform C) lay down on top of platform B and sliding along the edge pair intersection points of platform.

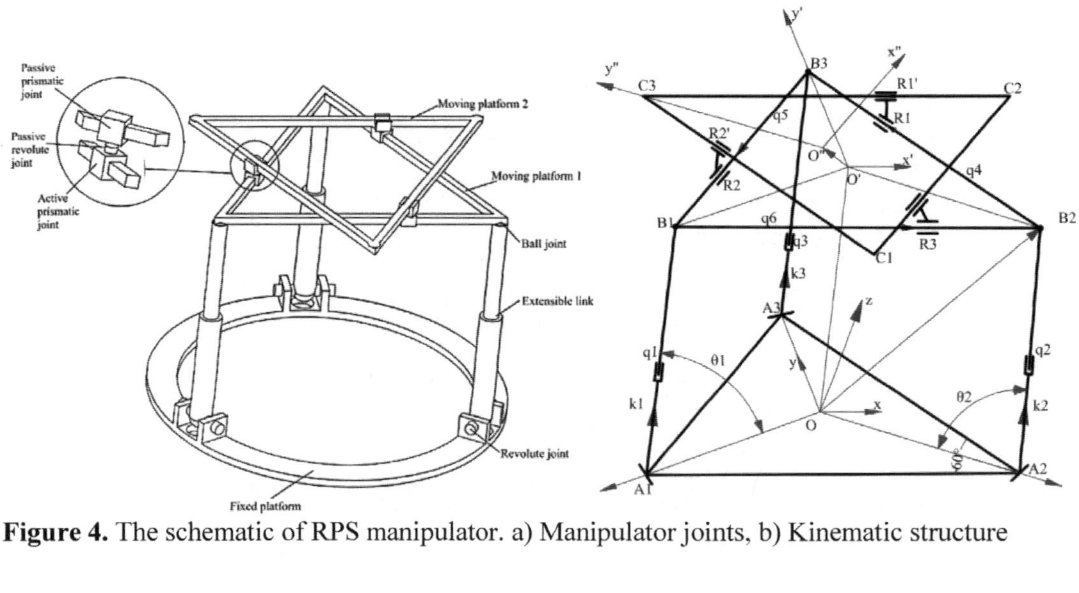

Figure 4. The schematic of RPS manipulator. a) Manipulator joints, b) Kinematic structure

Assuming the end-effector of the manipulator coincides with the center point $O'_{x'y'z'}$ and $O''_{x''y''z''}$ of the moving upper platform. Base coordinate system of x, y, z, laying on the center point O of the base platform with the plane xy coinciding with the base platform. The joint coordinates are the lengths q_1, q_2 and q_3 of the platform legs. The x'y'z' and x''y''z'' are the coordinate system attached to the moving platform with origin at point O' and O''. Considering the coordinates of the vertices B_1, B_2, B_3 of the upper platform of the manipulator. From the constraints applied by the rotational joints at points 1, 2, 3, the coordinates of the bottom platform points are obtained as follows:

The coordinates of the revolute joints relative to the original of the fixed platform:

$$\overrightarrow{OA_1} = \begin{bmatrix} -\frac{\sqrt{3}}{2}r \\ 1 \\ -\frac{1}{2}r \\ 0 \end{bmatrix} \overrightarrow{OA_2} = \begin{bmatrix} \frac{\sqrt{3}}{2}r \\ 1 \\ -\frac{1}{2}r \\ 0 \end{bmatrix} \overrightarrow{OA_3} = \begin{bmatrix} 0 \\ r \\ 0 \end{bmatrix}$$

The coordinates of the sphere joints in the moving platform O'x'y'z' frame:

$$\overrightarrow{O'B_1} = \begin{bmatrix} -\frac{\sqrt{3}}{2}r \\ -\frac{1}{2}r \\ 0 \end{bmatrix} \overrightarrow{O'B_2} = \begin{bmatrix} \frac{\sqrt{3}}{2}r \\ -\frac{1}{2}r \\ 0 \end{bmatrix} \overrightarrow{O'B_3} = \begin{bmatrix} 0 \\ r \\ 0 \end{bmatrix}$$

and the coordinates of C_1, C_2, and C_3 in frame O"x"y"z":

$$\overrightarrow{O''C_1} = \begin{bmatrix} -\frac{\sqrt{3}}{2}r \\ 1 \\ -\frac{1}{2}r \\ 0 \end{bmatrix} \overrightarrow{O''C_2} = \begin{bmatrix} \frac{\sqrt{3}}{2}r \\ 1 \\ -\frac{1}{2}r \\ 0 \end{bmatrix} \overrightarrow{O''C_3} = \begin{bmatrix} 0 \\ r \\ 0 \end{bmatrix}$$

The length of the individual links are denoted by q_i. In the fixed frame O-xyz, the position the spherical joints B_i can be expressed as:

$$\overrightarrow{OB_\iota} = \overrightarrow{OA_\iota} + \overrightarrow{A_\iota B_\iota} = \overrightarrow{OA_\iota} + q_i \cdot \overrightarrow{k_\iota} \tag{1}$$

and

$$\overrightarrow{OB_\iota} = \overrightarrow{OO'} + \mathbf{R}.\overrightarrow{O'B_\iota} = \overrightarrow{OO'} + \{0\} \cdot \overrightarrow{O'B_\iota} \tag{2}$$

where

$\overrightarrow{OO'} = \begin{bmatrix} x_{0'} \\ y_{0'} \\ z_{0'} \end{bmatrix}$ denotes the position of the origin of the center moving platform 1 in the fixed frame.

R is the orientation matrix of frame O'x'y'z' with respect to Oxyz with α, β, and γ being the RPY angles in the fixed frame

$$R = \begin{bmatrix} \cos(\alpha)\cos(\beta) & \cos(\alpha)\sin(\beta)\sin(\gamma) - \sin(\alpha)\cos(\beta) & \cos(\alpha)\sin(\beta)\cos(\gamma) + \sin(\alpha)\sin(\gamma) \\ \sin(\alpha)\cos(\beta) & \sin(\alpha)\sin(\beta)\sin(\gamma) + \cos(\alpha)\cos(\gamma) & \sin(\alpha)\sin(\beta)\cos(\gamma) - \cos(\alpha)\sin(\gamma) \\ -\sin(\beta) & \cos(\beta)\sin(\gamma) & \cos(\beta)\cos(\gamma) \end{bmatrix} \tag{3}$$

Herewith, there are 6 coordinates but three of them x0', y0', α dependent. Whereas β, γ, and z0' are three generalized coordinates. Similar approaches were used to define the coordinates of the moving platform 2. The velocity analysis and Jacobian matrices were calculate according to equation (1) and (2)

$$q_i \cdot \overrightarrow{k_\iota} = \overrightarrow{O'B_\iota} - \overrightarrow{OA_\iota} \tag{4}$$

where $\overrightarrow{k_\iota}$ is unit vector along three legs q_i.By substituting the equations (1) and (2) and differentiating equation (4), we found:

$$\dot{q}_i \cdot \overrightarrow{k_i} + q_i \cdot \left[\dot{\theta}_i \times \overrightarrow{k_i}\right] = \dot{\overrightarrow{OO'}} + \vec{\omega} \times \{0\} \cdot \overrightarrow{O'B_i} \quad (5)$$

where θ_i is the angular velocity of leg q_i and ω is the angular velocity of the moving platform 1 respectively to the fixed platform.

$$\dot{q}_i = \dot{\overrightarrow{OO'}} \cdot \overrightarrow{k_i} + \vec{\omega}\left(\overrightarrow{O'B_i} \times \overrightarrow{k_i}\right) \quad (6)$$

Set the Jacobian matrix

$$J_x = \begin{bmatrix} 1 & 0 & 0 \\ 0 & 1 & 0 \\ 0 & 0 & 1 \end{bmatrix} \quad (7)$$

$$J_y = \begin{bmatrix} \overrightarrow{k_1}^T & \overrightarrow{O'B_1} \times \overrightarrow{k_1} \\ \overrightarrow{k_2}^T & \overrightarrow{O'B_2} \times \overrightarrow{k_2} \\ \overrightarrow{k_3}^T & \overrightarrow{O'B_3} \times \overrightarrow{k_3} \end{bmatrix} \quad (8)$$

hence, arranged the equation (6) become:

$$J_x \cdot \begin{bmatrix} q_1 \\ q_2 \\ q_3 \end{bmatrix} = J_y \cdot \begin{bmatrix} \dot{x}_{0'} \\ \dot{y}_{0'} \\ \dot{z}_{0'} \\ \dot{\alpha} \\ \dot{\beta} \\ \dot{\gamma} \end{bmatrix} \quad (9)$$

4. Methods

The process of investigation synergetic of the combination 3D imaging fragments reduction technique and path planning a parallel robotic mechanism consist of three main consecutive steps reduction-fixation procedure is depicted in Figure 5.

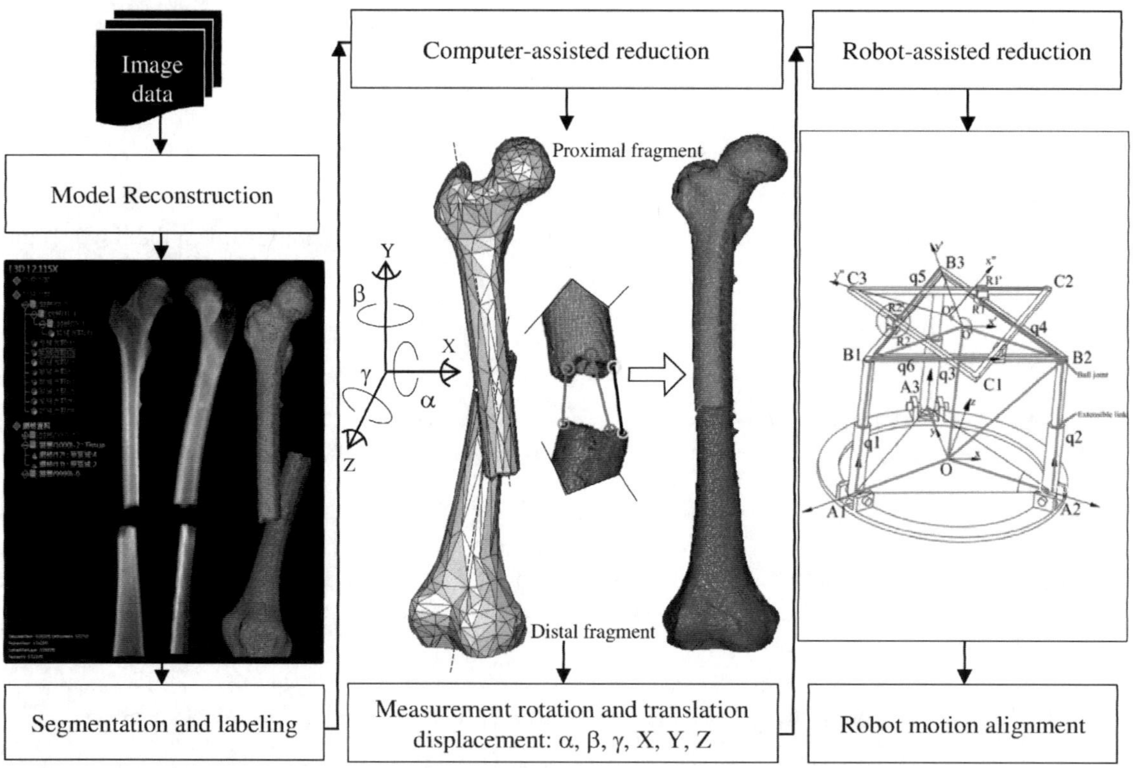

Figure 5. The main procedures of computer-aided preoperative with the planar parallel manipulator

These steps can be described as the following: (1) 3D model of the fractured bone are reconstructed from x-ray or CT-scan radiography. The fracture fragments segmentation and labeling are applied to the specifically desired fracture bone. Then, triangulation mesh is generated for ease manipulation reposition. The fracture fragment displacement in rotation and translation are measuring. (2) The virtual reduction is conducting with a paired point to point registration. The virtual fractured reduction simulation was reconstructed using the software PhysiGuide, version 2.7.2 (NCU Taiwan, bio-images, and clinical assistant laboratory) on a Desktop computer (Intel (R) Core (TM) i5-4440 CPU, 3.1 GHz processor, 4 GB RAM, Windows 7 operating system. All of the parameter measuring such as α, β, γ, X, Y, Z during the virtual reduction is recorded and used as the input for the next step. (3) The relative position of the fragments is determined according to virtual reduction data. Using inverse kinematic solution, an initial robot configuration (the length of the prismatic actuators) is determined to be fixed to the fragments, pull them apart and align them, with the final robot configuration of parallel moving and fixed platforms. The proximal and distal fragments are fixed to the fixed and moving platforms, respectively, by means of pins or wires. The prismatic pairs are actuated until the final configuration is obtained and then locked. The actuators are removed from the robot and the remaining frame acts as an external fixation device until the fracture heals. These steps were designed to be performed automatically under the supervision of the orthopedic surgeon who is able to slow down and/or stop the process. To encourage the fractured bone reduction assisted by robot mechanism, we proposed to develop the graphical user interface (GUI) for the clinician in a user-friendly environment. The clinician performs some measurements on these measured parameters along with certain known variables such as angular parameters, axial rotation, ring radius, legs length to the GUI. The parameters currently outcomes from the result of the virtual fractured bone reduction computer-assisted simulation. Those data will input to GUI for manipulating the reduction deformed bone robot movements. The GUI developed with Unity 5.4 that involves 6 main input box which is Δγ, Δβ, Δz OB, Δx, Δy, as depicted in Figure 6. The coding program is accomplished by the NCU Taiwan, medical assistance robotic system laboratory, with C# programing language refer to coordinate transformation as described in equation 1 to 9 in a windows environment. The simulation is running on a personal computer (Microsoft Windows 7 operating system with Intel Core i5, 3.6 GHz, and 8 GB RAM).

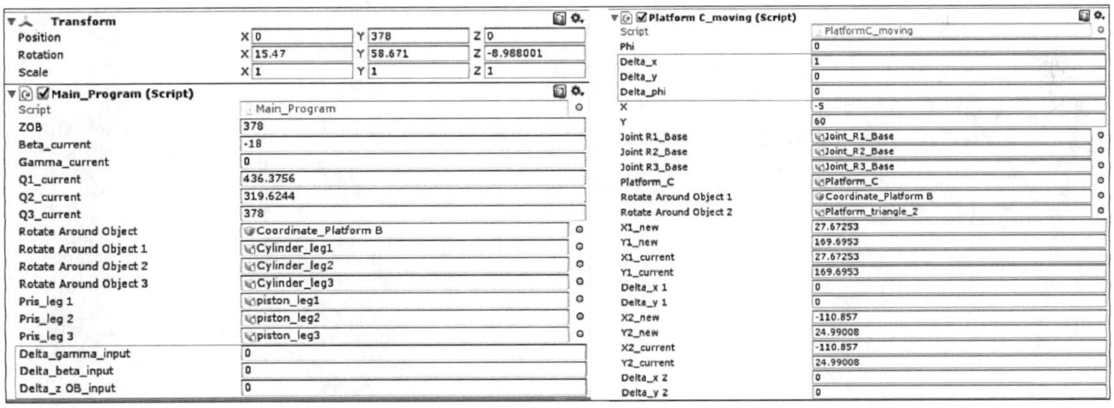

Figure 6. GUI windows displaying the operator interface of robot movement software program

5. Results and Discussion

To evaluate feasibility study of virtual fractured bone reduction and 3-RPS manipulator, we measure displacement of bone fragments and simulate bone reduction using paired point registration. Figure 7 show the measurement result of fractured bone displacement. Then, α, β, γ, ΔX, ΔY, ΔZ values were input into robotic model graphical user interface to control manipulation of arm length, drove fractured fragments back to their original position. The measurement of displaced fragments in translation and rotation were computed. The root mean square deviation was used to evaluate fragments alignment error.

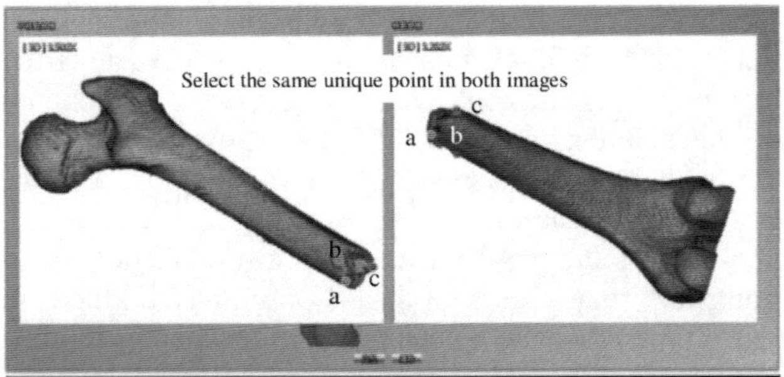

Three translation degree of freedom (d.o.f)	Three rotational degree of freedom (d.o.f)
$\Delta X = 24.67 \approx 25$ mm	$\alpha = 22.30°$
$\Delta Y = 27.13 \approx 27$ mm	$\beta = 13.45°$
$\Delta Z = 70.82 \approx 71$ mm	$\gamma = 0°$

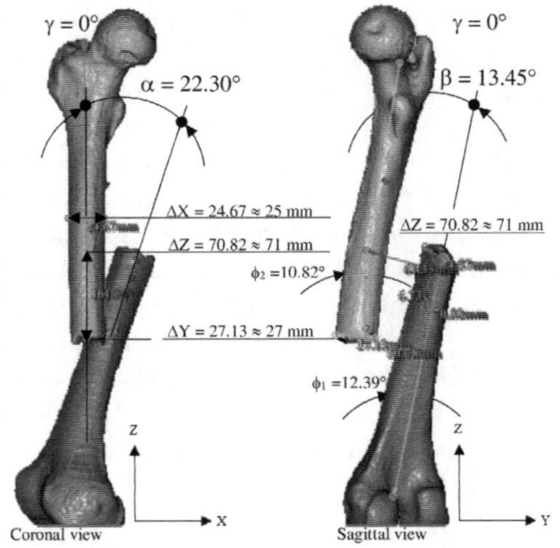

Figure 7. The measurement result of fractured fragments dislocation

The relative length between two centroid bounding box coordinates is measured. The grey colour fragment is fixed as reference, and the brown fragment move to match the reference as shown in Figure 8. Both fragments are realign closest possibly as their anatomy shape respectively to ground truth.

Parameter	Results	
	Fragment 1 (reference)	Fragment 2
$d_s - x$ (m)	0	0.03203
$d_t - y$ (m)	0	0.03411
$d_l - z$ (m)	0	0.02820
α (°)	0	55.17°
β (°)	0	34.3°
γ (°)	0	90°

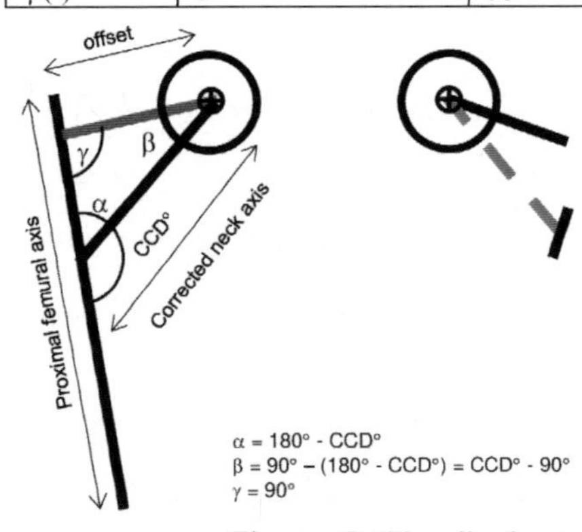

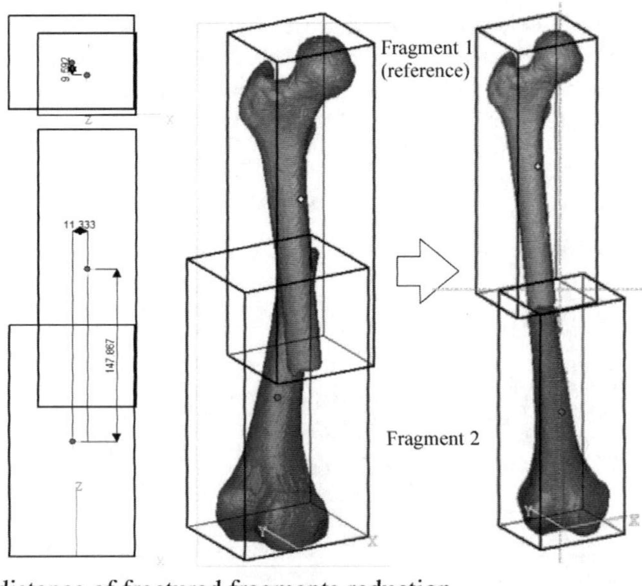

Figure 8. The displacement distance of fractured fragments reduction

The external fixator simulation was conducted to relocate back bone fragments. Actuators lengths, rotation angle of universal and spherical joints were calculated respectively from virtual bone fracture reduction planning. The coordinate transformation of the reduction movement was performed between two adjacent fragments. Using the same setting-up of fragment transformation was applied to execute the manipulator. The moving fragment was drove into the fixed platform to align the bone fragment. In first step, moveable platform was manipulated in such a way trajectory to attain coincidence axis of both fragments, as shown in Figure 9a-e. Then, fragments displacement and miss-orientation from other view was evaluated, as depicted in Figure 9f-j. Finally, the fractured fragments are validated refer to the healthy bone geometry and femoral mechanical axis. Figure 9j-k show 3D pose orientation view.

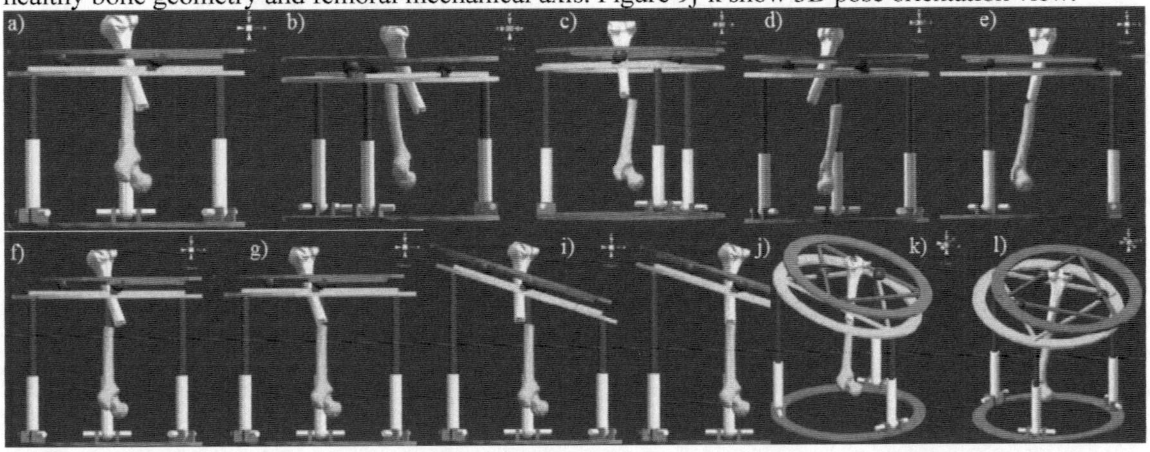

Figure 9. The graphical stages of fracture reduction with sequential adjustment and deformation correctness evaluation. a-e) reduction in anterior view, f-j) posterior view, k-l) iso view

6. Conclusion

In this study, we have developed the virtual model of fractured bone reduction graphical user interface to integrate with 3-RPS manipulator for repositioning the fracture bone into anatomically original position. The proposed integrated system was shown potentially improving alignment visibility of bone reduction. The outcomes simulation of virtual bone reduction were employed to control and manipulate external mechanism actuator to match bone fragments. Due to limitation in accuracy and unpredicted soft tissue, the intuitive supervision is suggested with high skill and experience human operator.

References

[1] Irwansyah, Lai J Y, Lee P Y and Chung C Y 2016 *Proc. of XIV Int. Symp. on 3D Analysis of Human Movement* July 18-21 Taipei Taiwan pp 141-144

[2] Lee P. Y, Lai J. Y, Hu Y. S, Huang C. Y, Tsai Y. C and Ueng W. D 2012 *Biomedical Engineering: Applications, Basis, and Communications* 24 3 pp. 1250007-1 - 1250007-18

[3] Tockus L, Joskowicz, Simkin A and Milgrom C 1998 *Int. Conf. on Medical Image Computing and Computer-Assisted Intervention-MICCAI* pp 29-38

[4] Avsar E and Un K 2016 *Informatics in Medicine Unlocked* 2 pp. 78-91

[5] Mukhherjee S, Rendsburg M and Xu W.L, 2005 *1st Int. Conf. on Sensing Technology*, November 21-23, Palmerstone North, New Zealand pp 78-84

[6] Graham A E, Xie S Q, Aw K C, Xu W L, and Mukherjee S Robotic Long Bone Fracture Reduction Medical Robotics Book edited by Vanja Bozovic ISBN 978-3-902613-18-9 pp.526 I-Tech Education and Publishing Vienna Austria

[7] Westphal R, Winkelbach S, Gosling T, Oswald M, Hufner T, Krettek C, Wahl F. M 2009 *Advances in Robotics Research* Springer Berlin, Heidelberg pp 251-262

[8] Du H, Hu L, Wang T, Zhao L, Li Y, Liu DZhang L, He C, Zhang L, Hou H, Zhang L, Tang P 2015 *The Int. J.l of Medical Robotics and Computer Assisted Surgery* 11 pp. 348-359

[9] Majidifakhr K, Kazemirad S and Farahmand F 2009 *Studies in Health Tech. and Inf.* pp 142-177

[10] Avsar E, Akcali I. D, Aydin A, Un M. K, Mutlu H, Ibrikci T, Ozkan C, Bicer O. S, 2015 *Cukurova Universiti Muhendislik Mimarlik Fakultesi Dergisi* 20 (2) pp 93-101

[11] Jimenez-Delgado J. J, Paulano-Godino F, Pulido Ram-Ramirez R, Jimenez-Perez J. R 2016 *Medical Images Analysis* 30 pp 30-45

[12] Lee P. Y, Lai J. Y, Yu S. H, Huang C. Y, Hu Y. S and Feng C. L 2013 *JMBE* 34 4 pp 368-376.

[13] Lai J. Y, Chen K. J 2007 *The Int. J. of Advanced Manufacturing Technology* 32 11 pp 1188–1200

[14] Lee K. M and Arjunan S. A 1991 *IEEE Transactions on Robotics and Automation*, 7 5 pp 634-641
[15] Daniali M, Zsombor-Murray H. P, Angeles P. J, 1993 *Computational Kinematics*, pp 153-164

IOP Conf. Series: Materials Science and Engineering **308** (2018) 012011 doi:10.1088/1757-899X/308/1/012011

Detection of active noise control on the standard motorcycle exhaust Supra X 125 D using PVC pipe technique form Y

I Isranuri[1], Alfisyahrin[*2], A R Nasution[3]

[1]Department of Mechanical Engineering, University of Sumatera Utara, Medan 20155, Indonesia
[2]Department of Electrical Engineering, University of Syiah Kuala Darussalam, Banda Aceh 23111, Indonesia
[3]Department of Mechanical Engineering, University of Sumatera Utara, Medan 20155, Indonesia

[*]E-mail : ikhwansyah@usu.ac.id; alfisyahrin@unsyiah.ac.id

Abstract. This detection aims to obtain noise reduction on the supra X 125D motorcycle exhaust by using the Active Noise Control Method. The technique is done using a Y-shaped PVC pipe to be bolted on the exhaust, which then branch Y PVC is placed loudspeaker with impermeable conditions. The function of this loudspeaker is as a secondary noise to counter the primary noise of the sound of exhaust motorcycle Supra X 125D. The sound generator in this study is the ISD 4004 module, which serves to generate noise to counter the source noise. How this ISD 4004 module works is by recording source noise then recording the source noise and then reversed the phase 180° by phase reversing circuit. So that, the noise generated by the sound generator will hit the source noise and encounter or such as addition of two different phase of sound will result in noise reduction when detected at the end of the Y-shaped PVC pipe. Inverted phase reversed using feed-back resistor 1 kΩ and 2 kΩ input resistors, 16V capacitor 2500µf and as amplifier using ICL 7660 and TL 702 CP. Test results on the highest 1000 rpm rotation engine speed on the Z axis of 2 dB, and at the highest 2000 rpm rotation engine speed also occurs on the Z axis of 1.5 dB.

1. Introduction

Comfort and tranquility in the environment is a thing that is very desired by many people, but unfortunately at this time the noise level of environment has not made comfort in human hearing again. It relate to the development of motor fuel engines, gas turbines, fan, blower and so on that are widely used in factories and also on transportation equipment that emit noise to the environment.

Commercial motorcycle exhaust (made of manufacturer) still emit noise significantly in the environment. The conventional method of silencing is done passively by placing dampers around the noise source or in the damped space. In this study, an active damping method is created by generating different phase of noise i.e 180° with the existing noise source. When both sources of sound are reunited, then both will weaken each other. But all experts are understandable in practice there is no ideal condition, meaning it is not possible zero conditions can be achieved. Therefore, the authors create the design of the Active Noise Control device, which will be tested on the muffler standard Supra X 125D motorcycle.

From some previous research which has been done above, has been reported that is applying active noise control on simulation media of PVC tube pipe by playing sound of speaker 1 and speaker sound 2 with same frequency and seen noise reduction happened to oscilloscope. While the method of damping with carpet made on PVC tube and tested directly on motorcycle exhaust by comparing the noise of exhaust noise before given carpet damping and noise after given carpet damper by calculating the absorbing coefficient of the sound.

In this research, an electronic device capable of generating sound signal to fight the sound of standard exhaust that is same frequency to be considered and make phase shift 180° so that noise reduction can occur. The method that is done, a signal generator ISD 4004 records the sound of the exhaust. A phase shifter function increase the phase of counter sound then translated through the speakers with the phase upside down and encounter with the sound of the exhaust directly.

2. Method

This research is conducted with two stages as follows: (1) Experiment was carried out on a PVC tube that a sound signal and a sound opponent signal was taken through two speakers coming from the same sound source. Here, from an amplifier without a phase reversal device. The distance between the two speakers is adjusted and the measurement direction using the sound pressure level is also set up. The experimental scheme can be seen in Figure 1 to prove whether there is a noise reduction [1].

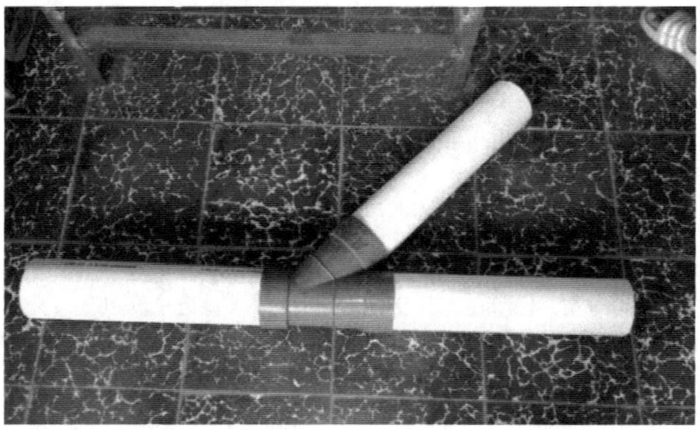

Figure 1. Experimental design using PVC tubes.

Caption: (1) The length of the pipeline is 45 cm; (2) the overall pipe length is 120 cm; (3) diameter of pipe 11.5 cm (4 inch); (4) Pipe angle 45 degree angle.

At the time of the test, occurred echoes on PVC tubes. Acoustical material has been added as anti-echo sound absorbers in the form of carpets. The ratio of the sound energy absorbed by a material to the sound energy coming on the surface of material is defined as the coefficient of sound absorption or the absorption coefficient (α). It can be explained in the following formula.

$$\alpha = \frac{W_a}{W_i} \quad (1)$$

Where:
α : Coefficient of sound absorption.
Wa: the sound energy absorbed.
Wi: the sound energy that comes on the surface of material.

From the test, it is obtained that the absorption coefficient of sound before the carpet damper installed around 78.4 dB and after installed carpet damper around 78.2 dB. The absorption coefficient of sound:

$$\alpha = \frac{0.2\ dB}{78.4\ dB} \tag{2}$$

$$\alpha = 0.0025\ dB$$

So, the carpet absorption coefficient is 0.0025 dB
The form of carpet damper mounted on PVC tube can be seen in Figure 2 below

Figure 2. Carpet Silencer on PVC Pipes.

Regarding the experiment on PVC tube, the test result is reported as follow:
1. The amount of sound generated by speaker 1 is 78.2 dB.
2. The amount of sound generated by speakers 2 is 78.2 dB.
3. Measurement results with Sound Pressure Level of the speaker 1 and speaker 2 is 77,1 dB. So the noise reduction that occurred in this experiment was 78.2 dB - 77.1 dB = 1.1 dB.

A prototype of active noise control with phase inverting is designed to obtain a greater noise reduction. For more details set-up of noise data retrieval on the exhaust, sketched picture can be seen in Figure 3.

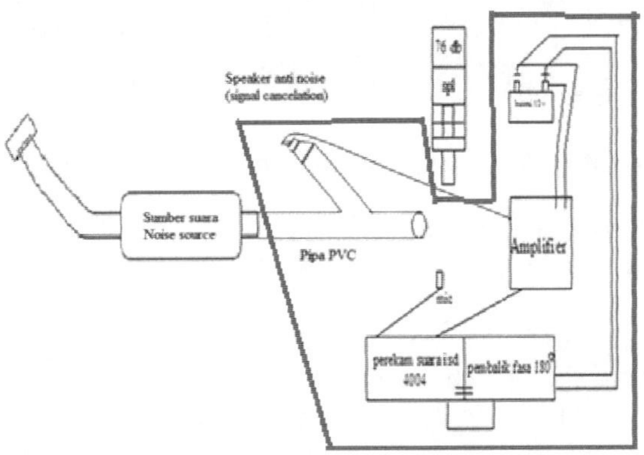

Figure 3. Set-up of noise data retrieval on exhaust

Set-up of the exhaust sound data retrieval of the active noise control device, the design of experimental equipments look as with red circled parts i.e amplifiers, ISD 4004 sound recorders, phase inverters, speakers and PVC pipe tubes.

2.1. Sketch of Measurement Direction on the Muffler

Measurements were made using a half-sphere method with (1) vertical directions (Z); (2) horizontal direction (-X, + X); and (3) axial direction (Y). The measuring has taken from the back, side and top

IOP Conf. Series: Materials Science and Engineering **308** (2018) 012011 doi:10.1088/1757-899X/308/1/012011

of the exhaust using the sound pressure level gauge. Variations of measurement directions are performed to obtain various reduction values that occur, as in Figure 4 below:

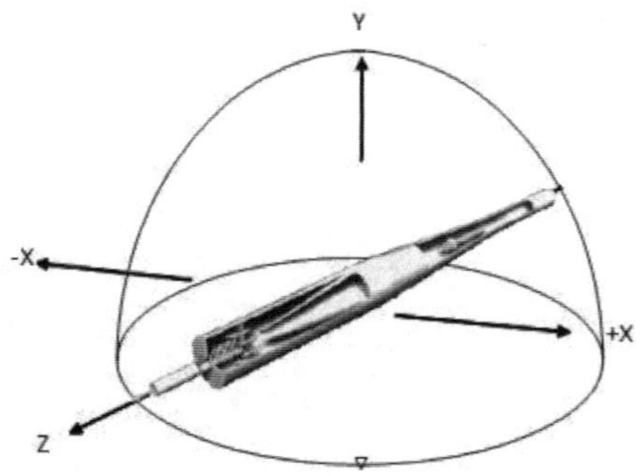

Figure 4. Sketch of direction noise measurement of a half-sphere method [3].

2.2 Use of Amplifier and Speaker

Selection of amplifiers in this study is the thing that affects the resulting sound to generate a sound signal that is useful to adjust the size of the sound that will be generated. Amplifier that will be applied to this tool has a fitting such as potentiometer that serves to adjust the amount of frequency released from the amplifier to be equal to the amount of frequency that is removed from the exhaust noise signal.

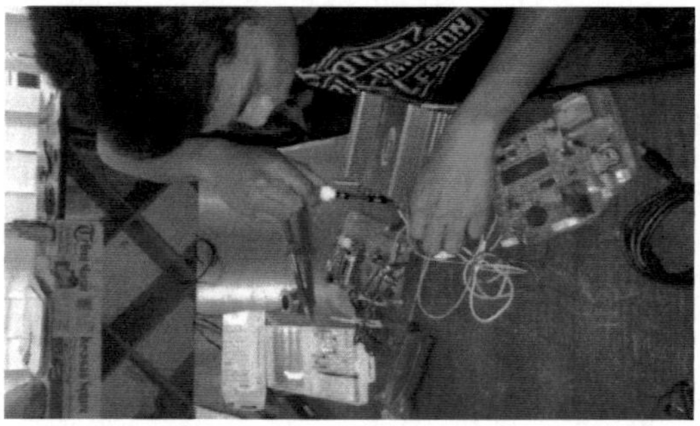

Figure 5. Amplifier & Speaker installation process specification amplifier

Caption: (1) Brand: Clarin; (2) Type: Cl 3002 2 Channel and 3 variable potensio; (3) Dimension: Length 17 cm, width 13 cm and thick 4 cm; (4) Weight: 0.25 kg; (5) Current source: DC 12V.

In this active noise control research, the speakers are used as final voice translator output after going through the recording process. The selected speakers are considered quality, so the sound is produced also better. It should have capability to reproduce power of amplifiers as well as possible without losses. Sensitivity is assessed by dB unit. The impedance value will determine the output power of the amplifier. Smaller impedance, greater of power out from the amplifier.

Figure 6. Speaker And Carbon Fiber Heat Resistant.

Specifications of speakers: (1) Brand: Alpine series Mk- 820 f; (2) Power: 45 W; (3) Impedance: 4 Ohm; (4) Dimensions: 4 inches

2.3 Inverting Phase Shift Design

To design angle 180° as phase shifter or called inverting circuit by using electronic component, consisting of capacitor, resistor, ICL7660 and Tl 072 CP. The circuit can shift the phase by 180° so that the output of the inverting circuit will be opposite 180° phase with its input. Angles 180°, 2250, 2700, and 3150 are obtained from each of the 0, 450, 900, and 1350 phase angles which are inputs for the inverting circuit.

In the inverting op-amp circuit has the following values: (1) The feedback resistance = 1 kΩ; (2) Input resistance = 2 kΩ; and (3) Input voltage = 12V.

Acquisition voltage (Av), output voltage (Vout) and power supply voltage (Vcc), can be seen in the following Figure 7.

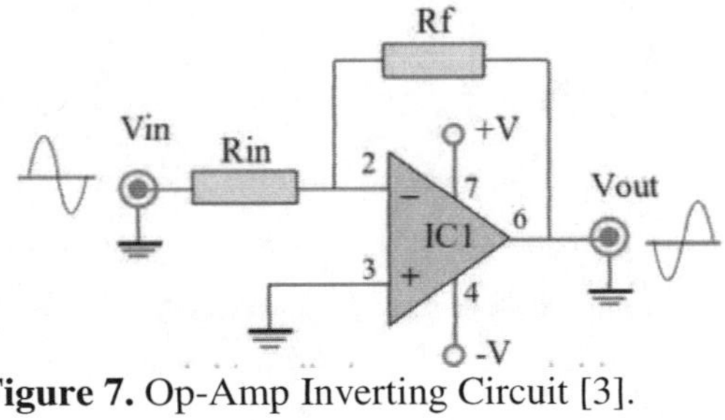

Figure 7. Op-Amp Inverting Circuit [3].

Where:

- Rf = 1 kΩ = 1000 Ω
- Rin = 2 kΩ = 2000 Ω
- Vin = 12 V

$$A_v = \frac{-R_f}{R_{in}} \quad (3)$$

$$A_v = \frac{-1000\Omega}{2000\Omega} = -0.5\Omega$$
$$V_{out} = A_v . V_{in} = -0.5\Omega \ . \ 12V = -6V$$

If the given input is 12V, then the resulting output is - 6 V. This assumes that the power supply voltage Vcc used allows the moving output to reach that value. A power supply ± 6V is too small for that, therefore requires a power supply with a voltage rating of at least ± 9V (or approximately ± 150%

× Vout), to amplify the input voltage of 12V. So obtained: Av = -6 V; Vout = - 5.61 V; Vcc = ± 8 V; and 2500 μF and 3k resistor. The assembly process and the result of the phase shifting component can be seen in Figure 8.

Figure 8. Assembling Process and Phase Slider Results.

2.4 Active Noise Control ISD4004 Module

After going through the stages of the process of active design of noise control, starting from the selection of materials to consideration of calculation of signal generator and phase shifter. Next step is how to combine all the components as a system that can be seen in Figure 9 as follow.

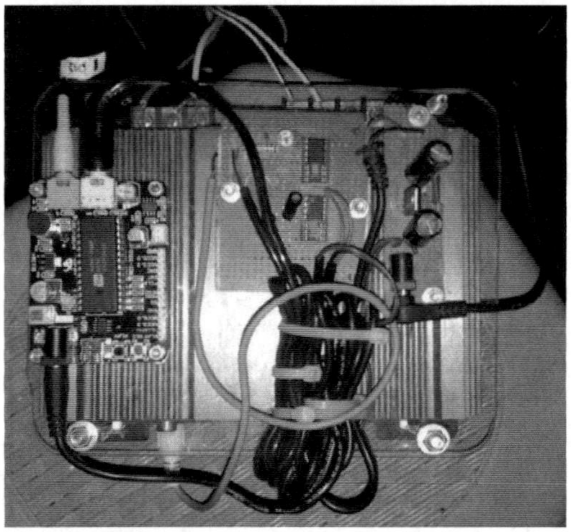

Figure 9. Active Noise Control Circuits.

2.5 Media Focusing Voice

Extraction of exhaust sound signal data is done by using PVC pipe tube media. It serves to focus the sound to be unidirectional. In this case, the diameter of the PVC pipe tube affects against the fast ripple of sound. The picture of PVC tube media can be seen in Figure 10 below.

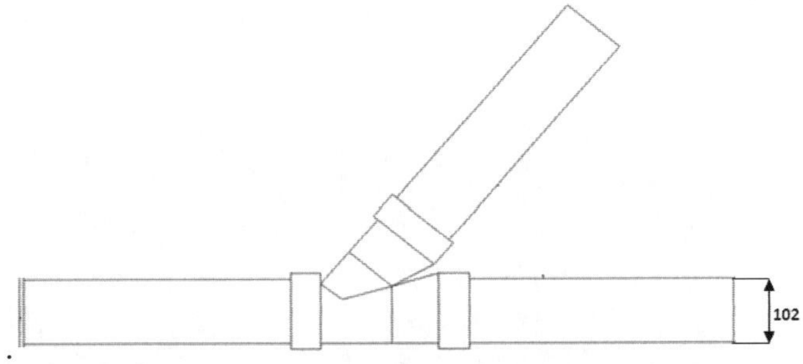

Figure 10. Media Focusing Voice.

Where: D = 102 mm; R = 51 mm; π= 3.14; f = 16.67 Hz; c = 343 m/s

The acoustic wave propagation in the tube affected by its diameter can be calculated by the formula:

$$c^1 = c\left(1 - \frac{0.76}{2\,x\,r\,\sqrt{\pi}}\right) \text{x} \frac{1}{\sqrt{f}} \quad (4)$$

$$c^1 = c\left(1 - \frac{0.76}{2\,x\,51\,\sqrt{3.14}}\right) \text{x} \frac{1}{\sqrt{16.67}}$$

$$= 343\left(1 - \frac{0.76}{102\,\sqrt{1.77}}\right) \text{x} \frac{1}{\sqrt{4.08}}$$

$$= 343\left(1 - \frac{0.76}{180.54}\right) \text{x } 0.245$$

$$= 343\,(1 - 4.20\,x\,10^{-3}) \text{ x } 0.245$$

$$c^1 = 83{,}68 \text{ m/s}$$

Thus, the acoustic wave propagation in a 102 mm diameter tube is 83.68 m/s. While the acoustic wave propagation in air is 343 m/s. So it can be concluded that the greater the acoustic wave propagation media, the sooner the sound is also faster.

The next research process continue after the active series of noise control is ready to be assembled. It puts directly testing on Motorcycle Supra X 125 D, in order to prove whether there is noise reduction already can be produced using this tool. The testing done at 02.00 a.m for reason to get a quiet atmosphere. Preparation is done by setting all tools ranging from the installation of speaker-microphone and current supply from batterei to generate anti noise signal. The process of taking exhaust sound data can be seen in Figure 11.

Figure 11. Setting Active Noise Control device for taking exhaust sound data.

IOP Conf. Series: Materials Science and Engineering **308** (2018) 012011 doi:10.1088/1757-899X/308/1/012011

After installing the Active Noise Control device and setting the SPL distance to the exhaust, the measuring activity held and can be seen in Figure 12 below.

Figure 12. SPL distance setting on the muffler.

When installing all equipment and set the SPL distance to the exhaust according to standard accomplished, then the exhaust sound data collection can be taken. It can be seen in Figure 13 as follow.

Figure 13. Process of Exhaust Sound Data Retrieval.

Finished with the initial exhaust data retrieval process, then tested continue with the exhaust sound generator. Here, with the same engine rotation but the phase wave has been already shifted by 180° as shown in Figure 14.

Figure 14. Sound Exhaust Testing Process with Anti Noise Phase Shift 180°.

3. Result and Discussion

3.1 Test results of Active Noise Control

Using SPL equipment, the testing accomplished by measuring directly the noise generated by Supra X 125 D motorcycle and comparing its result when using the Active Noise Control device. The engine speeds set-up at 1000 and 2000 rpm, the measuring distance is 1 meter with three measuring direction i.e X, Y and Z. The orientation of measurement can be seen in Figure 15 as follows:

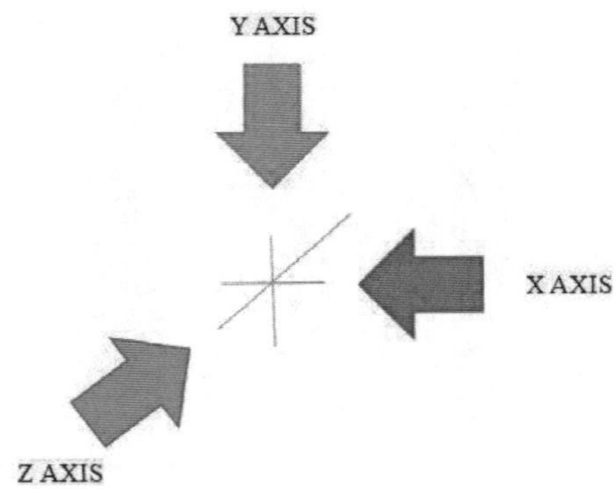

Figure 15. Measurement direction and orientation

To change the engine speed in the scale of rpm to frequency (Hz), can be calculated with the following formula:

- Engine Speed (n) = 1000 rpm
  where : $\omega \frac{2\pi}{60}\ x\ n = \frac{6.28}{60}\ x\ 1000 = 104.667$
  Thus Frequency : $f = \frac{\omega}{2\pi} = \frac{104.667}{6.28} = 16.667$ Hz

- Engine Speed (n) = 2000 rpm
  where : $\omega \frac{2\pi}{60}\ x\ n = \frac{6.28}{60}\ x\ 2000 = 209.33$
  Then the frequency : $f = \frac{\omega}{2\pi} = \frac{209.33}{6.28} = 33.33$ Hz

The following table is the results of noise measurement on exhaust of Supra X 125D without using active noise control and when using active noise control using sound level meter equipment as on Tabel 1. Given, engine rotation variation of 1000 rpm.

Table 1. Exhaust noise data without active noise control and when using active noise control at 1000 rpm engine speed

Rotation (rpm)	Frequency (Hz)	Measurement Distance	Measurement Direction	Measurement Result	
				Without ANC	Using ANC
1000 rpm	16,67	1 Meter	Z	66.5 db	64.5 db
1000 rpm	16,67	1 Meter	Y	65.5 db	64 db
1000 rpm	16,67	1 Meter	X	65 db	64 db
1000 rpm	16,67	1 Meter	-X	66 db	64.2 db

IOP Conf. Series: Materials Science and Engineering **308** (2018) 012011 doi:10.1088/1757-899X/308/1/012011

According to Tabel 1, the measurement results have shown the reduction of the sound signal of motorcycle muffler when using the active noise control device. Detection of sound signal data with measurement distance of 1 meter with engine rotation 1000 rpm, the reduction that happened is as follows: Z-axis: 66.5 dB - 64.5 dB = 2 dB; Y-axis: 65.5 dB – 64 dB = 1.5 dB; X axis: 65 dB – 64 dB = 1 dB; Axis -X: 66 dB - 64.2 dB = 1.8 dB. From the noise reduction data that occurs above, the comparison of the standard exhaust noise level without using active noise control and when using active noise control device at 1000 rpm engine speed can be seen in Figure 16 below.

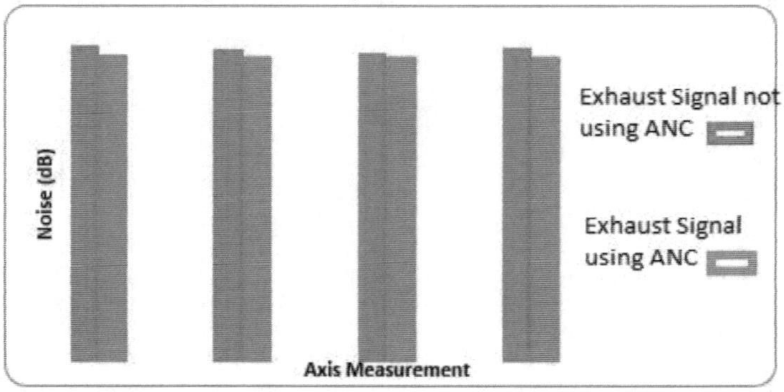

Figure 16. Comparison between without ANC and using ANC at 1000 rpm rotation of engine.

Highest noise reduction at 1000 rpm engine speed is on the Z axis. Using the active noise controller device, the noise reduction achieve up to 2 dB. The noise measurement results at 2000 rpm engine speed can be seen on Table 2 below.

Table 2. Exhaust Noise Data without active noise control and using active noise control at 2000 rpm engine speed.

Rotation (rpm)	Frequency (Hz)	Measurement Distance	Measurement Direction	Measurement Result	
				Without ANC	Using ANC
2000 rpm	33.3	1 Meter	Z	71 db	69.5 db
2000 rpm	33.3	1 Meter	Y	69.9db	68.9db
2000 rpm	33.3	1 Meter	X	69.3 db	68.3 db
2000 rpm	33.3	1 Meter	-X	68.8 db	68.2 db

Reduction calculation of sound signal data for measurement distance 1 meter with engine rotation at 2000 rpm come about as follows: Z-axis: 71 dB - 69.5 dB = 1.5 dB; Y axis: 69.9 dB - 68.9 dB = 1 db; X axis: 69.3 dB - 68.3 dB = 1 dB ; Axis -X: 68.8 dB - 68.2 dB = 0.6 dB. From the noise reduction data that occurs above, the comparison of the standard exhaust noise level without using active noise control and when using active noise control device at 2000 rpm engine speed can be seen in Figure 17 as follow.

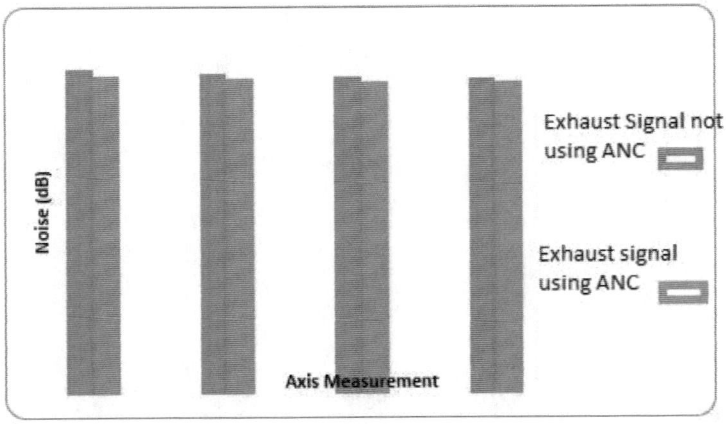

Figure 17. Comparison between without ANC and using ANC at 2000 rpm rotation of engine.

Highest noise reduction at 2000 rpm engine speed is on the Z axis. Using the active noise controller device, the noise reduction achieve up to 1.5 dB. As already known between source sound and counter noise, in implementation the sound waves are very random and difficult to meet them at zero. Illustration of the randomness of signals on the Z, Y, X axes and the phase shift signal waveform on the illustrated Z axis can be seen in Figure 18.

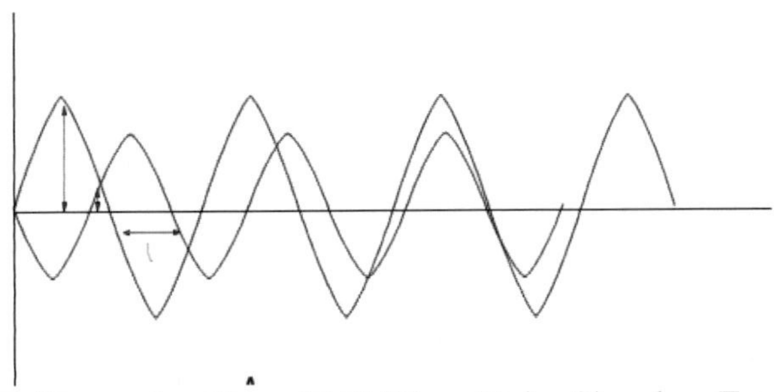

Figure 18. Illustration Phase Shift Wave Noise Signal on Z-axis [3].

At 4000 rpm engine speed or 66.67 Hz, frequency has increased higher than the previous frequency. In this case, the noise is not only caused from the exhaust but already there is influence of engine vibrations which cause noise. It can be seen in Figure 19 and 20.

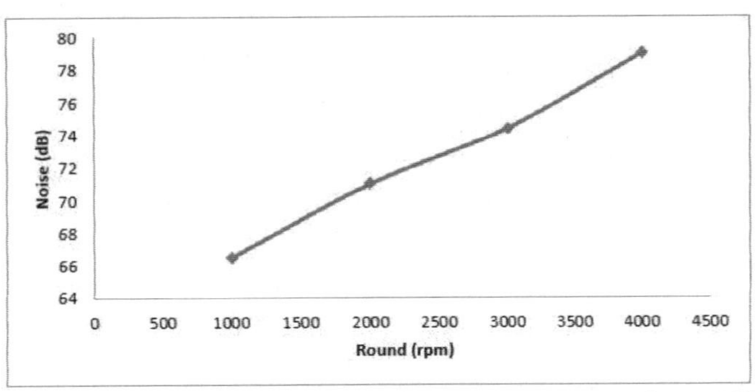

Figure 19. Graph of Z-axis exhaust noise without using ANC.

The reduction results between without active noise control and using active noise control device on the exhaust, it can be seen in Figure 19 and 20. The highest noise reduction find out on the Z axis.

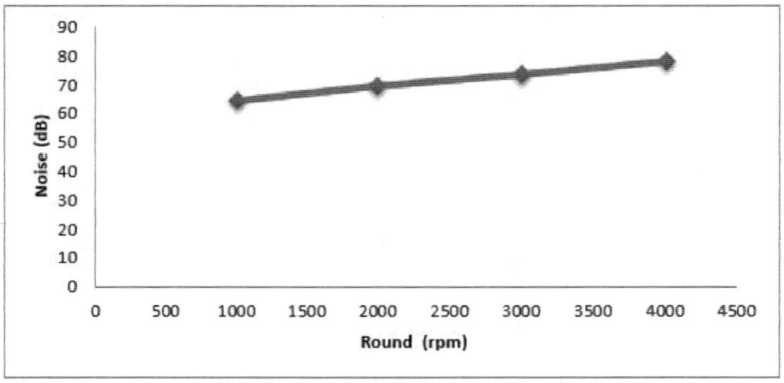

Figure 20. Graph of Z-axis exhaust noise using ANC.

Based on Figure 21, we can see that the noise reduction from engine rotation of 1000 rpm to 4000 rpm decreases unequally compared to Figure 18. Acording to different waveforms and amplitude that must be adjusted with the anti-signal generated from Tools assembled. The noise reduction that occurs on the Z axis ranges from 1.2 dB to 2 dB. Illustration of the randomness of signals on the Z, Y, X axes and the phase shift signal waveform on the illustrated Y axis can be seen in Figure 21 as follow.

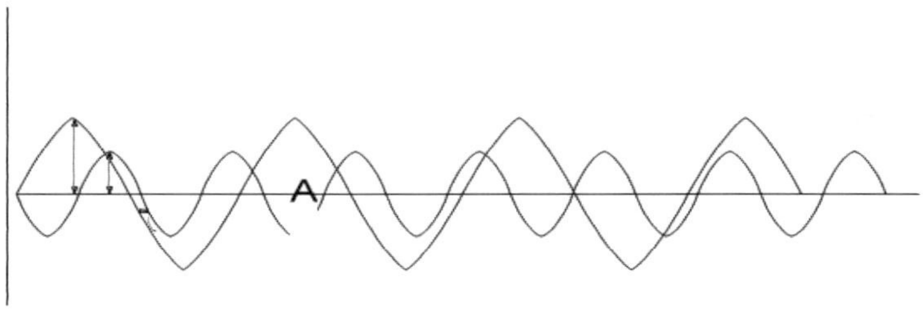

Figure 21. Illustration Phase Shift Wave Noise Signal on Y-axis [3].

The result of noise measurements for standard exhaust Supra X 125 D on Y axis, it can be seen in Figure 22 below.

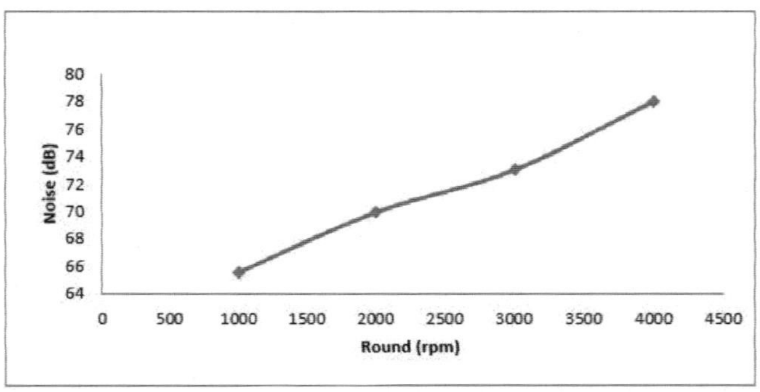

Figure 22. Graph of Y-axis exhaust noise without using ANC.

Based on Figure 22, we can see the level of standard exhaust noise without active noise control starting from 65.5 dB to 78 dB. At 3000 rpm engine speed, the exhaust noise level drops by 1.2 dB. In general, because of the highly variable random signal fluctuation and sound absorption properties at

50 Hz frequency corresponding to the exhaust acoustic material properties and noise absorption corresponding to the sound focusing medium. This influence of PVC tube coated with carpet silencer.

According to the test results of the effectiveness of the active noise control device on the Y axis, it can be seen in Figure 23 below.

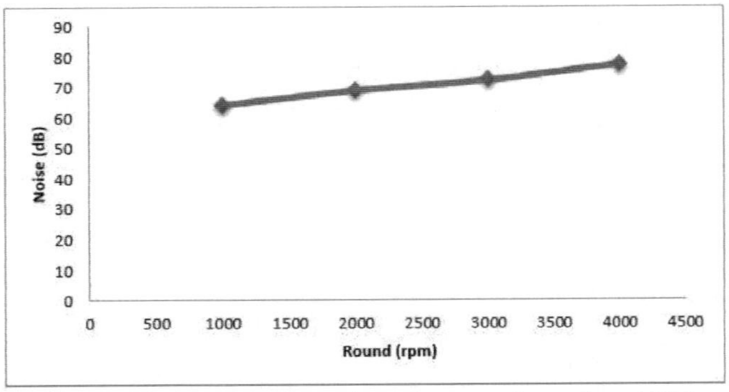

Figure 23. Graph of Y-axis exhaust noise using ANC.

On the Y-axis the noise reductions that occur from the installation of the noise control device are smaller at 1dB to 1.5 dB. This is influenced by the direction of SPL to the exhaust parallel or in line with the Z-axis. On the Y-axis and the X-axis, the direction of SPL is not aligned but from the side. In this case, the measured exhaust noise with this tool becomes different. The highest exhaust noise that can be reduced on the Y-axis by testing is 1.5 dB.

The waveform of phase shift measurements on the illustrated X-axis can be seen in Figure 24 below.

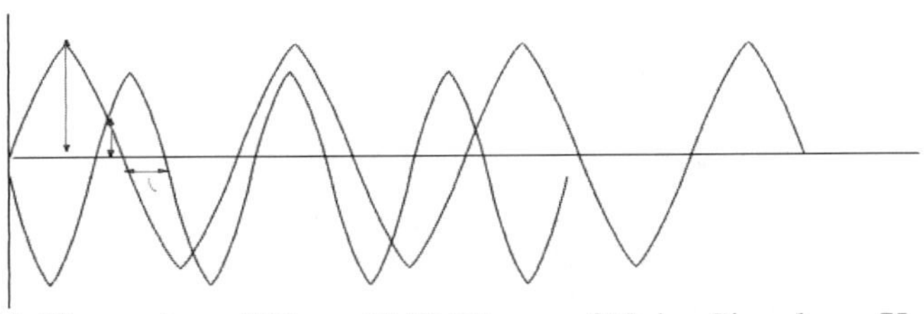

Figure 24. Illustration of Phase Shift Waves of Noise Signals on X-axis [3].

Based on the measurement results of motorcycle muffler standard without active noise control at engine speed 1000 to 4000 rpm can be seen in Figure 25 below.

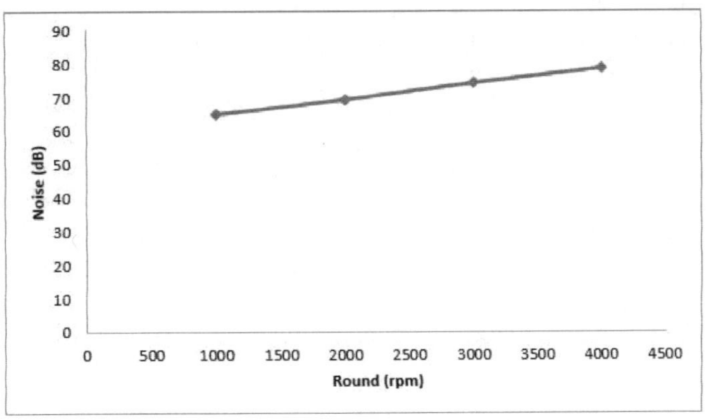

Figure 25. Graph of X-axis exhaust noise without using ANC.

By applying the above graph based on the position of the sound level on the X-axis, the engine will affect the measurement results. On the Z-axis, the influence will be greater against the measurement result because parallel to the exit direction of the exhaust sound. Therefore we can see the level of exhaust noise on the X-axis rise almost the same in every engine speed. Based on the result of noise level reduction test which can be obtained by active noise control device can be seen in the following Figure 26.

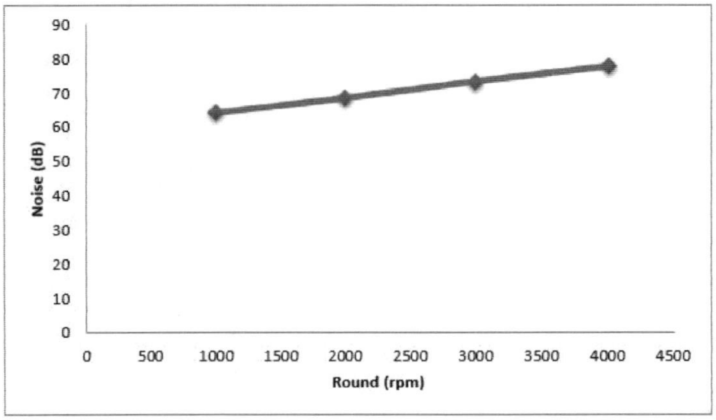

Figure 26. Graph of X-axis exhaust noise using ANC.

Of the various measurements performed on the X-axis, we have the same reduced noise level such as in Figure 26 above. The highest noise can be reduced on the X-axis of 1 dB, and will increase with the high frequency.

4. Conclusion

a. Preparation of simulation media in the form of PVC pipe tube with Y-shaped modeling, with the addition of carpet in pipe layer to avoid echo during sound testing. The absorption coefficient of the carpet is 0.0025 dB. The result of noise reduction test is 1.1 dB.
b. Inverted phase reversed using feedback resistor 1 kΩ and 2 kΩ input resistors, 16V capacitor 2500μf and as amplifier using ICL 7660 and TL 702 CP works very well. All series are assembled into one by using acrylic glass that is attached to the amplifier body by bolting.
c. In accordance to objective of this study is to identify the noise reduction that occurs from the use of tools designed in the form of active noise control device on the exhaust motorcycle Supra X 125D. Test results on the highest 1000 rpm engine speed on the Z-axis of 2 dB, and at the highest 2000 rpm engine speed also occurs on the Z-axis of 1.5 dB.

References

[1] Lim R, Patrick M J 2002 Reduksi Noise Akustik Secara Aktif [Active Noise Reduction] *Journal of Electrical Engineering University of Kristen Petra* 2:2
[2] Dewanto RA, Asep S, Aradea 2008 *Active Noise Control For Damping Electric Generator with Neurofuzzy Algorithm* (STMIK-DCI Tasikmalaya)
[3] Alfisyahrin 2008 *Simulasi Pemrosesan Sinyal Suara Untuk Kendali Kebisingan Aktif Pada Knalpot [Thesis:* Voice Signal Processing Simulation For Active Noise Control On Muffler] (USU)
[4] Doelle, Leslie L 1986 *Environment Acoustic* (Montreal: Erlangga)
[5] Leo B L, Ver I L 2015 *Animation and Acoustic Wave* (ISVR - Institute of Sound and Vibration Research)

[6] Jhon F E K, Reinhold V N 1990 *Sound Analysis and Noise Control* (New York)
[7] Jadine T, Cheryl P, Elaina M, Meghan W, Kay T, Devies H *Active Noise Control Communication Headsets For The Entertainmentm Industry*
[8] Graham J B, Harris C M 1991 *Handbook of acoustical measurements and noise control*
[9] Cyril H M 1957 *Handbook of Acoustical Measurements and Noise Noise* (New York Colombia University)
[10] Michael B, Christopher C, Melissa K, Kimberly L, Richard R, Kevin S *Engineering Silence, Active Noise Cancellation*
[11] Wilson, Charles E 1989 *Noise Control Measurement, Analysis, and Control of Sound Vibration* (New York Harper & Row Publisher)

IOP Conf. Series: Materials Science and Engineering **308** (2018) 012012 doi:10.1088/1757-899X/308/1/012012

Accuracy and stability analysis of the modified Lax numerical method in solving the sea waves refraction equation

S Arif

Department of Mathematics, Faculty of Mathematics and Sciences, Syiah Kuala University, Banda Aceh - 23111, Indonesia

E-mail: salmawaty@unsyiah.ac.id

Abstract. We analyse the accuracy and stability of the modified Lax numerical method used in solving the two dimensional sea waves refraction equation. The refraction phenomenon on the beach was observed by the fact that waves hit the shore line in perpendicular way. Angles of the incoming waves from offshore keep changing due to the smaller depth of water, and therefore become important to be determined. The accuracy was evaluated by means of the Taylor series expansion and the stability was analysed by the Von Neumann method. The analysis showed that the method is stable for a certain interval although it is only first order accurate.

1. Introduction

In this paper we discuss a numerical method performance for the sea waves refraction equation. We investigate analytically the accuracy and the stability of the selected numerical scheme, in this case the modified Lax method. The modified Lax scheme was chosen due to its stability for the simple wave equation. It generally improves stability because of the presence of the factor in the method to manage the calculation, although does not improve accuracy [1]. Another advantage of this method is its simplicity which is characterized by explicit form and only one computation stage needed at a point. The discretization of this equation was done by using finite difference method which is suitable for the numerical scheme.

The accuracy was evaluated by means of the Taylor series expansion. The truncation error of the method can be shown by substituting Taylor expansion to the equation. By assuming the increment intervals for horizontal and vertical axes are in the same order, then the order of accuracy of the method is determined [1].

The stability was analysed by the von Neumann method by using the Fourier expansion applied to the equation. The stability of the method can be achieved when the value of amplification factor or ratio of two successive steps is not larger than one [1].

Waves propagate from offshore to the beach following a line which is called a wave ray. Wave refraction can be seen as a bend of a ray of waves at an oblique angle when it enters the less depth of water [2]. Therefore angles of the incoming waves become important to be determined as input for obtaining other coastal quantities such as long shore currents [3], and wave energy [4]. However, the angle distributions solution will not be discussed here.

IOP Conf. Series: Materials Science and Engineering **308** (2018) 012012 doi:10.1088/1757-899X/308/1/012012

2. The modified Lax method

Before implementing the modified Lax method to the refraction equation, we first consider the method in its general form. In a two dimensional horizontal xy-plane, x refers to on-offshore axis and y is axis parallel to the shoreline.

Consider a rectangular grid of resolution Δx and Δy where i and j denote the indices in x- and y-direction, respectively ($0 \leq i \leq N_x$, $0 \leq j \leq N_y$). The parameter of the method α should satisfy $0 < \alpha < 1$. Let t represents the time and the asterisk $*$ indicates evaluation at the next time step for an interval Δt. The general formulation of the modified Lax method is defined for one dimensional case determining derivative $\frac{dq}{dt}$ of a quantity q, as

$$\frac{q_i^* - \frac{1}{2}\alpha(q_{i+1} + q_{i-1}) - (1-\alpha)q_i}{\Delta t} = \text{right hand side} \tag{1}$$

For $\alpha = 1$, it returns back to the Lax method. For $\alpha = 0$, it returns to forward Euler [1]. The scheme is expanded to the two dimensional in spatial form as it is applied to the refraction equations.

2.1. Wave refraction

For the purpose of modelling the refraction equation, let θ represents incident wave angle with respect to the positive x-axis, ω is the angular frequency, and k is wave number. Figure 1 shows the definition sketch for the equation.

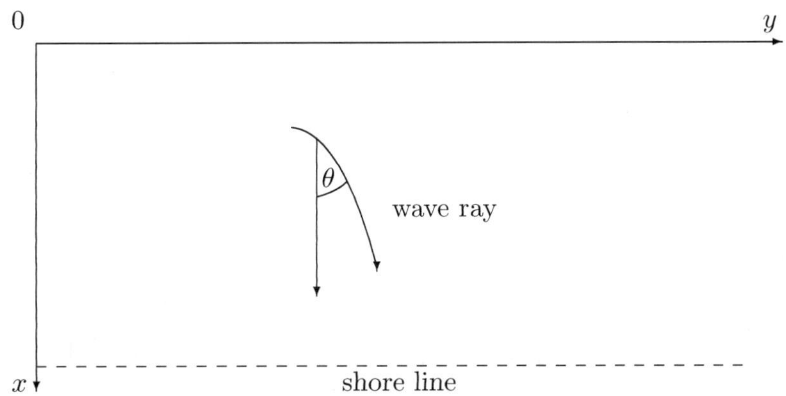

Figure 1. Plan definition sketch.

A wave ray is defined as a line tangent to the wave number vector at every point [2]. When the wave direction makes an angle with the beach normal, the wave number oriented in the direction is a normal vector of the wave phase function. This fact represents irrotational condition of the wave number vector. This implies that wave length and wave angle are reduced when the waves enter the shallow water. This property is expressed according to the following consideration.

For a scalar phase function $\Omega = kx - \omega t$, define the wave number vector $\mathbf{k} = (k_x, k_y)$ as $\mathbf{k} = \nabla\Omega$. Since

$$\nabla \times \mathbf{k} = 0, \tag{2}$$

it follows that [5]

$$\begin{aligned} \partial_x k_y - \partial_y k_x &= 0, \\ \partial_x(k \sin\theta) - \partial_y(k \cos\theta) &= 0. \end{aligned} \tag{3}$$

IOP Conf. Series: Materials Science and Engineering **308** (2018) 012012 doi:10.1088/1757-899X/308/1/012012

These equations lead to another shoaling related property on the coastline, where y variations of the variables are zero, (3) turns to

$$\begin{aligned}\partial_x(k\sin\theta) &= 0, \\ k\sin\theta &= \text{constant}.\end{aligned} \tag{4}$$

2.2. Numerical scheme

In the computational plane, the central difference applies for the scheme with the points i, j are positions for k and θ. For particular discretization purpose, it is helpful to define the indices $c = (i, j)$, $w = (i, j-1)$, $e = (i, j+1)$, and $s = (i+1, j)$.

Employing Lax method (1) in forward x-direction which is the 'timelike' direction, the discretized form for refraction equation (3) is

$$k_s \sin\theta_s - \frac{\alpha}{2}(k_e \sin\theta_e + k_w \sin\theta_w) - (1-\alpha)k_c \sin\theta_c = \frac{\Delta x}{2\Delta y}(k_e \cos\theta_e - k_w \cos\theta_w). \tag{5}$$

Given k_s, then θ_s is determined, as shown by the corresponding molecule depicted in Figure 2.

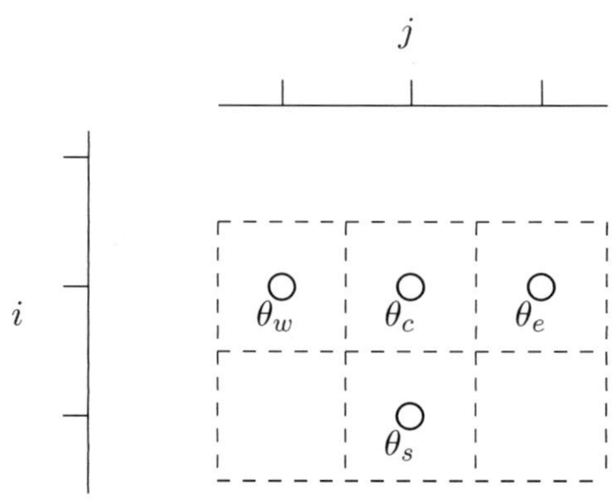

Figure 2. Discretization molecule for wave refraction. At θ location, k is also defined.

3. Numerical analysis

3.1. Accuracy

In order to investigate the accuracy for the modified Lax method (1), consider Taylor expansion for the solution q with respect to x around a point i:

$$q_{i\pm1} = q_i \pm \Delta x \partial_x q_i + \frac{1}{2!}(\Delta x)^2 \partial_x^2 q_i \pm \frac{1}{3!}(\Delta x)^3 \partial_x^3 q_i + O(\Delta x)^4, \tag{6}$$

and let

$$\widehat{q}_i = \frac{1}{2}\alpha(q_{i-1} + q_{i+1}) + (1-\alpha)q_i. \tag{7}$$

By using the Taylor expansion in spatial and temporal dimensions, the following relations are needed to be used in the accuracy analysis [1].

$$\widehat{q}_i = \alpha(q_i + \frac{1}{2}(\Delta x)^2 \partial_x^2 q_i) + (1-\alpha)q_i + O(\Delta x)^4,$$

$$q_{i+1} - q_{i-1} = 2\Delta x \partial_x q_i + \frac{2}{3!}(\Delta x)^3 \partial_x^3 q_i + O(\Delta x)^5,$$
$$q_i^* - q_i = \Delta t \partial_t q_i + \frac{1}{2}(\Delta t)^2 \partial_t^2 q_i + O(\Delta t)^3.$$

The truncation error for the two dimensional equation is determined in the same way as the one in one dimensional problem. Term with respect to x-direction was added to replace the term with respect to t. Here we derive the accuracy for the wave refraction equation.

$$\frac{k_s \sin\theta_s - \frac{\alpha}{2}(k_e \sin\theta_e + k_w \sin\theta_w) - (1-\alpha)k_c \sin\theta_c}{\Delta x} = \frac{k_e \cos\theta_e - k_w \cos\theta_w}{2\Delta y}.$$

Apply Taylor expansion around $k_c \sin\theta_c$, the left hand side becomes

$$\begin{aligned} \frac{k_s \sin\theta_s - \frac{\alpha}{2}(k_e \sin\theta_e + k_w \sin\theta_w) - (1-\alpha)k_c \sin\theta_c}{\Delta x} &= \frac{\Delta x \partial_x (k_c \sin\theta_c)}{\Delta x} \\ + \frac{\frac{1}{2}(\Delta x)^2 \partial_x^2 (k_c \sin\theta_c) + O(\Delta x)^3}{\Delta x} &- \frac{\frac{\alpha}{2}(2k_c \sin\theta_c + (\Delta y)^2 \partial_y^2 (k_c \sin\theta_c) + O(\Delta y)^4) - \alpha k_c \sin\theta_c}{\Delta x} \\ = \partial_x (k_c \sin\theta_c) + \frac{1}{2}\Delta x \partial_x^2 (k_c \sin\theta_c) &- \frac{\alpha}{2}\frac{(\Delta y)^2}{\Delta x}\partial_y^2 (k_c \sin\theta_c) + O(\Delta x)^2, \end{aligned} \quad (8)$$

and the right hand side becomes

$$\begin{aligned} \frac{k_e \cos\theta_e - k_w \cos\theta_w}{2\Delta y} &= \frac{2\Delta y \partial_y (k_c \cos\theta_c) + \frac{2}{3!}(\Delta y)^3 \partial_y^3 (k_c \cos\theta_c) + O(\Delta y)^5}{2\Delta y} \\ &= \partial_y (k_c \cos\theta_c) + \frac{1}{3!}(\Delta y)^2 \partial_y^3 (k_c \cos\theta_c) + O(\Delta y)^4 \end{aligned} \quad (9)$$

The truncation error for refraction equation is

$$\begin{aligned} \partial_x (k_c \sin\theta_c) - \partial_y (k_c \cos\theta_c) = &- \frac{1}{2}(\Delta x)^2 \partial_x^2 (k_c \sin\theta_c) + \frac{\alpha}{2}\frac{(\Delta y)^2}{\Delta x}\partial_y^2 (k_c \sin\theta_c) \\ &+ \frac{1}{3!}(\Delta y)^2 \partial_y^3 (k_c \cos\theta_c) + O(\Delta x)^2. \end{aligned} \quad (10)$$

Assume Δx and Δy are in the same order, the method is first order accurate.

3.2. Stability

The von Neumann stability analysis was used to evaluate the stability of the method after being used used for the equation. θ was written in the Fourier series [6]: $\theta_j = \widetilde{\theta} e^{iky_j}$.

The stability of the method will be achieved when the amplification factor ρ (modulus of the comparison of the two consecutive steps) is less than 1. Stability of refraction equation in the 'timelike' direction is determined.

$$\partial_x (k \sin\theta) - \partial_y (k \cos\theta) = 0.$$

Let $P = \sin\theta, Q = \cos\theta$, the equation and its discretization becomes

$$\partial_x (kP) - \partial_y (kQ) = 0, \quad (11)$$

such that

$$\begin{aligned} k_{i+1,j}P_{i+1,j} - \frac{\alpha}{2}(k_{i,j+1}P_{i,j+1} + k_{i,j-1}P_{i,j-1}) - (1-\alpha)k_{i,j}P_{i,j} \\ = \frac{\Delta x}{2\Delta y}(k_{i,j+1}Q_{i,j+1} - k_{i,j-1}Q_{i,j-1}). \end{aligned} \quad (12)$$

Consider a Fourier mode, $P_{i,j} = \widetilde{P}_i e^{iny_j} = \sin\theta_0 e^{iny_j}$, and let $\rho = \frac{\widetilde{k}_{i+1}\widetilde{P}_{i+1}}{\widetilde{k}_i \widetilde{P}_i}$ to be applied to the discretized form,

$$\begin{aligned} \rho - \frac{\alpha}{2}(e^{in\Delta y} + e^{-in\Delta y}) - (1-\alpha) - \frac{\Delta x}{2\Delta y}(e^{in\Delta y}\frac{\widetilde{Q}}{\widetilde{\widetilde{P}}} - e^{-in\Delta y}\frac{\widetilde{Q}}{\widetilde{\widetilde{P}}}) &= 0, \\ \rho - 1 + \alpha(1-\cos n\Delta y) - i\frac{\Delta x}{\Delta y}\frac{\widetilde{Q}}{\widetilde{\widetilde{P}}}\sin n\Delta y &= 0. \end{aligned} \quad (13)$$

By replacing back $\frac{\widetilde{Q}}{\widetilde{\widetilde{P}}} = \cot\theta_0 = C_0$ constant, and defining $\sigma = \frac{\Delta x}{\Delta y}$, stability condition is determined by the following relations,

$$\begin{aligned} \rho &= 1 - \alpha(1-\cos n\Delta y) + iC_0\sigma\sin n\Delta y, \quad (14) \\ |\rho|^2 &= (1-\alpha(1-\cos n\Delta y))^2 + (C_0\sigma\sin n\Delta y)^2, \\ &= 1 + \alpha^2(1-\cos n\Delta y)^2 - 2\alpha(1-\cos n\Delta y) + (C_0\sigma\sin n\Delta y)^2 \leq 1. \end{aligned}$$

$$\begin{aligned} \alpha^2(1-\cos n\Delta y)^2 - 2\alpha(1-\cos n\Delta y) + (C_0\sigma)^2(1-\cos^2 n\Delta y) &\leq 0, \\ \alpha^2(1-\cos n\Delta y) - 2\alpha + (C_0\sigma)^2(1+\cos n\Delta y) &\leq 0. \end{aligned} \quad (15)$$

This is a linear function in $\cos n\Delta y$, so all values of the left hand side are in the interval for $\cos n\Delta y = -1$ and $\cos n\Delta y = 1$. Thus the interval is

$$C_0^2\sigma^2 < \alpha < 1. \quad (16)$$

The solution of the above inequation gives the limits to choose the factor for stability criteria of the method in refraction equation.

4. Conclusion

The contribution of one dimensional model leads to an easier way to investigate the performance of the two dimensional equation. The modified Lax method shows that it is an appropriate method for doing the integration of the refraction equation due to its stability and simple computations. Although the method has only first order accuracy, the stability property of the method allows the simulation to maintain the stable execution until it is accomplished.

References

[1] Vreugdenhil C B 1989 *Computational Hydraulics* (Springer-Verlag)
[2] Dean R G and Dalrymple R A 1991 *Water Wave Mechanics for Engineers and Scientists* (Singapore: World Scientific)
[3] Shen Y M, Tang J, and Huang W 2008 Numerical simulation of long-shore currents induced by regular breaking wave *J. Coastal Res. (Special Issue - Surface Water Modeling)* ed Huang and Chen **52** pp 215-222
[4] Arif S 2014 Mathematical study on energy of coastal waves *Proc. 9th Annual Int. Workshop and Expo on Sumatra Tsunami Disaster and Recovery, October 22-24, 2014* (Banda Aceh: Syiah Kuala University) pp 129-136
[5] Watanabe A 1982 Numerical model of nearshore current and beach deformation *Coastal Engineering in Japan* **25** pp 147-161
[6] Pletcher R H, Tannehill J C and Anderson D A 2012 *Computational Fuid Mechanics and Heat Transfer, 3rd edition* (CRC Press)

Human face recognition using eigenface in cloud computing environment

S T M Siregar, M F Syahputra, R F Rahmat

Department of Information Technology, Faculty of Computer Science and Information Technology, University of Sumatera Utara, Medan, Indonesia

Email: tr.sintong@students.usu.ac.id |fadlysyah@gmail.com |romi.fadillah@usu.ac.id

Abstract. Doing a face recognition for one single face does not take a long time to process, but if we implement attendance system or security system on companies that have many faces to be recognized, it will take a long time. Cloud computing is a computing service that is done not on a local device, but on an internet connected to a data center infrastructure. The system of cloud computing also provides a scalability solution where cloud computing can increase the resources needed when doing larger data processing. This research is done by applying eigenface while collecting data as training data is also done by using REST concept to provide resource, then server can process the data according to existing stages. After doing research and development of this application, it can be concluded by implementing Eigenface, recognizing face by applying REST concept as endpoint in giving or receiving related information to be used as a resource in doing model formation to do face recognition.

1. Introduction

Face recognition system has become one of the research topics in image processing and computer vision applied in various systems, starting from attendance systems, security systems such as cctv, identity recognition, emotional recognition, granting system permissions and also applied to the robotics field as a way for the robot to recognize someone.

Face detection itself is a pre-processing stage in facial recognition. In doing face recognition of a person, it will take a lot of face samples with various expressions, angle of retrieval and general differences attached to the face like glasses and moustache [1]. To do face recognition of one face does not take a long time, but if we do it on the attendance system or security system on companies that have many faces to be recognized, it will take a long time. In addition, in some cases, there are deficiencies in some of the previous studies, such as face recognition not running in real time, using public datasets in testing, and computing processes that take a long time causing the computer cannot be used during the process.

To solve those problems, several studies have using cloud computing technology. Cloud computing is a computing service that is done not on a local device, but on an internet connected to a data center infrastructure. The system of cloud computing also provides a scalability solution where cloud computing can increase the resources needed when doing larger data processing. Because computing is not done on a local device, then it is done between client communication with cloud server by utilizing web-service communication. Generally using REST-API or SOAP. Because it does not use local computing capabilities, then cloud computing is mostly applied to mobile devices that have the limitation of computing on different devices. In addition, with cloud computing, eliminating the limits

IOP Conf. Series: Materials Science and Engineering **308** (2018) 012013 doi:10.1088/1757-899X/308/1/012013

of programming language that must be in doing face recognition. This is because the web service provides a response in the form of XML files or JSON.

In performing face recognition, it needs to perform several stages starting from face detection, feature extraction then doing face recognition. During such stages require adequate and available hardware resources at all times. If you want to do the local device, it will take a long time and besides, the device cannot be used for some time. Therefore, an approach is required to be able to perform face recognition without using the capabilities of local computing devices. This research proposed the implementation of Eigenface algorithm in performing face recognition that will be applied in the cloud computing environment. The contribution is we can perform face recognition in cloud computing requirement, with limited resources, and good image processing strategies.

Some researches have been done in the area of cloud computing which related to face recognition, one of the research is utilizing facial recognition service belongs to one of 'face.com', an online face recognition website and using 'facebook.com' as identity provider by utilizing Graph API owned by facebook.com [2]. A research in the robotics field also takes advantage of cloud computing as a computing medium for the Robot Operating System to be able to recognize humans [3]. Further research is done by developing a real time attendance system by utilizing local infrastructure network. Although it is not applying cloud computing, but research has applied the concept of client-server, in addition in this study only conducted trials on ten subjects only. [4]. In addition, a larger study was also conducted. The study aims to perform face recognition on large-scale face databases using the Extreme Learning method [5]. Furthermore, a research on face recognition by using cloud computing by improving the data security aspect that found in public-cloud [6].

2. Methodology

This research aims to provide face recognition service using API implemented in Cloud Computing environment with Software-as-a-Service (SaaS) model. To perform face recognition Eigenface method will be applied and applying the concept of REST API as a link between client and server.

Several assumptions have been made to limit the scope of the discussion in this study, namely:

1. Using front face only.
2. Using Face with no expression and smile.
3. Using photos of at least 15 people.
4. Not paying attention to lighting conditions or distance taking image.
5. The use of REST API does not have a limit on the number of uses.
6. The image used has a size of 320 x 240 pixels.

The general architecture of the system to be built can be seen in Figure 1.

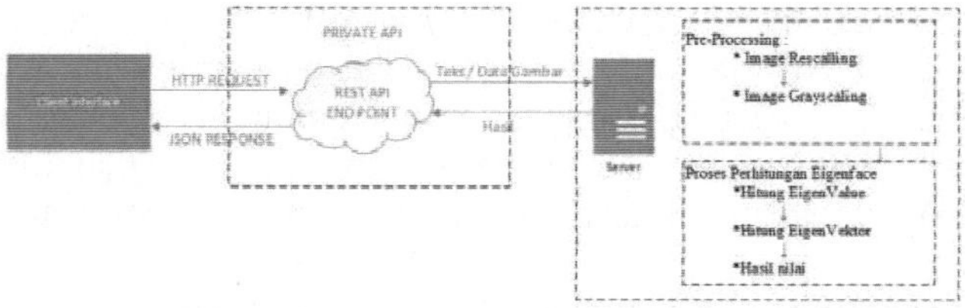

Figure 1. General Architecture

In this study, the data used in the form of taken images using a web camera. Several things to be considered when the data is acquired, they are:

A. Distance of image shooting
B. Image resolution
C. Image format

IOP Conf. Series: Materials Science and Engineering **308** (2018) 012013 doi:10.1088/1757-899X/308/1/012013

In doing this research, we use an external web camera that can produce images up to 1920 x 1080 pixels resolution. In addition, the resulting image is expected to have a resolution of 320 x 240 pixels with the format jpg / jpeg / png / gif.

In addition, in this study HTTP Request data and HTTP Response data has been used. Both of the datas play a role as a communication medium that occurs between the client and server. In this research, HTTP Request will be delivered with the following specifications:

A. Using the POST method
B. Multipart / form-data as encode type / content type

For more details on the HTTP Request structure can be seen in Figure 2.

```
POST regis.dev/upload HTTP/1.1
Host: 202.0.107.144
Content-Type: multipart/form-data; boundary=--------------------------
98215567712697562581705625561
Content-Length: 562

---------------------------98215567712697562581705625561
Content-Disposition: form-data; name="file"; filename="101402042_1.jpg "
Content-Type: multipart/form-data
```

Figure 2. HTTP Request Structure

2.1. Haar Feature

Haar feature extraction has take place in the system because the processing time of features is more quickly than image processing per pixel. The search for human face objects is done by searching for some features that have a high degree of differentiator. This is done by evaluating each feature of the training data by using the value of the feature. In this study authors use haarcascade_frontalface_default.xml specially to the detection of the human head from the front view.

2.2. Resizing Process

Resize process is one of the stages in pre-process which is by doing uniformity of each image data that will be used so they have the same size [8]. In this research, the image will be resized to 320 x 240 pixels.

2.3. Grayscale Process

The Grayscale process is performed to obtain similiar image data. Such similiarity is achieved by converting the RGB image into a gray image [7]. Here is a pseudocode to do the Grayscale process while the result of the grayscale process can be seen in Figure 3.

```
FOREACH pixel p IN image :
  luminosity = (p.red + p.green + p.blue) / 3
  p.red   = luminosity
  p.green = luminosity
  p.blue  = luminosity
```

Figure 3. Grayscale process

IOP Conf. Series: Materials Science and Engineering **308** (2018) 012013 doi:10.1088/1757-899X/308/1/012013

2.4. Principal Component Analysis

The PCA or Principal Component Analysis process has the following steps:

1. Initialize the face image of all existing face data.
2. Changing into the covariant matrix of any existing face image data.

$$\begin{bmatrix} 0 & 4 & 3 \\ 1 & 4 & 2 \\ 0 & 0 & 0 \end{bmatrix} \qquad \begin{bmatrix} 2 & 2 & 1 \\ 3 & 2 & 4 \\ 0 & 0 & 0 \end{bmatrix} \qquad \begin{bmatrix} 1 & 4 & 2 \\ 2 & 3 & 0 \\ 0 & 0 & 0 \end{bmatrix}$$

Citra A Citra B Citra C

Figure 4. Covariance Matrix of All Faces Image Data

3. Calculate mean value of the covariance matrix.

$$\Psi = \frac{1}{3}\sum_{n=1}^{3} \Gamma_n = \frac{1}{3}\left(\begin{bmatrix} 0 & 4 & 3 \\ 1 & 4 & 2 \\ 0 & 0 & 0 \end{bmatrix} + \begin{bmatrix} 2 & 2 & 1 \\ 3 & 2 & 4 \\ 0 & 0 & 0 \end{bmatrix} + \begin{bmatrix} 1 & 4 & 2 \\ 2 & 3 & 0 \\ 0 & 0 & 0 \end{bmatrix}\right)$$

$$\Psi = \frac{1}{3}\begin{bmatrix} 1 & 3 & 2 \\ 2 & 3 & 2 \\ 0 & 0 & 0 \end{bmatrix} \tag{1}$$

4. Reduced the covariance matrix against mean to get adjusted data or normal data.

$$\Phi_1 = \Gamma_1 - \Psi = \begin{bmatrix} 0 & 4 & 3 \\ 1 & 4 & 2 \\ 0 & 0 & 0 \end{bmatrix} - \begin{bmatrix} 1 & 3 & 2 \\ 2 & 3 & 2 \\ 0 & 0 & 0 \end{bmatrix} = \begin{bmatrix} 0 & 1 & 1 \\ 0 & 1 & 0 \\ 0 & 0 & 0 \end{bmatrix}$$
$$\Phi_2 = \Gamma_2 - \Psi = \begin{bmatrix} 2 & 2 & 1 \\ 3 & 2 & 4 \\ 0 & 0 & 0 \end{bmatrix} - \begin{bmatrix} 1 & 3 & 2 \\ 2 & 3 & 2 \\ 0 & 0 & 0 \end{bmatrix} = \begin{bmatrix} 1 & 0 & 0 \\ 1 & 0 & 1 \\ 0 & 0 & 0 \end{bmatrix}$$
$$\Phi_3 = \Gamma_3 - \Psi = \begin{bmatrix} 1 & 4 & 2 \\ 2 & 3 & 0 \\ 0 & 0 & 0 \end{bmatrix} - \begin{bmatrix} 1 & 3 & 2 \\ 2 & 3 & 2 \\ 0 & 0 & 0 \end{bmatrix} = \begin{bmatrix} 0 & 1 & 0 \\ 0 & 0 & 0 \\ 0 & 0 & 0 \end{bmatrix} \tag{2}$$

$$C = \frac{1}{M}\sum_{n=1}^{M} \Phi_n \Phi_n^T$$

$$C = AA^T \qquad A = [\Phi_1, \Phi_2, \dots, \Phi_M]$$

$L = AA^T$, dimana $L_{m,n} = \Phi_n \Phi_m^T$

$$L = \left(\begin{bmatrix} 0 & 1 & 1 & 1 & 0 & 0 & 0 & 1 & 0 \\ 0 & 1 & 0 & 1 & 0 & 1 & 0 & 0 & 0 \\ 0 & 0 & 0 & 0 & 0 & 0 & 0 & 0 & 0 \end{bmatrix} \begin{bmatrix} 0 & 0 & 0 \\ 1 & 1 & 0 \\ 1 & 0 & 0 \\ 1 & 1 & 1 \\ 0 & 0 & 0 \\ 0 & 1 & 0 \\ 0 & 0 & 0 \\ 1 & 0 & 0 \\ 0 & 0 & 0 \end{bmatrix}\right) \tag{3}$$

IOP Conf. Series: Materials Science and Engineering **308** (2018) 012013 doi:10.1088/1757-899X/308/1/012013

$$L = \begin{bmatrix} 4 & 2 & 1 \\ 2 & 3 & 1 \\ 1 & 1 & 1 \end{bmatrix} \tag{4}$$

a. After that, the obtained eigenvalue decomposition of data adjust,
b. After that, get eigenvektor of eigen value decomposition.
c. Sorting Eigenvectors.
d. The final data is obtained as the final result of the eigenvector adjusted data.

3. Result and Discussion

In the face image collection testing is done to show that the application can take pictures with web camera and make the process pengunggahan eight photos into the server with the right filename format after the API rules that have been set. Uploaded towards endpoint "/ upload". When the face image collection page is accessed it will look like in Figure 5 To perform a face image capture, users are required to enter a student ID (NIM) in the textbox section as shown in Figure 5.

Figure 5. Application view

In performing face image capture, an image shooting with expression criteria as can be seen in Table 1.

Table 1 Facial Expression List

Step	Expression type
1	Normal
2	Smile
3	Sad
4	Upset / Angry
5	Normal with Glasses
6	Smile with Glasses
7	Sad with Glasses
8	Upset / Angry with Glasses

Every time a user uploads an image, the application will send the image to the server, and before going to the next stage or table. Any image uploaded to the server will be saved in NIM_NoUrutFoto.jpg format, if the user is using an existing student ID (NIM) then the old image will be replaced with a newer image. Here are some images that have been successfully uploaded into the server that can be seen in Figure 6.

Figure 6. Some of the uploaded images

Because of the process is by utilizing the Cloud Computing concept with the Software-as-a-Service model, the facial recognition process is performed on the server side, where the face recognition application is only an interface application that does not perform the logical process of recognition using the Eigenface method. Therefore, the server as a service provider will provide results from the existing process. The face recognition results are resubmitted by the server in JSON format, where the server will only tell whether the NIM of the user has a match to the face contained in the dataset.

In addition, because the response given by the server is in the JSON format, it is necessary to have a mediation by the face recognition application to read out the response and make it Human-Readable by taking the object from JSON and parsing the information for use in the interface. The Interface Display means is the status bar that will display the information object from JSON and display it into status. The results of facial recognition can be seen in Figure 7.

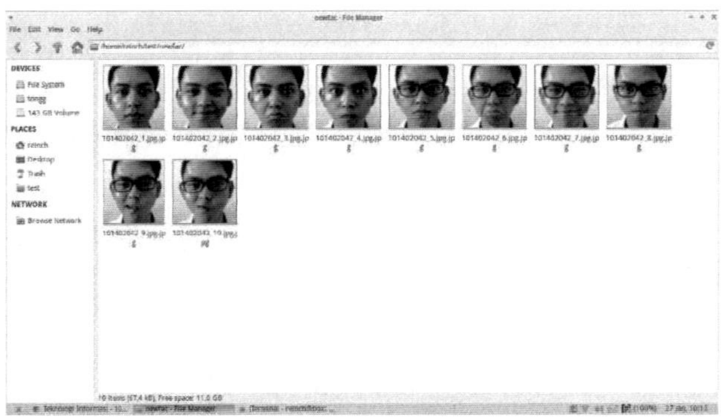

Figure 7. Face Recognition Result

In this study, an eigen image is produced which is a composite of each eigen image of the image used in the introduction. The combined image of the eigen can be seen in figure 8.

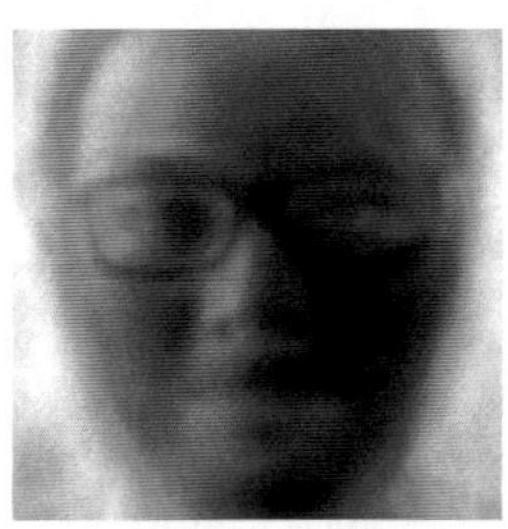

Figure 8. Imagery of the collected Eigenface

In this test, the server is hosted on the local domain which is regis.dev:2527, where the domain and port used can be set in the configuration file, so if it used on a server with a domain or Public IP and a particular port, it can be configured easily without having to configure networks on the server.

In this research, there are two endpoints that can be used as route or address in API usage. The first endpoint is "/ upload" and the second endpoint is "/ introduction". API built by using Python programming language with the help of Flask web-framework. As for the interface of facial image collection or facial recognition page using only HTML and Javascript. The impact of using the web-framework Flask resulted in the endpoint in access from the browser, the server will receive a request in the form of "GET" method while in the API development, the request received is only a request with the "POST" method. This also applies to the "/" endpoint, where the server will return the Response by rendering the index page, but since this study does not require an index page, so also if accessing the endpoint of the browser then the server will respond by rendering the index page stored in Directory upload or directory introduction. To overcome this, the endpoint API can only be used if the method used is a POST, where in most cases the POST method is only used in the data transmission of the form, this is because the GET method is the default method sent to be used by the browser. Therefore, testing of the Request received by the API, and also against the Response provided by the API.

4. Conclusion

It can be concluded that by implementing Eigenface we can apply REST concept in face recognition process by giving or receiving related information to be used as resource in forming model to do face recognition. Another conclusion is pre-processing phase must be done to get a similar image source. Furthermore, in this research has the possibility of development by doing uniformity of photo source, whether in the form of background, controlled light source. It can also take advantage of information security or data fields as a development of authentication in API usage.

To improve service performance of facial recognition system, further research can be done by applying scheduling or applying Job Management which can perform thread control that work to do introduction if the introduction process done by several client at once.

References

[1] Zhang, X., & Gao, Y. (2009). Face Recognition Across Pose: a Review. Pattern Recognition, 42(11), 2876-2896.Chandra, K. & Singh, S. 2014. Firefly algorithm to solve two dimensional bin packing problem. *International Journal of Computer Science and Information Technologies* 5(4): 5368-5373.

[2] Sari, R.F. & Indrawan, P. 2013. Cloud Computing Services in Mobile Devices Using Android Face Detector API and Rest Communication, p. 529-537, The Third International Conference on Digital Information Processing and Communications (ICDIPC 2013), Feb 2013.

[3] Bistry, H., & Zhang, J. 2010. A cloud computing approach to complex robot vision tasks using smart camera systems. In Intelligent Robots and Systems (IROS), 2010 IEEE/RSJ International Conference on (pp. 3195-3200). IEEE.

[4] Putra, I. N. T. A. 2014. Perancangan dan Pengembangan Sistem Absensi Realtime Melalui Metode Pengenalan Wajah. Jurnal Sains dan Teknologi, 3(2).

[5] Vinay, A., Shekhar, V. S., Rituparna, J., Aggrawal, T., Murthy, K. B., & Natarajan, S. 2015. Cloud Based Big Data Analytics Framework for Face Recognition in Social Networks Using Machine Learning. Procedia Computer Science, 50, 623-630.

[6] Haghiat, M., Zonouz, S., & Abdel-Mottaleb, M. 2015. CloudID: Trustworthy cloud-based and cross-enterprise biometric identification. Expert Systems with Applications, 42(21), 7905-7916.

[7] R. F. Rahmat, T. Chairunnisa, D. Gunawan, and O. S. Sitompul, “Skin color segmentation using multi-color space threshold,” in *2016 3rd International Conference on Computer and Information Sciences, ICCOINS 2016 - Proceedings*, 2016, pp. 391–396.

[8] Seniman, D. Arisandi, R. F. Rahmat, William and E. B. Nababan, "Chinese chess character

recognition using Direction Feature Extraction and backpropagation," *2016 International Conference on Data and Software Engineering (ICoDSE)*, Denpasar, Indonesia, 2016, pp. 1-6.

IOP Conf. Series: Materials Science and Engineering **308** (2018) 012014 doi:10.1088/1757-899X/308/1/012014

Adaptive neuro-fuzzy inference system for forecasting rubber milk production

R F Rahmat[1], Nurmawan[1], S Sembiring[1], M F Syahputra[1] and Fadli[2]

[1] Department of Information Technology, Faculty of Computer Science and Information Technology, Universitas Sumatera Utara, Medan, Indonesia
[2]Faculty of Management and Business, Universitas Sumatera Utara, Medan, Indonesia

Email: romi.fadillah@usu.ac.id

Abstract. Natural Rubber is classified as the top export commodity in Indonesia. Its high production leads to a significant contribution to Indonesia's foreign exchange. Before natural rubber ready to be exported to another country, the production of rubber milk becomes the primary concern. In this research, we use adaptive neuro-fuzzy inference system (ANFIS) to do rubber milk production forecasting. The data presented here is taken from PT. Anglo Eastern Plantation (AEP), which has high data variance and range for rubber milk production. Our data will span from January 2009 until December 2015. The best forecasting result is 1,182% in term of Mean Absolute Percentage Error (MAPE).

1. Introduction

Rubber is one of the leading plantation commodities that contribute significantly to non-oil exports. The increase in rubber exports is very significant, of export volume of 1.496 million tons valued at USD 1038 million in 2002 increased to 2.35 million tons valued at USD 7330 million in 2010 or an increase of 57%. The export contribution of plantation sector above is expected to be the driver of national economic development in the future [1]. The increased demand for rubber spurred an increase in the rubber milk production in rubber producing countries in the world, including Indonesia. Rubber is a vital need for human life, and it is associated with the mobility of human and goods that require components made of rubber such as tires, conveyor belt, transmission belt, etc. Rubber industries have to prepare raw materials in accordance with market demand to meet those needs. Rubber milk production is paramount to rubber industries. Rubber companies expect the accurateness of rubber milk production in term of its amount and time. Therefore, state-owned or private plantation companies require a system that can predict the production precisely.

Forecasting is an attempt to predict future circumstances through examination of the past circumstances [2]. One of the quantitative forecasting techniques is time series data forecasting, a set of observations built in sequence within a specified time in which data can be data daily, weekly, monthly, etc. [3]. Time series data forecasting has been developed in soft computing such as fuzzy method, genetic algorithm, neural networks and hybrid method.

ANFIS is a method that uses a neural network to implement fuzzy inference system. In other words, ANFIS is the incorporation of fuzzy inference system mechanisms depicted in artificial neural network architecture. In statistical modeling, ANFIS is applied to classification, clustering, regression, and forecasting in time series data [10].

IOP Conf. Series: Materials Science and Engineering **308** (2018) 012014 doi:10.1088/1757-899X/308/1/012014

Several kinds of research have been conducted. In 2013, Agus Harianto did a research on the forecast of rubber milk production using Autoregressive Integrated Moving Average (ARIMA) method [4]. The best ARIMA model for rubber milk production forecasting is ARIMA model (2,1,1) with an estimated R2-squared parameter of 93.20, AIC value (Akaike Information Criterion) of 25.64, SIC value (Schwarz Information Criterion) of 25.84 and significant probability close to zero. With the measurement result of forecasting error with MAPE (Mean Absolute Percentage Error) parameter of 1.37, MAD (*Mean Absolute Deviation*) value of 380,5, MSE (*Mean Squared Error*) value of 5.194.584.081 and MPE (*Mean percentage Error*) value of 0.01. The analysis of production forecasting shows that production of 2012 to 2013 is predicted to be increasing year by year with a small error rate.

In 2015, Hani'ah did adaptive neuro-fuzzy inference system (ANFIS) implementation for water use forecasting at Tirta Moedal Semarang drinking water company [5]. Ria in 2014, did weather forecasting using case-based reasoning and adaptive neuro-fuzzy inference system methods [6]. Gernowo and Yenny in 2013 used ANFIS to predict the variability of Indonesian Rupiah exchange value against US dollar. The result shows that the proposed method is capable in predicting the rise and fall of exchange rate fluctuations [7]. Saputra in 2012 applied ANFIS to analyze time series data [8]. Widyapratiwi in 2012 did short-term power load forecasting in Bali using adaptive approach neuro-fuzzy inference system (ANFIS) [9]. Several other methods to do forecasting also has been analyzed such as exponential smoothing method for palm oil production [11], evolving multilayer perceptron [12][16], distributed and adaptive neural network [13] [14] [17] and fast learning algorithm [15].

A company needs a precise forecasting of rubber milk production to be used as a reference to compose a business plan. The forecasting of rubber milk production done manually usually experiences a problem in the form of production prediction results that are not in accordance with the actual production results, so it required a proper method that generates a précised forecasting of the production to be used as a reference to compose a business plan of plantation company.

2. Methodology

Forecasting system of rubber milk production using adaptive neuro-fuzzy inference system (ANFIS) is a system that delivers results in rubber milk production forecasting based on time series data in the past. The system will receive input in the form of monthly production data. Afterwards, the data will be processed using adaptive neuro-fuzzy inference system method to generate the forecast result. The general flowchart of the proposed method can be seen in Figure 1. The principal work of the forecasting system of rubber milk production to be built is designed as follows:

1. Input CSV file data containing rubber milk production data through input form.
2. Set the input parameters in the form of with maximum epoch and maximum error.
3. Form network structure of ANFIS.
4. Classify rubber milk production data inputted into two clusters using fuzzy c-means algorithm and determine the type of *membership* function to be used. The system will use generalized bell membership function, as follows:

$$\mu(Z) = \frac{1}{1 + \left|\frac{Z - c}{a}\right|^{2b}}$$

Where: $\mu(Z)$ = membership degree
Z = the amount of production
c = premise parameter, the initial value is determined using mean cluster
a = premise parameter, the initial value is determined using the standard deviation of the cluster
b = premises parameter value 1

The output value of generalized bell membership function is layer 1 output of ANFIS.

5. Compute layer 2 output by multiplying weight values in the corresponding layer 1.
6. Compute layer 3 output by dividing the weight of each w1 or w2 by the amount of weights w1 + w2 in the second layer.
7. Compute output layer 4 by multiplying the output weights of layer 3 3 w_1f_1 and w_2f_2 also set the consequent parameter values using LSE Recursive.
8. Compute layer 5 output, in the form of output prediction which error value can be calculated by subtracting the actual output value with the value of output prediction. If the error is smaller than the maximum error specified in the input form, the system will stop and generate an output prediction.
9. The system will store the production data into a database along with all premise and consequent parameters.

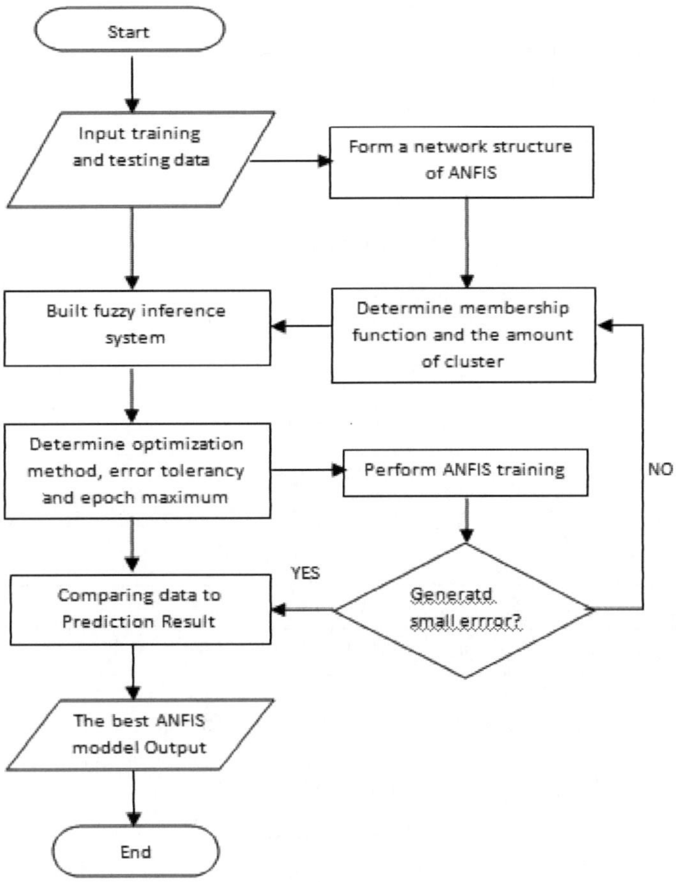

Figure 1. Flowchart of the System

2.1. ANFIS Simulation

The implementation of ANFIS calculation in the prediction of rubber milk production can be simulated as follows. Supposed the given time series data of 8 rubber milk production data are shown in Table 1

Table 1. Rubber milk Production Data's

t	1	2	3	4	5	6	7	8
Z_t	3875	3672	2697	2226	2364	2498	3111	4849

Given ANFIS input in the form of Z_{t-1} and Z_{t-2} while the output is in the form of Z_t as shown in Table 2.

Table 2. Input and Output Data of ANFIS

Z_{t-2}	Z_{t-1}	Z_t
3875	3672	2697
3672	2697	2226
2697	2226	2364
2226	2364	2498
2364	2498	3111
2498	3111	4849

Since the number of clusters is 2, then Sugeno model of ANFIS rule is as follows:

$$If\ Z_{t-2}\ is\ A_1\ and\ Z_{t-1}\ is\ B_1\ then\ Z_t = p_1Z_{t-2} + q_1Z_{t-1} + r_1$$
$$If\ Z_{t-2}\ is\ A_2\ and\ Z_{t-1}\ is\ B_2\ then\ Z_t = p_2Z_{t-2} + q_2Z_{t-1} + r_2$$

Afterwards, clustering process will be performed using fuzzy c-means clustering. Clustering result can be seen in Table 3.

Tabel 3. FCM Cluster Data

Membership Degree		Cluster Tendency	
Z_{t-1}	Z_{t-2}	*Cluster 1*	*Cluster 2*
0.0567545497136	0.9432454502863		*
0.1624564377419	0.8375435622581		*
0.9366214242975	0.0633785757024	*	
0.9778185681115	0.0221814318884	*	
0.9972202565087	0.0027797434912	*	
0.8155083611785	0.1844916388214	*	

Then input the value of clustering to ANFIS network, as follows:

Layer 1:

Since the membership function used is generalized bell:

$$\mu(Z) = \frac{1}{1 + \left|\frac{Z-c}{a}\right|^{2b}}$$

Determined the value of b=1 also as the initial value of a and c, mean value and standard deviation were used can be seen in Table 4.

Table 4. Mean and Standard Deviation

Data to-	*Cluster* 1		*Cluster* 2	
	Z_{t-1}	Z_{t-2}	Z_{t-1}	Z_{t-2}

1			3875	3672
2			3672	2697
3	2697	2226		
4	2226	2364		
5	2364	2498		
6	2498	3111		
m	2446.25	2549.75	3773.5	3184.5
sd	200.68943	390.29764	143.54267	689.42911

Then neurons of layer 1 can be calculated 1 as follows:

$$\mu_{A1}(x) = \frac{1}{1 + \left|\frac{Z_{t-2} - 2446.25}{200.689436692}\right|^2}$$

$$\mu_{A2}(x) = \frac{1}{1 + \left|\frac{Z_{t-2} - 2549.75}{390.297642831}\right|^2}$$

$$\mu_{B1}(x) = \frac{1}{1 + \left|\frac{Z_{t-1} - 3773.5}{143.542676580}\right|^2}$$

$$\mu_{B2}(x) = \frac{1}{1 + \left|\frac{Z_{t-1} - 3184.5}{689.429111656}\right|^2}$$

The result of layer 1 output in the form of membership degree of each data can be seen in Table 5.

Table 5. Membership Degree

Cluster 1		*Cluster* 2	
μ_{A1}	μ_{A2}	μ_{B1}	μ_{B2}
0.006685503	0.02289485	0.00144492	0.04477173
0.006685503	0.02289485	0.00144492	0.04477173
0.390456057	0.87539817	0.00853060	0.34096172
0.453631295	0.59239523	0.01026481	0.41384322
0.856188908	0.81532900	0.01250648	0.50212881
0.937653233	0.98272332	0.044840082	0.98876204

Layer 2:
Layer 2 output can be computed as follows:

$$w_1 = \mu_{A1}(x).\mu_{B1}(X); \quad w_2 = \mu_{A2}(x).\mu_{B2}(X);$$

The calculation result of layer 2 output can be seen in Table 6.

Table 6. Layer 2 Output

Layer 2 output	
w_1	w_2

9.6600532144628E-6	0.0010250424471626
9.6600532144628E-6	0.0010250424471626
0.0033308282542147	0.29847727756169
0.004656439238402	0.24515875557234
0.010707917147754	0.4094001885337
0.042044448460943	0.97167952783949

Layer 3:
Layer 3 output can be computed as follows:

$$\overline{w}_1 = \frac{w_1}{w_1 + w_2}; \qquad \overline{w}_2 = \frac{w_2}{w_1 + w_2}$$

The calculation result of layer 2 output can be seen in Table 7.

Table 7. Layer 3 Output

Layer 3 output	
$\overline{w}_1$	$\overline{w}_2$
0.0093360683007364	0.99066393169926
0.0093360683007364	0.99066393169926
0.011036245183707	0.98896375481629
0.018639535685288	0.98136046431471
0.025488480233879	0.97451151976612
0.041475243206127	0.95852475679387

Layer 4:
Layer 4 output can be computed as follows:

$$\overline{w}_1 f_1 = (\overline{w}_1 Z_{t-2})p_1 + (\overline{w}_1 Z_{t-1})q_1 + (\overline{w}_1)r_1$$
$$\overline{w}_1 f_2 = (\overline{w}_2 Z_{t-2})p_2 + (\overline{w}_2 Z_{t-1})q_2 + (\overline{w}_2)r_2$$

The calculation result of layer 4 output can be seen in Table 8.

Tabel 8. Layer 4 Output

$\overline{w}_1 Z_{t-2}$	$\overline{w}_1 Z_{t-1}$	$\overline{w}_1$	$\overline{w}_2 Z_{t-2}$	$\overline{w}_2 Z_{t-1}$	$\overline{w}_2$
36.17	34.28	0.0093	3838.82	3637.71	0.99
34.28	25.17	0.0093	3637.71	2671.82	0.99
29.76	24.56	0.0110	2667.23	2201.43	0.98
60.25	44.06	0.0186	2184.50	2319.93	0.98
60.2547	63.6702	0.0254	2303.74	2434.32	0.97
103.6051	129.0294	0.0414	2394.39	2981.97	0.95

The estimation of consequent parameters (p_1, p_2, r_1, p_2, q_2, and r_2) can be computed using LSE recursive method as follows:

$$\theta = (A^T A)^{-1} A^T y$$

Where A is the matrix of layer 4 output, y is production data output and θ is the consequent parameter. Therefore, the parameters value obtained are:

p_1 = 205.2208783291, p_2 = -8.7312130543796,
r_1 = -415469.33821946, p_2 = -2.1247969600763,
q_2 = 0.60963523398073dan r_2 = 5440.6762735378.

Layer 5:
layer 5 output can be computed by multiplying the obtained consequent parameters with layer 4 output (A matrix), layer 5 output can be seen in Table 9.

Table 9. Layer 5 Output

Z_t target	$Z_t output$	*Error*
2697	2696.9999999889	1.1070369E-8
2226	2225.9999999885	1.1506019E-8
2364	2363.9999999906	9.4355527E-9
2498	2497.9999999881	1.1867541E-8
3111	3110.9999999842	1.5775185E-8
4849	4848.9999999772	2.28219505-8

The estimation of premise parameters (a,c) is using error gradient descent propagation model.

3. Experimental Results

Data used for the testing in this research is monthly rubber milk production data of private company period January 2009 - December 2015. Production data consists of 696 data. The composition of training data amounted to 588 data and data testing amounted to 108 data.

The training result of ANFIS method can be tested using mean absolute percentage error (MAPE) as follows:

$$Error = \frac{\sum_1^n (\frac{|y - p|}{y})}{n} \times 100\%$$

Where y is the actual data, p is the prediction data, and n is the amount of data. From the result, obtained MAPE value of 1.182%.

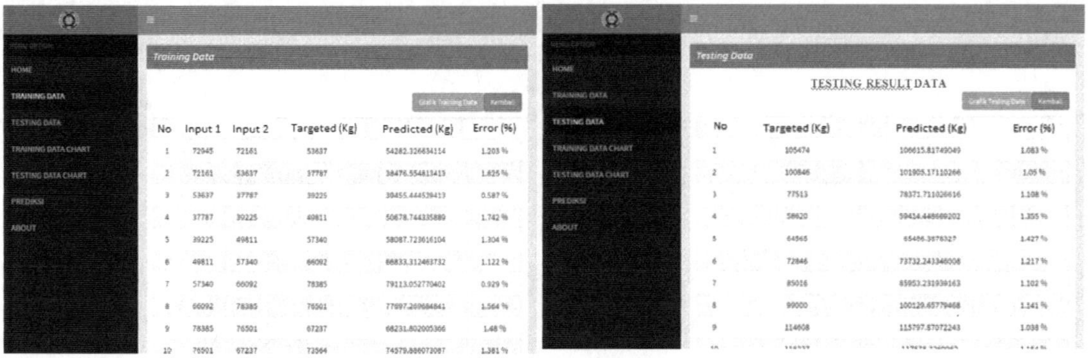

Figure 2. Training and Testing Results from the System

Data training aims to find the premise and consequent parameter values to be to test the data that do not train (See Figure 2 and 3). Training data used is rubber milk production data period January 2009 – December 2014 with total data of 588. Graphic of rubber milk production training result can be seen in Figure 3.

Figure 3. Graphic of rubber milk production training result

After training result was obtained, continued by performing a test on 108 data testing of rubber milk production data period January 2015 – December 2015. Based on training data, produced the smallest error value during training process which is 1.198%. The testing aims to test the accuracy of ANFIS architecture in identifying time series data patterns. From the testing results, obtained the smallest MAPE error value of 1.182%. The testing result of rubber milk production prediction that generates MAPE error value of 1.182% can be seen in Table 10, Figure 2 and Figure 4.

Table 10. Result of data testing

No.	Targeted Output (Kg)	Predicted Output (Kg)	MAPE Error (%)
1	105474	106615.81749049	1,182 %
2	100846	101905.17110266	
3	77513	78371.711026616	
4	58620	59414.448669202	
5	64565	65486.3878327	
6	72846	73732.243346008	
7	85016	85953.231939163	
8	99000	100129.65779468	
9	114608	115797.87072243	
10	116237	117578.3269962	

IOP Conf. Series: Materials Science and Engineering **308** (2018) 012014 doi:10.1088/1757-899X/308/1/012014

11	105235	106521.90114068	
12	121414	122975.05703422	

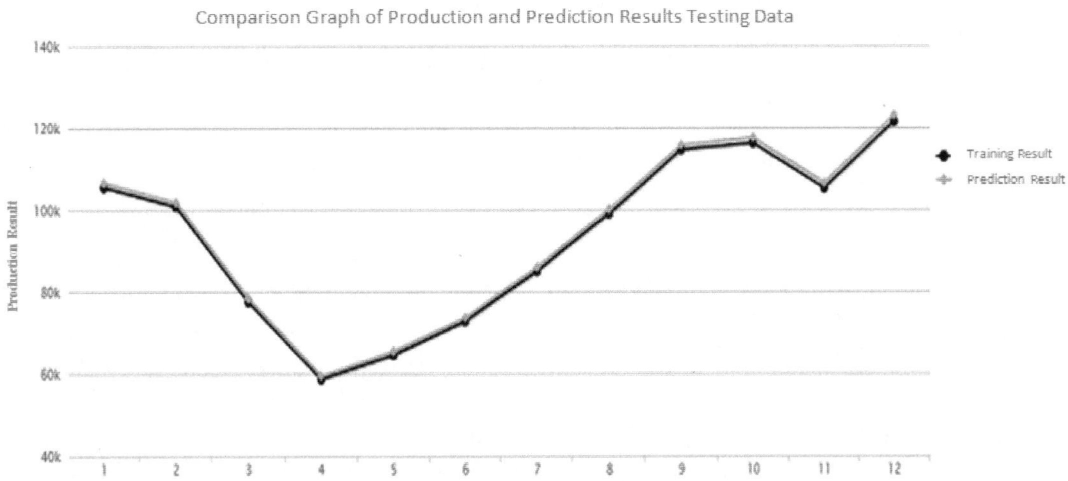

Figure 4. Graphic of testing result

The prediction testing is conducted to find out the result of rubber milk production for the next 12 periods. The testing result will show the predictive production data and prediction testing graphic. In Figure 4.5. Can be seen the results of production predictions and the graphs for production period January 2016 - December 2016.

PREDICTION RESULT

No.	Prediction Result (Kg)
1	104456.5659164
2	99873.209003215
3	76765.286173633
4	58054.533762058
5	63942.186495177
6	72143.305466238
7	84195.909967846
8	98045.01607717
9	113502.45659164
10	115115.74276527
11	104219.87138264
12	120242.80385852

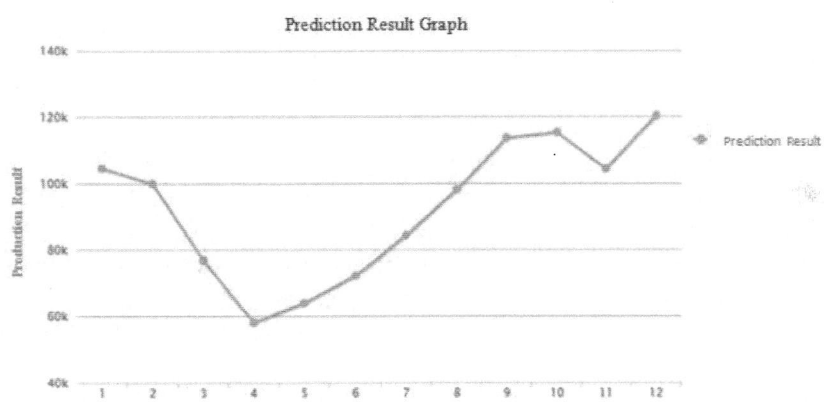

Figure 4.5. 2016 production prediction result

Based on the results of data training and data testing that have been conducted, it can be seen that the ANFIS prediction of rubber milk production generated MAPE error of 1.182%. When compared to Harianto's research in 2013 which researched rubber milk production using ARIMA method that produces MAPE error of 1.370%, the prediction of rubber milk production using ANFIS generates relatively smaller error compared to ARIMA method.

IOP Conf. Series: Materials Science and Engineering **308** (2018) 012014 doi:10.1088/1757-899X/308/1/012014

4. Conclusion

Based on the result, ANFIS is successfully implemented in a system to predict the production of rubber milk. The error rate of prediction result using ANFIS is relatively smaller, which is MAPE error value of 1,182%, if compared to previous research done by Harianto using ARIMA with MAPE error value of 1,370%.

For further research, we can apply other fuzzy determination methods to determine the value of fuzzy in the early stage of adaptive neuro-fuzzy inference system method. It is expected that adaptive neuro-fuzzy inference system method can be analyzed and compared to other methods in similar research data, to get more accurate prediction results by comparing the mean error between those methods.

5. References

[1] General Directorate of International Trade Cooperation. 2011. *ASEAN Community in a Global Community of Nations*.

[2] Rambe M F 2002 Market Demand and Sales Prediction Analysis. Scientific Journal of Management and Business, Management Department, Faculty of Economy **1** pp 1-12

[3] Halim S 2006 Time series analysis. Diktat. Universitas Petra Christian.

[4] Harianto A 2013 Forecasting Analysis of Rubber Production in PT Perkebunan Nusantara IX(PERSERO) Batujamus Plantation of Karanganyar Regency. *Agribusiness Review* **1** pp 45-64.

[5] Hani'ah U 2015 *Adaptive Neuro-Fuzzy Inference System* (ANFIS) Implementation for water use forecasting at Tirta Moedal Semarang drinking water company. Thesis. Universitas Negeri Semarang.

[6] Ria 2014 Weather Prediction Using Case-Based Reasoning and Adaptive Neuro-Fuzzy Inference System Methods. *Journal of Informatics* **12**(2): 90-95.

[7] Gernowo R dan Ernitawati Y 2013 Fuzzy Inference System Of The Variability of Indonesian Rupiah Exchange Value against US Dollar Prediction on Adaptive Neural Network as Economical Analysis. *Journal of Science and Mathematics* **21**(2): 39-41.

[8] Saputra A H 2012 Time series Data Analysis Using Adaptive Neuro-Fuzzy Inference System Method. Thesis. Universitas Diponegoro.

[9] Widyapratiwi L K 2012. Short-term Power Load Forecasting In Bali Using Adaptive Neuro-Fuzzy Inference System (ANFIS). Thesis. Universitas Udayana.

[10] Jang, JSR., CT Sun, dan E Mizutani. 1997. *Neuro-Fuzzy and Soft Computing: A Computational Approach to Learning and Machine Intelligence*. London: Prentice-Hall, Inc.

[11] Siregar B, Butar-butar I A, Rahmat R F, Andayani A and Fahmi F 2017 Comparison of Exponential Smoothing Methods in Forecasting Palm Oil Real Production *Journal of Physics: Conference Series* **801** p. 12004.

[12] Rahmat R F, Rizki A, Alharthi A F and Budiarto R 2016 Big Data Forecasting Using Evolving Multi-layer Perceptron *4th Saudi International Conference on Information Technology (Big Data Analysis) (KACSTIT)* pp. 1–5.

[13] Pasha M F, Rahmat R F, Budiarto R, and Syukur M 2010 A distributed autonomous neuro-gen learning engine and its application to the lattice analysis of cubic structure identification problem *International Journal of Innovative Computing, Information and Control* ***6*** pp. 1005–1022

[14] Rahmat R, Pasha M F, Syukur M, Budiarto R 2015 A gene-regulated nested neural network *International Arab Journal of Information Technology* ***12*** pp. 532–539

[15] Alwajeeh T, Alharthi A F, Rahmat R F and Budiarto R 2017 Fast Learning for Big Data Using Dynamic Function *IOP Conference Series: Materials Science and Engineering* **190** p. 12015

[16] Nababan E B, Iqbal M and Rahmat R F 2016 Breast Cancer Identification on Digital Mammogram Using Evolving Connectionist Systems *International Conference on Informatics and Computing (ICIC)* pp. 132–136

[17] Aaron, Sitompul O S and Rahmat R F 2014 Distributed autonomous Neuro-Gen Learning Engine for content-based document file type identification *International Conference on Cyber and IT Service Management* pp. 63–68

IOP Conf. Series: Materials Science and Engineering **308** (2018) 012015 doi:10.1088/1757-899X/308/1/012015

The morphological classification of normal and abnormal red blood cell using Self Organizing Map

R F Rahmat[1], F S Wulandari[1], S. Faza[1], M A Muchtar[1], I. Siregar[2]

[1] Department of Information Technology, Faculty of Computer Science and Information Technology, Universitas Sumatera Utara, Medan, Indonesia

[2] Department of Industrial Engineering, Faculty of Engineering, Universitas Sumatera Utara, Medan, Indonesia

Email: romi.fadillah@usu.ac.id,

Abstract. Blood is an essential component of living creatures in the vascular space. For possible disease identification, it can be tested through a blood test, one of which can be seen from the form of red blood cells. The normal and abnormal morphology of the red blood cells of a patient is very helpful to doctors in detecting a disease. With the advancement of digital image processing technology can be used to identify normal and abnormal blood cells of a patient. This research used self-organizing map method to classify the normal and abnormal form of red blood cells in the digital image. The use of self-organizing map neural network method can be implemented to classify the normal and abnormal form of red blood cells in the input image with 93,78% accuracy testing.

1. Introduction

Along with the times, technology in the medical field is currently growing. Technology itself has become a necessity in all areas, especially biomedical. Many techniques have been developed that can help doctors in identifying a particular disease. Digital image processing can be implemented to detect uterine tumor or cancer, to determine diseases such as lungs disease, liver disease and bone disease, bone segmentation of other muscles, teeth classification and microscopic image analysis [1]. To identify a disease can be tested through a blood test, one of which can be seen from the form of red blood cells. The normal and abnormal morphology of the red blood cells of a patient is very helpful to doctors in detecting a disease [2]

Blood is an essential component of living creatures in the vascular space, due to its role as an intercellular communication media to various parts of the body with the outside world because of its function of carrying oxygen from the lungs to the tissues and carbon dioxide of the tissues to the lungs to be removed, bringing nutrients from the gastrointestinal tract to the tissue and then deliver the rest of the metabolism through organ secretions such as kidney, providing hormones from blood clotting materials [3].

Previous research that discuss the image of red blood cells are: determining the morphology of red blood cells using Artificial Neural Network (ANN) method [2], the calculation of overlap red blood cells using Morphology Operation method [4], calculating the number of red blood cells using Multilayer Perceptron [5], Detecting and calculating the amount of normal and abnormal form of red blood cells using the object feature extraction based on the shape [6].

Based on previous research we are encouraged to investigate the detection of normal and abnormal red blood cells by implementing the Self-Organizing Map method. In this study, we use Self-Organizing Map (SOM) or better known as Kohonen method to perform normal and abnormal red blood cells pattern recognition of an image. SOM is a method often used in clustering process. SOM included in competitive learning, where the output nodes will compete to be the winning node and become the only node to be activated. Also, the transformation can create group data automatically, similar data will be stored in the nearest cluster. Since the initial weights are determined randomly, the SOM can determine itself to enter a particular group. SOM also takes only a short time to complete the learning process. However, the success of the system using SOM depends on the size of the output value of neurons number, radius neighbors, and the pace of learning [7]. Related research on SOM or image classification includes: Classification of nutritional status of children [7], Vectors quantization on a 24 -bit bitmap image compression [8], Rainfall prediction [9], character recognition [12], classification of ribbed smoke sheet using LVQ and shape feature extraction for its features [13] and character recognition using directional feature extraction [15] .

We hope the system to be built will result in better accuracy and is able to help doctors to distinguish normal and abnormal red blood cells automatically. Then the analysis of normal and abnormal forms of red blood cells performed by a doctor is not always the same with other doctors. The accuracy and concentration of doctors significantly affect the analysis result, therefore required a system that can identify normal and abnormal forms of red blood cells with better accuracy and in a short time.

2. Research methodology

The general architecture of our proposed method is shown in figure 1.

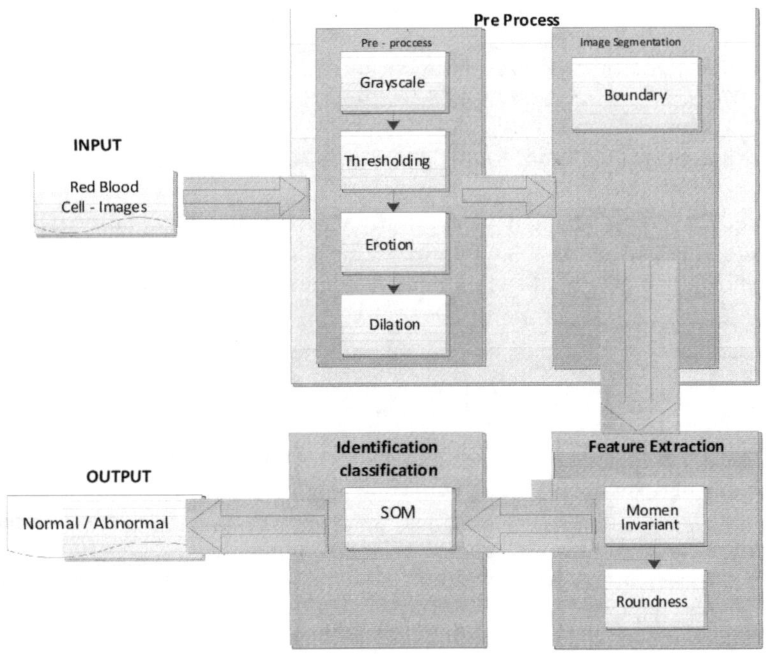

Figure 1. General Architecture

2.1. Red Blood Cells

The most abundant cells in the blood membrane are red blood cells or also known as erythrocytes. Red blood cells in the form of biconcave discs with a diameter of about 7.5 microns, the thickness of the edges of 2 microns and the central part of 1 micron or less, are composed of very thin membranes that are very susceptible to diffusion of oxygen, carbon dioxide and cytoplasm, but do not have a cell nucleus [10]. Viewed from the side, erythrocytes look like discs or biconcave with central acromial approximately ⅓ - ½ cell diameter [2]. Conventionally red blood cells are calculated by diluting a few drops of blood in a diluent solvent. The diluted blood is placed in the calculation space based on the known amount, and the number of cells presents calculated microscopically. Calculation of cells is assumed as the number of cells in one liter of blood. Calculation of red blood cells is considered as a number multiplied by 10^{12} per litter (i.e., 5 x 10^{12}/ l) [10].

2.2. Data Pre-processing

2.2.1. Data Acquisition. In this research, we used several images which contain red blood cells inside. This image has been taken from Prodia Laboratory and several pathology doctors with their permission. The images taken from the laboratory has been classified into two general images. Images which contain abnormal red blood cells or we called it as anemia, and images which contain normal red blood cells. We collect 28 images that consist of 24 images which contain abnormal red blood cells and 4 images which contain normal red blood cells.

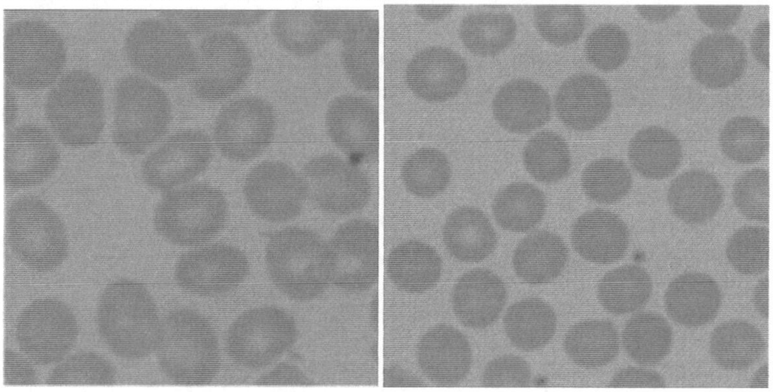

2a 2b

Figure 2a. Abnormal Red Blood Cells, **2b.** Normal Red Blood Cells

2.2.2. Greyscaling. It is the process of changing the pixel values from color (RGB) to gray-level or grayscale. The greyscale process is done by equalizing the pixel values of the three RGB values into one value [8][11][14]. The process of converting color image to grayscale image can be done using equation (1)

$$I(x,y) = \frac{R+G+B}{3} \quad (1)$$

Where $I(x,y)$ is grey level in position (x,y). While R, G, and B declare the color component value of each color pixel value in position (x,y) respectively.

2.2.3. Thresholding. Thresholding will generate binary image, an image that has two levels of gray which is black and white. In general, thresholding process of a grayscale image to generate binary image explained in equation (2):

$$g(x,y) = \begin{Bmatrix} 1 \; if \; f(x,y) \geq T \\ 0 \; if \; f(x,y) < T \end{Bmatrix} \quad (2)$$

where g(x,y) is binary image of grayscale image f(x,y), and T is the value of threshold. T value holds significant role in thresholding process [6].

2.2.4. Erosion and Dilation. The erosion process is executed by comparing each pixel of input image to SE center value by superimposing the SE with the image so that the SE center is precisely at the position of the processed image pixel. Erosion process will generate smaller object. The hole in the object will also appear to enlarge as it narrows the object boundary. The dilation process is executed by comparing each pixel of input image to SE center value by superimposing the SE with the image so that the SE center is precisely at the position of the processed image pixel.

2.2.5. Boundary. The boundary is the process to do segmentation on the images. We did background separation, and we fill the objects with different colors to differentiates them. In this process, we also do notation on the objects.

2.3. Feature Extraction

Invariant moment feature is beneficial for declaring objects by taking into account the object area. This feature uses central moment normalized basis. The moment generated can be implemented to handle translation, scaling and image rotation [13]. In this method, we took all 7 invariant moments as object features extracted from the image.
Another feature that we consider as a shape parameter is a roundness. Roundness describes the level of cellular roundness. Roundness is the ratio between object area and squared perimeter, which can be calculated using equation (13). The result is a value of ≤ 1. Value 1indicates R object as a circle.

$$R = (4\pi)\left(\frac{A(R)}{P^2(R)}\right) \tag{13}$$

Where:

R= Roundness
A = Area
P = Perimeter

2.4. Self Organizing Maps (SOM)
Self-organizing map (SOM) method or known as Kohonen, introduced by Teuvo Kohonen, is a competency-based neural network system capable of unsupervised learning due to its self-organizing ability. This neural network will study the distribution of set patterns without any previous class information [1].

3. Result and discussion

3.1. *Image Processing Results* – Here we describe the result from our proposed method based on our proposed general architecture in figure 1 above. Figure 3 shows that the process starts from original image, greyscale it down, then we did threshold to differentiate the background, then the morphological operator such as erosion and dilation are performed. The boundary process will be shown in figure 5. Figure 4 shows the extraction of the objects into 7 invariant moments and roundness values.

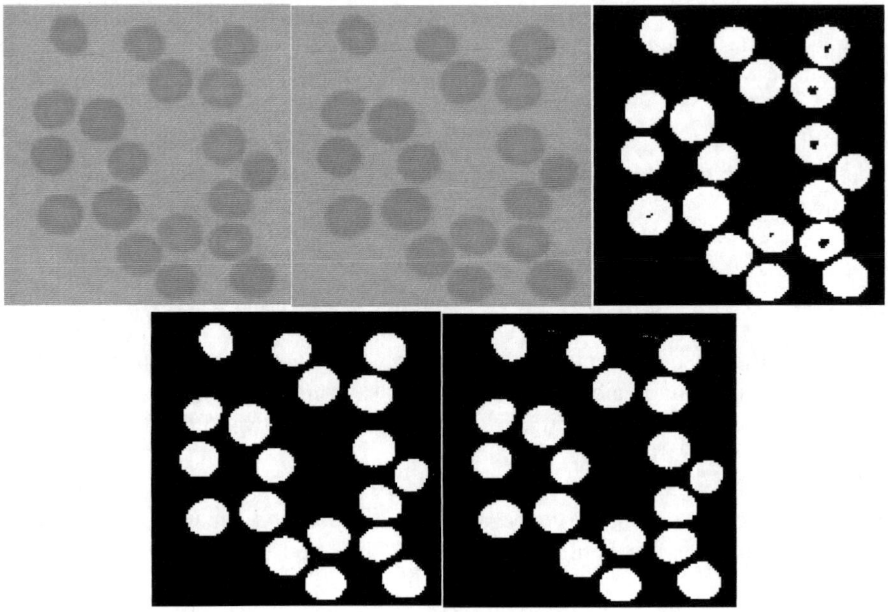

Figure 3. (from left up to right down). a.Original Image, b. Grayscaled Image, c. Threshold Image, d. Erosion Image, e. Dilation Image

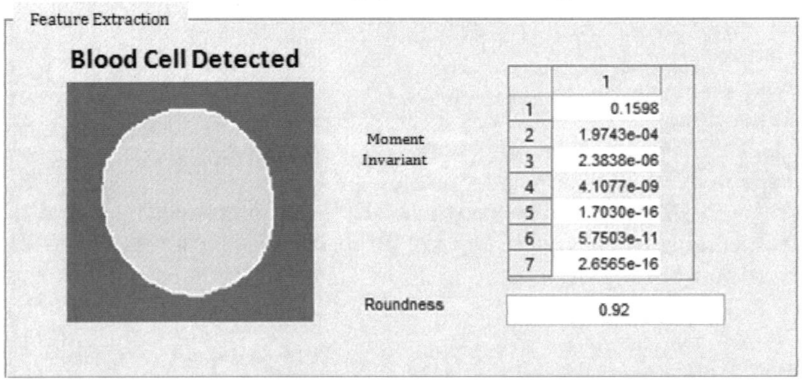

Figure 4. Invariant Moment and Roundness Extraction

3.2. System Overview

The result of this research is in the form of image that displays the cells considered as normal and abnormal along with information on the number of red blood cells detected. From the identification results shown in Fig 5, the tested cell is considered to be normal that contains roundness value of 0.90. Red blood cells considered to be normal if the roundness value is approaching 1. From the sample shown in Fig 5 contained result of 7 invariant moment values:

1. 0.1602
2. 3.1938e-04
3. 1.2415e-06
4. 4. 7320e-09
5. 3.9522e-16
6. 7.4814e-11
7. -3.2081e-16

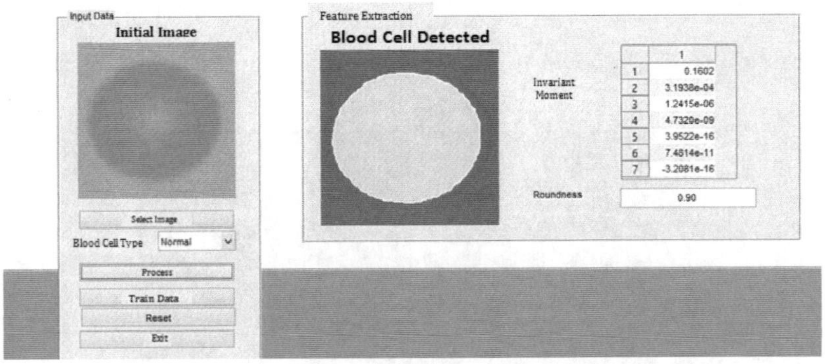

Figure 5. Red blood cell identified as normal cell

From the identification results shown in Fig 6, the abnormal cells contained roundness value of 0.86. The abnormal red blood cells tend to have a roundness value smaller than 0.9. The result depends on the input image.

From the sample shown in Fig 6 contained result of 7 invariant moment values:

1. 0.1689
2. 0.0024
3. 2.0364e-04
4. 4.9047e-06
5. -3.3754e-11
6. -1.0413e-07
7. 7.2269e-11

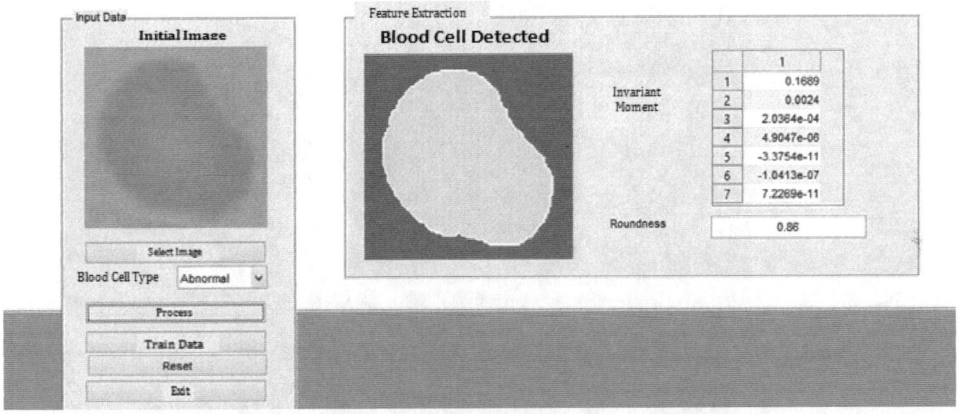

Figure 6. Red blood cell identified as abnormal cell

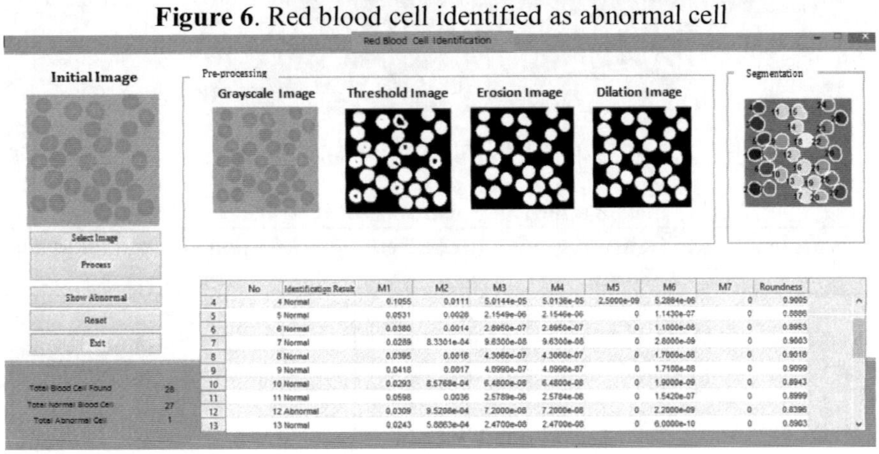

Figure 7. Proposed System

IOP Conf. Series: Materials Science and Engineering **308** (2018) 012015 doi:10.1088/1757-899X/308/1/012015

3.3. Testing Results

From the identification results shown in Fig 8, we present 5 samples from 15 testing images that we conducted before.

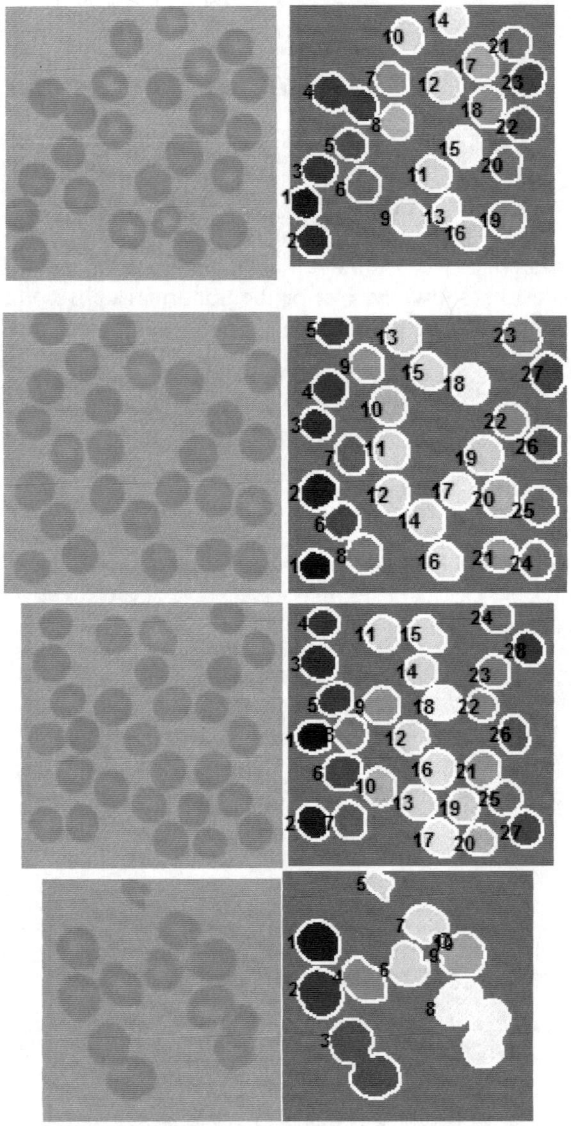

Figure 8. Identification Results

Table 1. Accumulation Results

Image Number	Cells in Total	Identified	Failed	Accuracy
1	23	22	1	95.65%
2	27	27	0	100%
3	28	26	2	92.85%
4	10	7	3	70%
5	16	10	6	62.5%
6	23	21	2	91,30%

7	27	25	2	92,59%
8	19	18	1	94,74%
9	25	25	0	100%
10	28	27	1	96.42%
11	19	19	0	100%
12	19	17	2	89,47%
13	22	22	0	100%
14	20	20	0	100%
15	16	16	0	100%

We used 322 cell partitioned images from 15 big cell images. Based on our experiment shown in Table 1, it shows that the system successfully identified 302 cell images. From this result, we can accumulate overall accuracy from our system which is shown below.

$$Percentage\ of\ System\ Accuracy = \frac{302}{322} \text{ x } 100\% = 93.78\%$$

4. Conclusion

Based on the data testing done can be concluded that self-organizing map neural network method can be implemented to classify normal and abnormal forms of red blood cells with 93.78% accuracy. We also can conclude that the image pre-processing stage has not been able to separate the cells that coincide or overlap effectively. Therefore, we suggest several things for further research:

1. Self-Organizing Map (SOM) method can be compared with other methods in similar conducting research to get better results.
2. We built the system to identify normal and abnormal forms of red blood cells limited to the shape; we hope in future research, system will be developed to identify normal and abnormal forms of red blood cells using other two components which are based on size and color.
3. Several other methods are required in the pre-processing process to generate non-stacked blood cells. Thus, the next process can produce better data information.
4. There are still some cells considered as one cell, but in fact, there are two or more cells, for that issue, it will require a different method that can solve this problem.

5. References

[1] Putra D 2009 *Digital Image Processing.* Yogyakarta: Andi

[2] Warni E 2009 The morphological operation of red blood cells (erythrocytes) based on image processing and artificial neural networks *Scientific Journal Electrical Engineering*

[3] Tarwoto and Wartonah 2008 *Medical Nursing Surgery of Hematology System Disorders.* Jakarta: Trans Info Media.

[4] Usman K 2008 Calculation of Stacked Red Blood Cells Based on Digital Image Processing and Morphological Operation. *National Seminar of Informatics*, 31-38.

[5] Poomcokrak J and Neatpisarnvanit C 2008 Red Blood Cell Extraction And Counting. *The Third International Symposium on Biomedical Engineering*, 199-203

[6] Kartikasari D 2013 Abnormal and Normal Red Blood Cell Detection and Calculation System Based on Digital Image Processing Using Matlab. *Thesis*. Universitas Komputer Indonesia

[7] Zunairoh U, Khotimah B K, and Rachmad A 2012 Classification of Children Nutrition Status Using Kohonen Self Organizing Map. *Journal of Bachelor of Informatics Engineering*.

[8] Tae G F, Suwarno S, and Hapsari W 2010 Kohonen Self Organizing Map Implementation in Vectors Quantization on 24-bit Bitmap Image Compression. *Jurnal Informatika*, **6** pp 15-23.

[9] Septiadi D 2008 Soft Computing Application of Rainfall Prediction in Kalimantan. *Journal*

of Meteorology and Geophysics **9** pp 65 – 77

[10] Bain B J 2004 *A Beginner's Guide to Blood Cells. 2nd Edition*, pp. 9 - 22.

[11] Rahmat R F, Chairunnisa T, Gunawan D and Sitompul O S 2016 Skin color segmentation using multi-color space threshold *International Conference on Computer and Information Sciences, ICCOINS 2016 – Proceedings* pp. 391-396.

[12] Gunawan D, Arisandi D, Ginting F M, Rahmat R F and Amalia A 2017 Russian character recognition using Self-Organizing Map *Journal of Physics: Conference Series* **801**, p. 12040

[13] Rahmat R F, Pulungan A F, Faza S and Budiarto R 2017 Image classification of ribbed smoked sheet using Learning Vector Quantization *Journal of Physics: Conference Series* **801**, p. 12050.

[14] Nababan E B, Iqbal M and Rahmat R F 2016 Breast Cancer Identification on Digital Mammogram Using Evolving Connectionist Systems *International Conference on Informatics and Computing (ICIC)* pp. 132–136

[15] Seniman, Arisandi D, Rahmat R F, William and Nababan E B 2016 Chinese chess character recognition using direction feature extraction and backpropagation *International Conference on Data and Software Engineering 2016*

Boyer-Moore Algorithm in retrieving deleted Short Message Service in Android Platform

R F Rahmat, D F Prayoga, D Gunawan and O S Sitompul

Department of Information Technology, Faculty of Computer Science and Information Technology, Universitas Sumatera Utara, Medan, Indonesia

Email: romi.fadillah@usu.ac.id, dian.fajar.prayoga@students.usu.ac.id, danigunawan@usu.ac.id, opim@usu.ac.id.

Abstract. Short message service (SMS) can be used as digital evidence of disclosure of crime because it can strengthen the charges against the offenders. Criminals use various ways to destroy the evidence, including by deleting SMS. On the Android OS, SMS is stored in a SQLite database file. Deletion of SMS data is not followed by bit deletion in memory so that it is possible to rediscover the deleted SMS. Based on this case, the mobile forensic needs to be done to rediscover the short message service. The proposed method in this study is Boyer-Moore algorithm for searching string matching. An auto finds feature is designed to rediscover the short message service by searching using a particular pattern to rematch a text with the result of the hex value conversion in the database file. The system will redisplay the message for each of a match. From all the testing results, the proposed method has quite a high accuracy in rediscovering the short message service using the used dataset. The search results to rediscover the deleted SMS depend on the possibility of overwriting process and the vacuum procedure on the database file.

1. Introduction

From the beginning of digital cellular telephony era, SMS has become one of the services to do communication. Since then, it is also likely to facilitate the communication of criminal activity. Therefore, SMS has become a significant object in the criminal investigation until now. In Indonesia, the importance of digital evidence has been stated explicitly in Law No. 11 In 2000 Article 5 on Information and Electronic Transaction. It shows that digital evidence such as SMS is crucial for a criminal investigation [1]. There are four phases of digital evidence investigation, namely digital evidence identification, digital evidence storage, digital evidence analysis, and digital evidence presentation. Each of them has their function to reveal the data from digital evidence [2]. Several examples of digital evidence feature and services in the mobile phone include Subscriber Identity Module (SIM), internal memory, memory card and network service providers [3].

In the Android operating system, SMS will be stored in SQLite Database file, which consists of two parts, namely the table storage and its record. It means one single SMS will be stored in a record in the table storage. In the case of SMS deletion and removal, the termination of the bit in memory does not occur, which means deleted SMS will be removed to empty space in the memory. This process is usually called as vacuum procedure [4]. From this procedure, we can assume that SMS retrieval becomes possible if no overwriting process occurred in memory. If so, then SMS object will be permanently deleted from memory. Thus, we can define that overwriting process is a process of adding new SMS in

IOP Conf. Series: Materials Science and Engineering **308** (2018) 012016 doi:10.1088/1757-899X/308/1/012016

the memory space of preceding SMS [5]. The other aspect that can determine the existence of deleted SMS is how often the database file executes the vacuum procedure.

To find the deleted SMS, we need to build a system that can recover deleted SMS with string matching in its hexadecimal bytes. The algorithm for string matching requirement is Boyer-Moore Algorithm proven as a good and fast string-matching algorithm [6]. Previous work has been conducted in term of device forensic such as file-based content type identification [7], file undelete [8] and file type identification [9]

2. Research methodology

In figure 1, we describe the general architecture of our proposed method. It starts from connecting Android smartphone with our system. Then several steps of pre-processing are conducted, namely doing image making to get another copy of data from the smartphone, extracting the image file to get database file, which contains SMS file from the Android root directory and generating hex values from database file to get hexadecimal bytes pattern. Next, input generalization is used as hex pattern in the searching process. This specific hex pattern will be used in Boyer-Moore Pre-processing phase to count the shifting values along with generated hex string. The last processes are the searching process using Boyer-Moore Algorithm and filtering process.

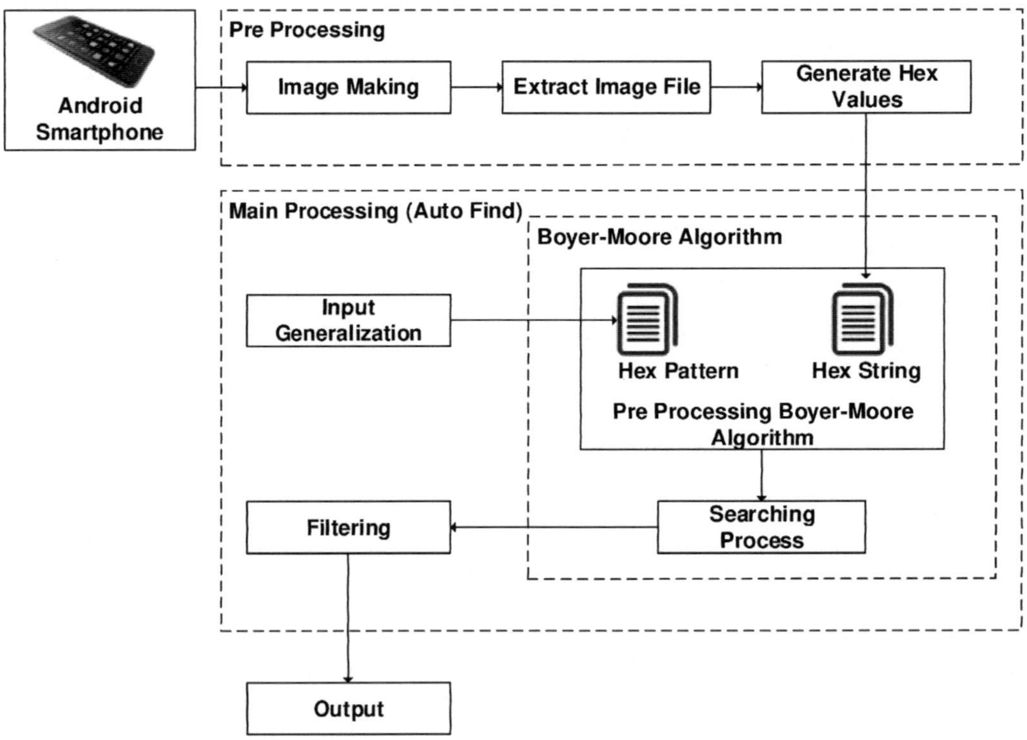

Figure 1. General Architecture

2.1. Dataset

The used dataset in this research is divided into two sets. Both of them are mmssms.db SQLite database file. These files are taken from /data/data/com.android.providers.telephony/databases/ Android directory files. The first dataset is mmssms1.db, mmssms2.db, mmssms3.db that consist of 10-30 SMS in one database, while mmssms4.db, mmssms5.db, and mmssms6.db have 0 SMS because it has been deleted before. The last set is mmssms7.db, mmssms8.db, mmssms9.db, and mmssms10.db which has unknown SMS existed before and unknown deleted SMS. The dataset properties can be seen in table 1.

Table 1. Dataset properties.

File	SMS	Deleted SMS
mmssms1.db	10	0
mmssms2.db	20	0
mmssms3.db	30	0
mmssms4.db	0	*Unknown*
mmssms5.db	0	*Unknown*
mmssms6.db	0	*Unknown*
mmssms7.db	*Unknown*	*Unknown*
mmssms8.db	*Unknown*	*Unknown*
mmssms9.db	*Unknown*	*Unknown*
mmssms10.db	*Unknown*	*Unknown*

2.2. Image Making and Extracting Image File

In this step, we will copy the data in the Android mobile phone root directory where the SMS database files contained. The purpose of image making is to copy entire SMS file to a safe container and to prevent any damage, modification, or loss of the file.

Next step is to extract image file to get file folder from image making process. In this folder, we can get the system or user's applications data including SMS messages. The output of this process is SMS database.

2.3. Generating Hex Values

After SMS database file is found, the next step in this proposed method is generating hex values to get hexadecimal bytes value from the database file. In this hexadecimal bytes values, we can get deleted SMS. The output of this process is hex string which is saved inside a .txt file. In this hexadecimal byte representation, SMS will have SMS file pattern, containing a header, sms_number, sms_date, and sms_body. The structure of SMS file pattern can be shown in figure 2.

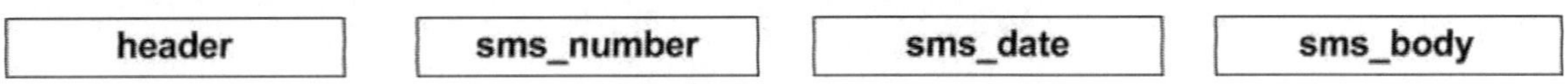

Figure 2. SMS File Pattern

2.4. Input Generalization

In this step, we generalize input string that will be searched in hex string. This input string will be used as a hex pattern in the searching process. Generalization is aimed to get the desired output. In the implementation, our proposed system uses hexadecimal the SMS header pattern (see Figure 2). 08 08 08 bytes from this header will be used as hex pattern for the next process.

2.5. Auto Find

Auto finds process is designed to retrieve deleted SMS file automatically. In the implementation, Boyer-Moore algorithm is used to support searching process. This feature assigned to search hex pattern 08 08 08. The result from this feature will give a whole content of SMS that has been found including its supporting attributes inside hexadecimal bytes.

2.5.1. Boyer-Moore Pre-processing Algorithm. Generated hex values create hex string, while input generalization creates hex pattern, these two values are used to calculate next shifting value based on bad character shift and good suffix shift functions. Thus, those two-shifting value is stored in two different table BmBc and BmGs. Shifting value of BmBc and BmGs can be described in Table 2 and Table 3 respectively.

Table 2. Shifting value BmBc in auto find feature

char	BmBc[char]
0	1
8	2
0	1
8	2
0	1
8	2

Table 3. Shifting value BmGs in auto find feature

i	Char	BmGs[char]
0	0	2
1	8	2
2	0	4
3	8	4
4	0	6
5	8	1

2.5.2. Searching Process. Our proposed system will do pattern searching using two shifting tables bad-character shift and good-suffix shift, wherein every unmatched character, the system will choose the biggest summation of shifting value for the next shifting. The pseudo code of Boyer-Moore Algorithm in the searching process can be shown below:

```
procedure BM(input pattern, text: array of char)
Variable Declaration
       i, j: integer
       bmGs : array [0..XSIZE] of integer
       bmBc : array [0..ASIZE] of integer
Algorithms
       /* Preprocessing */
       preBmGs(pattern, BmGs)
       preBmBc(pattern, BmBc)

       /* Searching */
       i = 0
       while (i <= text.Length - pattern.Length);
       for (j = pattern.Length - 1; j >= 0 and pattern[j]=text[i+j]; --i)
       if (j < 0)
              arr.Add(i);
              i ← i + BmGs[0];
       else
              i ← i + MAX(BmGs[j],BmBc[text[i+j]]- pattern.Length+1+j)
```

As has been described before, hexadecimal bytes from generated hex values is transformed into hex string, then the searching of hex pattern from input generalization will become a string matching process. When the criteria matches, the system will take some bytes from the file structure that contains SMS number pattern, SMS date, and SMS body. This method refers to the fact that undamaged SMS has similar pattern in their representation of hexadecimal bytes. The pattern for this auto finds method is shown in figure 3.

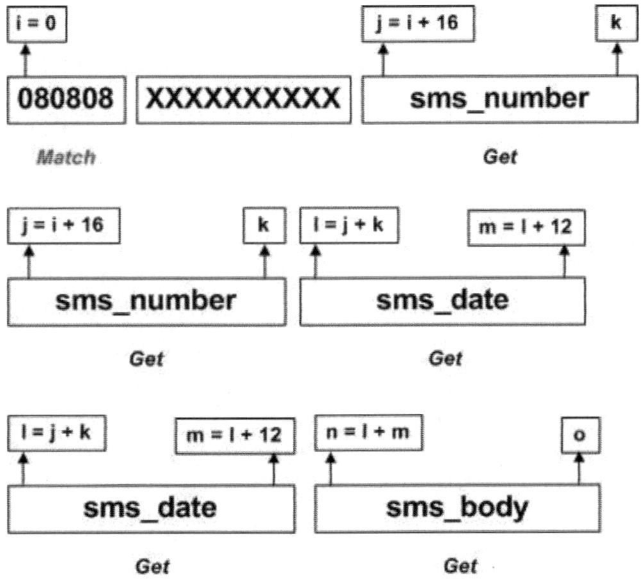

Figure 3. Auto Find Pattern Features

2.6. Filtering Process

In auto find step, there is possibility that system will display the message repeatedly. Therefore, an additional process is required to overcome it. Filtering is aimed to reduce the duplication of the data that frequently occurs in auto find. We use date of message as a comparator variable for implementing the filtering process. When the date of message is compatible, SMS will be stored in the system then it will be matched again. If the date of message cannot be found later, SMS will be displayed. Otherwise, it will not be displayed if the date of message is found. Figure 4 shows the filtering process.

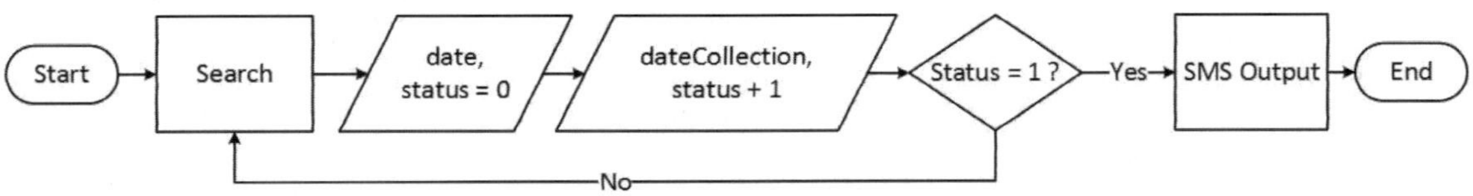

Figure 4. Filtering Process

2.7. Target Output

Target output in our proposed method is that system can summarize and retrieve the deleted SMS and identify if it is still intact or removed. The system will display the number, date, content and the status of SMS whether it is still integrated or deleted. For the message that can be found in the system and is still stored in the database file, then it can be identified that it is still integrated, otherwise, if it cannot be found in the database file, it has been deleted. The date of message will be used as epoch time and will be matched again with the database file for receiving the status of SMS. Table 4 below shows the proposed target output.

Table 4. Target output

Position	Number	Date	Body	Status
111222	+6281311112222	1 May 2015, 17:32:15	reference item (partial)	Exist
222333	+6282300005555	abstract body	abstract heading (also in Bold)	Delete

333444	+628112012012		level-2 heading, level-3 heading, author affiliation	Exist
...	...	...	...	...

3. Result and discussion

3.1. Generating hex values

The generated hex values will be stored in .txt file before being used in auto find feature. Figure 5 shows the generated hex values, and table 5 describes Byte Frequency Distribution (BFD) from mmssms7.db.

Table 5. BFD Table mmssms7.db dataset.

Byte (hexadecimal)	Byte (decimal)	Frequency
00	0	24980
01	1	1795
02	2	723
03	3	482
04	4	237
...	...	...
FD	253	0
FE	254	1
FF	255	31

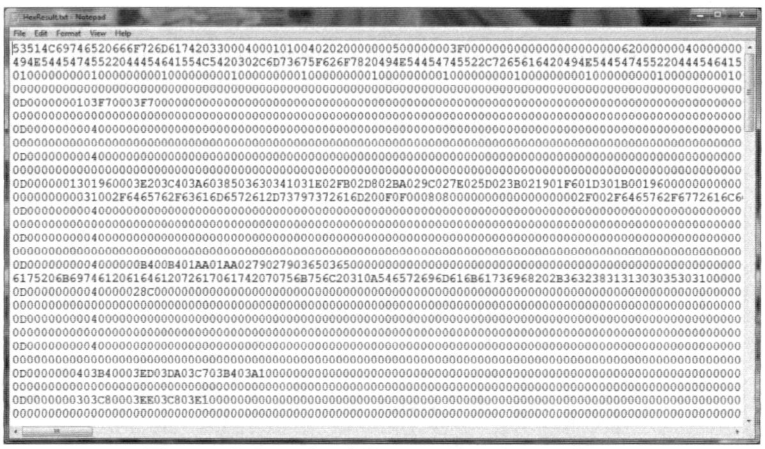

Figure 5. Result of Generating Hex Values

3.2. Auto Find

In this feature, the system will do searching by using hex pattern and hex string in each dataset that contains the generated hex values. The system will display the message when hex pattern and hex string matches. Several parameters that affect the searching process, include:

- Length of pattern used by default in auto finds feature is header byte of SMS data that consists of 6 bytes of hexadecimal, 080808.
- Length of character counted based on the result of generated hex values in each file of SMS database.
- The number of matching pattern and text in generated hex values counted for each file of SMS database.

- The number of call number found in generated hex values, where the default number is 24 bytes of hexadecimal or 12 ASCII characters.
- The number of the integrated message found and still stored in database file.
- The number of deleted message found and not stored in database file.
- Unreadable message caused by vacuum procedure or not integrated.

The sample result of auto finds feature using mmssms7.db dataset is shown in figure 9 and table 6. It shows that there are 29 match messages (23 integrate messages and 6 deleted messages).

Table 6. Result of auto find

Parameter	Value
Length of pattern	6
Length of character	151552
Match message	29
Number of call number as SMS sender	3
Number of integrating message	23
Number of deleted message	6

3.3. Testing Result

Table 7, table 8 and table 9 show the testing result using testing dataset and training dataset. The obtained accuracy for finding integrate message is 100% as shown in Table 4. While for the deleted message in training dataset, the system only can find 26 deleted SMS (around 43.33%), and in testing dataset, the system can discover 87 integrate message and 20 deleted message. The finding of deleted message depends on several processes before, namely overwriting or vacuum procedure in the database file.

Table 7. Testing result using training dataset for integrate file.

File	Integrate Messages	Found Messages	Accuracy
mmssms1.db	10	10	100%
mmssms2.db	20	20	100%
mmssms3.db	30	30	100%
Total	60	60	100%

Table 8. Testing result using training dataset for deleted file.

File	Integrate Messages	Deleted Messages	Total
mmssms4.db	0	9	9
mmssms5.db	0	9	9
mmssms6.db	0	8	8
Total	0	26	26

Table 9. Testing result using testing dataset.

File	Integrate Messages	Deleted SMS	Total
mmssms7.db	22	6	28
mmssms8.db	20	4	24
mmssms9.db	31	5	36
mmssms10.db	14	5	19
Total	87	20	107

4. Conclusion

From all testing results, there are several obtained conclusion for retrieving the deleted SMS in Android mobile phone. First, our proposed method can be used for retrieving the deleted SMS in Android mobile phone with 100% integrate message and 43.33% deleted message in the training dataset. In testing dataset, system can find 87 integrate message and 20 deleted message. The result of finding depends on the entire contents of database file. If the database file has overwritten and vacuum procedure, then it will be difficult to find because the database file has the different pattern. However, our proposed method cannot be used to retrieve message permanently deleted by overwriting and limited to specific digital

IOP Conf. Series: Materials Science and Engineering **308** (2018) 012016 doi:10.1088/1757-899X/308/1/012016

device and platform. Another additional approach is required to identify the message in other device and platform.

For further research, we suggest considering record recover method to retrieve SMS data deleted permanently. To get higher accuracy, the data searching should use a different pattern, and the data should be more and more various. Another research possibly done is retrieving data in SQLite database using the same method.

5. References

[1] Carrier B 2003 Defining Digital Forensic Examination and Analysis Tools Using Abstraction Layers *Int. J. Digital Evidence* **1** 1 – 12

[2] Casey E 2011 *Digital Evidence and Computer Crime: Forensic Science, Computers and the Internet* (Waltham: Elsevier)

[3] Zareen A and Baig S 2010 Mobile Phone Forensics: Challenges, Analysis and Tools Classification *Proc. of the 2010 Int. Workshop on Systematic Approaches to Digital Forensic Engineering* pp 47-55

[4] Stahlberg P, Miklau G and Levine B N 2007 Threats to Privacy in the Forensic Analysis of Database Systems *Proc. of the 2007 ACM SIGMOD Int. Conf. on Management of Data* pp 91 - 102

[5] Hoog A 2011 *Android Forensic: Investigation, Analysis, and Mobile Security for Google Android* (Waltham: Elsevier)

[6] Boyer R S and Moore J S 1977 A Fast String Searching Algorithm *Communications of the ACM* **20** 762 – 772

[7] Aaron, Sitompul O S and Rahmat R F 2014 Distributed autonomous Neuro-Gen Learning Engine for content-based document file type identification *International Conference on Cyber and IT Service Management* pp. 63–68.

[8] Sitompul O S, Handoko A, and Rahmat R F 2016 A file undelete with Aho-Corasick algorithm in file recovery *International Conference on Informatics and Computing (ICIC)* pp. 427–431.

[9] Rahmat R F, Nicholas F, Purnamawati S and Sitompul O S 2017 File type identification of file fragments using longest common subsequence (LCS) *Journal of Physics: Conference Series* **801** p. 12054.

IOP Conf. Series: Materials Science and Engineering **308** (2018) 012017 doi:10.1088/1757-899X/308/1/012017

The association rules search of Indonesian university graduate's data using FP-growth algorithm

S Faza, R F Rahmat*, E B Nababan, D Arisandi, S Effendi

Faculty of Computer Science and Information Technology, Universitas Sumatera Utara, Medan, Indonesia

*E-mail: romi.fadillah@usu.ac.id

Abstract. The attribute varieties in university graduates data have caused frustrations to the institution in finding the combinations of attributes that often emerge and have high integration between attributes. Association rules mining is a data mining technique to determine the integration of the data or the way of a data set affects another set of data. By way of explanation, there are possibilities in finding the integration of data on a large scale. Frequent Pattern-Growth (FP-Growth) algorithm is one of the association rules mining technique to determine a frequent itemset in an FP-Tree data set. From the research on the search of university graduate's association rules, it can be concluded that the most common attributes that have high integration between them are in the combination of State-owned High School outside Medan, regular university entrance exam, GPA of 3.00 to 3.49 and over 4-year-long study duration.

1. Introduction

Association rules mining is a data mining technique to determine the integration of the data or the way of a data set affects another set of data [1]. By way of explanation, there are possibilities in finding the integration of data on a large scale.

University graduates data is one of the large-scale data of an educational institution consisting of identity data of the students that were collected during the freshman year and academic data that were collected throughout the years in the university. From the existing graduates' data, some attributes can deduct student's length of studies such as previous school, entrance exam types, faculties, and GPA.

The attribute varieties in university graduate's data, e.g., previous school, entrance exam types, faculties, and GPA, have caused frustrations to the institution in finding the combinations of attributes that often emerge and have high integration between attributes of university graduate's data. Therefore, it is necessary to have an alternative to obtain the combination of common attributes that have a high integration between attributes using association rules mining. Association rules mining can generate high-integrated attributes combination by searching for attribute combinations that often appear first.

Frequent Pattern-Growth (FP-Growth) Algorithm is an algorithm that can be used in association rules mining technique to determine the frequent itemset in a set of data [2]. The algorithm only performs the database scanning process twice to determine the frequent itemset [3] and represents the transaction using FP-Tree data structure [4]. The result is expected to generate numerous rules that have high integration on the attributes in university graduates' data.

IOP Conf. Series: Materials Science and Engineering **308** (2018) 012017 doi:10.1088/1757-899X/308/1/012017

Research on university graduates has been performed using various methods. In 2011, Siregar analysed the integration determinant between student and length of study data using multiple linear regression algorithms [5]. This research generates a result at 61% of the integration of student's data toward study period data. Parameters set as the integration determinants are GPA, average national exams score, amount of credit hours, and parents background education. The research produces a prediction without yet knowing which combination of parameters which frequently emerge.

Huda, in 2010, used Apriori algorithm to show the information of student graduation rates [6]. The result of data mining process in the research can be applied as consideration in taking further decisions on the factors that affect the graduation rates, especially the one in student's primary data of a faculty. The student's primary data in the process of mining include data admission process, data from school, student city data, and study program data. The research advocates future researcher to use FP-Growth algorithm since the proposed method required numerous iteration of searches to the database.

In 2013, Ma'ruf also used Apriori algorithm to determine the relationship of admission process with student graduation rates [7]. The research generates information on the origin of student admission achievement which set as a reference in maximizing the advertisement in particular area. Apriori algorithm performed candidate generate of each item, so it required many iterations of searches to the database.

The attribute varieties in university graduate's data, e.g., previous school, entrance exam types, faculties, and GPA, have caused frustrations to the institution in finding rules that consist of the combinations of attributes that often emerge and have high integration between attributes. Therefore, an alternative is required to generate association rules for university graduate's data.

2. Methodology

The proposed method to search the association rules on university graduate's data consists of several phases, i.e., data pre-processing by determining the attributes, namely school types, university entrance types, GPA, and length of study, then classifying the attributes into several groups. The next phase is data mining process to obtain association rules. Association rules can be obtained by two steps [3], which are the search of the frequent itemset using FP-Growth algorithm then followed by the creation of association rules to search the integration of each item in an item set.

In determining association rules, there is an interestingness measure obtained from the data processing with specific calculations [3], namely support as the supporting value and confidence as an absolute value. Both will be applied to determine the interesting association rules, by comparing the threshold specified by the user. The limit usually contains minimum support and confidence. According to [8], association rules mining aims to determine all rules that have the value of support >= min_support and confidence >= min_confidence.

Prediction matching will be performed based on the obtained association rules to generate suggestion as a consideration for the faculties and universities to take further decisions about factors affecting the passing rate. The general architecture of the proposed method is shown in figure 1.

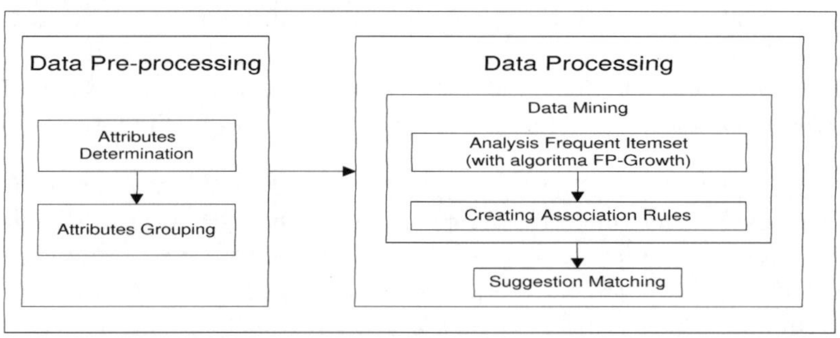

Figure 1. General architecture of the proposed method

2.1. Data pre-processing

Data pre-processing is a data selection process that aims to obtain the data being used in this research by eliminating some attributes that will not be analyzed since it is not suitable to be the factor of student length of study.

2.1.1. Attributes determination

In university graduates data, there are several attributes such as student ID, alumnus ID, name, gender, place, and date of birth, religion, city, zip code, previous school, faculty, major, university entrance types, GPA, length of study, and graduation period. The attributes that will be analyzed are school types, faculties, university entrance types, GPA and length of study, while the rest of the attributes will not be analyzed.

There are various types of formal education institutions taken before entering universities such as state or privately owned high schools which located in the city or outside the city. University entrance types are varied to enter the university, as for the first type is only entitled to those who have excellent grades throughout high school years. GPA (Grade Point Average) is the accumulated score obtained by students from each semester. Length of study is the time spent to finish the bachelor degree. The standard time for a student to graduate in four years. However, numerous students spent over 4 years to achieve the bachelor degree, yet the number of students who spend less than 4 years to obtain the degree is quite numerous.

2.1.2. Attributes grouping

There are several item groups in each attribute, which are:

1. School types
   1.1. State-owned high school (HS)/Islamic high school (IHS) in Medan
   1.2. State-owned HS/ IHS outside Medan
   1.3. Private-owned HS/ IHS in Medan
   1.4. Private-owned HS/ IHS outside Medan
   1.5. Other HS/ IHS
2. University entrance types
   2.1. High school scores based entry (HSSBE)
   2.2. Regular country-subsidized entry (RCSE)
   2.3. Independent entry (IE)
   2.4. Independent country-subsidized entry (ICSE)
3. GPA
   3.1. GPA < 3.00
   3.2. GPA = 3.00-3.49
   3.3. GPA > 3.49
4. Length of study
   4.1. < 4 years
   4.2. 4 years
   4.3. > 4 years

2.2. Data processing

There are several phases in processing the data to generate the desired data, which is a combination of rules or association rules of the university graduates.

2.2.1. Data mining

Data mining is a process of collecting and using the historical data to find the regularity of patterns and relationships in a large dataset [9]. This phase is performed to obtain association rules. Below in Table 1 is the list of 30 university graduates data that will be used to search the association rules.

Table 1. University graduates data

Student ID	School type	Entrance type	GPA	Length of study
1	State-owned HS/ IHS outside Medan	RCSE	2.97	> 4 Years
2	Private-owned HS/ IHS in Medan	RCSE	2.92	> 4 Years
3	State-owned HS/ IHS outside Medan	RCSE	3.29	4 Years
4	State-owned HS/ IHS in Medan	RCSE	3.41	< 4 Years
5	State-owned HS/ IHS in Medan	IE	2.92	< 4 Years
6	State-owned HS/ IHS in Medan	RCSE	2.78	> 4 Years
7	State-owned HS/ IHS in Medan	IE	3.05	> 4 Years
8	State-owned HS/ IHS in Medan	RCSE	3.22	< 4 Years
9	State-owned HS/ IHS outside Medan	RCSE	3.30	< 4 Years
10	State-owned HS/ IHS outside Medan	RCSE	3.10	< 4 Years
11	State-owned HS/ IHS outside Medan	RCSE	3.36	< 4 Years
12	Other HS/ IHS	RCSE	3.07	> 4 Years
13	State-owned HS/ IHS outside Medan	RCSE	3.02	> 4 Years
14	State-owned HS/ IHS in Medan	IE	3.06	< 4 Years
15	State-owned HS/ IHS in Medan	RCSE	3.29	< 4 Years
16	State-owned HS/ IHS outside Medan	RCSE	3.44	< 4 Years
17	State-owned HS/ IHS outside Medan	RCSE	3.24	< 4 Years
18	State-owned HS/ IHS in Medan	RCSE	3.08	< 4 Years
19	Private-owned HS/ IHS in Medan	RCSE	3.03	> 4 Years
20	State-owned HS/ IHS in Medan	RCSE	3.46	4 Years
21	State-owned HS/ IHS in Medan	IE	2.97	> 4 Years
22	Other HS/ IHS	RCSE	3.59	4 Years
23	State-owned HS/ IHS in Medan	IE	2.79	> 4 Years
24	State-owned HS/ IHS outside Medan	RCSE	2.95	> 4 Years
25	Private-owned HS/ IHS in Medan	RCSE	3.02	4 Years
26	State-owned HS/ IHS outside Medan	RCSE	2.99	4 Years
27	Private-owned HS/ IHS outside Medan	RCSE	3.04	4 Years
28	State-owned HS/ IHS outside Medan	RCSE	3.54	4 Years
29	Private-owned HS/ IHS in Medan	RCSE	2.90	> 4 Years
30	State-owned HS/ IHS outside Medan	RCSE	3.06	4 Years

According to Table 1, 30 university graduates have unique attribute groups individually. Therefore, each attribute item is given a code to simplify the search process rules as in table 2.

Table 2. Code for each attribute item

Attribute	*Item*	Code
School types	State-owned HS/ IHS in Medan	A
	State-owned HS/ IHS outside Medan	B
	Private-owned HS/ IHS in Medan	C
	Private-owned HS/ IHS outside Medan	D
	Other HS/ IHS	E
Entrance types	HSSBE	F
	RCSE	G
	IE	H
	ICSE	I
GPA	GPA < 3.00	J
	GPA = 3.00-3.49	K
	GPA > 3.49	L

	< 4 Years	M
Length of study	4 Years	N
	> 4 Years	O

Based on table 2, the item set of university graduates data or code based item set can be seen in table 3.

Table 3. University graduates data with *item set*

Student ID	*Item set*	Student ID	*Item set*	Student ID	*Item set*
1	BGJO	11	BGKM	21	AHJO
2	CGJO	12	EGKO	22	EGLN
3	BGKN	13	BGKO	23	AHJO
4	AGKM	14	AHKM	24	BGJO
5	AHJM	15	AGKM	25	CGKN
6	AGJO	16	BGKM	26	BGJN
7	AHKO	17	BGKM	27	DGKN
8	AGKM	18	AGKM	28	BGLN
9	BGKM	19	CGKO	29	CGJO
10	BGKM	20	AGKN	30	BGKN

This stage has the purpose of searching the item combinations that meet the minimum requirements of the support value in the database. Support value of each item can be obtained using equation 1 below:

$$Support(A) = \frac{\text{Total Transaction A}}{\text{Total Transaction}} \; x \; 100\% \tag{1}$$

Meanwhile, the support value of 2 items can be retrieved using the equation below:

$$Support(A.B) = \frac{\text{Total Transaction A and B}}{\text{Total Transaction}} \; x \; 100\% \tag{2}$$

Based on Table 3, the frequent itemset can be determined, suppose given a minimum value of support = 10% and then search the support value of each item with the equation 1 as in table 4.

Step 1: Finding frequent item

Table 4. *Support value* per *item*

No	Code	Total	Support
1	A	11	11/30 * 100% = 36%
2	B	12	12/30 * 100% = 40%
3	C	4	4/30 * 100% = 13%
4	D	1	1/30 * 100% = 3%
5	E	2	2/30 * 100% = 6%
6	F	0	0/30 * 100% = 0%
7	G	25	25/30 * 100% = 83%
8	H	5	5/30 * 100% = 16%
9	I	0	0/30 * 100% = 0%

10	J	9	9/30 * 100% = 30%
11	K	19	19/30 * 100% = 63%
12	L	2	2/30 * 100% = 6%
13	M	11	11/30 * 100% = 36%
14	N	8	8/30 * 100% = 26%
15	O	11	11/30 * 100% = 36%

Based on the data in table 4, the common items that meet the minimum support value of 10% are school types A, B, and C, entrance types G and H, GPA J and K, and length of study M, N, and O.

Step 2: Building FP-Tree

FP-Tree is a data storage structure formed by a null-labeled root [3]. FP-Tree constructed consists of frequent items on every student, so the rare items are omitted. University graduates data with dataset can be seen in table 5.

Table 5. University graduates item set data

Students ID	*Item set*	Students ID	*Item set*	Students ID	*Item set*
1	BGJO	11	BGKM	21	AHJO
2	CGJO	12	GKO	22	GN
3	BGKN	13	BGKO	23	AHJO
4	AGKM	14	AHKM	24	BGJO
5	AHJM	15	AGKM	25	CGKN
6	AGJO	16	BGKM	26	BGJN
7	AHKO	17	BGKM	27	GKN
8	AGKM	18	AGKM	28	BGN
9	BGKM	19	CGKO	29	CGJO
10	BGKM	20	AGKN	30	BGKN

FP-Tree that can be constructed from the existing item set can be seen in figure 2.

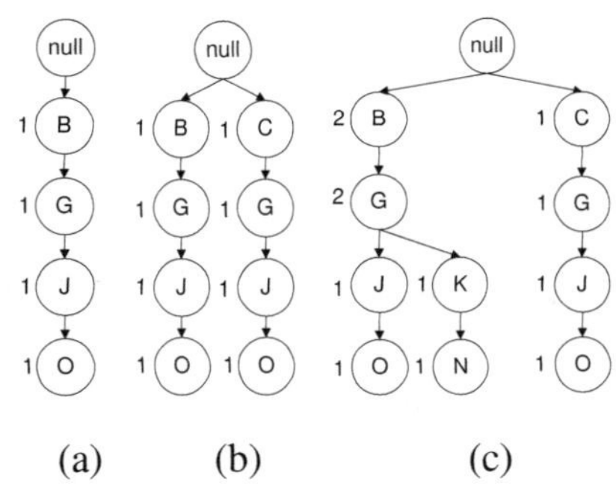

IOP Conf. Series: Materials Science and Engineering **308** (2018) 012017 doi:10.1088/1757-899X/308/1/012017

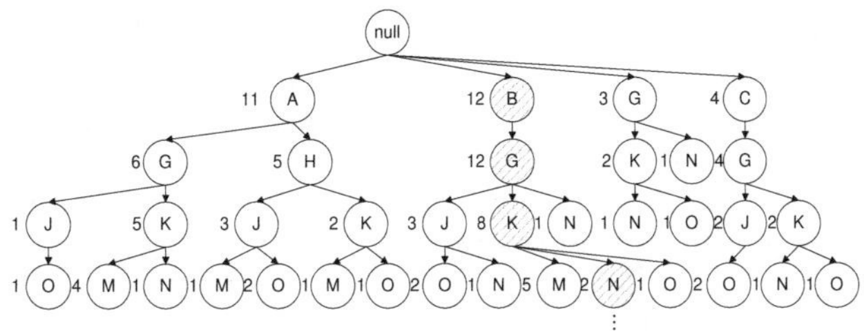

(d)

Figure 2. (a) Tree after reading itemset-1 (BGJO), (b) Tree after reading itemset-2 (CGJO), (c) Tree after reading itemset-3 (BGKN), and (d) Tree after reading itemset-30 (BGKN).

Step 3: Searching frequent item set

Frequent itemset search is conducted through several stages [10].

1. Conditional FP-Tree evocation, which only contains the same item suffix on each item set as in figure 3.

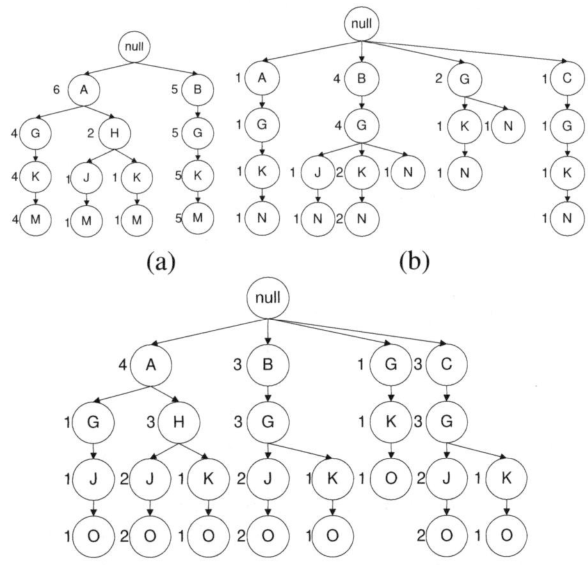

(c)

Figure 3. Conditional FP-Tree (a) suffix M, (b) suffix N, (C) suffix (O).

2. Checking whether the support value of the item set suffix is higher or less than the minimum support of 10% by the equation (1), if it meets the requirements, the second check will be performed on the second suffix continuously until the prefix using equation (2).

From the calculation of frequent itemset search obtained AGKM and BGKM item set that meets the minimum support as in table 6.

Table 6. *Item set* meets minimum *support*

No	Suffix	Total	Support
1	AGKM	4	4/30 * 100% = 13%
2	BGKM	5	5/30 * 100% = 16%

IOP Conf. Series: Materials Science and Engineering **308** (2018) 012017 doi:10.1088/1757-899X/308/1/012017

2.3. Creating association rules

Creating association rules aims to find the associative rule A → B that meets the minimum requirement of confidence value, i.e., to look for a connection between items in an item set. The confidence value is obtained using equation 3 below:

$$\underset{A\rightarrow B}{Confidence} = \frac{\text{Total Transaction containing A \& B}}{\text{Total Transaction containing A}} x\ 100\% \qquad (3)$$

This research looks at the relations between a student length of study to school types, entrance types, and GPA. Suppose that given a 60% confidence value, the search for confidence can be calculated using the equation (3) and can be seen in table 7.

Table 7. *Confidence value*

No	*Item set*	*A → B*	*Confidence*
1	AGKM	$AGK \rightarrow M$	5/8 * 100% = 62%
2	BGKM	$BGK \rightarrow M$	4/5 * 100% = 80%

Based on Table 7, AGKM and BGKM item sets have confidence values that meet the minimum confidence value. Therefore, AGKM and BGKM can be used as the rules of a university graduates data, which have a high degree of emergence and connectivity in a university graduates data. Association rules obtained are:

1. Students originated from State-owned HS/ IHS in Medan with entrance type RCSE and GPA range of 3.00 s/d 3.50, will spend <4 years to obtain the degree.
2. Students originated from State-owned HS/ IHS outside Medan with entrance type RCSE and GPA range of 3.00 s/d 3.50, will spend <4 years to obtain the degree.

2.4. Suggestion Matching

Suggestion matching can for university or faculty be conducted based on the obtained association rules to see the suffix on the rules that are by the length of study. The advice given can be useful for universities to accelerate student's length of study, or to reduce the student's rate with over 4 years-long studies.

Suggestion matching can be identified from the obtained rules based on the length of study, if the length of study is < 4 years or = 4 years, then the initial two combinations of the obtained rules, namely school, and entrance types. It can be selected as the suggestion for the faculty. The faculty can provide information to prospective students from particular schools using obtained rules for school types. Faculty can also offer numerous opportunities to enter the faculty to prospective students with particular entrance types that based on the obtained rules.

If the length of study is > 4 years, the rules suggested the faculty to avert or limit the promotion to students with school types specified by the rules, as for the entrance type rules, it is wise for the faculty to limit the opportunities for the prospective students with specified entrance types to enter the faculty.

Suggestion matching also considers the GPA from the obtained rules. However, it is not in accordance with the length of study attribute. If GPA is lower than 3.00, then it is implied that faculty needs to give extra attention and guidance to the students about the courses to be taken for the thesis. If the GPA range between 3.00 to 3.49 that means faculty should guide the students to improve their grades while those who have GPA above 3.50 should be advised to maintain their excellent work by the faculty.

3. Experimental Results

The test of the association rules search was conducted in every faculty. It generated rules with support value range of 1 to 23% depends on the faculty. The smaller the support value due to numerous attribute combinations, the less the combination appearance rate of a combination. Meanwhile, the confidence value with the highest support value in all faculties, has an average value at over 70%. It is implied that the obtained association rules have high integration between each attribute. The test result of the highest support value can be seen in table 8.

Table 8. Highest support value

Faculty	***Item set***	**Highest support value**	***Confidence value***
Medical	BGKM	6,12	98,59
Dentistry	CGJO	5,9	57,81
Pharmacy	CGJO	11,83	96,88
Public health	BGKO	8,98	64,13
Nursery	BGKM	11,42	85,29
Mathematics & science	BGKO	10,76	80,73
Computer science	AGKO	12,01	92,42
Engineering	BGKO	14,37	91,8
Psychology	AGJO	8,21	97,06
Agriculture	BGJO	12,13	99,29
Law	BGKO	5,62	69,9
Cultural literature	BGKO	9,7	77,1
Economy	BGKM	7,17	58,04
Social and politics	BGKO	8,08	57,14

According to Table 8, the faculty of Medical, Nursery and Economy attained BGKM item set (State-owned HS/ IHS outside Medan, RCSE, GPA range of 3.00 to 3.49, length of study < 4 years), while Dentistry and Pharmacy faculties obtained CGJO item set (Private-owned HS/ IHS in Medan, RCSE, GPA < 3.00, length of study > 4 years). The faculty of Public Health, Mathematics and Science, Engineering, Law, Cultural Literature and Social Politics obtained BGKO item set (State-owned HS/ IHS outside Medan, RCSE, GPA range of 3.00 to 3.49, length of study > 4 years). AGKO item set (State-owned HS/ IHS in Medan, RCSE, GPA range of 3.00 to 3.49, length of study > 4 years) were acquired for Computer science faculty and Psychology obtained AGJO item set (State-owned HS/ IHS in Medan, RCSE, GPA < 3.00, length of study < 4 years). Lastly is faculty of Agriculture which attained BGJO item set (State-owned HS/ IHS outside Medan, RCSE, GPA < 3.00, length of study < 4 Years).

Based on the item set, the system can deliver recommendations to faculty or university. Below is the suggestion for the faculty of Medical, Nursery, and Economy with BGKM item set:

a. The three faculties should provide information or persuade the prospective students originated from State-owned HS/ IHS outside Medan to choose them.
b. Universities should offer numerous opportunities to the prospective students with RCSE entrance type to enter the three faculties.
c. These faculties should guide the students who have GPA range between 3.00 to 3.49 to improve their grades and to start thinking and deepen the courses to be taken for the thesis.

Suggestions for Dentistry and Pharmacy faculties with CGJO item set are:

a. These faculties should provide information or persuade the prospective students originated from Private-owned HS/ IHS in Medan to choose them.
b. Universities should offer numerous opportunities to the prospective students with RCSE entrance type to enter these faculties.

c. These faculties should give extra attention to the students who have GPA lower than 3.00 improve their grades and to start thinking and deepen the courses to be taken for the thesis.

These are the suggestion obtained for Public Health, Mathematics and Science, Engineering, Law, Cultural Literature and Social Politics faculties with BGKO item set:

a. These faculties can provide information or persuade the prospective students originated from all high school types excluding the State-owned HS/ IHS outside Medan to choose them.
b. Universities should limit the available opportunities to the students with RCSE entrance type to enter these faculties.
c. These faculties should guide the students who have GPA range between 3.00 to 3.49 to improve their grades and to start thinking and deepen the courses to be taken for the thesis.

Computer Sciences faculty with AGKO item set is presented with suggestions, which are:

a. Computer Sciences faculty should provide information or persuade the prospective students originated from State-owned HS/ IHS in Medan to choose the faculty.
b. Universities should offer numerous opportunities to the prospective students with RCSE entrance type to enter the faculty.
c. This faculty should guide the students who have GPA range between 3.00 to 3.49 to improve their grades and to start thinking and deepen the courses to be taken for the thesis.

Suggestions for Psychology faculty with AGJO items set are:

a. Psychology faculty should provide information or persuade the prospective students originated from all high school types excluding the State-owned HS/ IHS in Medan to choose the faculty.
b. Universities should offer numerous opportunities to the prospective students with RCSE entrance type to enter the faculty.
c. This faculty should give extra attention to the students who have GPA lower than 3.00 improve their grades and to start thinking and deepen the courses to be taken for the thesis.

At last is the suggestions for Agriculture faculty with BGJO item set which are:

a. Agriculture faculty should provide information or persuade the prospective students originated from all high school types excluding the State-owned HS/ IHS outside Medan to choose the faculty.
b. Universities should offer numerous opportunities to the prospective students with RCSE entrance type to enter the faculty.
c. This faculty should give extra attention to the students who have GPA lower than 3.00 improve their grades and to start thinking and deepen the courses to be taken for the thesis.

4. Conclusion and Future Research

Based on the result of the association rules search process using the FP-Growth algorithm, it can be concluded that the use of the FP-Growth algorithm to search association rules can generate association rules using minimum support and confidence values as the reference values. The most frequent attribute combination to appear which has high inter-attribute connectivity in all faculties is school type of State-owned HS/ IHS outside Medan, RCSE entrance types, GPA range of 3.00 to 3.49, and length of study > 4 Years.

From the result, the faculties can determine which school types that are suitable to promote the faculties to the prospective student. Moreover, the university can see which entrance type that has a significant impact on passing rate to determine the number of accepted students proposed by the faculties.

For further research, authors are expected to predict the student's length of study using prediction method to support the result of this research. It can be done using Evolving Connectionist Method that has been implemented in [11] with fast learning [12] or using Distributed Adaptive Engine [13-15]. Further research can also add more parameters, such as national high school exams score and high school major which can be used as the factor of student's study period.

5. References

[1] Said A M, Dominic P D D and Abdullah A B 2009 A Comparative Study of FP-growth Variations *IJCSNS International Journal of Computer Science and Network Security* **9**

[2] Yongmei L and Yong Guan 2008 FP-Growth Algorithm for Application in Research of Market Basket Analysis *IEEE 6th International Conference on Computational Cybernetics* Stara Lesna Slovakia

[3] Han J and Kamber M 2006 *Data Mining: Concepts and Techniques* Second Edition (Elsevier: USA)

[4] Han J, Pei J, and Yin Y 2000 Mining Frequent Patterns without Candidate Generation. *Proceedings of the ACM SIGMOD International Conference on Management of Data* pp 1-12

[5] Siregar A S 2011 Application Development Of Interrelationship Determinant Between Student Data And Length Of Study Using Multiple Linear Regression Algorithm (STT Harapan Medan Students Case Study) *Thesis* Universitas Sumatera Utara: Medan

[6] Huda N M 2010 Data Mining Application To Show Information of Students Graduation Rates. *Thesis* Universitas Diponegoro: Semarang

[7] Ma'ruf F A 2013 Data Mining Application To Identify The Relationship Of Students Graduation Rates and Admission Type (Case Study: STMIK AMIKOM Yogyakarta) *Journal* STMIK AMIKOM: Yogyakarta.

[8] Avadh K S, Ajeet K, Ashish K 2014 An Empirical Analysis and Comparison of Apriori and FP- Growth Algorithm for Frequent Pattern Mining *IEEE International Conference on Advanced Communication Control and Computing Technologies (ICACCCT)*

[9] Pel W A, Chunhong A and Lei W 2014 An Improved Algorithm For Mining Association Rule In Relational Database. *Proceedings of the 2014 International Conference on Machine Learning and Cybernetics*

[10] Tan P N, Steinbach M and Kumar V 2005 *Introduction to Data Mining* (Addison-Wesley: United States)

[11] Rahmat R F, Rizki A, Alharthi A F and Budiarto R 2016 Big Data Forecasting Using Evolving Multi-layer Perceptron *4th Saudi International Conference on Information Technology (Big Data Analysis) (KACSTIT)* pp 1–5

[12] Alwajeeh T, Alharthi A F, Rahmat R F and Budiarto R 2017 Fast Learning for Big Data Using Dynamic Function *IOP Conference Series: Materials Science and Engineering* **190** p 12015

[13] Rahmat R, Pasha M F, Syukur M, Budiarto R 2015 A gene-regulated nested neural network *International Arab Journal of Information Technology* ***12*** pp 532–539

[14] Pasha M F, Rahmat R F, Budiarto R, and Syukur M 2010 A distributed autonomous neuro-gen learning engine and its application to the lattice analysis of cubic structure identification problem *International Journal of Innovative Computing, Information and Control* ***6*** pp 1005–1022

[15] Aaron, Sitompul O S and Rahmat R F 2014 Distributed autonomous Neuro-Gen Learning Engine for content-based document file type identification *International Conference on Cyber and IT Service Management* pp 63–68

IOP Conf. Series: Materials Science and Engineering **308** (2018) 012018 doi:10.1088/1757-899X/308/1/012018

Grip Analysis of Road Surface and Tire Footprint Using FEM

M Sabri*, S Abda
Computational & Experimental System Mechanics Research Centre,
Mechanical Engineering Department, Universitas Sumatera Utara, Medan, Indonesia

*E-mail: sabrimesin@gmail.com

Abstract. Road grip involve a touch between road pavement and the tire tread pattern. The load bearing surface, which depends on pavement roughness and local pressures in the contact patch. This research conducted to develop a Finite element model for simulating the experimentally testing of asphalt in Jl. AH Nasution Medan, North Sumatera Indonesia base on the value of grip coefficient from various tire loads and the various speed of the vehicle during contact to the road. A tire model and road pavement are developed for the analyses the geometry of tire footprint. The results showed that the greater the mass of car will increase grip coefficient. The coefficient of grip on the road surface contact trough the tire footprint strongly influence the kinetic coefficient of friction at certain speeds. Experimentally show that Concrete road grip coefficient of more than 34% compared to the asphalt road at the same IRI parameters (6-8). Kinetic friction coefficient more than 0.33 was obtained in a asphalt path at a speed of 30-40 Km/hour.

1. Introduction
The automobile tires base of Pneumatic work principle began to be manufactured since 1895. They have replaced the solid tires, which incriminate the mechanical principle of the vehicle and cause discomfort for passengers as the driving force and speed become larger. However, the refinement to be achieved is not only the greater Comfort, but the better grip of the pneumatic tire also needs to be corrected.

Part of the kinetic energy developed by a vehicle has to be absorbed by the suspension system, the brakes and the tires during cornering and braking. Where the car meets the road, there are only the vehicle's tires to ensure the ultimate contact patch. The aim of this study is to identify the grip coefficient of tire-road contact using Finite Element Method. The mechanics of grip are to be explained by the astonishing visco-elastic properties of the tire's rubber [2] which within the contact patch produce a host of physical phenomena that strive to counteract any untimely skidding over the road surface.

2. Tire/road interface model
The model requires essentially a contact numerical models (which includes a contact condition and a friction model) and a propagation model. This approach allows for the development of the individual models independently from the other two, without affecting the entire model structure.

The tire is modeled by a 2D flexible ring model, as described in [1]. A contact condition is used, which states that no part of the tire may penetrate the road surface (the condition is actually an inequality or constraint equation). The friction model is based on Coulomb's friction. The calculated structural vibrations of the tire are used as input for a 3D Finite element code (FEM), to calculate the grip coefficient of tire-road interface.

A 2D flexible ring model is used to explain the suggested approach. The model is shown in figure (1). The ring is assumed to be a thin elastic body of thickness h and radius R, the thickness being small compared to the radius and displacements of the ring. Stress stiffening due to the pressurized air in the tire is accounted for. The radial displacement, $x_r(\theta)$, is assumed to be constant over the thickness, whereas the tangential displacement $x_\theta(\theta)$ varies linearly over the thickness. The material is assumed to be isotropic and Hooke's law is used to relate strains to stresses (Young's modulus E). Rotary inertia is neglected. The ring is supported by an elastic foundation (a distributed radial and tangential stiffness (k_r and k_θ, respectively)), representing the side-wall stiffness.

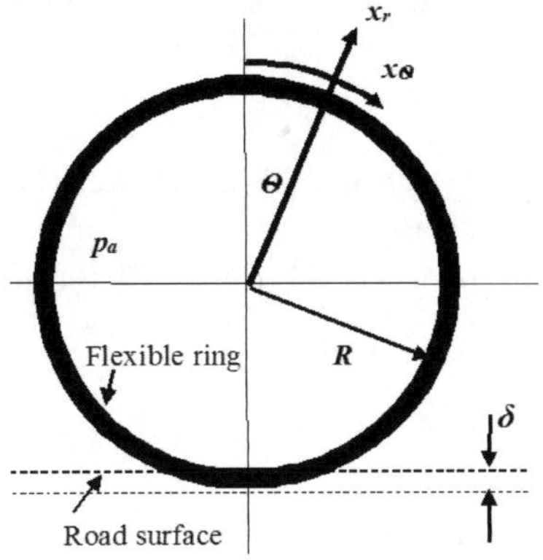

Figure 1. Ring model of tire. Degrees of freedom x_r and x_θ.

The dynamic equations for the flexible tire are, see [1] and [2]:

$$\rho h\ddot{x}_r + d\dot{x}_r + \frac{EI}{R^4}\left(\frac{\partial^4 x_r}{\partial\theta^4} - \frac{\partial^3 x_\theta}{\partial\theta^3}\right) + \frac{Eh}{R^2}\left(x_r + \frac{\partial x_\theta}{\partial\theta}\right) + \frac{p_a}{R}\left(x_r - \frac{\partial^2 x_r}{\partial\theta^2} + 2\frac{\partial x_\theta}{\partial\theta}\right) + k_r\, x_r = f_r \quad (1)$$

$$\rho h\ddot{x}_\theta + d\dot{x}_\theta + \frac{Eh}{R^2}\left(\frac{\partial^2 x_\theta}{\partial\theta^2} - \frac{\partial x_r}{\partial\theta}\right) + \frac{EI}{R^4}\left(\frac{\partial^3 x_r}{\partial\theta^3} + \frac{\partial^2 x_\theta}{\partial\theta^2}\right) + \frac{p_a}{R}\left(x_\theta - \frac{\partial^2 x_\theta}{\partial\theta^2} + 2\frac{\partial x_r}{\partial\theta}\right) + k_\theta\, x_\theta = f_\theta \quad (2)$$

Where ρ is the density, d the damping coefficient, $I = h^3/12$ denotes the area moment of inertia per unit width, p_a is the air pressure and f_r and f_θ are the distributed forces per unit width in the, respectively, radial and tangential direction.For convenience, the differential equations (1) and (2) are rewritten as:

$$L_r(x_r,x_\theta) = f_r \quad (3)$$

$$L_\theta(x_r,x_\theta) = f_\theta \quad (4)$$

Essential in the development of a fast numerical solution procedure is that the tire model equations can be used to solve the unknown displacements x_r, x_θ if the forces f_r, f_θ are known, but also that they can be used to calculate the unknown forces f_r, f_θ if the displacements x_r, x_θ are given. This is the case for the position of points on the tire which are in the footprint, as these positions are equal to the position of the road surface (or, if friction is included, can be calculated from the position of the road and the friction model used).

For simplicity, we use a coordinate system which is fixed to the tire. This implies that, in the transient calculations, we let the road surface rotate around the tire. This is however not essential to the method.

3. Contact model

The tire will deform when in contact with the road surface. The footprint of the tire in obviously depends on the position of the road and also related to the friction forces between the tire and the road surface, as well as the tire model (note that the radial displacements x_r, Tangential displacement x_θ should also

IOP Conf. Series: Materials Science and Engineering **308** (2018) 012018 doi:10.1088/1757-899X/308/1/012018

satisfy the differential equations in the footprint!). The normal and friction force will cause the tire to adhere to the road surface and, as already indicated, the position of the tire in the footprint equals the position of the road surface. If the friction force exceeds a maximum friction force at some points in the footprint (as in Coulomb's friction) these points will slip and their position, i.e. the displacements x_r, x_θ, is determined by the geometry of the road surface, the maximum friction force and the tire model.

The contact condition is simply a constraint equation, specifying that the tire cannot penetrate the road surface. For simplicity, we discuss a road surface with no texture, but we have extended the algorithm to various types of texture. For the tire model given above and a smooth road surface, this implies that the radial and tangential displacements of the tire should satisfy:

$$f(x_r, x_\theta) = (R + x_r)\cos\theta - x_\theta \sin\theta + R - \delta \geq 0 \tag{5}$$

where δ denotes the deflection approach of the tire to the road surface, see figure (1). Thus, if $f(x_r,x_\theta) < 0$ the tire penetrates the road surface and the contact condition is violated. On the other hand, in the contact region or footprint the displacements of the tire satisfy the equality:

$$f(x_r,x_\theta) = 0, \tag{6}$$

Which, as is discussed below, is one of the equations for the two unknown displacements x_r,x_θ in the footprint.

The contact forces f_r, f_θ and displacements x_r, x_θ are, besides the contact condition, determined by the friction model. For given displacements x_r, x_θ of any point of the tire, one can calculate the actual forces needed to keep that point at that position. Hence, one can also calculate the (normal) normal force f_n and shear stress τ. For a non-textured road surface, f_n and τ follow from:

$$f_n = f_r \cos\theta - f_\theta \sin\theta \tag{7}$$

$$\tau = f_r \sin\theta + f_\theta \cos\theta \tag{8}$$

As the displacements x_r,x_θ are known, we can use equations 3 and 4 to calculate, respectively, the forces f_r and f_θ. If the contact is assumed to be frictionless, i.e. if $\tau = 0$, this implies

$$L_r(x_\theta,x_r)\sin\theta + L_\theta(x_\theta,x_r)\cos\theta = 0 \tag{9}$$

and together with $g(x_r,x_\theta) = 0$, one can solve for x_r and x_θ. If friction is accounted for $|\tau| \leq \mu|fn|$ when there is no slip and x_r,x_θ can, initially, be determined from the position of the road surface. If $|\tau| > \mu|f_n|$ slip occurs and, besides the contact condition, $\tau = \mu f_n$. Substitution of equations 7 and 8 yields:

$$L_r(x_\theta,x_r)(\sin\theta \mp \mu\cos\theta) + L_\theta(x_\theta,x_r)(\cos\theta \pm \mu\sin\theta) = 0, \tag{10}$$

where the sign ($\pm$) depends on the sign of τ.

4. Numerical model

The vibrations of the tire, calculated by the tire/contact model, are used as input for the 3D Finite element model. Hence, phenomena like the horn-effect are accounted for roughness absorption of the road surface can be included as well. As in the FEM code the source mesh is fixed, we need to transform the vibrations of the rotating tire to normal vibrations on a fixed source mesh. The average shape of the tire, as calculated by the tire/contact model, is used as the 3D source mesh, see figure 2.

Figure 2. Finite element source mesh

This research is focused on grip coefficient analysis of road surface influenced by road surface roughness (IRI) and road surface deflection (dL). Fig. 3 show the deformation of the contact surface area between tire footprint and road should satisfy the differential equation (or element matrix equations). For all points in the footprint, the position is given and one can calculate the contact forces needed to put the tire at that position. We use this procedure to iterate the solution. The average value of IRI According to the experimental measurement in Abdul Haris Nasution Street Medan [3] are between 30 – 50 mm.

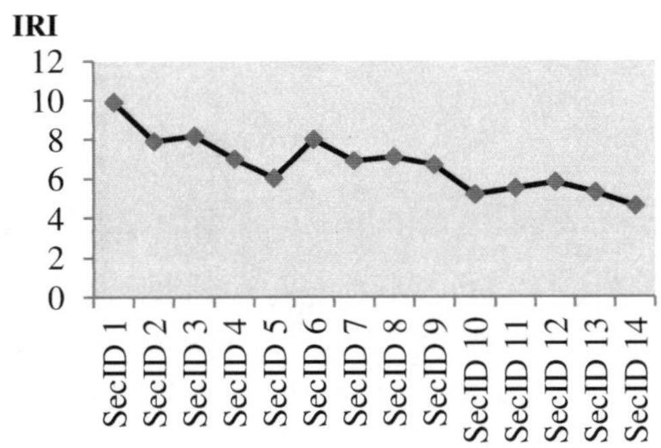

Figure 3. IRI along tire-road contact surfaces

With respect to the current 3D tire model, we assume all points to vibrate along the entire width of the tire. The vibrations are subsequently interpolated to this fixed source mesh. Furthermore, the vibrations are show in fig. 4 in the finite element simulation.

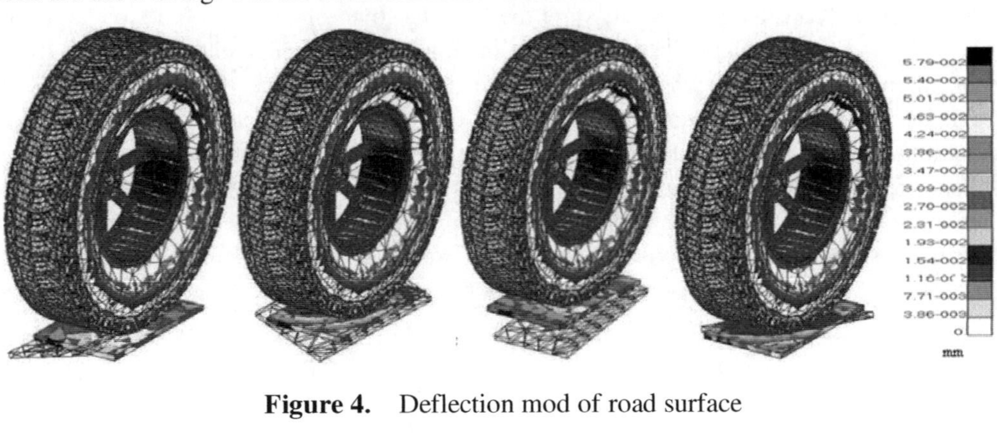

Figure 4. Deflection mod of road surface

Transforming the time signal to the Fourier domain seems trivial. However, for some point's on the source mesh close to the contact region, the actual displacement can be such that they come in and out of contact (figure 5).

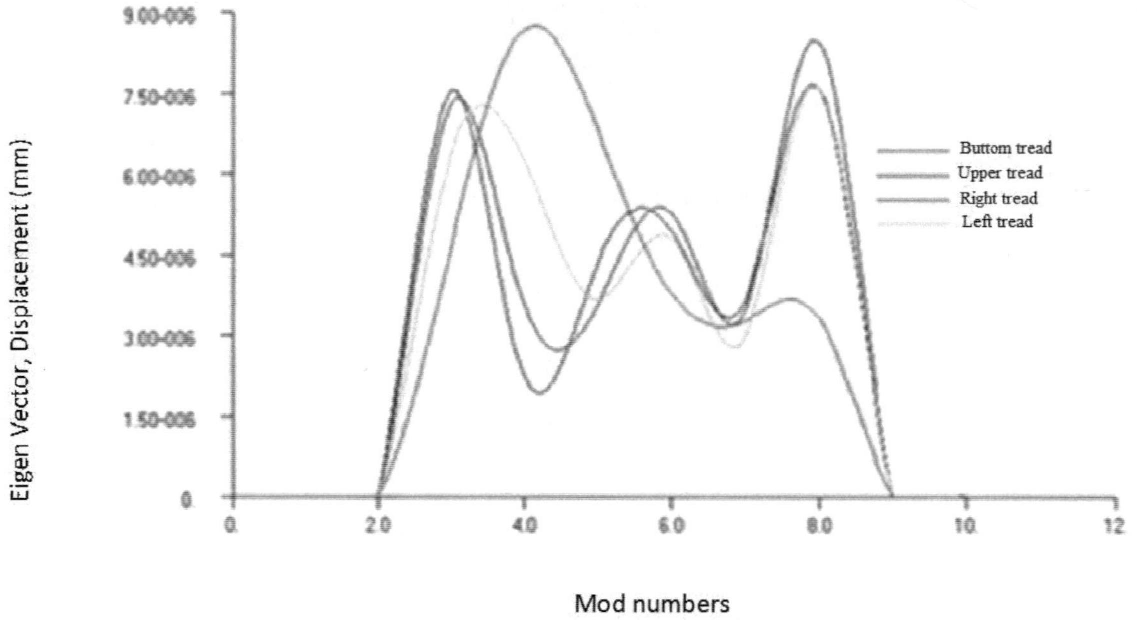

Figure 5. Eigen Vector tire-road deflection contact

Once these points are in contact, they should not be able to radiate the friction at all, as the position of that point in the contact is assumed to be fixed. Therefore, near the contact region, we keep track of the volume (between the tire and the road surface) in the very small region between the footprint and those points which are never in contact. The change of the volume in time is then converted to a harmonic velocity and this value is set to the velocity of an element radiating from the contact region into the 'horn'. Note that a similar approach can be used to simulate air-pumping.

5. Grip on road surfaces

The Vehicles cannot be able to move, if there were no grip between tire and road. a moving vehicle has to deal with natural forces, such as the banking, the slope or the unevenness of the road, or rolling resistance, which are constantly trying to slow the vehicle down or push it off its path. As the only contact point between the vehicle and the road, the tire ensures two fundamental functions. It gives the vehicle its directional stability, which the driver needs to steer it. The tire acts as a transmission component for brake and drive torque.

The maximum values of longitudinal grip coefficient μmax and transversal grip coefficient τmax cannot therefore be reached simultaneously. the tire-road interface represents a global grip potential delineated approximately by a circle, and available to provide longitudinal or transversal grip or both. To obtain the results from the purpose of this study, initial research includes measurement of IRI, measurement of dL and grip coefficient measurement (F, Ff, $\mu k = f$ (IRI), $\mu k = f$ (dL)}.

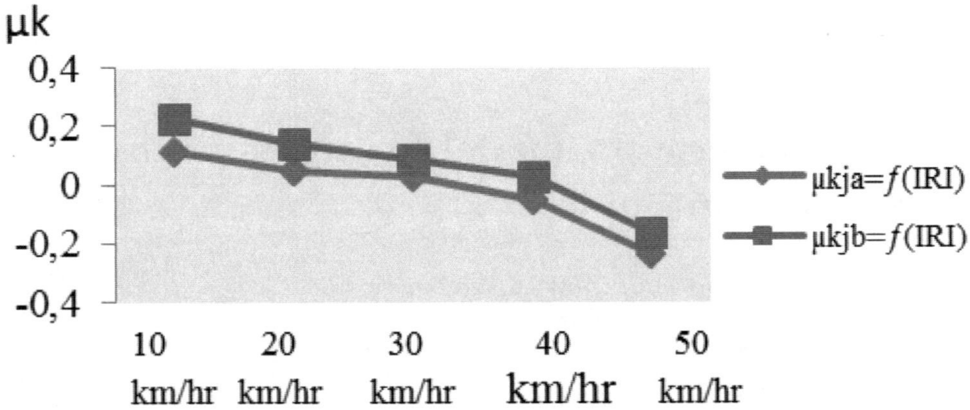

Figure 6. Simulation μ_{kja} = f(IRI) compare to experimental μ_{kjb} =f(IRI) Grip coefficient of tire-road contact.

Using the finite difference equations, the discrete analogue of the contact condition and the friction model calculated for each of the discrete points of the tire. Before proceeding with the next point, we check whether the contact condition is violated. If so, the position of that discrete point is fixed to the road surface (fig. 6). The forces required to keep that point at that position is calculated (the normal and shear forces f_n and τ). It is then checked whether $|\tau|$ is smaller than $\mu|f_n|$. If so, the forces are correct and the algorithm can proceed with the next point. If not, the position of that point of the tire is calculated from solving the displacements from the equations $g(x_r, x_\theta) = 0$ and equation 10. For points already fixed to the road surface, one does need to check whether the normal force becomes negative.

6. Conclusion

The correlation between road surface roughness and the pattern of tire tread will generate the grip coefficient of tire-road contact. The Effect of $\mu_k = f$ (IRI) for every increase speed, will decrease kinetic grip coefficient (μ_k).

Reference

[1] Sabri M 2016 Displacement of a Free Tyre in Natural Frequencies to Determine Noise Structure *Applied Science and Engineering Conference (AASEC 2016)*

[2] Sabri M at al., 2010 Free Vibration of Tire/Road Contact 2010 Fracture and Strength of Solids VII *Key Engineering Materials* **462-463**:2

[3] Nuhudawi M P, Sabri M 2016 Analisa Interaksi Beban Dinamik Mobil Terhadap Karakter Struktur Permukaan Jalan [Automated Dynamic Load Interaction Analysis Of The Road Surface Structure Character] *Proceeding of Yearly National Engineering IV* ***600*** *ISBN 978-602-96269-2-6 ITB Bandung*

[4] Karakus M, Cavus A, Colakoglu M 2017 Vibration analysis of a tire in ground contact under varied conditions *Journal of Theoretical and Applied Mechanics* **47**:1 pp 3-17

[5] Sabri M, et al. 2011 Free Vibration Tire Road Contact *Key Engineering Materials 462* (10.4028/www.scientific.net/KEM.462-463.) pp 843-848

IOP Conf. Series: Materials Science and Engineering **308** (2018) 012019 doi:10.1088/1757-899X/308/1/012019

Reliability enumeration model for the gear in a multi-functional machine

M. K. M. Nasution[1,*] and H. Ambarita[2]

[1]Teknologi Informasi, Fasilkom-TI, Universitas Sumatera Utara, Padang Bulan 20155 USU, Medan, Indonesia

[2]Sustainable Energy Research Centre, Faculty of Engineering, Universitas Sumatera Utara, Padang Bulan 20155 USU, Medan, Indonesia

E-mail: mahyuddin@usu.ac.id

Abstract. The angle and direction of motion play an important role in the ability of a multi-functional machine to be able to perform the task to be charged. The movement can be a rotational action that appears to perform a round, by which the rotation can be done by connecting the generator by hand through the help of a hinge formed from two rounded surfaces. The rotation of the entire arm can be carried out by the interconnection between two surfaces having a jagged ring. This link will change according to the angle of motion, and any yeast of the serration will have a share in the success of this process, therefore a robust hand measurement model is established based on canonical provisions.

1. Introduction

The arm or wrists in multi-functional machine play an important role in replacing dangerous human tasks like lever and lifting heavy loads, or resolve welding at a car manufacturer, defusing bombs, chemicals management, paint spraying and assembling. To perform such a task, the robotic arm or wrist must be capable of rotating within a range the same as that of a human, as a gear mechanism [1].

Most of the gear are modelled round with a serrated gaze transmitted in such a way that the spherical shape facilitates movement within the specified range. The canonical lever has been designed and known to be a discrete gear distributed on a round surface so that the gear direction is not continuous. In order for the serrations to be produced, the spindle-shaped teeth are designed by distributing serrations around the ring in a continuous manner [2]. Therefore, the gear movement has a probability, but the encounter of two jagged surfaces is interdependent and thus requires an enumeration model. This paper intend to reveal a model associated with the enumeration of the gear on the hand.

2. A Concept of Model

The arm and wrists of the lever of machine are connected to two rounded surfaces, and the wrists will rotate when the arm is also spinning or vice versa. The rotation due to the mechanism on the arm is transformed to the wrist through the relationship between two round surfaces, which have several rings with the gears attached in such a way, there are some rings that are not connected between two unrelated surfaces, depending on the angle of motion formed by the arm and wrist. Space created by a rounded ringed surface of the serration causes a lever to move his hand in a straight or crooked state without losing its power.

Each ring on a round surface, depending on the presence of its teeth, performs the performance of the lever [3]. A performance can be measured based on the interaction results of the teeth of two related surfaces, and loss of capture power of one gear from a surface against the gears on the other surface will cause the gear not to operate properly. Performance measurement of reliability refers to the probability that a system will carry out a stated mission that is based on the odds caused by the number of teeth [4, 5]. A mission is expressed in term of quantity, quality, time and others, whose operations can be assessed. Assuming that the extent to which expansions are expressed from this mission can be described with the Boolean function with all of the assumptions that have been widely described about a lever, the multi-functional machine performance measurement model is established and the enumeration.

3. An Approach

As already disclosed, a round-shaped gear is formed by producing a ring around the ring on a round surface. The rounded gear mechanism can be used to order two degrees of freedom, based on the direction of the rounded gear motion [6]. A rack-cutting vector function in the $S_1(o_1, x_1, y_1)$ reference system can be represented by $\mathcal{R}_1(\beta)$, β representing the shelf-cutting curve parameter. The rack-cutting set can be expressed by a vector function

$$\mathcal{R}_2 = \mathcal{R}_2(\beta, \phi_2) \tag{1}$$

ϕ_2 is a motion direction parameter. Thus, the point on the rack-cutting set wrapper simultaneously satisfies the above equation and the following equation

$$\frac{\partial \mathcal{R}_2}{\partial \beta} \times \frac{\partial \mathcal{R}_2}{\partial \phi_2} = 0 \tag{2}$$

Based on the gear theory, the design of the gear-shaped surface is rounded by a ring around the ring, which is a mechanism involving a spherical Σ^1, Σ^2 and a rack cutter for a multi-functional machine. Coordinate the system $S_1(o_1, x_1, y_1)$ is rigidly attached to the rack cutters. Coordinate system $S_2(o_2, x_2, y_2, z_2)$ is rigidly attached to the cross section of the Σ^1 gear. The coordinate system of $S_3(o_3, x_3, y_3, z_3$ and $S_4(o_4, x_4, y_4, z_4)$ is stacked rigidly against Σ^1 and Σ^2 as a lever. The S_3 coordinate system is introduced to rotate at the angle ϕ_1 and ϕ_2 each axis: y_3 and x_3. In the same way, ϕ_4 and ϕ_5 are rotating angles of the rounded gear Σ^2 which each rotates around the axis y_4 and axis z_4 in a lever. For a round-shaped gear Σ^1 has a ring around gear Σ^1 has a ring around the ring, the angle of ϕ_1 and ϕ_2 are the same. So, it is stated that $\phi_1 = \phi_2 = \phi$. In the same way the ϕ_4 and ϕ_5 annotations are equal to θ_1. The relationship between ϕ and θ_1 is $r_{p1}\phi = r_{p2}\theta_1$ [5]. The relationship between ϕ and θ_1 is $r_{p1}\phi = r_{p2}\theta_1$, with which r_{p1} and r_{p2} are the radius of relations of each gear Σ^1 and Σ^2. Therefore, the coordinate transformation matrix from S_1 to S_2 can be parsed as

$$\mathcal{M} = \begin{pmatrix} \cot\phi_2 & -\tan\phi_2 & r_{p1}\cot\phi_2 + S\tan\phi_2 \\ \tan\phi_2 & \cot\phi_2 & r_{p1}\tan\phi_2 - S\cot\phi_2 \\ 0 & 0 & 1 \end{pmatrix} \tag{3}$$

System reliability is the probability of system success, expressed as an assessment. The success of a system depends on the success of its building components, in this case a ring with a round-shaped gear of the arm and wrist at lever. Success is expressed for each teeth as the performance of a given set of specifications for a period of time. So, it is a probability of success applied directly to where the A event (total sample space) is the total number of experiments. Therefore, reliability A, $\mathcal{R}(A)$, expressed by

$$\mathcal{R}(A) = \frac{N(A)}{N(S)} = \frac{\text{success number}}{\text{total tried}} \tag{4}$$

Functional relationships should be established to express the reliability of the system as a function of component reliability by series and parallel terms. The series of relationships with each component must be successful in order for the system to succeed [8]. Failure of any one component causes failure in the system. For system with n series components, the Boolean function AND relationship is used to connect between system components [7], i.e.

$$T = A_1 A_2 \ldots A_n \tag{5}$$

If the component fails freely, then the equation that can be applied to the reliability function is

$$\mathcal{R}(T) = \prod_{i=1}^{n} \mathcal{R}(A_i) \tag{6}$$

The parallel connection resulted in the system to be successful if one of the components of any component succeeded. Success in parallel for n component is represented by an operator OR in Boolean that can express the following truth value [7]

$$T = A_1 + A_2 + \ldots + A_n \tag{7}$$

In order for the system to fail otherwise the whole component must fail. Hence $\overline{T}$ denotes system failure, and $\overline{A}_i$ signifies the failure of the i component, the following relation is derived from Eq. (5)

$$\overline{T} = \overline{A}_1 \overline{A}_2 \ldots \overline{A}_i. \tag{8}$$

It can be transformed into probability function relationships $P(\overline{A}_1)P(\overline{A}_2)\ldots P(\overline{A}_n)$, it may be expressed in the following parts of the system and component reliability

$$\mathcal{R}(T) = 1 - \prod_{i=1}^{n} [1 - \mathcal{R}(A_i)] \tag{9}$$

Determination of reliability as a general approach consist of two steps

(i) Write a Boolean expression for reliability configuration, by series, parallel or a mixture of both. This expression will relate the truth value of the system to the value of the success of the component.

(ii) Translate this phrase in algebraic form for the probability of system success as a function of the reliability of the components.

4. Enumeration

The motion direction of the gear mechanism by which assembly is used which describes the surface models of the Σ^1 and Σ^2. First (left-handed gear) can rotate around the y_3 and z_3 at the same time based on the frame. If the left-handed gear rotates around the y_3 when associated with the second (right-handed gear), the rotation of the z_3 is determined. Similarly, if the first part rotates around the x_3 with the right gear, the rotation of the y_3 is determined. Based on that it can be stated that if there are n rings from the center to the edge of two rounded surfaces,

IOP Conf. Series: Materials Science and Engineering **308** (2018) 012019 doi:10.1088/1757-899X/308/1/012019

there will be m ringed rings, so $m < n$ corresponds to the established angle that is applicable for the transformation of the Eq. (3).

The relationship illustrates that for the angles formed and according to the spherical surface there are two rings that catch each other on the teeth, while the other rings will do the same in pairs according to the angle sequence constructed by two surfaces. Therefore, referring to the preceding part descriptions can be modeled that a series of Boolean functions will represent a link between the rings of two surfaces, whereas the parallel shaped Boolean function represents a sequence of m of meshed ring of two surfaces. Suppose there are n rings on a surface, written c_{S1i}, $i = 1, 2, \ldots, n$, on the other surface expressed by c_{S2i}, $i = 1, 2, \ldots, n$. If there are m interlocked rings of both surface then in general

$$T = \Sigma_{i=1}^{m} c_{S1i} c_{S2i} \tag{10}$$

From the middle ring of the surface to the edge rings, the m coupling of the linked pair can use the above equation. Based on the model formed, it can be stated that this pairing resulted in an incision on each pair of links, resulting in a series relationship between the sequences that are formulated into

$$T = \prod_{j=0}^{n-m} \Sigma_{i=1+j}^{m+j} c_{S1i} c_{S2i} \tag{11}$$

As example, $n = 5$ rings on each surface and there is a sequence of $m = 2$ interlocked rings, serial and parallel series can be modeled as follows

$$\begin{aligned} T &= \prod_{j=0}^{3} \Sigma_{i=1+j}^{2+j} c_{S1i} c_{S2i} \\ &= (c_{S11}c_{S21} + c_{S12}c_{S22})(c_{S12}c_{S22} + c_{S13}c_{S23})(c_{S13}c_{S23} + c_{S14}c_{S24}) \\ &\quad (c_{S14}c_{S24} + c_{S15}c_{S25}) \\ &= T_0 T_1 T_2 T_3 \end{aligned} \tag{12}$$

i.e. it is $T_j = \Sigma_{i=1+j}^{2+j} c_{S1i} c_{S2i}$, $j = 0, 1, 2, 3$.

Enumeration of each T_j can be done by stating the configuration in progress, for example for T_1 as follows

(i) Write the success function for configuration, done by enumerating success path. Both c_{S11} and c_{S21} or c_{S12} and c_{S22} work, thus

$$T = (c_{S11}c_{S21} + c_{S12}c_{S22}) \tag{13}$$

or

$$T = vw + xy. \tag{14}$$

(ii) Map the success functions for the multi-functional machine.

(iii) Rearrange and merge the function of success into an event of mutual. It produces the function

$$T = vw + \overline{v}xy + v\overline{w}xy. \tag{15}$$

(iv) Determine the reliability functionality based on Eqs. (4), (6) and (10), i.e. the reliability function for the system

$$\mathcal{R}(T) = \mathcal{R}(vw) + \mathcal{R}(\overline{v}xy) + \mathcal{R}(v\overline{w}xy). \tag{16}$$

Then, this equation is changed by tribe depending on whether the component fails freely or is bound. Assume that the dependency and subsequently raised in the following facts

$$P_x = \mathcal{R}(x) \tag{17}$$

and

$$Q_x = 1 - \mathcal{R}(x) \tag{18}$$

Thus the reliability function of (13) can be rewritten as

$$\mathcal{R}(T) = p_v p_w + q_v p_x p_y + p_v q_w p_x p_y. \tag{19}$$

Reliability values can be expressed as the opportunity of each yeast to successfully perform the interconnection between the two rings in the same direction of movement, and all the values with the rings with the rings can be substituted into the above equation. To obtain the overall system reliability from the same application robot hand from Eq. (13) to Eq. (15) can be repeated as much as T_j, $j = 0, 1, 2, \ldots, nm$.

5. Conclusion

The lever's reliability model of the multi-functional engine is formulated on the basis of the interface between the rings of two surfaces, the direction of motion, and the angle formed between two rounded surfaces. A freely two-degree canonical form represents the interconnection between rings occurring on two rings on each surface, resulting in series and parallel relationships, thus formulations can be expressed in terms of truth values and then can be amenable after determining their combinations based on mapping.

References

[1] M. D. Byrne, K. K. Greene, and S. P. Everett 2007 Usability of voting systems: baseline data for paper, punch cards, and lever machines *Proceedings of the SIGCHI Conference on Human Factors in Computing Systems*.
[2] J. Qimi, Z. and L. Huamin 1998 The design of quasi-ellipsoidal gear ratio and pitch curved surface, *ASME Journal of Mechanical Design* **120**.
[3] R. Jin, and G. Agrawal 2005 A methodology for detailed performance modeling of reduction computations on SMP machines *Performance Evaluation* **60**.
[4] E. W. Brehm, R. T. Goettge, and F. W. McCaleb 1995 START/ES: an expert system tool for system performance and reliability analysis *Performance Evaluation* **22**.
[5] S. C. Yang 2002 Mathematical model of a ring-involute-teeth spherical gear with a double degree of freedom, *The International Journal of Advanced Manufacturing Technology* **20**.
[6] L. Wuxing, P. W. Tse, Z. Guicai, and S. Tielin 2004 Classification of gear faults using cumulants and the radial basis function network *Mechanical Systems and Signal Processing* **18(2)**.
[7] E. Mendelson 1970 Boolean algebra and switching circuits *McGraw-Hill Book Company*, New York.
[8] M. Harahap and M. K. M. Nasution 2005 Model and enumerasi pengukuran kinerja antarmuka aplikasi Kabupaten Road Management System *al-Khawarizmi: Journal of Computer Science* **1(2)**.

IOP Conf. Series: Materials Science and Engineering **308** (2018) 012020 doi:10.1088/1757-899X/308/1/012020

Interaction models on sand surface of natural adsorbent with adsorbate Cd^{+2} metal ions in solution with batch operation

B Haryanto*[1], Siswarni MZ[1], CH Chang[2], AT Kuo[2], WB Singh[3]

[1]Department of Chemical Engineering, Universitas Sumatera Utara, Padang Bulan, Medan 20155, Indonesia
[2]Department of Chemical Engineering, National Cheng Kung University, Tainan 701, Taiwan
[3]Department of Chemistry, Manipur University, Canchipur, Imphal, Manipur 79503, India

*E-mail: bode.haryanto@usu.ac.id

Abstract. The interaction type of metal ion with the adsorbent surface in adsorption operation was evaluated in this study. The binary metal ions in solution was mixed with the sand then shaken in 100 rpm. The kinetics adsorption capacity was than measured with the pseudo order model to evaluate the interaction type from the r^2 indicator. The result found that the Cd ions with r^2 =0.9656 and r^2= 0.9989 for first and second order respectively, for black sand and r^2 =0.9070 and r^2= 0.999 for first and second order respectively, for white sand. It was indicated that the interaction type of metal ions on sand surface was occurred together as physical and chemical interaction.

1. Introduction

Heavy metals have ability to harm the physiology and human biological system, if it exceeds the tolerance level; it causes a number of diseases [1,2]. Sand has reported as adsorbent [3,4]. It has used after modified as reported by Gusain et al. (2013) [5] of sands of the Ganges-Varanasi, modified with Fe ion from India, Han [6] of sand modification with FeS from Korea; used naturally as reported by Haryanto [7], with natural sand from Taiwan [8,9,], Shi et al. (2014) [10] with the natural sand, Ruba Al-Khali from Saudi Arabia; Thambavani dan Kavitha (2014) [11] of sand from river in Tamil Nadu (Hindi).

The reason to used sand because it has macro-porous and meso-porous structures [12]. Pores contribute a more surface area and a higher selectivity on reaction and adsorption interaction [13]. Charles W, 1994 [14] has reported that sand with porous characteristic creates the intra-particle porosity. Sand as the tetrahedral silica with hydroxyl functional groups on the surface, it can be represented as =Si–OH [15]. The functional interaction with the primary bonding force in inner-sphere complexes is a coordinate-covalent bonding in contrast to electrostatic bonding in outer-sphere complexes [16-18].

There are several purification processes for water contaminated by heavy metals such as: adsorption process [19], coagulation and precipitation chemistry, electro-flotation [19,20], ion exchange and membrane separation [21]. Adsorption process is more efficient and less expensive than other technology entrapment of heavy metals.

Equation (1) uses in measuring the adsorption capacity until it reaches equilibrium; is the amount of metal ion adsorbed in sand at equilibrium (qe). The qt is the amount of metal ion adsorbed in sand at certain time t, when one run on contact time (Equation 2). Equation 3 (%R) is to determine the percentage of metal ion adsorbed removed from water, when using the adsorption operation.

$$q_e = \frac{(C_0 - C_e)V}{m_{ads}} \quad (1)$$

$$q_t = \frac{(C_0 - C_t)V}{m_{ads}} \quad (2)$$

$$R\% = \frac{(C_0 - C_e)}{C_0}.100\% \quad (3)$$

Where: C_0 is the initial metal concentration (mg/L), Ct is the metal concentration at time t (mg/L) and Ce is the metal concentration at equilibrium (mg/L); qe, qt and qr is (mg/g). V is the volume of solution (L) and m_{ads} is the mass of adsorbent (g) [7,19,22,23,24,25,26].
Adsorption kinetic model can be evidenced by the coefficient correlation (r^2). First order (4) and second order (5) equations of respectively expressed as follows [1,24,26,27]:

$$\log(q_e - q_t) = \log q_e - \frac{k_1}{2,303}t \quad (4)$$

$$\frac{t}{q_t} = \frac{1}{k_2 q_e^2} + \frac{1}{q_e}t \quad (5)$$

The aims of this investigation is to apply the pseudo order 1st and order 2nd equations to identify the type of adsorption interaction of metal ion Cd^{2+} on two type sand surface base on correlation coefficient (r^2). The empirical data of adsorption process on single substance of Cd^2 as adsorbate and sands white and black as adsorbent with batch operation were used.

2. Methodology

Two types sand were used in this study as shown in Figure 1. The white sand Figure 1A was obtained from Pantai Cermin district and the black sand (Figure 1B) from Namorambe district, Sumatera Utara, Indonesia. The size of sand used was 40 mesh. Sand was cleansing then continues to adsorption proses. All sand was washed until the washing water has constant pH as the confirmation that sand already has the same condition before used as natural adsorbent.

A. White sand B. Black sand

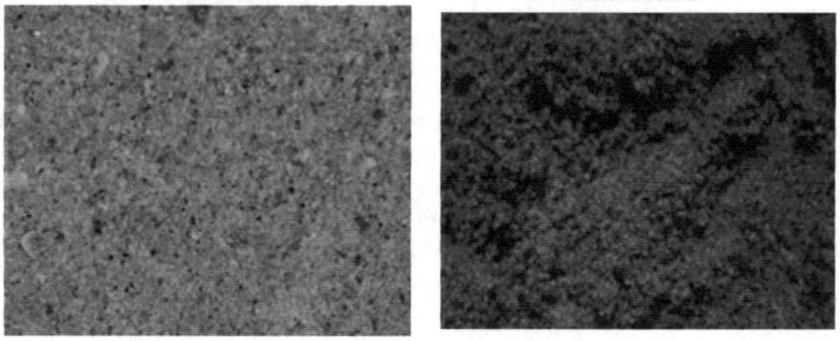

Figure 1. White and black sand natural adsorbents

In this experiment, Figure 2 shows every section in adsorption process. Section A prepare 50 ppm $CdCl_2$ dissolved by pure water at pH 4.5. Then continue to section B mix 100 ml $CdCl_2$ 50 ppm with 100g sand. Section C**,** each Erlenmeyer was shacked for 6 hours with 150 rpm revolutions per minute. 0.5 ml samples were taken every 5 minutes to be analyzed by AAS. To confirm the presence of Cd^{2+} on sand surface was analyzed by FTRI. SEM was used to identify the sand surface characteristic. Equations 2 and 3 were used to measure the adsorption kinetic. The equations 4 and 5 were used in measure to predict the adsorbent interaction type with adsorbate Cd^{2+}. Correlation interaction the r^2 was used to predicts interaction type on both adsorbents surface with contaminant Cd^{2+} ion.

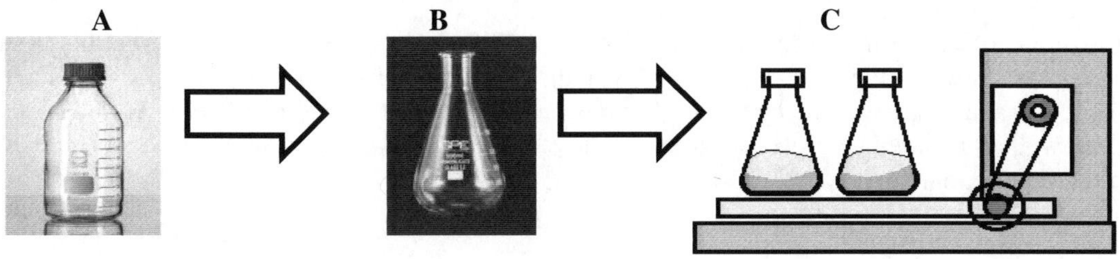

Figure 2. Adsorption process

3. Result and Discussion

3.1. Adsorption Kinetics

Adsorption kinetics is demonstrated at figure 3. The contact time of adsorption process was conducted for 360 min. The absorption capacity of both adsorbent increased with initial concentration of metal ion Cd 50 ppm. From the results for, in the first 5 minutes, the amount of metal ions Cd^{2+} adsorbed was only 13.66%. Cd^{2+} adsorbed achieves optimum point in the minute-200, where the %R adsorbed was 39.31% for white sand (Figure 3A). For the black sand has optimum point in the minute-200 with the %R 25% as shown in Figure 3B. The adsorption capacity was changes largely in the first 10 minutes. It was lower in comparing the ability of white sand. After some time, the ability of the adsorption capacity starts to decline and relatively constant.

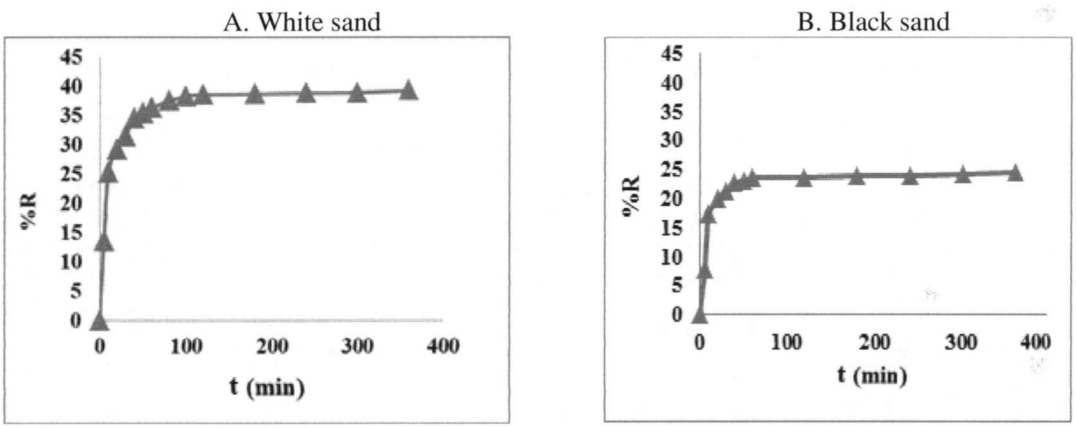

Figure 3. Adsorption kinetic of Cd ion on white sand (A) and black sand (B).

The adsorption equilibrium can be achieved by measuring the contact time between the adsorbent with the adsorbate Cd^{2+}. For the white sand, adsorption capacity continues to increase to adsorb Cd^{2+} ion until the time of 300 minutes with the concentrations of adsorbate Cd^{2+} consecutively: 25.35%, 29.22%, 31.44%, 34.55%, 35.41%, 37.51 %, 38.35%, 38.61%, 38.89%, and 38.96%. After 100 minutes interaction, adsorption of Cd^{2+} by white sand had approached a constant point. This show, that the adsorption process has achieved an equilibrium state. The equilibrium time is to find out when an adsorbent already reached saturation so that the adsorption process has been completed [11]. For the black sand, the equilibrium time was after 80 minutes.

In this investigation, the data of adsorption kinetics of Cd ion on white sand and black sand are empirically derived using models pseudo first order (Equation 4) and pseudo second order (Equation 5). This modelling is required to describe and evaluate adsorption mechanism and identify an average rate of adsorption on the adsorbent sands for single solution [1]. The models can also be used to categorize whether during the adsorption process chemical reactions and physical interaction occur or not in the adsorbent based on the correlation coefficient. The (r^2) correlation coefficient obtained by plotting the data of adsorption capacity (qt) versus time using the above equation.

3.2. Pseudo First Order

The equation of Pseudo First Order have value of $r^2 = 0.907$ and $r^2 = 0.9606$ for white and black sand respectively as shown in Figure 4. Figure 5 shows the equation of Pseudo second order for white (A) and black (B) sand. The results have value of $r^2 = 0.999$ for white sand and $r^2 = 0.9989$ for black sand. These show that modeling of pseudo first and second order makes the data more representative adsorption interaction.

A. White sand B. Black sand

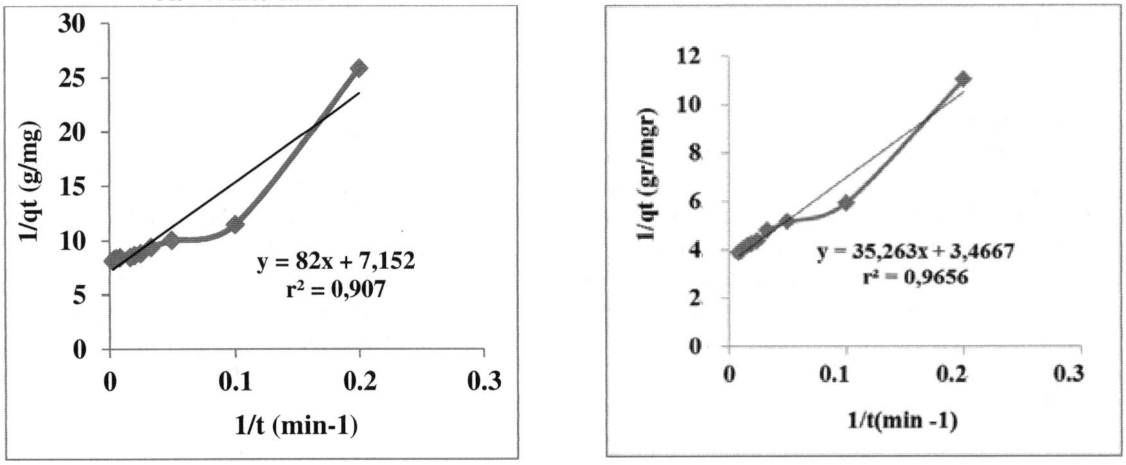

Figure 4. Pseudo First Order model with Cd^{2+} 50 ppm and speed 150 rpm

3.3. Pseudo Second Order

From the results of theoretical calculations, the correlation coefficient (r^2) for first order is lower than second order as shown at Figure 4 and 5. The equation of first order and second order, each has the value of correlation coefficient (r^2) related quite similar. From these data it can be determined that the adsorption process was involving a chemical interaction (chemisorption) and physical interaction (physical sorption) between the adsorbate and adsorbent on the surfaces occur together [1,24,26,27].

A. White sand **B. Black sand**

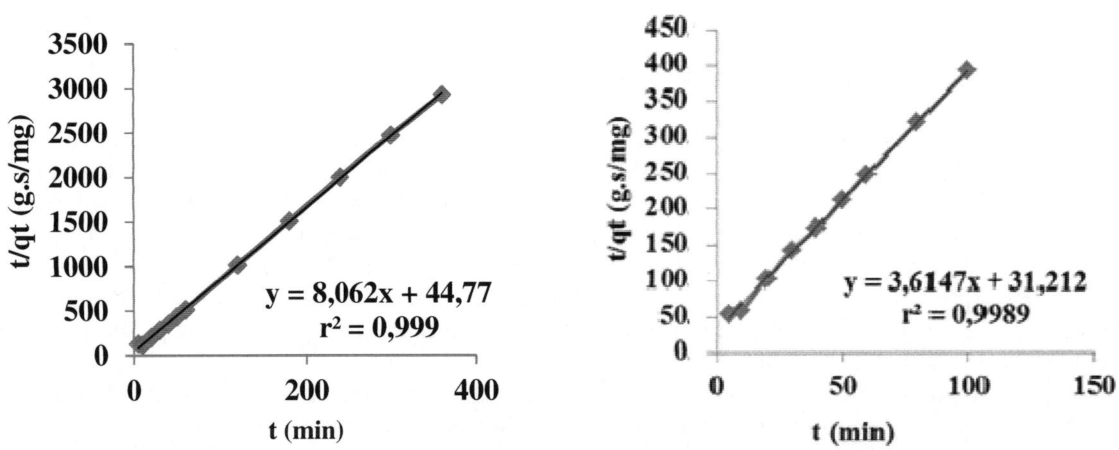

Figure 5. Second Order Pseudo with Cd^{2+} 50 ppm and speed 150 rpm

Cleanse Sand (A) and contaminated sand (B) was confirmed of the changing the effect of adsorption on sands surface by using FTIR as shown in Fig. 6. Galo et al. (2004) [28] have reported the range wavenumber (cm^{-1}) for Si-OH is in the range 500 to 1000. As shown in Figure 6 within the wavenumber range FTIR analysis, the interaction of on sand surface to interact with Cd^{2+} after and before contaminated of the white sand. In this study, the interaction of Cd^{2+} with functional Si-OH on sand surface have confirmed as shown on variation of wavenumber in the range 500 to 1000 (cm^{-1}). Figure 7 shows the surface characteristic the white sand with 1500x magnitude. The SEM image shows the porous characteristic of sand. The surface characteristic of white sand with ability to interact with Cd ion is about 40 ppm. The possible diffusion adsorbed metal ion goes to inter particle area on adsorption operation. It is impact to the interaction on sand surface with porous characteristic as chemical and physical interaction on sand surface.

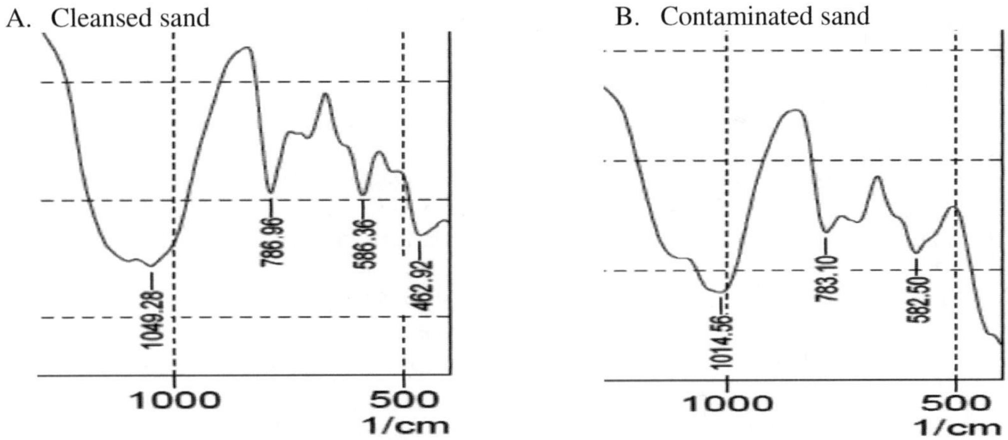

Figure 6. FTIR confirmation cleansed white sand (left) and contaminated white sand (right)

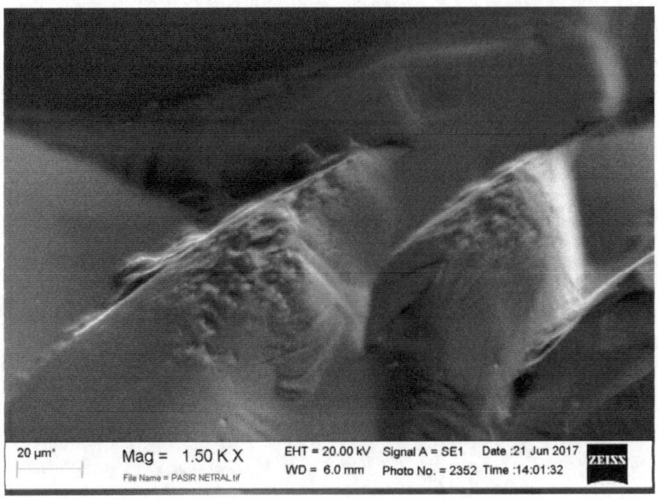

Figure 7. SEM surface of white sand

4. Conclusions

The adsorption capacities of the white sand were higher than black sand. The investigation was to apply the pseudo order 1st and order 2nd base on the adsorption kinetic empirical data for both sand types. It was obtained the trend similar on correlation coefficient (r^2). The results found that the Cd^{2+} ions with r^2 =0.9070 and r^2 =0.9656 for first order for white and black sand respectively. For the second order found that the r^2= 0.999 and r^2= 0.9989 for white and black sand respectively, The result was indicated that the interaction type of Cd^{2+} ions on sand surface was occurred together as physical and chemical interaction. Analyze with using BET is needed to confirm the difference pore size area of both sand types that impact to adsorption capacity of white sand higher than black sand.

ACKNOWLEDGMENT
The authors wish to express sincere gratitude to Lembaga Penelitian Universitas Sumatera Utara on the TALENTA Project 2017, No: 299/UN5.2.3.1/PPM/KP-TALENTA USU/2017 for the financial support for this research project; also to Professor Robison Tarigan to help and for providing comments on this manuscript.

Reference
[1] H Liu, C Wang, J Liu, B Wang and H Sun 2013 *Journal of Environmental Management* **128** 727-734
[2] KG Akpomie and FA Dawodu 2015 *Journal of Advanced Research* **6** 6 1003-1013
[3] M A Awan, I A Qazi and I Khalid 2003 *Journal of Environmental Sciences* **15** 3 413-416
[4] Sangiumsak N and Punrattanasin P 2014 Pol. J. Environ. Stud. **23** 3 853-865
[5] Gusain D, Srivastava V and YC Sharma 2013 *Journal of Industrial and Engineering Chemistry* **20** 841-847
[6] YS Han, AH Demond, TJ Gallegos and KF Hayes *Chemosphere* 2014 **134** 499-503
[7] B Haryanto, JS Chang and CH Chang 2014 *Surfactants Detergents* **51** 5 407-414
[8] B Haryanto and CH Chang 2014 *Journal of the Taiwan Institute of Chemical Engineers* **45** 2170-2175
[9] B Haryanto and CH Chang 2015 *Journal of Oleo Science* **64** 2 161-168
[10] Shi S, Gondal MA and Shen K 2014 *Res. Chem. Intermed.* **41** 6999-7013
[11] DS Thambavani and B Kavitha 2014 *International Journal of Research* **1** 4 513-625
[12] H Chen, J Wang, Z Rahman, J G Worden, X Liu, Q Dai and Q Huo 2007 *J. Mater. Sci.* **42** 6018-6026
[13] K Kaneko 1994 *Journal of Membrane Science* **96** 59-89
[14] CW Fetter 1994 Applied hydrogeology New York Maxwell Macmillan International
[15] Sparks D L 2003 Environmental Soil Chemistry San Diego Academic Press
[16] Koretsky C 2000 *Journal of Hydrology* **230** 127-171
[17] Goldberg S and C T Johnston 2001 *Journal of Colloid and Interface Science* **234** 204-216
[18] Bradl H B 2004 *Journal of Colloid and Interface Science* **277** 1-18
[19] W Liu, T Wang, AGL Borthwick, Y Wang, X Yin, X Li, and J Ni 2013 *Journal Science of The Total Environment* **456-457** 171-180
[20] DI Da Mota, AJ De Castro, CR De Goes, J De Oliveira and A Gomes 2014 *Journal of Materials Research and Technologi* **4** 2 109-113
[21] R Mahmoudkhani, A Torabian, HA Hassani and R Mahmoudkhani 2014 *Journal Procedia APCBEE* **10** 79-83
[22] L Zheng, Z Dang, X Yi and Z Hui 2010 *Journal of Hazardous Materials* **176** 650-656
[23] S Vafakhah, ME Bahrololoom, R Bazarganlari and M Saeedikhani. 2014 *Journal of Environment Chemical Engineering* **2** 356-361
[24] M Arshadi, MJ Amiri and S Mousavi 2014 *Water Resources and Industry* **6** 1-17
[25] Q Hu, Z Xiao, X Xiong, G Zhou and X Guan 2014 *Journal of Environmental Sciences* **27** 207-216
[26] B Haryanto, WB Singh, ES Barus, A Ridho and M R Rawa 2017 *Journal of Physics: Conference Series* **801** 1-7
[27] S Chen, Q Yue, B Gao, Q Li and X Xu 2011 *Journal Chemical Engineering* **168** 909-917
[28] GJAA Soler-Illia, EL Crepaldi, D Grosso and C Sanchez 2004 *J. Mater. Chem.* **14** 1879-1886

IOP Conf. Series: Materials Science and Engineering **308** (2018) 012021 doi:10.1088/1757-899X/308/1/012021

Loading time and channeling effect in removing copper ions from sand surface by surfactin solution in flushing column

B Haryanto*[1], WB Singh[2], CH Chang[3] A Chairunnisa[1], FR Butarbutar[1]

[1]Department of Chemical Engineering, Universitas Sumatera Utara, Medan 20155, Indonesia

[2]Department of Chemistry, Manipur University, Canchipur, Impal, Manipur-795003, India

[3]Department of Chemical Engineering, National Cheng Kung University, Tainan 701, Taiwan

*E-mail: bode.haryanto@usu.ac.id

Abstract. The variations of loading time (t) were used to indicate the channeling effect of in removing copper ions from sand surface by surfactin solution in continuous flushing column technique. Results trend shows that by increasing loading time will increase the interaction with the contaminant on sand surface. In shorter time showed that the removal efficiency of water was higher than surfactin if flushing without foam. The micelle structure within the surfactin solution has ability to move into intra particle and tend to flow down which decrease the loading time in comparing with water molecule. Flushing with foam, the bubbles gas penetrates and spreads the surfactin micelles to the whole sand surface and spend shorter time in loading the sand column. Types of flushing with and without foam with the removal efficiency result are possible to predict the channeling affect.

1. Introduction

Flushing technique was used to remove metal ion from contaminated sand in column remediation. Flushing approaches with surfactant solution with and without foam were applied. Flushing without foam made the solution difficult to search the surface area uniformly in the packed column. The formations of channels provide low contact areas between the solution and the contaminated material. The solution has tendency to flow through certain channels [1,2].

Figure 1. The foam inhibits the channeling effect [3].

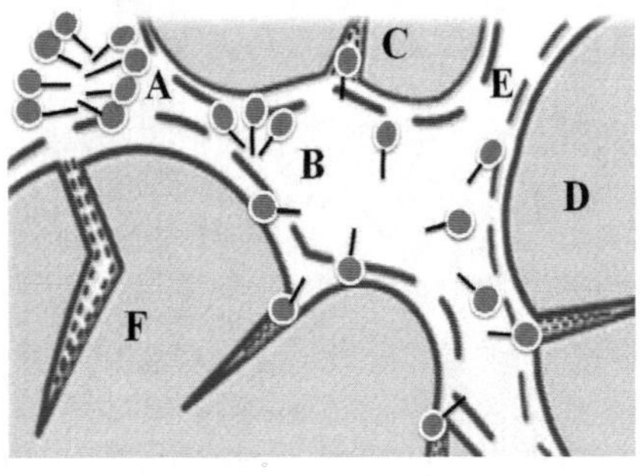

Figure 2. Illustration of the sand particles with inter and intra particle area [4].

Figure 1 shows flushing the solution without foam and with foam. The sand-packed column was mostly composed of dry sands. Flushing without foam, demonstrating the pronounced channeling effect (Fig.1, left). Flushing with foam could improve the solution movement in the sand column and then flow more evenly through the column medium (Fig. 1, center). Increasing the surfactant concentration is possible to wet all sand surfaces and this significantly reduced channeling effect (Fig. 1, right). With the inhibition of the channeling effect, it has been reported that the removal efficiency was apparently increased.

As shown in the illustration, flushing with foam is possible to spread the surfactant molecules to intra particle areas. Without foam within the micelle structure, it is difficult to reach the intra particle area that impacts metal ions removal from the contaminated sand medium. In Figure 2 is shown that flushing with foam enhanced solution technique in the inside of the sand-packed column. A is the foam lamella; B is monomers spread onto inter-particle sand pores; C is penetration into the intra-particle sands pores. D is sand particle, E is inter-particle pores and F is intra particle pores [4].

The objective of this study is to observe the time needed between the first drops of solution into the column, until the first drop left the column (t). The loading time of water and surfactin solution was investigated and then compared with surfactin solution with foam flushing the column. Adsorbed metal ions in sands were prepared by an adsorption process and by putting them in oven to get the residual type. The loading time with the removal efficiency was used to indicate the channeling effect.

2. Experimental Method

Surfactant was produced from Bacillus subtilis ATCC 21332 with purity about 90%, [5]. Surfactin solution was dissolve in a 10^{-3} M phosphate buffer with a pH value of 8.0 [4,6]. Copper (II) sulfate pentahydrate (Cu2SO4.5H2O) was chosen as the model metal ion contaminant and was purchased from Showa Chemical Co., Ltd., Japan. Purified water with a resistivity of 18.2 MΩ-cm was produced by Milli-Q plus purification system, Millipore, USA and was used in all experiments. Contaminated sands were provided by mixing the sands and the copper ion-containing solution for 24 hours at a rate of 150 rotations per minute. 24 hours waiting period of was allowed for the system to reach adsorption equilibrium, there after the sands were collected. An atomic absorption spectrometer (Sens AA Dual, GBC, Australia) was used to analyze the metal ion concentration in the solution, and then dried by putting it in oven at 50°C. As the residual type, the adsorption density of metal ions on sands with the inner-sphere interaction and precipitation were then evaluated.

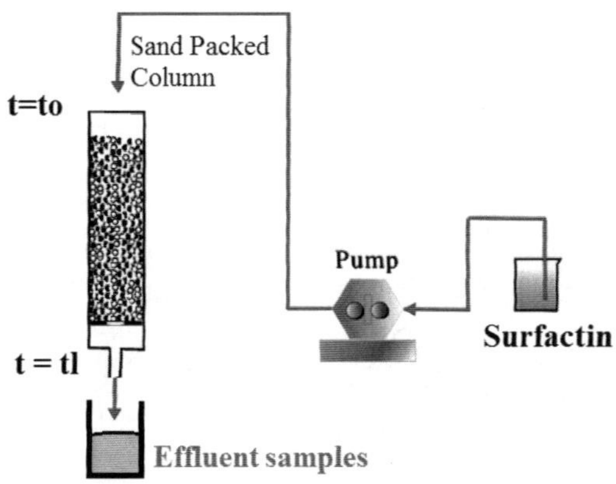

Figure 3. Illustration of instrument used for surfactant solution flushing operation without foam [4]

The instrument used for surfactant solution flushing operation without foam is as shown in Figure 3. The loading time was measured by timer for t=t0 to t=tl as shown at the sand column. Contaminated sand of 10.15 g was put to the column. A glass column with a length of 7.5 cm and with outside diameter of 1.5 cm was used in the column experiments. The pore volume was about 2.2 ml and the sand height is about 3.8 cm. Solution was pumped by peristaltic pump 2 ml/min into the column. Flushing solution was collected after 4 pore volume. A peristaltic pump (Masterflex, Model 7518-10, Cole-Parmer Instrument Company, Barrington, USA) was used to create continuous flow of surfactin solution [7,8] . The instrument used for surfactant solution flushing operation with foam is similar as to that been used by Huang and Chang, 2000 and Haryanto et al. 2014 [7,9]. AAS was used to analyze metal ions removal from the concentration in the effluent solution.

3. Results

Flushing with water only was used in comparing with surfactin solution. The impact of surface properties of Surfactin with ability to decrease the air water surface tension till 29mN/m, provides shorter loading time in comparing to water with surface tension 72mN/m. Increasing the flow rate will decrease the loading time needed. The volume of solution to flow through the column decreases the time needed when increasing the rate as shown in Figure 4. The surfactin micelles at 5x cmc in solution tend to flow out than to interact with sand interface then produce channeling effect. Water molecule is possible to wet the sand surface then interact with the chemical compound on sand surface which tend to delay the loading time.

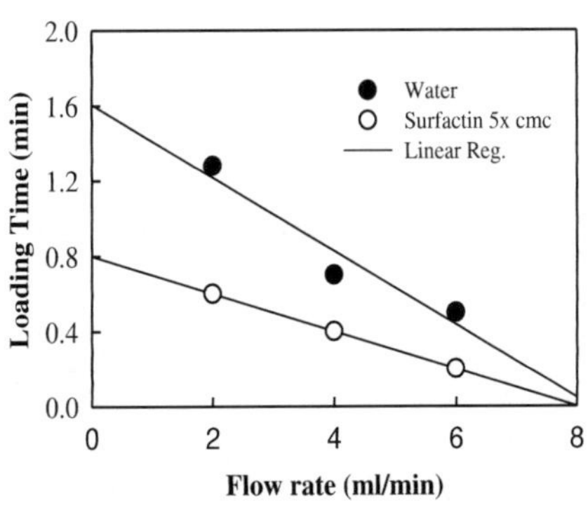

Figure 4. Time loading at different water flow rates and surfactin solution

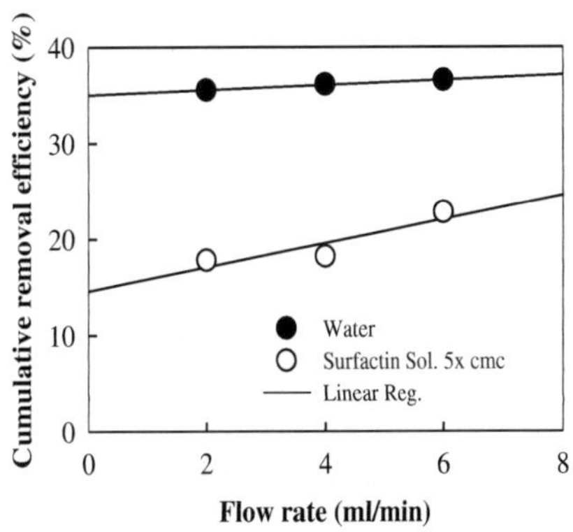

Figure 5. Removal efficiency at different water flow rate and surfactin solution of sand with residual type.

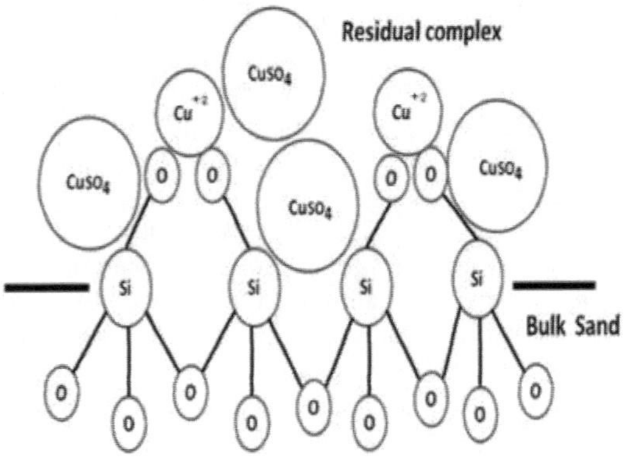

Figure 6. The residual sand type [4,10]

The removal efficiency of the flushing operation with, flowrate variation is shown in Figure 5. Sand with residual type model was used to confirm the flushing with surfactin solution, in comparing with water. Water with longer loading time was possible to remove the copper ion contaminant from sand surface. The trend removal efficiency of surfactin solution is also possible to increase the copper ion removal as increasing the flow rates. Surfactin molecules in micelles structure may be more affected by increasing the flow rate rather than by decreasing the loading time. Residual type of contaminant occupied the inter particle sand surface area (Figure 6). In the presence of water molecule, it is possible to desorb the adsorbed then the Surfactin molecule was trapped in their micelle. The micelle have to break then interact with adsorbed on sand surface and desorb the adsorbed to solution phase. The ability of Surfactin to decrease interface tension of water and solution was limited by sand surface areas which tend to create channeling effect.

Flushing characteristic by water is shown in Figure 7. The removal efficiency at each 4 pore volume flushing solution was collected in the flushing column with flow rate variations. The removal efficiency of adsorbed was not impacted by the variation of flow rate. It tends that water molecules have the same way to interact on sand surface. The loading time impacts to wet the sand surface with the longer loading time make it possible to desorb the adsorbed but with limited surface area interaction as shown at the first 4 pore volume result. The removal efficiency each of 4 pore volume has the same result even by increasing the flow rate.

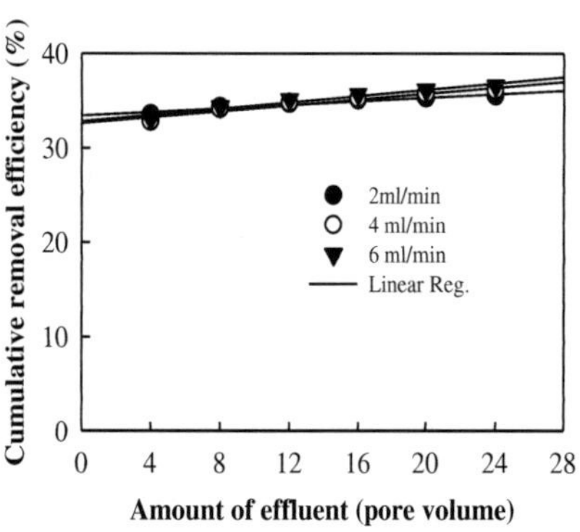

Figure 7. Effects of flow rates variation to the removal efficiency of sand with residual type by flushing with water

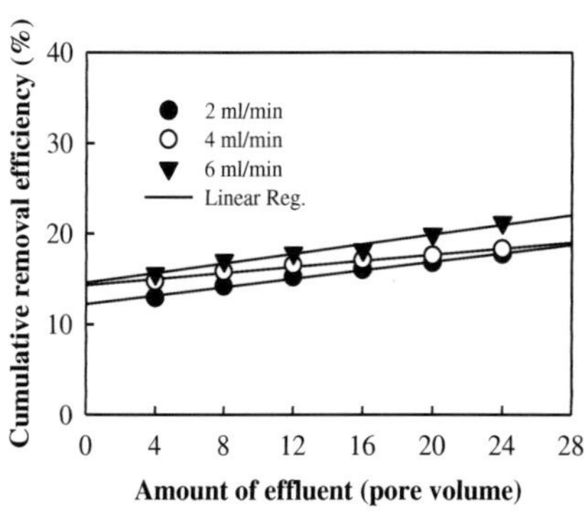

Figure 8. Effects of flow rates variation to the removal efficiency of sand with residual type by flushing with Surfactin 5x cmc

Flushing by 5x cmc surfactin is shown in Figure 8. The removal efficiency at each 4 pore volume flushing solution was increased with the increasing of flow rate. The removal efficiency of adsorbed was impacted by the variation of flow rate. The surfactin molecules in the micelles with higher flow rate are possible to increase interaction in the surface area then increased the adsorbed to desorb. The shorter loading time but with higher flow rate was possible to desorb continuously with different higher removal efficiency. The removal efficiency is tending to increase for each 4 pore volume by adding the flushing pore volume.

Flushing column with foam possible to increase the removal efficiency of the copper ion as shown in Figure 9 with the flow rate was 2 ml/min. Surfactin foam with variation concentrations are possible to spread within the surfactin molecules in the foam structure to interact with copper ion at sand surface. The foam is possible to spread within the surfactin molecule into the intra particle on sand then increases the removal efficiency of the adsorbed. The loading time with foam flushing is relatively shorter but with foam surfactin, it is more possible to reach surface area of sand then increases the removal efficiency.

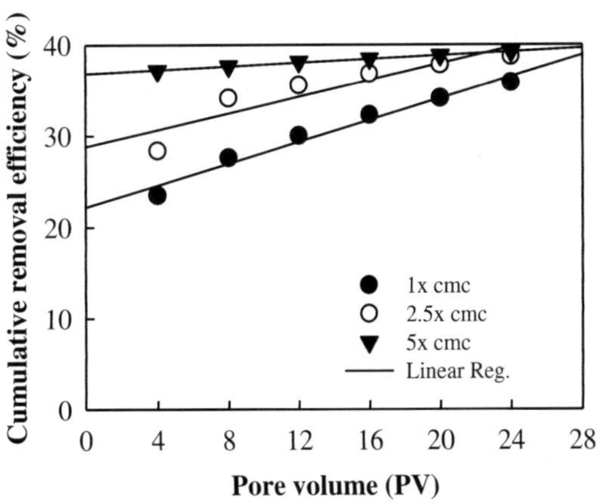

Figure 9. Effects of Surfactin concentration variation to the removal efficiency of sand with residual type by flushing with foam solution with flow rate 2 ml/min.

4. Conclusions

Loading time was impacted by flow rate variation. For surfactin molecule with micelles structure in solution tend to flow out faster in comparing with water molecule. The removal efficiency was influenced by at least 3 factors: flow rate variation, concentration and flushing by surfactin solution or foam. The channeling effect was occurred and influenced by at least 3 factors flow rate, concentration and flushing type.

References

[1] Chowdiah P, Misra B R, Kilbane J J, Srivastava V J, Hayes T D 1998 *Journal of Hazardous Materials* **62** 265-280

[2] RK Rothmel, RW Peters, E S Martin E S and MF Deflaun 1998 *Environ. Sci. Technol.* **32** 1667-1675

[3] B Haryanto and CH Chang 2015 *Journal of Oleo Science* **64** 2 161-168

[4] B Haryanto 2014 PhD Thesis Chem. Eng. Department, NCKU Taiwan

[5] MS Yeh, YH Wei and JS Chang 2005 *Biotechnol. Prog.* **21** 1329-1334

[6] CN Mulligan, RN Yong, BF Gibbs, J Susan, and HP Bennett 1999 *Environ. Sci. Technol.* **33** 3812-3820

[7] B Haryanto and CH Chang 2014 *Journal of the Taiwan Institute of Chemical Engineers* **45** 2170 -2175

[8] B Haryanto, JS Chang, and CH Chang 2014 *Tenside Surf. Det.* **5** 1-5

[9] CW Huang, and CH Chang 2000 *Colloids and Surfaces A: Physicochemical and Engineering Aspect* **173** 171-179

[10] C Koretsky 2000 *Journal of Hydrology* **230** 127-171

The Simulation Study of Horizontal Axis Water Turbine Using Flow Simulation Solidworks Application

H Prasetyo, EP Budiana, DDDP Tjahjana, S Hadi*

[1]Mechanical Engineering Department, SebelasMaret University.
Jl. Ir. Sutami 36 A, Surakarta, Central Java, Indonesia

*E-mail: syamsulhadi@ft.uns.ac.id

Abstract. The design of Horizontal Axis Water Turbine in pico hydro power plants involves many parameters. To simplify that, usually using computer simulation is applied. This research performs simulation process variation on turbine blade number, turbine blade curvature angle, turbine bucket angle and blocking system tilt angle. Those four variations were combined in order to obtain the best design of turbine. The study used Flow Simulation Solidworks application, and obtain data on turbine speed, pressure, force, and torque. However, this research focused on turbine torque value. The best design of turbine was obtained in the turbine with 6 blades, blade curvature angle of 65° and bucket angle of 10°, and blocking system tilt angle of 40°. In the best turbine, the produced torque value was 8.464 Nm.

1. Introduction

A water turbine as an alternative and a small-scale power plants was classified as pico hydro of power plants because it usually produces a maximum power output of 5 kW. Many researcher interests to do a study on this topic because of its advantages for the remote area which need small electric power even though producing a small power output. Many studies have been conducted to develop pico hydro power plants and involved so many parameters [1-3]. For example, Chen et al. developed a Vertical Axis Water Turbine (VAWT) using a drag type turbine which only involved the number of turbine blades and the form of blocking system. In order to obtain generated power equal to 88,2 W with a flow rate of water equal to 1,5 m/s the research need until three generation of the research [3]. Due to many parameter affect to the turbine performance, researches by simulation method in order to minimize experimental time and cost were also interesting topics [4-9]. Chen et al. also conducted a simulation study using ANSYS CFD for turbine blades and blocking system form. Another research designed and studied flow simulations on blade turbine propellers using the Flow Simulation Solidworks application. The study used a mass flow rate parameter of 1.499 m^3/s as input and resulted in a turbine speed distribution of 0 to 10,949 m/s [9].

A Horizontal Axis Water Turbine (HAWT) is proposed in order to generate electric power in the high rise building wich has the capability to be directly installed in the pipeline of collected waste and rainfall water. It is difficult to collect data directly using the experimental method with many used parameter. To facilitate the data collection, this study first conducted a computational simulation of fluid dynamics in turbine system with various variations such as blade number, blade curvature, turbine bucket angle and blocking system angle using simulation. All of those parameters will be varied to get the most efficient turbine system. In the implementation of simulation, an appropriate method was required to get the best result.

2. Method

Fluid dynamic simulation in this current study is carried out using flow simulation SolidWorks 2014 application which has the capability in analyzing the design of product particularly related to fluid flow. Through this application, data on water flow would be obtained including speed, pressure, flow rate, turbine torque, etc. Thus, those data used to analyze the best variation of turbine system based on morphological matrix method. Turbine blade number, blade curvature angle, bucket angle, and angle of blocking system are shown as figures 1, 2, 3, and 4.

(a) (b) (c) (d)

Figures 1. Turbine blade number of (a) 6, (b) 8, (c) 10, and (d) 12 blades

(a) (b) (c) (d)

Figures 2. Turbine blade curvature angle of (a) 65°, (b) 70°, (c) 75°, and (d) 80°

(a) (b) (c) (d)

Figures 3. Turbine blade bucket angle of (a) 5°, (b) 10°, (c) 15°, and (d) 20°

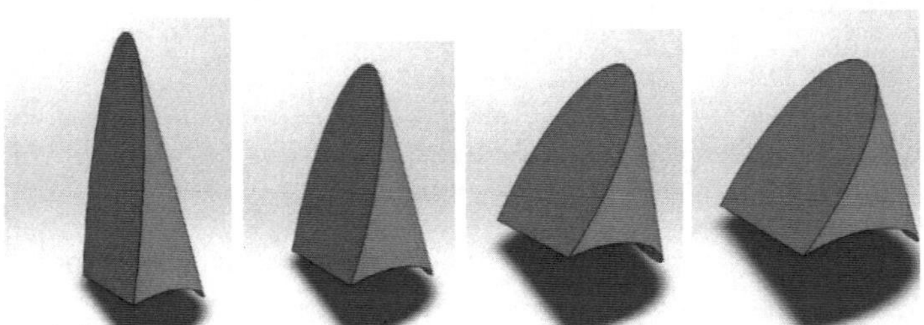

(a) (b) (c) (d)

Figures 4. Turbine angle of blocking system of (a) 20°, (b) 30°, (c) 40°, and (d) 50°

Morphological matrix is one of the creative technical tools to yield ideas, based on the potential characteristic variation of a problem. Morphological matrix is a random design that can be used individually or in the group. The result of this morphological matrix concept development is then poured into the simulation. The variation to be simulated in this research included turbine and blocking system variations. There are three components of designed turbine variation: turbine blade number, blade curvature angle and bucket angle. Meanwhile, for blocking system, there was one type, and it was only varied for its tilt angle. Here is the designed model of turbine variation and blocking system.

Before simulation of all obtained turbine variations from the morphological matrix, meshing validation was conducted first as a pre-study step in this research. Meshing validation was required to obtain the best mesh level out of eight mesh levels existing in Flow Simulation Solidworks of 2014. Meshing validation was carried out from 3rd level mesh to the best mesh. The best mesh level is the condition when there is no more cell number increase, or in other words, the number of the cell has been constant as shown in figure 5.

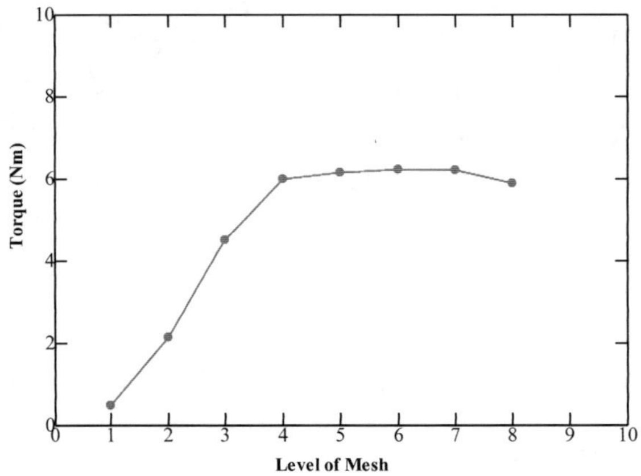

Figure 5. The graph of meshing validation.

3. Results and Discussions

In this subchapter, the effect of the four parameters on the turbine torque will be analyzed. The best turbine design dealing with highest torque value, because of the higher the torque value, the higher is the power produced.

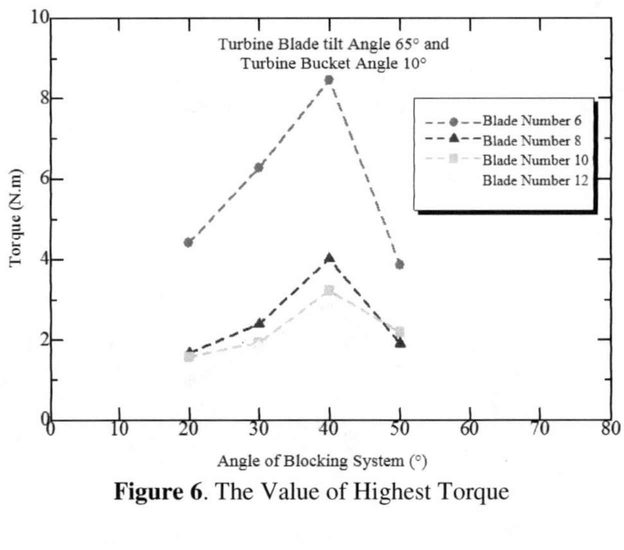

Figure 6. The Value of Highest Torque

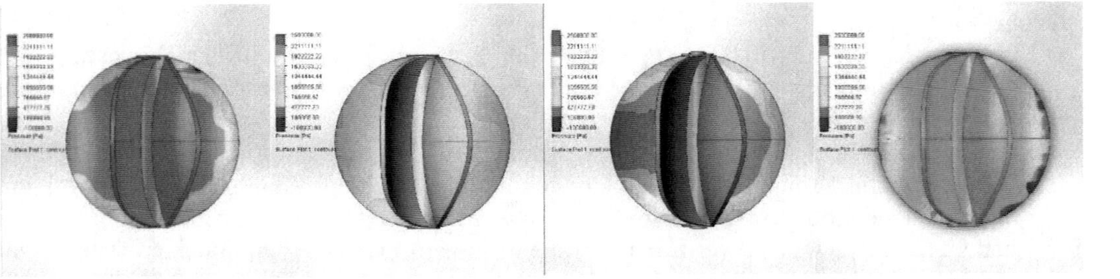

(a) (b) (c) (d)

Figures 7. Pressure plot on the turbine with the blades number of 6, the blade curvature angle of 65° and bucket angle of 10° at various blocking system angle of (a) 20°, (b) 30°, (c) 40°, and (d) 50°

Figure 6 is a chart of the best turbine design by varying the number of the turbine blade, turbine blade tilt angle and turbine bucket angle. The result of the analysis showed that turbine with 6 blades, the turbine blade tilt angle of 65° and turbine bucket angle of 10° producing the largest torque of 8.464 N·m. The turbine produced high torque value because it was affected by blocking system with the best tilt angle of 40°. Thus, combining those four variables will result in the best design of turbine using the produced torque.

Figures 7 shows the pressure obtained on the same turbine blade with a different angle of blocking. Turbine with 40° angle of blocking had direct flow towards the main blade of the turbine, so that in this turbine get big pressure and centered on the main blade. It can be concluded that in the turbine with the blades number of 6, the blade curvature angle of 65° and the bucket angle of 10 ° and with the blocking system angle of 40° can create great momentum and energy in the turbine.

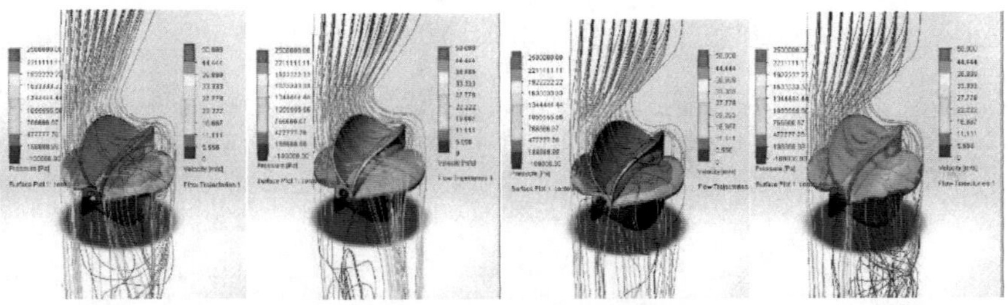

(a) (b) (c) (d)

Figures 8. Velocity plot on the turbine with the blades number of 6, the blade curvature angle of 65° and bucket angle of 10° at various blocking system angle of (a) 20°, (b) 30°, (c) 40°, and (d) 50°

Figure 8 shows the distribution of fluid flow in the turbine. As shown on a turbine with a blocking angle of 40°, the fluid flow velocity increases as it passes through the top of the blocking system and continues to increase until it reaches the turbine blade. After passing turbine blades the velocity of fluid flow tends to decrease. It shows that most of the fluid flow is right on the turbine blades so that it can produce high force and torque. The turbine surface area and the angle of the blocking system have a great effect on the results. So turbines with 65° curvature angles, 10° bucket angles and 40° blocking system slope angle are the best designs of all variations.

4. Conclusion

Research on horizontal axis water turbine (HAWT) simulation by flow simulation of SolidWorks application give the result that the best turbines which produce the highest torque of 8.464 Nm are generated with 6 number of blades, 65° blade curvature angle, 10° bucket angle, and 40° blocking system angle.

IOP Conf. Series: Materials Science and Engineering **308** (2018) 012022 doi:10.1088/1757-899X/308/1/012022

5. References

[1] Marthin, Shaleen, Sharma, Abhay Kumar 2014 Analysis of Rainwater Harvesting and its Utilization for Pico Hydro Power Generation *International Journal of Advanced Research in Computer Engineering & Technology (IJARCET)* **3**:6

[2] Sarma N, Biswas A, Misra R 2014 Experimental and computational evaluation of Savonius hydrokinetic turbine for low velocity condition with comparison to Savonius wind turbine at the same input power *Energy Conversion and Management* **83** pp 88-98

[3] Chen J, et al. 2012 A novel vertical axis water turbine for power generation from water pipelines *Energy* **54** pp 184 – 193

[4] Matsson, John 2014 *An Introduction to SolidWorks Flow Simulation 2014* (USA : Stephen Schoroff)

[5] Musa, Masjuri, Razak J A, Ayob M R, Rosli M A, Herawan S G, Sopian K 2011 CFD Analysis on Cost-effective Pico-hydro Turbine: A Case Study for Low Head and Low Flow Rate Condition *International Conference and Exhibition on Sustainable Energy and Advanced Materials (ICE-SEAM 2011) Solo-Indonesia*

[6] Nuantong W, Sirivit T 2009 Flow Simulations on Blades of Hydro Turbine, *International Journal of Renewable Energy* **4:**2 pp 61-66.

[7] Leila S, Mohammed H 2013 Numerical Simulation of the Flow Into a Rotating Pelton Bucket *International Journal of Emerging Technology and Advanced Engineering (Ijetae)* **3**:2

[8] Yang Z Y, Voke P R 2001 Large-eddy simulation of boundary-layer separation and transition at a change of surface curvature *J Fluid Mech* **439 pp** 305-33

[9] Myint Y W, Win H H 2014 Design and Flow Simulation of Runner Blade for Propeller Turbine *International Journal of Scientific Engineering and Technology Research (IJSETR)* 3:11 pp 2259-2262

IOP Conf. Series: Materials Science and Engineering **308** (2018) 012023 doi:10.1088/1757-899X/308/1/012023

Numerical Analysis of Flood modeling of upper Citarum River under Extreme Flood Condition

R I Siregar[1, 2]

[1]Civil Engineering Department, Faculty of Engineering, Universitas Sumatera Utara, Jl. Dr. Mansur Kampus USU, Padang Bulan, Medan 20155, Indonesia
[2]Sustainable Energy and Biomaterial Centre of Excellence, Universitas Sumatera Utara, Jl. Almamater Kampus USU, Padang Bulan, Medan 20155, Indonesia

E-mail: rizawaloed@yahoo.com

Abstract. This paper focuses on how to approach the numerical method and computation to analyse flood parameters. Water level and flood discharge are the flood parameters solved by numerical methods approach. Numerical method performed on this paper for unsteady flow conditions have strengths and weaknesses, among others easily applied to the following cases in which the boundary irregular flow. The study area is in upper Citarum Watershed, Bandung, West Java. This paper uses computation approach with Force2 programming and HEC-RAS to solve the flow problem in upper Citarum River, to investigate and forecast extreme flood condition. Numerical analysis based on extreme flood events that have occurred in the upper Citarum watershed. The result of water level parameter modeling and extreme flood discharge compared with measurement data to analyse validation. The inundation area about flood that happened in 2010 is about 75.26 square kilometres. Comparing two-method show that the FEM analysis with Force2 programs has the best approach to validation data with Nash Index is 0.84 and HEC-RAS that is 0.76 for water level. For discharge data Nash Index obtained the result analysis use Force2 is 0.80 and with use HEC-RAS is 0.79.

1. Introduction

Numerical analysis is as a technique to solve the problems in the field of engineering, which formulated in the form of mathematical equations. Mathematical equation in other conditions cannot solve by analytically but numerically. It has analytic approaches errors with value must be less than the provisions [1]. Flow pattern recognition using the physical modeling and numerical simulation helps designers to propose optimal shape for hydraulics structures. Optimal shape of hydraulic structure causes increasing their performance [2]. Need to be versed in all the formulas and theories related to the problem, analytical procedures and problems of scale and ignoring some parameters, physical and laboratory methods [3]. The model adopts two well-established empirical formulas to determine the flooding under extreme flood condition [4].

Flood events caused by the extreme flood conditions can have damaging consequences on buildings and infrastructure located in prone areas [5]. Flood simulation models can provide practitioners of flood disaster management with sophisticated estimates of floods. Despite the advantages that flood simulation modeling may provide, experiences have proven that these models are of limited use. Until now, this problem investigated by evaluations, which information demanded by decision-makers versus what models can actually offer [6].

Factors that affect flooding due to river floods when extreme flood conditions affected is river morphology. The amount of morphological change, erosion, and deposition related to peak flow conditions at the sites. The measured changes in bank full capacity were highly correlated with most hydraulic variables [7]. The aims of this study are to perform a numerical approach and modeling floods especially during extreme flooding condition. This study analyzed extreme flood condition in upper Citarum River used numerical analysis of flood modeling. This study used computation approach with Force2 programming and HEC-RAS, to investigate and model extreme flood condition.

2. Study Area and Materials

2.1. Characteristics of the River in Study Area

The upper Citarum watershed has upper Citarum River as a main river. The watershed consists of thirteen tributaries, where they form a sub-watershed. Upper Citarum River originated from well Cisanti, located at the foot of Mount Wayang, about 60 kilometers south of Bandung. Upper Citarum River has 37 kilometers in length and it flows through the southern of Bandung Regency up to Saguling Reservoir. The Figure 1 shows upper Citarum River in upper Citarum Watershed that lies in Citarum Watershed, West Java, Indonesia. The location map of river image shows at the top right of the figure 1.

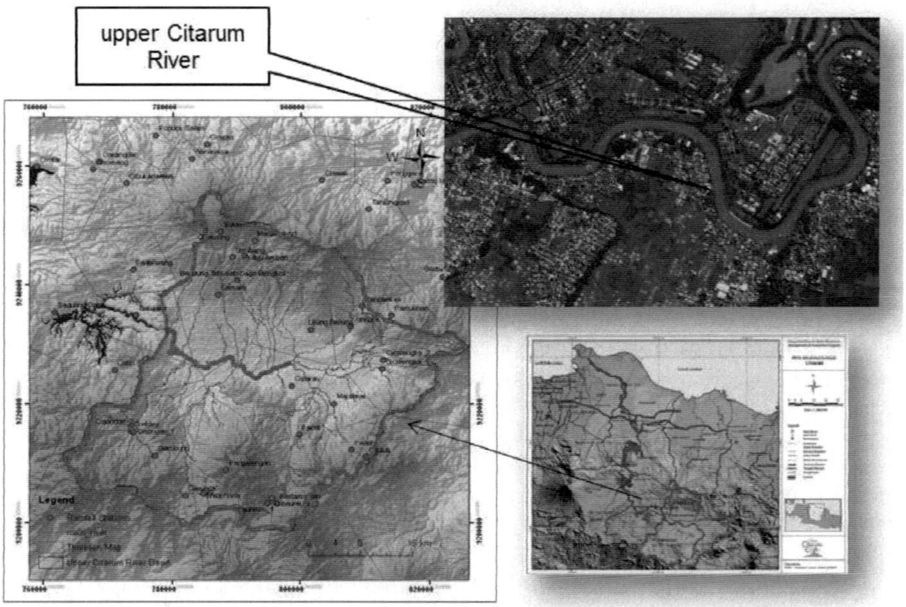

Figure 1. The location map of Citarum Watershed in West Java that shows the location of upper Citarum Watershed as a study area

2.2. Data Sources

The data required in this study comes from the Major River Basin Organization of Citarum, Public Work Ministry. These data include upper Citarum river flow discharge in years 2001 to 2012, daily rainfall data in years 2001 to 2012, and land cover data in years 2001 and 2010. In addition, river morphology such cross section and long section of upper Citarum River data and watershed characteristic are also required for modeling. Flood modeling also requires DEMs data for 2D modeling for extreme flood that occurs in upper Citarum River. The input parameter for analysis involved rainfall intensity, land cover, river cross and long sections, characteristics of watershed, and flood extreme discharge. All of these data run in Force2 programming and HEC-RAS to get information about what happened about the influences of this flood. As validation, we use discharge data and water level from AWLR data from the Major River Basin Organization of Citarum.

IOP Conf. Series: Materials Science and Engineering **308** (2018) 012023 doi:10.1088/1757-899X/308/1/012023

3. Methods

The stages of methodology involved two programs. They are Force2 and HEC-RAS. The numerical analysis that we have done base on momentum and continuity equation, run in Force2, this program help to iterate the discharge and water level in simulation time. The result of from that program will compare to the HEC-RAS value. In this study, HEC-RAS as a comparison to know extending of this model, because in numerical analysis only as one dimensional modeling, but in HEC-RAS analysis use two dimensional modeling.

3.1. Governing Equations

Finite Element Method is the most widely used method to solve the problem of flow because its formulation is relatively easy and efficient as well as giving satisfactory results. The properties of the conservative discretization increment used to control volume approach [8]. Subsequent analysis is to determine the elevation of the base and the water level of each section, determine the cross-sectional area A and a discharge Q using explicit discretization Lax-Wendroff [9]. Discretization shapes that given by Lax-Wendroff to solve the equations, which is in the form of conservative equation.

$$\frac{\partial W}{\partial t} + \frac{\partial F(W)}{\partial x} + T(W) = 0 \tag{1}$$

$$W_j^{n+1} = W_j^n + \Delta t\left(\frac{\partial W}{\partial t}\right)_j^n + \frac{\Delta t^2}{2!}\left(\frac{\partial^2 W}{\partial^2 t}\right)_j^n + \frac{\Delta t^3}{3!}\left(\frac{\partial^3 W}{\partial^3 t}\right)_j^n + \dots \tag{2}$$

$$\frac{\partial A}{\partial t} + \frac{\partial Q}{\partial x} = 0 \tag{3}$$

$$\frac{\partial Q}{\partial t} + \frac{\partial}{\partial x}\left(\beta\frac{Q^2}{A} + gI\right) + gAS_0 + g\frac{Q\,|Q|}{AC^2R} = 0 \tag{4}$$

3.2. Flood Modeling Method

This study performs numerical analysis with Finite Element Method. Numerical simulations derived out by the unstructured grid, Finite Element Method [11]. In the analysis phase, it will discretize the discharge equation and water level based on the fluid momentum and continuity equations. The equation solved with the help of computer programming that is Force2.

This method applied the Hydrologic Engineering Center River Analysis System (HEC-RAS) model to estimate the potential vulnerabilities of flood for different peak outflow scenarios with conclusions and recommendations [12]. All raster datasets used to create 1D and 2D HEC-RAS model. HEC-RAS is the most commonly used flood model tool in Indonesia. The HEC-RAS parameters kept unchanged for different DEMs because the goal is to investigate the sensitivity of inundation maps to topographic errors instead of trying to create new calibrated model for each topographic dataset [13].

The numerical results compared with a fully 2D model as well as with measurements in some gauge points giving satisfactory results. The simulation of a real flooding scenario in the upper Citarum River near the urban area then performed. A lateral coupling configuration provided, in which the flood wave propagation in the main channel simulated by means of a 1D model and the inundation of the riverside simulated by means of a 2D model [14]. Based on the results of numerical analysis and simulation results in HEC-RAS will be compared the results of flow discharge and water level to see the relationship between the two parameters in terms of flood modeling.

Base on discharge data in years 2001 to 2012, shows that the trend of increasing flow occurs in 2010. For further simulation, we use the flow in year 2010 as the most extreme flood events in upper Citarum River.

3.3. Flow Validation Method

In this study, the model use Nash–Sutcliffe efficiency (NI) for flow validation method. The result of comparison then determined the strength in predicting the modeling by Nash-Sutcliffe model of efficiency coefficient (E). The Nash–Sutcliffe efficiency (NI) is the two criteria most widely used for calibration and evaluation of hydrological models with observed data [15]. Nash–Sutcliffe Index can range from -∞ to 1. The value of 1 (NI=1) corresponds to a perfect match of modeled discharge to the observed data. The value NI = 0 indicates that the model predictions are as accurate as the mean of the observed data, if it is less than 0; it occurs when the observed is better prediction than the model.

4. Results and Discussions

4.1. Flow Simulation Equations

Based on the result of a decrease in the continuity equation and momentum, with incorporates discretization scheme results Lax-Wendroff, the equation set forth in Taylor series as follows:

$$W_j^{n+1} = W_j^n - \Delta t\left(\frac{\partial F}{\partial x} + T\right) + \frac{\Delta t^2}{2}\left(\frac{\partial}{\partial x}\left[R\left(\frac{\partial F}{\partial x} + T\right)\right] - \frac{\partial T}{\partial t}\right) = \{W_j^n\} - \frac{\Delta t}{2\Delta x}\{(F_{j+1}^n - F_{j-1}^n) + (2\Delta x)T_j^n\} + \frac{1}{2}\frac{\Delta t^2}{\Delta x^2}\left\{\left(R_{j+\frac{1}{2}}^n\left[F_{j+1}^n - F_j^n + \frac{\Delta x}{2}(T_{j+1}^n + T_j^n)\right] - R_{j-\frac{1}{2}}^n\left[F_j^n - F_{j-1}^n + \frac{\Delta x}{2}(T_j^n + T_{j-1}^n)\right]\right) - \frac{\Delta x^2}{\Delta t}(T_j^n - T_j^{n-1})\right\} \quad (5)$$

Definitely, W_j^{n+1} and W_j^n consists of cross sectional area A and discharge Q. The equation above can derive into the equation to determine the value of A and Q. The equation becomes Lax-Wendroff scheme that derived into the equation in the programming language. From these two equations then determined the parameters of other channels. The following is equation upstream and downstream boundary conditions.

$$\frac{Q_1^{n+1}}{h_1^{n+1}B} - 2\sqrt{gh_1^{n+1}} - u_R^n + 2c_R^n = -g\Delta t S_0 + g\frac{u_1^n|u_1^n|}{(C_z)^2 R_1^n}\Delta t \quad (6)$$

$$\frac{Q_N^{n+1}}{h_N^{n+1}B} + 2\sqrt{gh_N^{n+1}} - u_L^n - 2c_L^n = -g\Delta t S_0 + g\frac{u_N^n|u_N^n|}{(C_z)^2 R_N^n}\Delta t \quad (7)$$

$$Q_N^{n+1} + 2Bh_N^{n+1}\sqrt{gh_N^{n+1}} = P', \ dimana\ P' = Bh_N^{n+1}(u_L^n - 2c_L^n - g\Delta t S_0 + g\frac{u_N^n|u_N^n|}{(C_z)^2 R_N^n}\Delta t) \quad (8)$$

From the equation above can be obtained by the value of water depth h because h non-linear equations that can be solved with Newton Raphson method of interpolation.

4.2. Flood Modeling

In flood modeling with numerical analysis method, two parameters that observed, they are flow discharge and water level of upper Citarum River. Based on the purpose of this study is to perform a comparison of the numerical approach by generating an equation solved by FEM. Three things helped to analyze the model the flood in extreme conditions. There are iteration analysis with Force2 program and analysis with the help of HEC-RAS. Results of modeling with HEC-RAS can looked in Figure 2 below. The comparison result depicted in the graph as shown in Figure 3.

The figure 2 shows that the results shown based on simulations of extreme flood events with HEC-RAS 2D modeling. The picture shows the inundation area released by the upper Citarum River when overflowing. The resulting inundation area based on terrain data generated for 2D modeling in HEC-RAS. Based on the analysis, the amount of inundation based on the simulation is 75.26 square kilometers around the river.

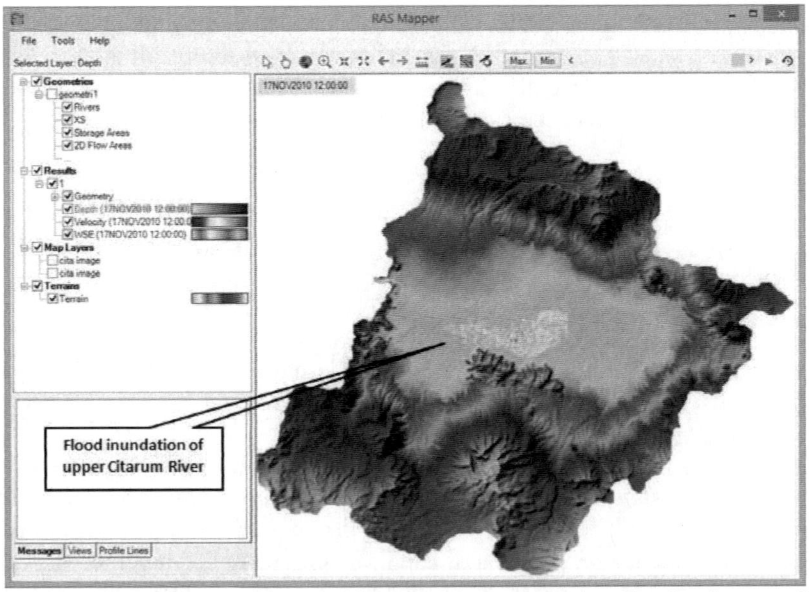

Figure 2. Flood inundation of upper Citarum River when overflowing in year 2010

The figure 3 shows that the results comparison with numerical analysis approaches. The pictures show the maximum value from simulation among Force2, HEC-RAS, and validation data. From the graph, we know that the change of Q and water level of H have differences between the results from Force2 and HEC-RAS in certain time simulation.

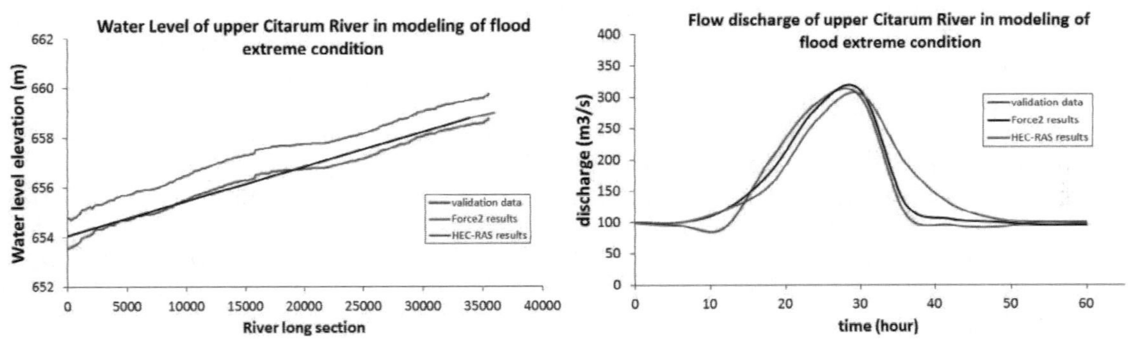

Figure 3. Water level and flow discharge of upper Citarum River when overflowing in year 2010

4.3. Flow Validation

Based on the validation of Nash Index of water level obtained the result analysis use Force2 is 0.84 and with use HEC-RAS is 0.76. For discharge data Nash Index obtained the result analysis use Force 2 is 0.80 and with use HEC-RAS is 0.79. The index values is more acceptable.

5. Conclusions

This study has shown that numerical modeling of the flood modeling has become unavoidable works in the hydraulic modeling. There is need for flood inundation models that efficiently and accurately applied over practical scales such as a city or regional flood plain. In this study, based on modeling results with HEC-RAS for 2D modeling, during the extreme flood event in 2010, several areas in the upper Citarum river were almost entirely flood by upper Citarum River overflow, and its inundation area is about 75.26 square kilometers. Comparing two-method show that the FEM analysis with

Force2 programs has the best approach to validation data with Nash Index is 0.84 and HEC-RAS that is 0.76 for water level. For discharge data Nash Index obtained the result analysis use Force2 is 0.80 and with use HEC-RAS is 0.79. Overall, this study showed that optimal design of the flood modeling leads to increase the performance of the flood prevention structures.

Acknowledgments

The author gratefully acknowledged that the research Universitas Sumatera Utara, Indonesia supported for the financial. Thanks to Major River Basin Organization of Citarum, Public Work Ministry, for providing the data and others.

References

[1] Zhou Z, Olabarrieta M, Stefanon L, Alpaos A, Carniello L, and Coco G 2014 A Comparative Study of physical and Numerical Modeling of Tidal Network Ontogeny *Journal of Geophysical Research* **119** 892-912
[2] Dehdar-behbahani S, Parsaie A 2016 Numerical modeling of flow pattern in dam spillway's guide wall. Case study: Balaroud dam Iran *Alexandria Engineering Journal* **55** 467-473
[3] Ershadi S, Arasteh M, Tajziehchi M 2013 *Numerical Modeling of Flow Pattern Changes in Tidal Inlet of TIYAB Port* Journal of Environmental and Earth Sciences **11** 691-702
[4] Marsolli R, Orton P M, Georgas N, Blumberg A F 2016 Three-Dimensional Hydrodynamic Modeling of Coastal Flood Mitigation by Wetlands *Coastal Engineering* **111** 83-94.
[5] Aureli F, Dazzi S, Maranzoni A, Mignosa P, and Vacondio R 2015 Experimental and Numerical Evaluation of the Force due to the Impact of a Dam-Break Wave on a Structure *Advances in Water Resources* **76** 29-42.
[6] Leskens J G, Brugnach M, Hoekstra A Y, Schuurmans W 2014 Why are Decisions in Flood Disaster Management so Poorly Supported by Information from Flood Models? *Environmental Modelling & Software* **53** 53-61
[7] Hooke J M 2016 Geomorphological Impacts of an Extreme Flood in SE Spain *Geomorphology* **263** 19-38
[8] Gravenkamp H, Man H, Song C, Prager J 2013 The Computation of Dispersion Relations for Three- Dimansional Elastic Waveguides Using the Scaled Boundary Finite Element Method *Journal of Sound and Vibration* **332** 3756-3771
[9] Fridrich D, Liska R, Wendroff B 2016 Some Cell Centered Lagrangian Lax-Wendroff HLL Hybrid Schemes *Journal of Computational Physics* **326** 878-892
[10] Lu J, Fang J, Tan S, Shu C, Zhang M 2016 Inverse Lax-Wendroff Procedure for Numerical Boundary Conditions of Convection-Diffusion Equations *Journal of Computational Physics* **317** 276-300
[11] Karna T, Baptista A M, Lopez J E, Turner P J, McNeil C, and Sanford T B 2015 Numerical Modeling of Circulation in High-Energy Estuaries: A Columbia River Estuary Benchmark *Ocean Modelling* **88** 54-71
[12] Dehdar-behbahani S, Parsaie A 2016 Numerical modeling of flow pattern in dam spillway's guide wall. Case study: Balaroud dam, Iran *Alexandria Engineering Journal* **55** 467-473
[13] Butt M J, Umar M, Qamar R 2013 Landslide Dam and Subsequent Dam-Break Flood Estimation Using Hec-Ras Model in Northern Pakistan *Natural Hazards* **65** 241-254
[14] Saksena S, Merwade V 2015 Incorporating the Effect of Dem Resolution and Accuracy for Improved Flood Inundation Mapping *Journal of Hydrology Elsevier* **530** 180-194
[15] Yucel I, Onen A, Yilmas K K, Gochis D J 2015 Calibration and evaluation of a flood forecasting system: Utility of numerical weather prediction model, data assimilation and satellite-based rainfall *Journal of Hydrology* **523** 49-66

IOP Conf. Series: Materials Science and Engineering **308** (2018) 012024 doi:10.1088/1757-899X/308/1/012024

Beeswax as phase change material to improve solar panel's performance

R Thaib[*1], S Rizal[1], M Riza[2], T M I Mahlia[3], T A Rizal[4]

[1] Department of Mechanical Engineering, Faculty of Engineering, University of Syiah Kuala, Darussalam, Banda Aceh 23111, Indonesia
[2] Department of Chemical Engineering, Faculty of Engineering, University of Syiah Kuala, Darussalam, Banda Aceh 23111, Indonesia
[3] Department of Mechanical Engineering, Faculty of Engineering, Universiti Tenaga Nasional, Malaysia
[4] Department of Mechanical Engineering, Faculty of Engineering, University of Samudra, Kota Langsa, Aceh, 24416, Indonesia

[*]E-mail: razalithaib@unsyiah.ac.id

Abstract. One of the main obstacles faced during the operation of photovoltaic (PV) panels was overheating due to excessive solar radiation and high ambient temperatures. In this research, investigates the use of beeswax phase change materials (PCM) to maintain the temperature of the panels close to ambient. Solar panels used in this study has 839 mm length, 537 mm wide, and 50 mm thick, with maximum output power at 50 W. During the study, there were two solar panels was evaluated, one without phase change material while the other one was using beeswax phase change material. Solar panels were mounted at 15° slope. Variables observed was the temperature of solar panel's surface, output voltage and current that produced by PV panels, wind speed around solar panels, and solar radiation. The observation was started at 07:00 am and ended at 06:00 pm. The research shows that maximum temperature of solar panels surface without phase change material is ranging between 46-49 °C, and electrical efficiency is about 7.2-8.8%. Meanwhile, for solar panels with beeswax phase change material, the maximum temperature solar panels surface is relatively low ranging between 33-34 °C, and its electrical efficiency seems to increase about 9.1-9.3%.

1. Introduction

In line with the increasing pace of development and the increasing pattern of life, energy consumption in Indonesia continues to increase. This increase occurred almost in all sectors that include industry, transportation, commercial, household, power generation and other sectors. The total national energy consumption in the period 2003-2013 has increased with an average growth rate of 4.1% per year. Total energy consumption increased from 117 million TOE in 2003 to 174 million TOE in 2013 [1].

The commercial sector consists of trade, hotels, restaurants, financial, government agencies, schools, hospitals, communications and others. Data for years 2004-2011 showed that the sector grew on average 8% per year. However, growth in the sector does not have the same pattern of the growth of energy consumption in the commercial sector. The growth of energy consumption in the commercial sector in the period 2004-2011 had an average growth of 4% per year. In 2014, this sector consumes the energy of 5:22 million TOE, or about 3% of total final energy consumption. Based on the master plan of national energy conservation, the commercial sector has the potential energy savings that can be achieved by 10-30%.

A decrease in energy consumption and its impact can be achieved in various ways, including the use of low carbon building materials, improved insulation, low-energy lighting, embedding renewable technology, intelligent control and use of low carbon fuels. The incorporation of renewable energy in the building component must be done in a way that ensures reliable performance and cost-effective energy saving. A Building Integrated Photovoltaics (BIPV) system consists of integrating photovoltaics modules into the building envelope, such as the roof or the façade. By simultaneously serving as building envelope material and power generator, BIPV systems can provide savings in materials and electricity costs, reduce the use of fossil fuels and emission of ozone depleting gases, and add architectural interest to the building [2].

However, BIPV faces a challenge with the rise in temperature, manifested as electrical efficiency loss and overheating. This efficiency loss is mainly due to a decrease in open-circuit voltage (V_{oc}), which has a negative temperature coefficient. Currently, however, only 15–20% of the solar energy incident on a PV panel can be converted to electricity, with the rest of the energy being transformed into heat. This heat can well be absorbed by the PV device itself, causing its working temperature up to 80°C. PV conversion efficiency is thereby decreased by 0.4–0.65 % for every increased degree [3].

Many researchers, therefore, are seeking to mitigate the influence of high temperature on PV conversion efficiency by rapidly removing heat from PV module surfaces to maintain as good a performance as possible, and hence better meet expectations. Much of the research has focused on natural or forced air circulation, heat pipes and hydraulic or refrigerant cooling [4]–[7].

Study of the use of the PC as a PV panel cooling material both experimentally and analytically also been widely reported [3], [8]–[10]. This system was named PV–PCM or PV/PCM, a hybrid technology integrating a PV panel and a PCM into a single module to achieve a higher module solar conversion efficiency than a PV panel acting alone. This is a promising integration, as the PV converts visible and ultraviolet components of the solar spectrum, while the PCM utilizes the infra-red components and the waste heat from the PV.

The PCM also reduces the PV temperature thereby increasing its efficiency. However, the PV–PCM system is still in the infant stages, with various issues unclear, presenting opportunities for further research [11]. The aim of these experiments is to study the behavior of the PV-PCM systems in outdoor realistic uncontrolled conditions to determine how effective the PV-PCM systems are in 'real' conditions. In addition, the PV-PCM systems were characterized in low latitude hot climate of Banda Aceh, Indonesia. Experiments were conducted outdoors of the Engineering Faculty, Syiah Kuala University, Banda Aceh, Indonesia (05.57 N, 95,37 E).

2. Methods

This study was conducted outdoors in the Faculty of Engineering, University of Syiah Kuala. The research was conducted from 08-07-2016 to 15-07-2016. Three PV panels are used. One PV panel as a reference, the PV panels with paraffin wax as a phase change material (PV-PCM_1), and a PV panel with beeswax as a phase change material (PV-PCM_2). The study was conducted from 07:00 am to 06:00 pm, where ambient temperatures, wind speed, solar radiation intensities, temperatures at front and back surfaces, V_{oc} and I_{sc} for the reference PV and the PV-PCM systems were measured. PV panel is placed facing north.

Figure 1 shows the experimental setup consisting of the reference PV panels without PCM, PV-PCM1, and PV-PCM_2 installed outdoors of Engineering Faculty. The temperatures and V_{oc} and I_{sc} were measured with data logger and the multimeters. PCM paraffin wax used had a melting temperature of 46.7°C, and beeswax is 51°C.

Figure 1. An outdoor experimental setup consisting of a reference PV, PV-PCM1 and PV-PCM2 installed.

According to the Cox and Raghuraman 1985, report, the insolation of wavelength above 1.1μm is transmitted through the silicon cell without any absorption and this is absorbed by the back sheet of PV module. Electric efficiency (η_e) of PV panel was determined using following equation:

$$\eta_e = \eta_o \left[1 - \beta\left(T_p - T_r\right)\right] \quad (1)$$

Standard electric efficiency (ηo) is calculated with:

$$\eta_o = \frac{V_{oc} I_{sc}}{G A_p} x\ 100\ \% \quad (2)$$

where ηo is the nominal electrical efficiency under standard condition, Ap is the area of the PV module, G is the irradiation and it is defined as 1000W/m2 for the standard condition, Voc is the PV voltage at maximum power point and Isc is the PV current at maximum power point. All the relevant data can be obtained from the specification of PV module shown in table 1, T_o is the temperature of the standard condition, 25°C, T_c is the cell temperature, β is the temperature coefficient of silicon cell, β=0.0045/°C.

Table 1. Specifications PV Panel

Type	:	Silicon Nitride Monocrystalline
Maximum power (Pmax)	:	50 W
Voltage at Pmax (Vmp)	:	17.5 W
Current at Pmax (Imp)	:	2.9
Warranted minimum Pmax	:	45W
Short-circuit current (Isc)	:	3.2A
Open-circuit voltage (Voc)	:	21.8
Dimensions	:	
Length		839mm
Width		537mm
Depth		50mm

3. Results and Discussion

Although experiments were conducted from 08-07-2016 to 15-07-2016, the weather conditions were not stable on all days. On certain days the solar radiation intensity and the ambient temperatures were not high enough due to overcast weather conditions in Banda Aceh which restricted temperatures at the PV surface to such low values that the PCM could not melt properly. For other days the ambient temperatures and the solar radiation were high enough to raise the PV temperature above the PCM melting point which allowed heat absorption and thus melting of PCM to occur.

IOP Conf. Series: Materials Science and Engineering **308** (2018) 012024 doi:10.1088/1757-899X/308/1/012024

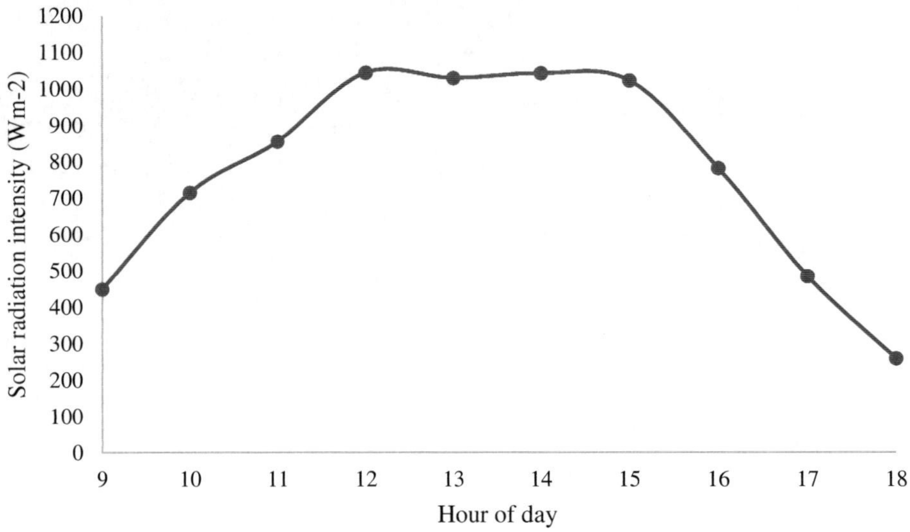

Figure 2. Solar radiation intensity, Banda Aceh on 9-07-2016

Figure 2 shows that solar radiation intensity ranged from 448 W/m2 at 9:00 am to 259 W/m^2 at 06:00 pm with a peak at 1042 W/m2 at 02:00 pm. The best case data with higher solar radiation intensities (1042 W/m^2) peak and ambient temperatures (34℃ peak) was obtained at 09-07-2016. The stable conditions resulted in higher temperatures at the reference PV front surface. This made more heat available at PV enabling higher heat harnessing by the PCM at the back of PV resulting more temperature reduction.

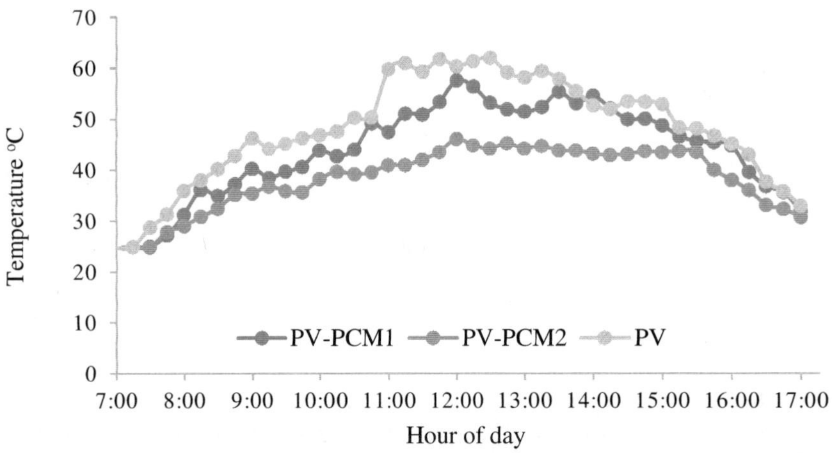

Figure 3. The average temperature at the front surface of reference PV, PV-PCM1, and PV- PCM2.

Figure 3 shows the average temperature evolution on the front surface of the reference PV, PV-PCM1 and PV-PCM2 measured. Temperature at the PV front surface increased rapidly from 24°C (at 07:00 am) to 40 °C in 1 hours and 30 minutes (at 08:30 am) and remained above 40 °C for 7 hours and 30 minutes (up to 04:00 pm) with a peak temperature of 629 °C at 12:30 pm. As shown in Figure 3 PV-PCM_1 and PV-PCM_2 followed the same trend in temperature rise as the reference PV temperature, however, the PV-PCM systems maintained lower temperatures than the reference PV throughout the experiment. After 12:00 pm the temperature of the reference PV dropped rapidly due to combined effect of (i) decreased incoming heat caused by decreasing solar radiation intensity and (ii) increased heat loss to ambient due to a rapid increase in wind speed. The temperature in the PV-PCM systems showed less of a decrease, primarily due to the stored latent heat and thermal mass of the PCM in the PV-PCM systems, which releases stored heat with small temperature changes.

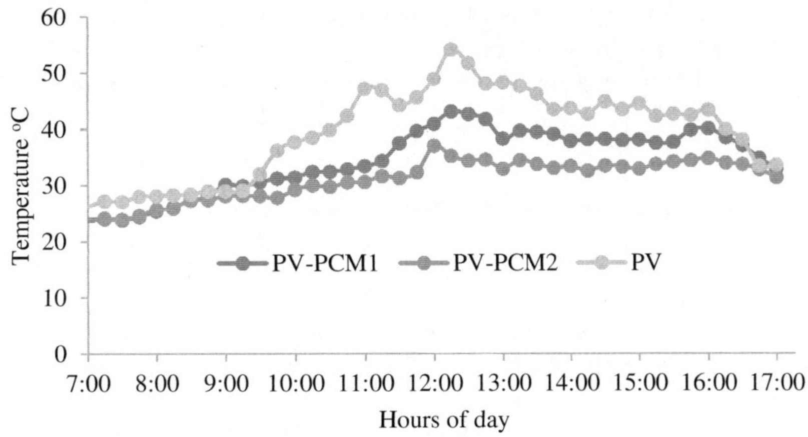

Figure 4. Average temperature at rear surface of reference PV, PV-PCM1, and PV- PCM2

Figure 4 shows the temperature evolution on the rear surface of the PV panels. It is seen that the rear surface of the PV reference temperature at 54 °C maximum occurred at 12:00 pm At that moment the front surface temperature is 62 °C, there is a difference 8°C. While the PV-PCM1 maximum temperature 43.1°C, while on the front surface 60.4°C, then the difference between 17°C, while the PV-PCM2 difference of 9.15°C.

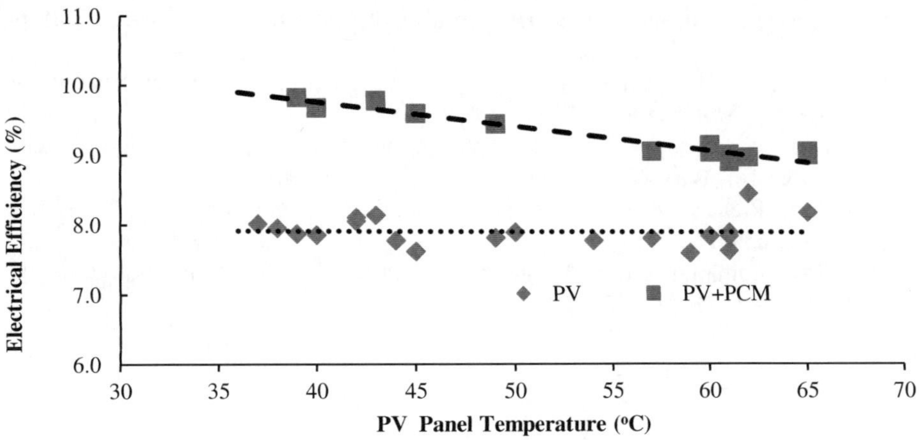

Figure 5. Relation of electrical efficiency with PV panels surface temperature

Figure 5 shows the impact of front temperature of PV panel on PV electricity efficiency. Electricity efficiency was determined using eq. (1). The graphic shows that electricity efficiency of PV panel without PCM is ranged between 6.1 % - 6.5 %. While for PV panel with PCM the efficiency is ranged at 7.0 % - 7.8 %. This proved that air cooling is capable of increasing the efficiency of PV panel.

4. Conclusion

The result of observation conducted on July 09 2016, that the changes of PV panel surface temperature, which solar radiation intensity for PV panel with and without PCM mechanism is shown in figure 4. The graphic shows that maximum intensity of solar radiation was achieved at 12.00 pm which is at 1024 W/m^2, followed by a maximum temperature of PV panel surface without PCM at 62°C. The significant impact of lowering surface temperature is shown during PCM cooling, where maximum temperature was only at 60.4°C. As a comparison, PV panel without PCM recorded that during 09:00 am to 02:00 pm, the temperature rose up to 28°C, while PV panels with PCM only 11°C. Beeswax has a better ability to absorb heat compared to the paraffin wax. This emphasized that test device is working properly. Both PCM evaluated showed promising potential for the desired application, with higher potential in low latitude hot in Banda Aceh.

References

[1] Dewan Energi Nasional 2014 *Outlook Energi Indonesia 2014*

[2] Sharma S, Tahir A, Reddy K S, and Mallick T K 2016 Solar Energy Materials & Solar Cells Performance enhancement of a Building-Integrated Concentrating Photovoltaic system using phase change material *Sol. Energy Mater. Sol. Cells* **149** pp 29–39

[3] Hasan A, McCormack S J, Huang M J, and Norton B 2010 Evaluation of phase change materials for thermal regulation enhancement of building integrated photovoltaics *Sol. Energy* **84**:9 pp 1601–1612

[4] S. Dubey S, G. S. Sandhu, and G. N. Tiwari, “Analytical expression for the electrical efficiency of PV / T hybrid air collector,” *Appl. Energy*, vol. 86, no. 5, pp. 697–705, 2009.

[5] Chauhan D and Agarwal S 2014 *Different Cases of Hybrid PV / T Air Collector : A Comparative Study* **1**:3 pp 29–34

[6] Tonui J K and Tripanagnostopoulos Y 2007 Air-cooled PV/T solar collectors with low-cost performance improvements *Sol. Energy* 81:4 pp 498–511

[7] K. A. Moharram 2013 Enhancing the performance of photovoltaic panels by water cooling *Ain Shams Eng. J.* **4**:4 pp 869–877

[8] Aelenei L, Pereira R, Gonçalves H, and Athienitis A 2014 Thermal performance of a hybrid BIPV-PCM : modeling, design and experimental investigation *Energy Procedia* **48** pp 474–483

[9] Huang M J 2011 The effect of using two PCMs on the thermal regulation performance of BIPV systems *Sol. Energy Mater. Sol. Cells* **95**:3 pp 957–963

[10] Thaib R, Hamdani, Irwansyah, Zaini 2014 Investigation Performance of Solar Water Heater System Using Paraffin Wax *ARPN J. Eng. Appl. Sci.*, **9** pp 1749–1752

[11] Pielichowska K and Pielichowski K 2014 Phase change materials for thermal energy storage *Prog. Mater. Sci.* **65** pp 67–123

[12] Cox C H and Raghuraman P 1985 Design considerations for flat-plate-photovoltaic/thermal collectors *Sol. Energy* **35**:3 pp 227–241

Techno-Economic analysis of solar photovoltaic power plant for small scale fish processing in Kota Langsa - a case study

S B Widodo[1], Hamdani[2], T A Rizal[1], N A Pambudi[3]

[1]Department of Mechanical Engineering, Faculty of Engineering, University of Samudra, Kota Langsa, Aceh, 24416, Indonesia
[2]Department of Mechanical Engineering, Faculty of Engineering, University of Syiah Kuala, Banda Aceh, Aceh, 23111, Indonesia
[3]Mechanical Engineering Education, Universitas Negeri Sebelas Maret, Jl. Ir. Sutami 36A, Surakarta 57126, Indonesia

*E-mail: hamdani_umar@yahoo.com

Abstract. In Langsa, fisheries are the sector leaders by fulfilling a capacity of about 6,050 tons per year and on the other hand, fish-aquaculture reaches 1,200 tons per year on average. The fish processing is conducted through catches and aquaculture. The facilities on which this processing takes place are divided into an ice factory unit, a gutting and cutting unit, a drying unit and a curing unit. However, the energy and electricity costs during the production process has become major constraint because of the increase in the fishermen's production and income. In this study, the potential and cost-effectiveness of photovoltaic solar power plant to meet the energy demands of fish processing units have been analysed. The energy requirements of fish processing units have reached an estimate of 130 kW, while the proposed design of solar photovoltaic electricity generation is of 200 kW in an area of 0,75 hectares. In this analysis, given the closeness between the location of the processing units and the fish supply auctions, the assumption is made that the photovoltaic plants (OTR) were installed on the roof of the building as compared to the solar power plants (OTL) installed on the outside of the location. The results shows that the levelized cost of OTR instalation is IDR 1.115 per kWh, considering 25 years of plant life-span at 10% of discount rate, with a simple payback period of 13.2 years. OTL levelized energy, on the other hand, is at IDR 997.5 per kWh with a simple payback period of 9.6 years. Blood is an essential component of living creatures in the vascular space. For possible disease identification, it can be tested through a blood test, one of which can be seen from the form of red blood cells. The normal and abnormal morphology of the red blood cells of a patient is very helpful to doctors in detecting a disease. With the advancement of digital image processing technology can be used to identify normal and abnormal blood cells of a patient. This research used self-organizing map method to classify the normal and abnormal form of red blood cells in the digital image. The use of self-organizing map neural network method can be implemented to classify the normal and abnormal form of red blood cells in the input image with 93,78% accuracy testing.

1. Introduction

Langsa is a municipality located in the province of Aceh, Northern Sumatra that covers 262.41 km2 of land and it's formed by 5 districts and 66 villages. The policy of the city government of Langsa is to boost the marine resources available through the improvement of the fishing industry, aquaculture, the processing industry and the marine industry, while relying on science and technology. Marine resources

IOP Conf. Series: Materials Science and Engineering **308** (2018) 012025 doi:10.1088/1757-899X/308/1/012025

are the backbone of the sustainable economic development in the region. The growth of the fisheries in the sector and the efforts to support them are constantly encouraged given that the local government is fully aware of the positive economic impact they can provide. In table 1, the production rate of the Langsa fisheries throughout the year 2015 can be observed.

Based on the data, it can be noted that the fisheries that generate the most production are located in the Langsa Timur and Langsa Barat district. Meanwhile, the largest aquaculture production belongs to Langsa Timur. In a fishing group there's a total of 80 members divided into 45 fishermen and 35 aquaculture business owners who specialize in milkfish aquaculture. Added value of the fisheries products has been a great issue for years. They haven't been successful to bring fish production into a wider market because ice factories and post-harvest fish processing technology are not available. Moreover, there is an energy shortage in the region that poses a constraint. Therefore, this paper will analyze how the solar energy is an option to deal with the energy shortage in the fish processing units.

Table 1. The production capacity of marine and inland fisheries in Langsa [2]

No.	Districts	Inland Fisheries (Ton)	Marine Fisheries (Ton)
1	Langsa Timur	420.50	3,019.00
2	Langsa Barat	80.51	2,022.00
3	Langsa Kota	74.03	1,009.00
4	Langsa Lama	245.05	-
5	Langsa Baro	380.05	-
	Sum	**1200.14**	**6,050.00**

The solar market for power generation is still relatively new and it's still being studied. There are a limited number of technology providers and the access to technical data is restricted. In addition, it's necessary to analyze the energy impact of these new technologies before considering using them on a large scale. A photovoltaic system converts solar energy into DC power and the system consists of a PV array, a battery, and elements for power conditioning. There are two types of PV systems: grid connected and standalone. The grid connected type feeds electricity directly into the electrical network and operates parallel to the conventional energy source. These types of grid systems create clean electricity without transmission or distribution losses and without a need for batteries; its performance relies on the local climate, the orientation and inclination of the PV array, and inverter performance. By contrast, a standalone system has no interaction with a network grid and it's directly connected to the load; the use of a battery becomes necessary. The battery stores energy when the power supplied by the PV modules exceeds the load demand and it returns it when the PV supply is deficient [9]. There are a large number of researches focused on the analysis of the feasibility of solar energy use for power plants with PV panels, i.e. [1,4-7,10]. One of these studies was conducted in Serbia [8] to figure out the possibilities of generating electrical energy through 1MW PV power plants by using the different types of solar PV modules. They concluded that using CdTe solar modules generates more electricity.

Chandel et. all [3] analysed the potential and cost-effectiveness of a solar photovoltaic power plant to fulfil the energy demand of a garment zone in Jaipur, India. It was estimated that the energy demands of the zone for the year 2011 were at 2.21 MW, and the design of the solar PV power plant had a capacity of 2.5 MW. This requires an approximate of 13.14 acres of land. The on-site solar PV power plant internal rate of return (IRR) is 11.88%, NPV at a 10% discount rate and of 119.52 million INR; the simple payback period is 7.73 years and the discounted payback period is at 10% in 15.53 years, while the off-site power plant represents an IRR of 15.10%, NPV of 249.78 million INR, simple payback period of 6.29 years and discounted payback period of 10.14 years. Levelized energy cost is Rs. 14.94 and Rs. 11.40 per kWh for on-site and off-site solar PV plants respectively at a 10% discount rate, which is quite attractive.

Grid connected and standalone photovoltaic systems can be analysed and dimensioned by using a wide array of tools. System designers and installers tend to use simpler tools for sizing the PV system. On the other hand, most scientists and engineers typically use more involved simulation tools for optimization. Software tools related to photovoltaic systems can be classified into pre-feasibility analysis, sizing, and simulation. PVSYST is a dedicated PC software package for PV systems developed by the University of Geneva. It integrates pre-feasibility, sizing and simulation support for PV systems. After defining the location and loads, the user selects the different components from a product database and the software automatically calculates the size of the system [9].

In this work, PVSYST 5.41 is used to perform the simulations of standalone solar PV systems by comparing the energy production, performance ratio, efficiency and cost. Throughout this project, the potential and the cost-effectiveness of a solar photovoltaic power plant is analysed for meeting the energy demand of fish processing at a small scale in Kota Langsa; this process makes use of an ice factory unit, a gutting and cutting unit, a drying unit, and a curing unit.

2. Method

Location: The geographic location where the project was conducted is Kuala Langsa which is situated at: 04° 31' 4.98" N, 98° 0' 54.3" E, and elevation at 2 m (a.s.l.) shows figure1. PVSYST includes its own solar data for some locations.he general architecture of our proposed method is shown in figure 1.

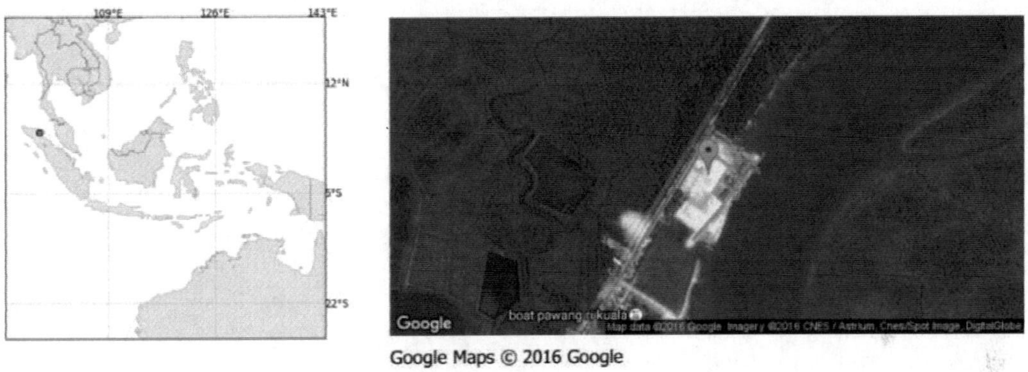

Figure 1. Location of small scale fish processing plant in Kota Langsa

In figure 2 the global horizontal irradiation and air temperature during a year can be observed. The global radiation components are direct, diffuse and reflected. The average daily sum of global irradiation on the horizontal surface is of 4.73 kWh/m^2 per day with a maximum value of 5.50 kWh/m^2 (March) and a minimum value of 3,852 kWh/m^2 (December). The diffuse component of radiation proved to be significant, especially during the months of Feb-Sept, while reflected radiation stayed relatively reduced throughout the rest of the year. The average yearly component of diffuse radiation is of 46.5 % of total global radiation.

Designing a standalone PV system doesn't entail many steps, but the preferred ones are as follows. Determination of the load defines the sum of electrical power used in the process of fisheries. Figure 3 shows the layout of equipment inside the fisheries' unit processing. Inside the unit, there are two unit freezers, one unit ice machine, one unit chiller, two units of fresh water pumps, a gutting and cutting unit, a drying unit, and a curing unit. The whole equipment will need 130 kW of electrical power, putting the daily average energy requirement at 3120 kWh.

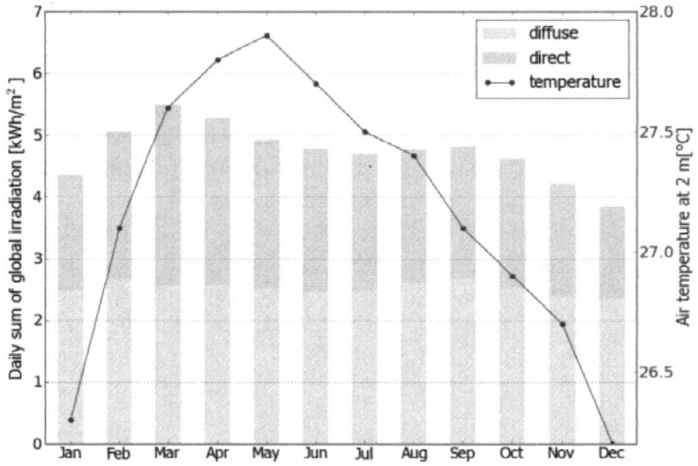

Figure. 2: Global horizontal irradiation on site

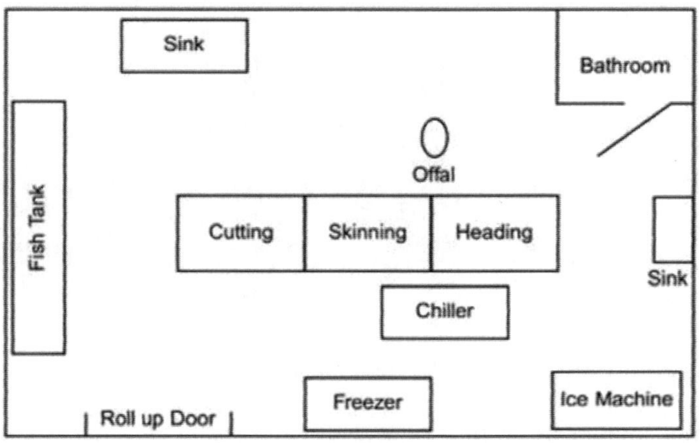

Figure 3. Layout of small-scale fish processing facility in Kota Langsa

The next step is to calculate the number of PV, battery and method of controller. The system defining parameter which was used is shown in table 2.

Table 2. System Defining Parameters

Module type	Standard
Technology	Poly-crystalline
PV module	120 Wp 28 V
Battery pack	Open 12 V/100 Ah

Using the aforementioned PVSYST software it was shown that the number of PV panels that should be provided to meet the energy demands was 935 modules, with the requirement of an area of about 0.75 hectares (1.85 acres).

Orientation: The panels are facing south (in this study), and the angle of the panels form an inclination with the ground. The energy usage between winter time and summer time is quite large at this latitude; on the contrary, the solar resource gap between winter and summer is not that considerable. That is the reason why the inclination needs to be optimized for the summer months. In this example, the plane tilt is about 150 and azimuth is about 00, as shown in figure 4.

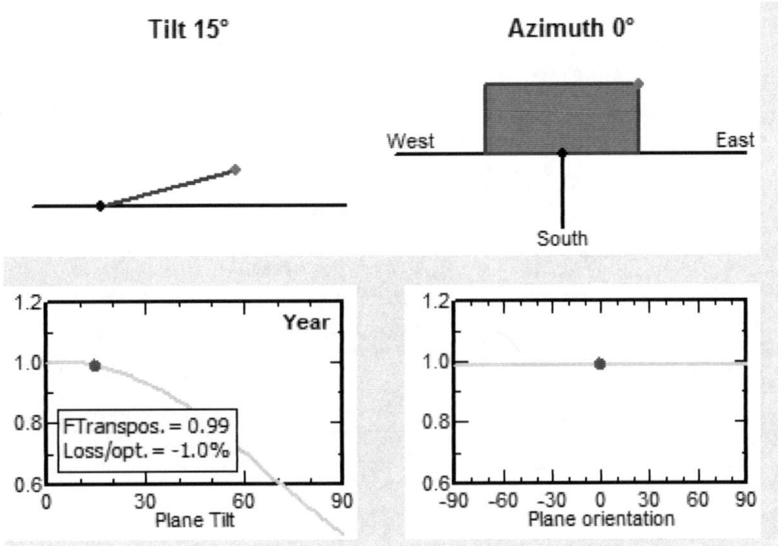

Figure 4. Orientation

The path of the sun in Kota Langsa throughout the year is represented in figure 5. The figure shows the terrain horizon, the module horizon, and an active area with civil and solar time. The module horizon, however, may have a shading effect on solar radiation.

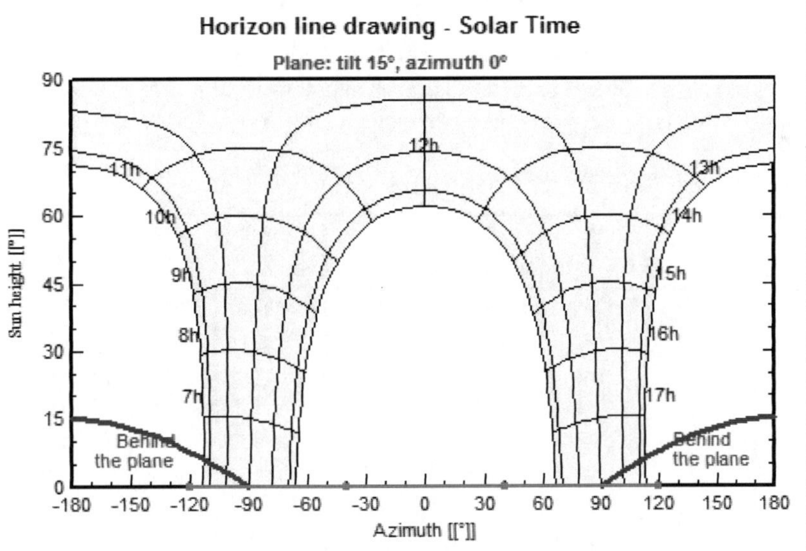

Figure 5. Insolation at location

System: A standalone PV system will depend on the demand of the user; the user has to input the desired nominal power, or the area that's available for installing PV modules. In the off-grid PV system, a battery needs to be chosen from the database. All the strings of PV modules should be homogeneously connected; this means that they should be identical modules, with the same number of modules in a series, the same orientation, etc. Figure 6 shows the schematic diagram of a standalone system. The diode shown here is the bypass diode used for the purpose of protecting and a back-up generator is used for emergency purposes.

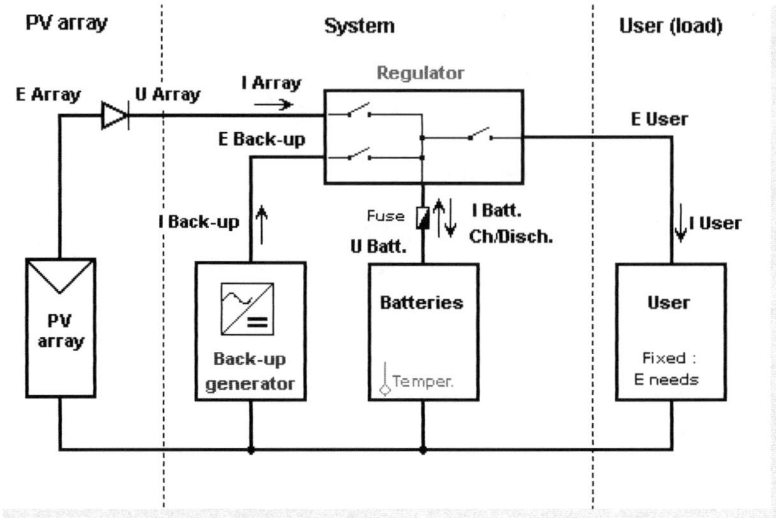

Figure 6. Schematic of the system

3. Result and discussion

This research is totally based on the PVSYST software and for modelling purposes. All the figures and tables depicted in this paper were generated during the simulation process. This paper only depicts the computational modelling, so that's why there's only the presentation of the results of the simulation rather than the description of it. Performance ratio (PR) is the ratio of the final PV system yield (Yf) and the reference yield (Yr). Table 3 shows the performance ratio of a grid connected system designed for the Kuala Langsa site. The yearly average performance ratio is 0.764.

Table 3. Normalized Performance Coefficients

	Yr kWh/m².day	Lu	Yu kWh/kWp/day	Lc	Ya kWh/kWp/day	Ls	Yf kWh/kWp/day	PR
January	4.30	0.139	4.30	0.956	3.35	0.240	3.11	0.722
February	4.68	0.444	4.68	1.392	3.29	0.182	3.11	0.664
March	5.11	0.433	5.11	1.463	3.65	0.545	3.11	0.607
April	4.51	0.306	4.51	1.180	3.33	0.224	3.11	0.689
May	4.20	0.091	4.20	0.910	3.29	0.187	3.11	0.739
June	4.05	0.016	4.05	0.776	3.27	0.166	3.11	0.767
July	4.14	0.000	4.14	0.788	3.36	0.249	3.11	0.750
August	4.16	0.000	4.15	0.792	3.36	0.257	3.11	0.748
September	4.29	0.236	4.29	1.061	3.23	0.120	3.11	0.724
October	4.61	0.253	4.61	1.166	3.44	0.483	2.96	0.642
November	4.13	0.112	4.13	0.883	3.24	0.137	3.11	0.753
December	3.72	0.053	3.72	0.749	2.97	0.359	2.61	0.702
Year	4.32	0.171	4.32	1.007	3.32	0.264	3.05	0.706

Table 4 shows the meteorological and incident energy of the PV system. The global horizontal irradiation (GlobHor) is of 1598.3 kWh/m2/year. The horizontal diffuse irradiation (DiffHor) is of 965.11 kWh/m^2. The overall global incident energy on the collector plane is of 1578.0 kWh/m^2.

Table 4. Meteor and Incident Energy

	GlobHor kWh/m²	DiffHor kWh/m²	T Amb °C	WindVel m/s	GlobInc kWh/m²	DifSInc kWh/m²	Alb Inc kWh/m²	DifS/GI
January	125.5	80.52	27.05	2.0	133.4	82.15	0.428	0.000
February	125.0	68.34	27.47	1.9	131.1	69.81	0.426	0.000
March	158.5	90.05	27.91	2.0	158.6	89.73	0.540	0.000
April	141.4	87.07	27.80	1.8	135.3	84.33	0.482	0.000
May	141.8	80.30	28.28	1.9	130.3	76.45	0.483	0.000
June	133.7	81.74	27.88	1.9	121.5	77.19	0.456	0.000
July	140.7	83.24	28.01	2.0	128.5	78.84	0.480	0.000
August	136.2	86.50	27.86	1.9	128.8	83.27	0.464	0.000
September	131.5	79.89	27.00	1.9	128.6	78.47	0.448	0.000
October	138.7	79.80	27.16	1.8	142.8	80.70	0.473	0.000
November	117.1	75.95	26.70	1.7	123.8	77.52	0.399	0.000
December	108.1	71.70	26.89	1.9	115.3	73.26	0.368	0.000
Year	1598.3	965.11	27.50	1.9	1578.0	951.72	5.446	0.000

Table 5 shows the detailed monthly average system losses in kWh. Module quality loss (ModQual) is of 33,515 kWh/year. Module mismatch loss (MisLoss) is of 13,071 kWh/year. Ohmic wiring loss (Ohm Loss) is of 15,543 kWh/ year. Array virtual energy at Maximum Power Point (MPP), EArrMPP is of 1,278,478 kWh/Year.

Table 5. Detailed system Losses

	ModQual kWh	MisLoss kWh	OhmLoss kWh	EArrMPP kWh	EArUfix kWh	EUnused kWh	EArray kWh
January	2845	1110	1319	108536	108536	4317	104219
February	2757	1075	1449	104989	104989	12495	92492
March	3338	1302	1710	127164	127164	13472	113692
April	2872	1120	1321	109554	109560	9205	100349
May	2761	1077	1224	105378	105378	2820	102557
June	2594	1011	1065	99072	99078	463	98609
July	2737	1067	1165	104502	104502	0	104502
August	2743	1070	1193	104728	104734	5	104723
September	2736	1067	1285	104357	104357	7112	97244
October	3016	1176	1524	114941	114947	7860	107081
November	2650	1033	1195	101116	101116	3367	97749
December	2467	962	1093	94141	94141	1644	92497
Year	33515	13071	15543	1278478	1278500	62762	1215714

Table 6 shows the balances and main results of the PV system employed. Yearly global horizontal irradiation is of 1,598.3 kWh/m2. The yearly global incident energy on the collector plane is of 1,521.1 kWh/m^2. Energy available at the output of the PV array is of 1.223.697 kWh. The energy supplied to the user is of 1,118,857 kWh. Energy the user needed = 1.138.971. SolFrac Solar fraction (EUsed / ELoad) = 98.2 %.

Table 6. Balances and main results

	GlobHor	GlobEff	E Avail	EUnused	E Miss	E User	E Load	SolFrac
	kWh/m²	kWh/m²	kWh	kWh	kWh	kWh	kWh	
January	125.5	128.9	103854	4317	0	96735	96735	1.000
February	125.0	127.0	100687	12495	0	87373	87373	1.000
March	158.5	153.2	122470	13472	0	96735	96735	1.000
April	141.4	130.3	105233	9205	0	93614	93614	1.000
May	141.8	125.2	100707	2820	0	96735	96735	1.000
June	133.7	116.4	94685	463	0	93614	93614	1.000
July	140.7	123.3	99838	0	0	96735	96735	1.000
August	136.2	123.9	100057	5	0	96735	96735	1.000
September	131.5	124.1	99920	7112	0	93614	93614	1.000
October	138.7	138.1	110069	7860	4707	92027	96735	0.951
November	117.1	119.4	96648	3367	0	93614	93614	1.000
December	108.1	111.3	89528	1644	15407	81328	96735	0.841
Year	1598.3	1521.1	1223697	62762	20114	1118857	1138971	0.982

Figure 7 shows a graphic the Performance Ratio of the Incident energy for the entire month. The average Performance ratio is 0.706.

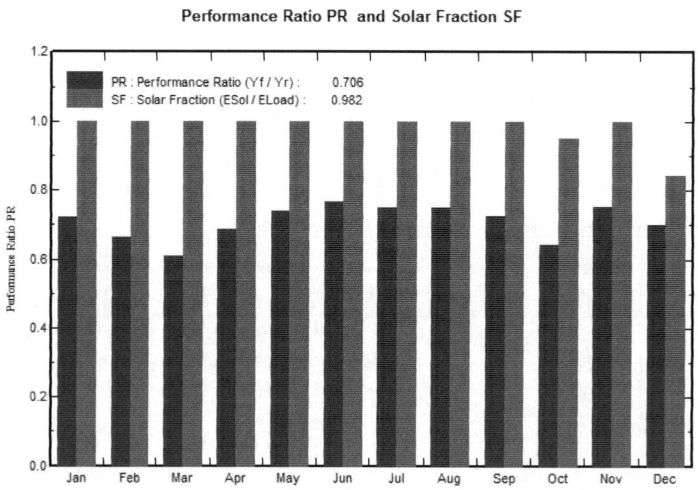

Figure 7. Performance Ratio PR and Solar Fraction

Figure 8 represents the overall system loss diagram for small scale fish processing at Kota Langsa. The horizontal global irradiation is of 1,598 kWh/m^2. The effective irradiation on the collector plane is of 1,521 kWh/m^2. Then the PV cell converts solar energy into electrical energy. After PV conversion, array nominal energy is of 1,528 MWh. Effective energy at the output of the array is of 1,216 MWh. After the inverter loss, the battery storage was of 1,161 MWh. Therefore the energy injected into the fish processing plant is 1,119 MWh.

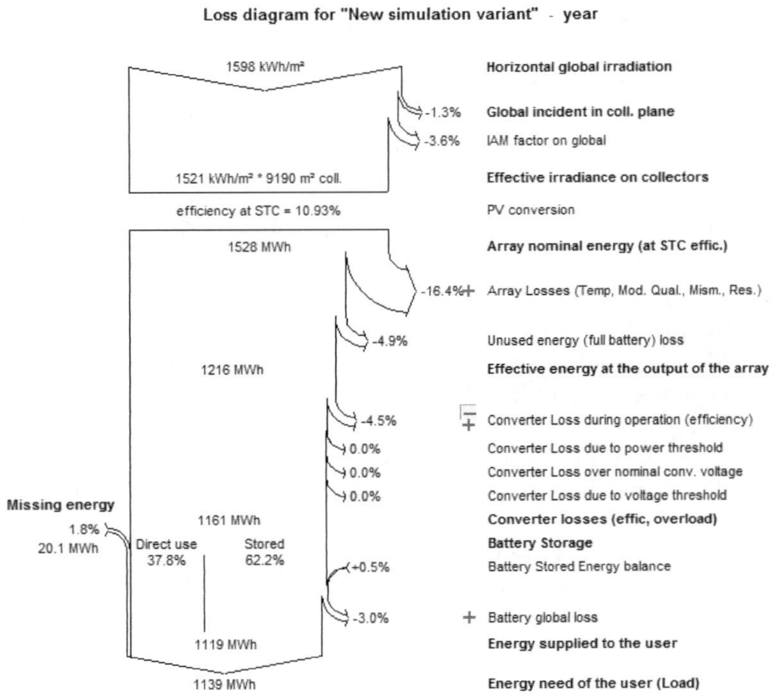

Figure 8. The overall system loss diagram for small scale fish processing at Kota Langsa.

4. Conclusion

Fisheries are leading sectors in the municipality of Langsa; the fish processing industry comprises an ice factory unit, a gutting and cutting unit, a drying unit, and a curing unit. However, the energy and electricity costs during the production process have become major constraints. In this paper, the solar energy as an option to deal with the energy shortage is studied. PVSYST 5.41 is used to perform the simulations of standalone solar PV system by comparing the energy production, performance ratio, efficiency and cost. The potential and the cost-effectiveness of a solar photovoltaic power plant can likewise be estimated.

The result of the analysis shows that for the installation of OTR, the levelized cost is IDR. 1,115 per kWh, taking into consideration the 25 years of life-span at a 10% of discount rate, with a simple payback period of 13.2 years. As for OTL, levelized energy was obtained at IDR.997.5 per kWh, with a simple payback period of 9.6 years.

Acknowledgement

Special thanks Ministry of Research and Technology and Higher Education for fully support this research through HI-link program.

References

[1] Bhoye H, Sharma G 2014 An Analysis of One MW Photovoltaic Solar Power Plant Design *International Journal of Advanced Research in Electrical, Electronics and Instrumentation Engineering* **3**(1) 6969–6973.

[2] BPS-Statistics of Langsa Municipality 2016 *Langsa Municipality in Figures 2016* (BPS-Statistics of Langsa Municipality **1**)

[3] Chandel M, Agrawal G D, Mathur S, Mathur A 2014 Techno-economic analysis of solar photovoltaic power plant for garment zone of Jaipur city *Case Studies in Thermal Engineering* **2** 1–7

[4] Ganguli S, Sinha S 2010 Design of a 9 kWp Grid Connected Solar Photovoltaic Power Plant Using 100 M2 Available Area In Patiala *Journal of Engineering Research and Studies* **II**

[5] Khelif A, Talha A, Belhamel M, Hadj Arab A 2012 Feasibility study of hybrid Diesel-PV power plants in the southern of Algeria: Case study on AFRA power plant *International Journal of Electrical Power and Energy Systems* **43**(1) pp 546–553

[6] Lan H, Liao Z, Yuan T, Zhu F 2012 Calculation of PV Power Station Access *Energy Procedia* **17** pp 1452–1459

[7] Mekhilef S, Saidur R, Safari A 2011 A review on solar energy use in industries *Renewable and Sustainable Energy Reviews* **15**(4) pp1777–1790

[8] Pavlović T, Milosavljević D, Radonjić I, Pantić L, Radivojević A, Pavlović M 2013 Possibility of electricity generation using PV solar plants in Serbia *Renewable and Sustainable Energy Reviews* **20** 201–218

[9] Ramoliya J V 2015 Performance Evaluation of Grid-connected Solar Photovoltaic plant using PVSYST Software *Journal of Emerging Technologies and Innovative Research (JETIR)* **2**(2) pp 372–378

[10] Tarigan E, Dwi F 2015 Techno-Economic Simulation of a Grid-Connected PV System Design as Specifically Applied to Residential in Surabaya, Indonesia *Energy Procedia* 65 pp 90–99

Performance analysis of low temperature heat source of organic Rankine cycle for geothermal application

A Pintoro[*1] , H Ambarita[1,2], T B Nur[1,2] and F H Napitupulu[1]

[1]Department of Mechanical Engineering, Faculty of Engineering, University of Sumatera Utara, Indonesia
[2] Sustainable Energy and Biomaterial Center of Excellence, Universitas Sumatera Utara, Padang Bulan, Medan 20155, Indonesia.

[*]E-mail: andi.pintoro1967@gmail.com

Abstract. Indonesia has a high potential energy resources from geothermal activities. Base on the report of Asian Development Bank and World Bank, the estimated of Indonesian hydrothermal geothermal resource considered to be the largest among the world. If it's can be utilized to produce the electric power, it's can contribute to increasing the electrification rates in Indonesia. In this study, an experimental studied of electric power generation, utilizing the Organic Rankine Cycle (ORC) system to convert the low level heat of hydrothermal as an energy source. The temperature of hydrothermal was modelled as hot water from water boiler which has a temperature range from 60 °C – 100 °C to heat up the organic working fluid of ORC system. The system can generated 1,337.7 watts of electricity when operated using R134A with hot water inlet temperature of 100 °C. Changing system working fluid to R245fa, the net power obtained increase to 1,908.9 watts with the same heat source condition. This study showed that the ORC system can be implemented to utilize low temperature heat source of hydrothermal in Indonesia.

1. Introduction

Electrical energy has a very important and strategic role to support the modern society and national development in Indonesia. Therefore the availability of electricity must be reliable, sustainable and environmentally friendly. In Indonesia, the demand of electrical energy is higher than the supply [1]. This situation has an effect on acceleration of development in Indonesia especially in remote area which is not reachable by electricity.

It has been known Indonesia has a lot of potential heat sources that can be developed into electrical energy, such as solar and geothermal energy. Among the various heat sources, a low-temperature heat source from geothermal is great interest to be developed as a solution to the fulfillment of energy needs in remote areas that possess it.

Organic Rankine Cycle (ORC) have widely been studied to harness low-temperature heat source that provide temperature ranging between 40°C and 200°C. An ORC has the same components as a conventional Rankine cycle, such as a pump to raise the pressure of the working fluid, an evaporator to absorbing the heat, turbine (or expander) to convert the heat energy contained in the working fluid to be shaft power and condenser to reject the heat from the working fluid. The major difference is the choise of the working fluid, ORC use organic component instead water. Lakewand Bolland [4] analyzed the power production capability and equipment size requirements for R134a, R123,

R227ea, R245fa, R290 and n-pentane. They used a subcritical Rankine cycle without superheating, specified the heat source temperature in the temperature range from 80°C to 200 °C and used the evaporator pressure as independent parameter. Their results show that the selection of working fluids depends on the type of heat source, the temperature level and the design objective. N-pentane can be either the maximum power output (turbine) or the smaller heat exchangers. It was shown that the maximum power is obtained for the optimal evaporator pressure. Furthermore, a working fluid may have the smallest turbine size factor but requires a large heat exchanger area. The authors concluded that an economic study is necessary to determine which working fluid is the most appropriate. Qiu et al. [2] gave an overview of expansion devices for micro-CHP ORC systems and concluded that both scroll and vane expanders are good choices for systems within the capacity range of 1-10 kW. In particular, scroll machines are well adapted to small-scale ORC applications and offer significant advantages such as reliability and robustness (reduced number of moving parts), as well as ability to handle high pressure ratios and can handle the presence of a liquid phase in the flow.

The objective of this experimental study is to analyse the performance of the designed electric power generation, utilizing the Organic Rankine Cycle (ORC) system to convert the low-temperature heat of geothermal in form of hydrothermal as an heat energy source to be shaft power, and the system performance will compared with using two different working fluid such as R134a and R245fa.

2. System Description

The designed Organic Rankine Cycle (ORC) system consist of three loops, as can be seen in Figure 1. The first loop is the refrigerant loop, in this loop, the liquid refrigerant as a working fluid (R134a or R245fa) is pressurized by the feed pump. The pressurized liquid than heated in the tubing of the evaporator by the hot water that circulate in the evaporator shell until change the phase to be saturated vapor. The fluid leaves the evaporator in dry saturated condition and enters the modified scroll compressor as an expansion device. The expansion process drives an electrical generator to produce electric power. Leaving the expander, the working fluid enter the condenser to be cooled by the cooling water until change the phase to be saturated liquid and than pump again to the evaporator. The second loop is the heat source loop, in here, the hot water from the water boiler (simulating of low-temperature hydrothermal heat source) circulate to the evaporator. And the third loop is cooling water loop, in here, the cooling water circulates to absorb heat from working fluid in condenser and than reject the heat in the cooling tower.

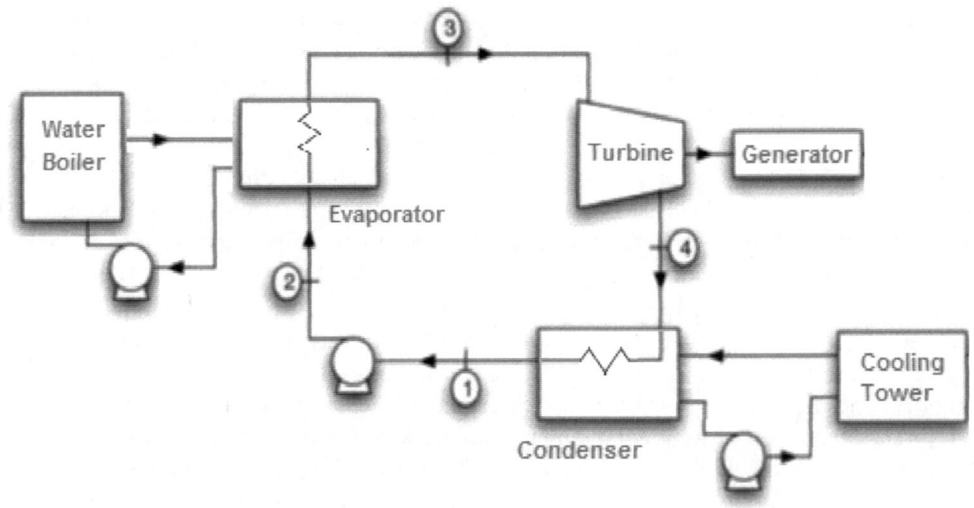

Figure 1. Schematic diagram of ORC System.

IOP Conf. Series: Materials Science and Engineering **308** (2018) 012026 doi:10.1088/1757-899X/308/1/012026

3. Methodology

3.1 Experimental Set-Up

Base on a required condition, an small-scale ORC power generating was designed, constructed and tested. Figure 2 is the schematic diagram of the experimental system. The experimental prototype is mainly comprised of an evaporator, a water boiler, a gas tank, an expander, a throttling valve, a condenser, a liquid tank, an feed pump, a cooling tower, two water pumps and the interrelated measurement and data acquisition system.

A vane pump is used to feed and pressurized the working fluid. It can provide maximum operating pressure of 28 bar and rotational speed of 1200 rpm. The working fluid mass flow rate set to be 450 kg/h for all level of heating temperature.

The water boiler was used to simulated the hydrothermal heat source. The hot water from water boiler was supplied to heat up the working fluid in the evaporator coil pipes use a hot water pump with constant mass flow rate of 2000 kg/h and the hot water temperature varying between 60 ^{o}C to 100 ^{o}C.

After heat up in the evaporator, the superheated working fluid vapor collect in to the gas tank temporarily whose volume is 10 litres and the allowable working pressure is 30 bar. The use of gas tank makes the expander inlet pressure much more stable.

In the expander, the superheated vapor of working fluid expand to gain the mechanical work at the shaft. The scroll expander is coupled to AC-generator by belt and use some incandescent bulb as working load.

After expansion process, the low-pressure working fluid vapor flows to condenser to be cooled , water as the cooler media circulate by the cooling water pump from condenser to cooling tower which have cooling capacity 100 kW. After condensing process, the liquid working fluid gathered in liquid receiver tank to pump again to evaporator.

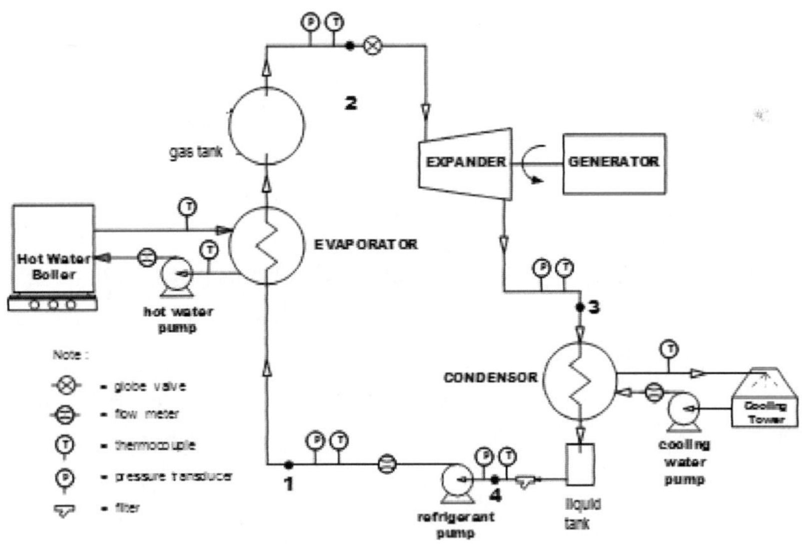

Figure 2. Diagram of experimental set-up.

3.2 The Measurement system

As shown in Figure 2, T-type thermocouples and pressure transmitters are installed at different positions in the prototype (stream 1 – 4), to measure the temperatures and pressures of the working fluid. And also the turbine flow meter used to monitor the mass flow rate of the working fluid, hot water and cooling water.

All the output signals of the experimental data can be automatically transported to the computer and recorded as functions of time in the computer, utilizing an Cole Palmer 18200-20 series data logger.

3.3 Thermodynamic Analysis of ORC System

Figure 3 illustrates the thermodynamic process for ORC system. A theoretical Rankine cycle consists of the following processes: 1 → 2: Compression (working fluid feed pump); 2 → 3: Heat supply (geothermal heat source); 3 → 4: Expansion (expander); 4 → 1: Heat rejection (condenser).

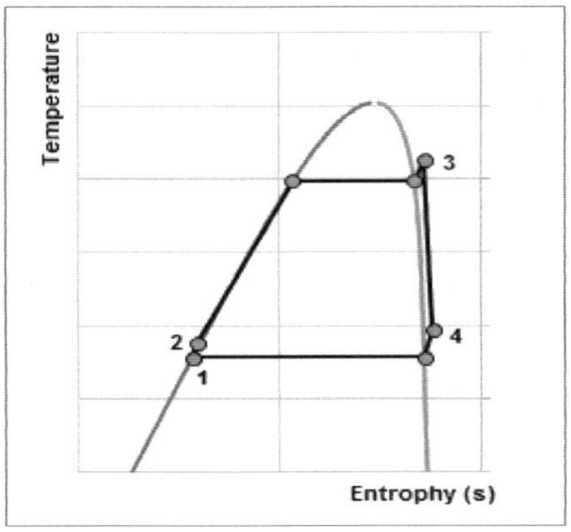

Figure 3. T-s diagram for ORC System.

The analysis of the cycle consists of applying mass and energy balances to each of the processes mentioned above. Process 1 → 1 is the work done by the working fluid pump:

$$W_p = (h_3 - h_4)/\eta_p \tag{1}$$

Process 2 → 3 is is the heating process in the evaporator where geothermal heat transferred to the working fluid:

$$Q_{evap..} = (h_4 - h_1) \tag{2}$$

Process 3 → 4 is the actual expansion of the working fluid through the turbine:

$$W = (h_1 - h_2).\eta_m.\eta_s \tag{3}$$

Process 4 → 1 is the condensation process which occurs within counter flow heat exchanger using cooling water at 28 °C so the heat transferred to the cooling water (heat out) is:

$$Q_{cond.} = (h_2 - h_3) \tag{4}$$

The net electrical power produced by the ORC unit is:

$$W_{cycle} = W_{exp} - W_p - W_{p1(hw)} - W_{p2(cw)} \tag{5}$$

The thermal efficiency of the ORC unit is:

$$\eta_{cycle} = W_{cycle}/Q_{evap} \tag{6}$$

4. Results and discussion

4.1 Measured experiment data

Base on the data from the experiment data and thermodynamic analysis, Figure 4 shows the impact of evaporator hot water inlet temperature variations from 60°C to 100°C to the cycle performance and expander power output. As the hot water inlet temperature increased, the expander power output and system efficiency also increased. This effect is the dominant effect accompanied by increasing the working fluid enthalpy mean while the work of the feed pump still remains ($\dot{m}_f$ set to be constant).

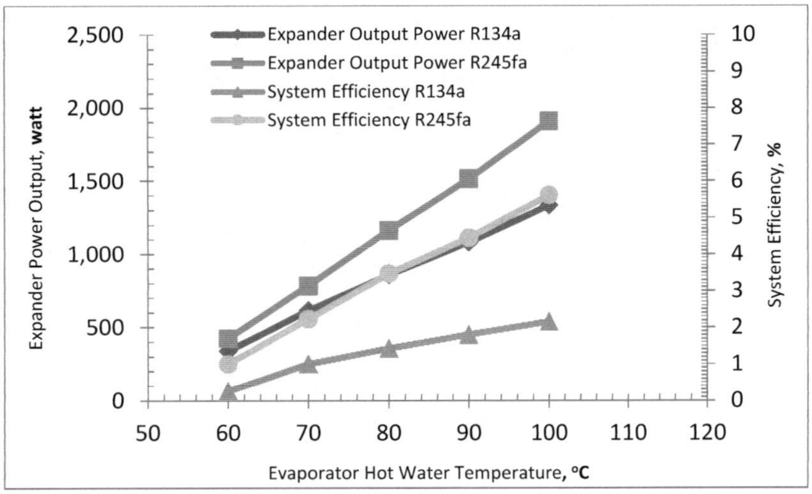

Figure 4. Impact of the hot water temperature variation to expander power output and system performance

From Figure 5 can be seen that with the same heating temperature, R134a and R245fa have almost the same expander inlet temperature but R245fa have much better system efficiency.

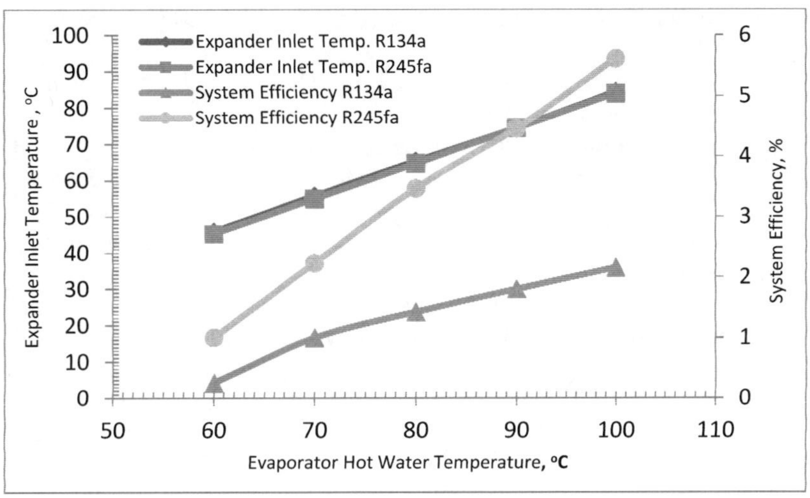

Figure 5. Impact of the hot water temperature variation to expander inlet temperature and system performance

4.2 Comparison of the Output for the Two WorkingFluids

The result of the thermodynamic calculation, base on experiment measured data, shows the ORC system used R245fa as working fluid, have a better performance than R134. The highest net output is $P_{out,\ net}$ = 1,908.9 watts with η_{cycle}= 5.62 %. The comparison of the two-working fluid can be seen on Table 1.

Table 1. Comparison of the cycle performance betwen R134a and R245fa.

No	Item	R-134a	R-245fa
1	Temperature of hot water (T_{hw} / °C)	100	100
2	Expander inlet temperature ($T_{in\ expd}$ / °C)	84,45	84,04
3	Expander outlet temperature ($T_{out\ expd}$ / °C)	39,13	55,16
4	Expander inlet pressure ($P_{in\ expd}$ / bar)	26,2	8,6
5	Expander inlet pressure ($P_{out\ expd}$ / bar)	8	2
6	System efficiency (η_{cycle} / %)	2,16	5,62
7	Expander efficiency ($\eta_{eksp.}$ / %)	67,37	84,13
8	Fluid mass flow rate ($\dot{m}_f$ / kg/h)	450	450
9	Hot water mass flow rate ($\dot{m}_{hw}$/ kg/h)	2000	2.000
10	Expander diffrence pressure (ΔP / bar)	18,2	6,6
11	Net Power / $P_{out,net}$ (watt)	1334,7	1908,99

5. Conclusion

It has been shown in this study that it is possible to use an organic Rankine cycle to produce electrical power from a low temperature geothermal source. A comparison result between two working fluids, R-134a and R-245fa, revealed that the refrigerant R-245fa had a better performance.

Increasing the temperature of heat source will increase the power output and system efficiency. However, many tasks should be done to see the effect of changing design parameters on the net power output and total system efficiency. Plans for the future work to thoroughly analyse and describe the sensitivity analysis for total ORC.

References

[1] BPPT 2014 *Indonesia Energy Outlook 2014* (Center for Energy Resources Development Technology - PTPSE)

[2] Qiu G, Liu H, Riffat S 2011 Expanders for micro-CHP systems with organic Rankine cycle *Appl. Therm. Eng.* **31** pp 3301-3307

[3] Lemort V, Declaye S, Quoilin S 2012 Experimental characterization of a hermetic scroll expander for use in a micro-scale Rankine cycle *Proc. IMechE Part J. Power Energy* **226** pp 126 - 136

[4] Lakew A A, Bolland O 2010 Working fluids for low-temperature heat source *Appl. Therm. Eng.* **30** pp 1262-1268

[5] Moran M J, Shapiro H N 1993 *Fundamentals of Engineering Thermodynamics, Second Edition*, SI Version (John Wiley & Sons. Inc. New York)

[6] Chang J C, Hung T C, He Y L, Zhang W. 2015 Experimental study on low-temperature organic Rankine cycle utilizing scroll type expander *Applied Energy Journal* **29** pp 167-171

[7] Badr O, O'Callaghan PW, Hussein M, Probert S D 1984 Multi-vane expanders as prime movers for low-grade organic Rankine cycle engines *Applied Energy* **16** pp 129–46

[8] Madhawa H D 2013 Optimum Design Criteria for an ORC using Low-Temperature Geothermal Heat Sources *Journal of Renewable Power Source* **1**:8 pp 143-175D

[9] Sylvain Q, Vincent L, Lebrun J 2010 Experimental study and modeling of an Organic Rankine Cycle using scroll expander *Applied Energy* **87** pp 1260–1268
[10] Wang J, Yan Z, Wang M, Ma S, Dai Y 2013 Thermodynamic analysis and optimization of an (organic Rankine cycle) ORC using low grade heat source *Energy* **49** pp 356–65

IOP Conf. Series: Materials Science and Engineering **308** (2018) 012027 doi:10.1088/1757-899X/308/1/012027

Effectiveness of a heat exchanger in a heat pump clothes dryer

A H Nasution[1], P G Sembiring[1], H Ambarita[1,2*]

[1]Mechanical Engineering, University of Sumatera Utara, Kampus USU Medan 20155, Indonesia
[2]Sustainable Energy Research Center, Faculty of Engineering University of Sumatera Utara, Kampus USU Medan 20155, Indonesia

*E-mail: himsar@usu.ac.id

Abstract. This paper deals with study on a heat pump clothes dryer coupled with a heat exchanger. The objective is to explore the effects of the heat exchanger on the performance of the heat pump dryer. The heat pump dryer consists of a vapor compression cycle and integrated with a drying room with volume 1 m^3. The power of compressor is 800 Watt and the refrigerant of the cycle is R22. The heat exchanger is a flat plate type with dimensions of 400 mm × 400 mm × 400 mm. The results show the present of the heat exchanger increase the performance of the heat pump dryer. In the present experiment the COP, TP and SMER increase 15.11%, 4.81% and 58.62%, respectively. This is because the heat exchanger provides a better drying condition in the drying room with higher temperature and lower relative humidity in comparison with heat pump dryer without heat exchanger. The effectiveness of the heat exchanger is also high, it is above 50%. It is suggested to install a heat exchanger in a heat pump dryer.

1. Introduction

Drying is a process to reduce moisture content by evaporating and uses amount of energy to overcome the latent heat evaporation. The drying process are used in several applications such as industrial drying, agricultural product drying, manufacturing drying and clothes drying. One of the biggest drying applications is clothes drier. This application consumes big amount of energy by using several energy sources such as fossil fuel and electricity. Meyers et al. [1] and Yadav and Moon [2] reported that in the USA 71 Terawatt hours (TWh) of electricity per year electricity is consumed for drying clothes. This energy is up to 9% of electricity consumption in the USA. To the best knowledge of the authors, there is no report on energy consumption of electricity for drying clothes in Indonesia. However, based on survey there is an increasing number of commercial laundry in big cities in Indonesia. These laundries use commercial drying instead of typical natural drying. Several types of clothes drier are found in literature and tumbler rotating drum and flowed by hot air of 40-60°C is the most used system [3].

One of the most promising artificial clothes drier is vapor compression heat pump drier. Here, the heat from the condenser of vapor compression cycle is used as a heat source for drying process. Minea [4] reported the review on the heat pump drier and system integration. It was shown that a lot of unworkable, unnecessarily complicated or doubtful improved drying heat pump concepts have been proposed. The true challenge is first to make heat pump drier and its combination feasible by providing relevant information about integration, schedules, control strategies, operating parameters and energy performance issues. Braun et al [5] has reported their study on energy efficiency analysis of air cycle heat pump dryers. Two types of typical cloth dryers were compared the first type is conventional tumbler dryer using electric heater and the second one is reversed Brayton cycle. Ameen and Bari [6] investigated the feasibility of drying clothes using waste heat from a condenser from a typical split-type residential Air-Conditioner (AC) typically used in high rise urban apartments. Three drying methods of clothes were compared, they are commercial dryer using 1 kW electric heater, indoor natural drying, and heat pump dryer. The results showed that the drying rate for the three methods commercial dryer, natural dryer, and heat pump dryer wer 0.319 kg/hr, 0.139 kg/hr, and 0.424 kg/hr, respectively. The drying time of the tested methods varied from 120 minute to 390 minutes. They proposed the parameters energy consumption (SEC) rate in kWh/kg and specific moisture extraction rate (SMER) in kg/kWh to compare the performance of the methods. Since this study, SEC and SMER typically used as a performance comparison in heat pump dryer.

Deng and Han [7] carried out an experimental study on clothes dryer using rejected heat from split-type residential AC. An experimental rig has been designed and set up to perform the study. The residential AC with specification of 6.4 kW of cooling capacity was used. It is a typical size of AC for a room of up to 30 m^2 in Hongkong. In the experiments, the initial weight of clothes dried was about 3 kg. Two drying methods were compared, electrical clothes dryer and their proposed method. The electric consumption and drying time from both methods were compared. The results revealed that there was an additional electric use of residential AC, but it was only 1.2% in comparison with AC without dryer. Mahlia et al [8] studied experimentally clothes dryer using heat wasted from split-type residential AC. They proposed a system which consists of a drying room and a residential AC which is moveable. The experimental data were used to compare the effectiveness of the system proposed with a conventional dryer in terms of drying time and specific energy consumption. It was shown that drying time varied from 70 minutes to 420 minutes. The drying rate varied from 0.56 kg/hr to 0.75kg/hr with residential AC, on the other hand it was 0.13 kg/hr for indoor drying and 0.18 kg/hr for out-door drying. In addition, SMER varied from 0.1809 kg/kWh to 0.2205 kg/kWh. The residential AC clothes dryer is claimed more efficient way to dry clothes. This system results in shorter drying time and high effectiveness. As a note in their study, the increase power due to additional fan was not taken into account. Suntivarakorn et al [9] studied on a clothes dryer using waste heat from split-type residential AC. The results showed that the drying rate was varied between 1.1 kg/hr to 2.26 kg/hr. Based on these rate, it was claimed that the system has a better performance than commercial and natural dryers. Ambarita et al. [10] studied experimental the performance of a clothes dryer by utilizing waste heat from a split-type residential AC. A drying room with a volume of 1 m^3 has been designed and fabricated. The waste heat from the condenser of a resdential AC with compressor power of 800W was utilized as a heat source. The tested clothes were made of pure cotton with initial weight varied 3.05kg, 5.25 kg, 6.21 kg, 8.22 kg, and 10.22 kg. Two different drying methods, single inlet and multi inlets, have been tested. The experiments showed that the drying time varies from 80 minutes to 410 minutes. The experiments with single inlet the averaged drying time, optimum initial weight, optimum drying rate and optimum SMER were 242 minutes, 6.21 kg, 0.868 kg/hr, and 2.345 kg/kWh, respectively. While with multi inlets, those parameters were 222 minutes, 8.22 kg, 0.922 kg/hr, and 2.492 kg/kWh, respectively.

The above reported studies showed that heat pump dryer can be used for drying with two different methods, pure heat pump system [5] and waste heat system [6-10]. However, studies are still needed to improve the effectiveness and the efficiency. Recently, Ambarita et al. [11] reported their study on the performance and characteristics of heat pump clothes dryer. In the study, a heat exchanger was employed in the system. However, the effect of the installed heat exchanger was not explored. In the present study, the effectiveness of the heat exchanger in a heat pump clothes dryer is investigated. The objective is to explore the effect of the heat exchanger to the heat pump dryer and the effectiveness of the heat exchanger. The results are expected to supply the necessary information in developing an efficient and effective artificial drying.

IOP Conf. Series: Materials Science and Engineering **308** (2018) 012027 doi:10.1088/1757-899X/308/1/012027

2. Methods

2.1. Experimental Apparatus

In this study, a lab scale model of heat pump drier with overall dimensions of 3000 mm × 1000 mm × 1800 mm has been designed and fabricated. The main components are the compressor, condenser, expansion valve, evaporator, fan, a flat plate heat exchanger, and drying chamber.The drying room has a volume of 1 m^3 with dimensions of 1400 mm x 1000 mm x 1080 mm. The heat pump is a modified split-type of residential AC with commercial made by Samsung and the model is AS09TUQX. The specifications of residential AC are shown in Table 1. The power of the fan is 105 W. The heat exchanger is a flat plate type with dimensions of 400 mm × 400 mm × 400 mm. It is made of zinc plate with a thickness of 0.35 mm, the distance between the flat plates is 4 mm, and the number of gap is 92.

Table 1. Specification of heat pump

Parameter	Value
Compressor Power (Averaged)	800 W
Cycle	Vapor Compression Cycle (VCC)
Input Voltage/Frequency	220 -240 V/50 Hz
Electric Current (Maximum)	4.7 Ampere
Dimension of the Condenser	660 mm × 470 mm × 240 mm
Dimension of the Evaporator	820 mm × 283 mm × 210 mm

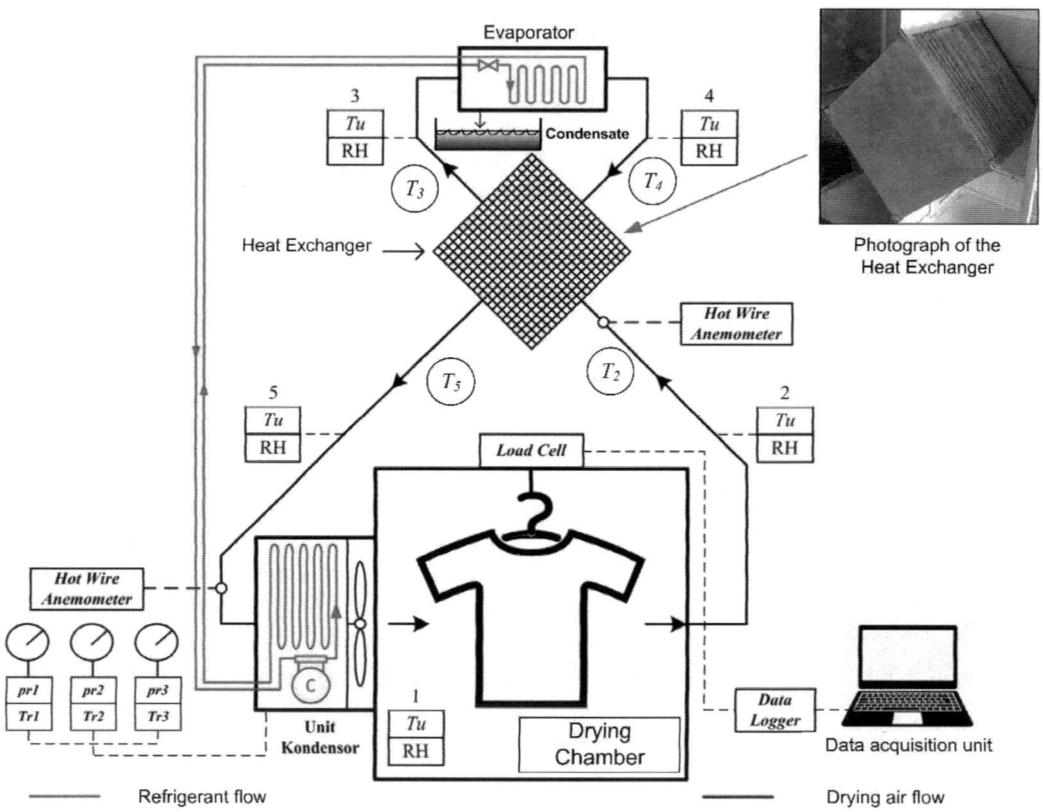

Figure 1. Experimental apparatus and data acquisition unit.

In order to perform the experiment, a data acquisition unit is installed on the experimental apparatus. The temperature and the humidity of the drying air are measured using a Portable Thermohygrometer with accuracy 1.0ºC and 5.0% RH. The temperature and humidity are measured at five locations. They are located in the drying room, at the inlet of the heat exchanger, at the evaporator entrance, at the exit of the evaporator, and at the heat exchanger exit. The velocity of the drying air is measured using a DT-8880 Hot wire anemometer. The specification of the anemometer are as follows, speed range and accuracy are 0.3 to 30 m/s and ± 3%, respectively. The weight history of the clothes dried is recorded using S-type Load cell with an accuracy of 1.0 gram. The load cell is placed on the top of the drying room. The pressure of the refrigerant is measured using a pressure gauge at three positions, at the compressor suction line, discharge line and the line after condenser. The schematic diagram of the experimental and the data acquisition unit are shown in Figure 1. In the figure, the photograph of the heat exchanger is shown.

Drying air flows into the drying room at point 1 and absorbs moisture from clothes.The air that contains high moisture content at point 2 is then directed past the heat exchanger (2 to 3) and then across the evaporator coil (3 to 4).Heat exchanger works as a heat recovery to air at 4 to 5 points.During the dehumidification process (the humidity) from point 3 to point 4, the first time the air conditioning cooled the sensible until the dew point (dew point). The cooling process will produce condensation of water vapor contained in the air dryer. Latent heat of evaporation of water is then absorbed by the evaporator to boil the refrigerant.Heat is recovered will be pumped to the condenser. The air cooled and lowered the humidity will absorb the heat that is released condenser from point 5 to point 1 as heating sensible to raise the temperature of the drying air.

2.2. Performance Parameters

During the experiments, the temperatures history, weights history, and air velocity in the drying room will be measured and recorded by the data acquisition system. In the vapor compression cycle, the voltage, electric current, and the pressure of refrigeration line will be measured. These measured data will be used to analyse the performance parameters. The performance parameters that used in the present study are explained as follows.

The first drying performance is specific moisture extraction rate (SMER). This parameter is defined as the ratio of moisture removed from the clothes to energy consumstion by the system. In the present system, the energy consumption for removing the moist from the clothes consists of heat released from the condenser and the energy rate to run the auxiliary fan ($\dot{W}_f$). Since the heat released from the condenser is the heat of vapor compression cycle as a byproduct of the system, the input power of compressor ($\dot{W}_c$) will be used to calculate the SMER [10]. Thus, SMER [kg/kWh] is given by the following equation:

$$\mathrm{SMER} = \frac{\dot{m}_d}{\dot{W}_c + \dot{W}_f} \tag{1}$$

Where $\dot{m}_d$ [kg/hr] is the drying rate of the clothes dried and it is calculated by the following equation:

$$\dot{m}_d = \frac{m_0 - m_{dr}}{\Delta t} \tag{2}$$

where m_0, m_{dr} and Δt are the mass of the clothes at intial time (wet) and when it is dry and drying time, respectively. In the analysis, non-dimensional (MR) weigh of the clothes is used. It is modeled using the below equation.

$$MR = \frac{m - m_{dr}}{m_0 - m_{dr}} \tag{3}$$

Heat pump dryer performance expressed in total achievement (TP), which is calculated by the below equation:

$$TP = \frac{\dot{Q}_k + \dot{Q}_e + \dot{Q}_{HE}}{\dot{W}_c + \dot{W}_f} \tag{4}$$

Where $\dot{Q}_k$ and $\dot{Q}_e$ are the amount of heat that is released by condenser and heat absorbed by the evaporator, respectively. And $\dot{Q}_{HE}$ is the total heat rate from hot flow to cold flow in the heat exchanger. The total heat flow rate in the heat exchanger consists of sensible and latent heat which can be calculated as follows.

$$\dot{Q}_{HE} = \dot{Q}_s + \dot{Q}_l = \dot{m}_h c_{p,h}\left(T_{h,i} - T_{h,0}\right) + \dot{m} h_{fg}\left(w_{in} - w_{out}\right) \tag{5}$$

Where $\dot{m}_h$ [kg/s] is the mass flow rate of air, $c_{p,h}$ [kJ/kg K] is the specific heat of the air, $T_{h,i}$ [K] and $T_{h,o}$ [K] are the temperature of air entering and leaving from the heat exchanger. The parameters h_{fg} [kJ/kg], w_{in} [kg/kg] and w_{out} [kg/kg] are the enthalpy vaporization, humidity of the air in and out, respectively.

The effectiveness of the heat exchanger is defined as the ratio of the actual heat transfer rate to the possible maximum heat transfer rate.

$$\varepsilon = \frac{\dot{Q}}{\dot{Q}_{max}} \tag{6}$$

As a note, the present heat exchanger is categorized a flat plate and cross flow type heat exchanger where the hot and cold fluids are unmixed. To the best knowledge of the author there is no empirical correlation of effectiveness can be found in the literature. Thus, the experimental measurement will be used to make the analysis.

The experimental method is explained in brief as follows. In the beginning the heat pump drier will be operated without any drying load or empty drying room for 30 minutes. This is known as idle operation time. After the idle operation time, the wet clothes which are pre-dried using washing machine are loaded to the drying room. The wet clothes are hanging inside the drying room using typical clothes hanger. The experiment is then started, and all of the temperatures, weight, current, and hot air velocity were measured and recorded using the data acquisition unit. In this study, the dryer load is the clothes made of pure cotton with initial weight of 245 gr.

3. Results and Discussions

The experiments have been carried out for the case heat pump drier without heat exchanger and the case where the heat pump drier with heat exchanger. The results will be discussed in the following sections.

3.1. *Temperature history*

Figure 2 shows the temperature history in the drying room and at several drying air lines. The temperature history only shown during the drying time only, it is about 20 minutes. The annotations refer to the positions as shown in figure 1. For instance, T1 shows temperature in the drying room and T2 shows temperature of the drying air leaving the drying room and entering the heat exchanger as a hot fluid. As expected, the figure shows that temperature in the drying room is higher than in all lines. The function of the system is to make temperature in the drying room is high and has low humidity. This condition is a perfect drying force for drying clothes. In the drying room, the moist is evaporated into vapor and it needs amount of heat to overcome the heat evaporation of water. The heat evaporation is provided by hot drying air. In other word, the temperature of the drying air when it is leaving the drying chamber will decrease. In the figure, temperature of the drying air leaving the drying room, named as T2, is lower than temperature in the drying room. This temperature difference provide energy to the drying process. In the figure, it can be seen clearly the temperature difference increase with increasing time. The reason for this is that at a high temperature, the heat loss to the surrounding increasing. Temperature of the drying air leaving the heat exchanger (shown by T3) is lower than T2. This because temperature of the cold fluid entering the heat exchanger (T4) cools the hot air. The trend of the temperature of the hot drying air leaving the heat exchanger and cold air entering the heat exchanger is similar. This fact reveals that the heat pump dryer can provide a good drying condition in the drying room.

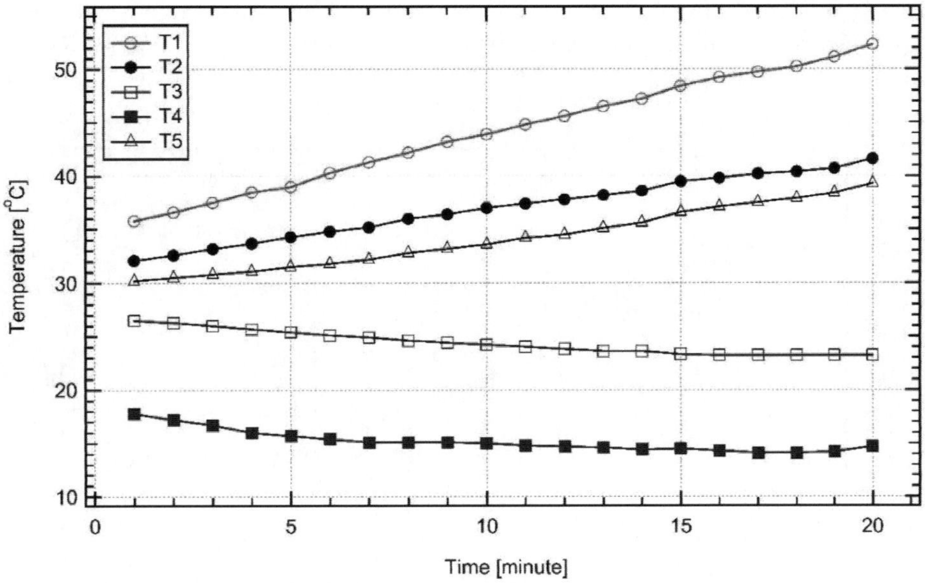

Figure 2. Temperature history in the experimental apparatus

3.2. Effects of drying medium velocity

In this experiment, three different speeds of drying air are tested. The hot air speed is measured in the condenser with surface area of 0.25 m^2. The experiments were carried out at low speed, medium speed and high speed with drying air velocity at condenser surface are 0.23 m/s, 0.41 m/s and 0.64 m/s, respectively. Figure 3 shows temperature in the drying room at different speeds.

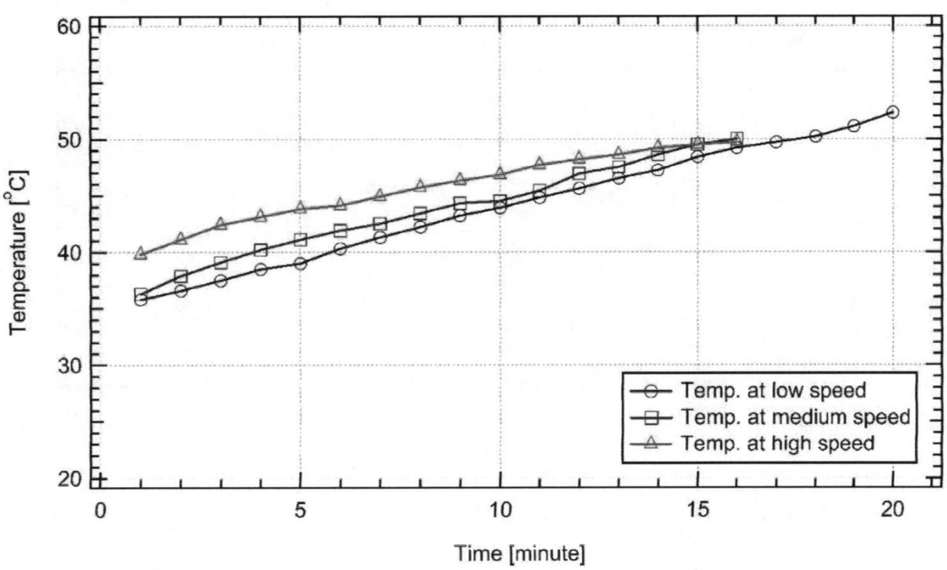

Figure 3. Temperature history in drying room at different hot air speed

The figure shows that higher hot fluid speed results in a higher temperature in the drying room. This is because at high speed results in lower temperature of the cold fluid leaving the evaporator. This will make temperature of the cold fluid entering the heat exchange will be higher. Thus, as a final product temperature of the hot air in the drying chamber will be higher. The higher temperature in the drying room will make drying process faster. This fact shown in figure 4. This is the reason why the temperature history at high speed is lower than temperature history at low speed. This fact suggest that the drying process should be carried out at high drying air speed. However, the effect of different humidity in the drying room is not investigated here. This will be an objective of the further research.

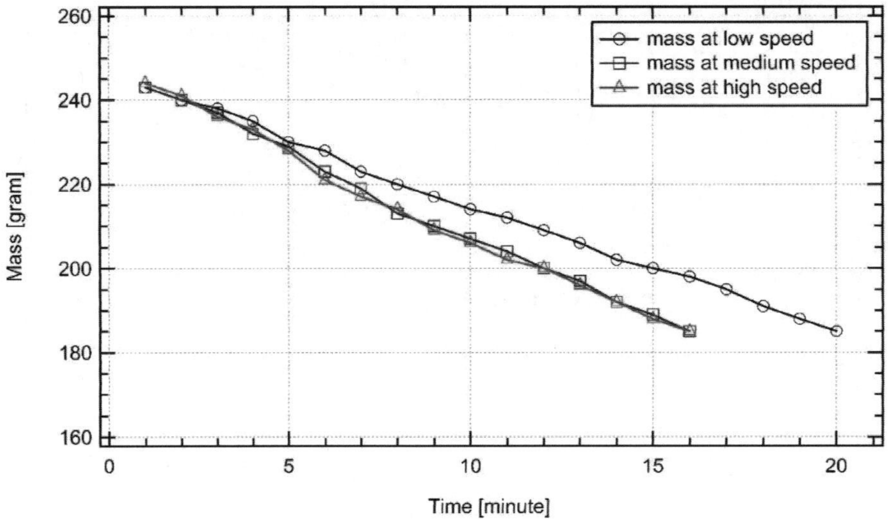

Figure 4. Mass history of clothes at different hot air speed

3.3. *Effect of the heat exchanger to the performance*

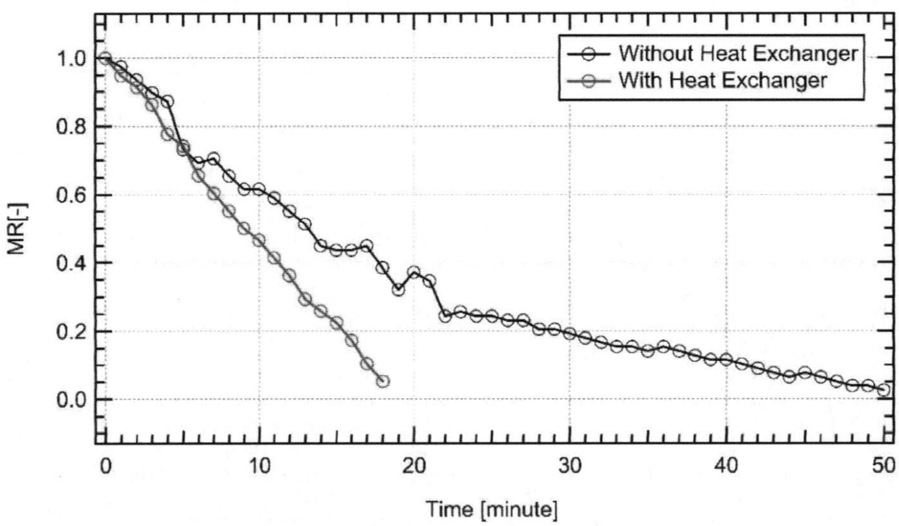

Figure 5. MR comparison for the heat pump dryer with and without heat exchanger

The performance of the present heat pump dryer, in term of drying time, with and without heat exchanger is shown in figure 5. In the figure, the non-dimensional weight (MR) in the drying room for heat pump dryer is shown. It can be seen that the drying rate in the drying room without heat exchanger is lower than the heat pump dryer with the heat exchanger. The gradient of the MR with heat exchanger is higher than without heat exchanger. As a result, the drying time for the heat pump dryer with heat exchanger is 50 minutes and with heat exchanger is only 18 minutes. This fact suggest that the present of the heat exchanger increase the performance of the heat pump dryer significantly. In term of drying time it can reduce the drying time up to 64%.

The reason of higher drying rate in the drying room of the heat pump dryer with the heat exchanger is explored using the temperature and relative humidity in the drying room as shown in Figure 6. In the case without heat exchanger, temperature in the drying room increase from 36°C to a maximum value of 47°C. In the same time, the RH decrease from 40% to a minimum value of 27%. For the heat pump with heat exchanger, the temperature in the drying room increase from 36°C to a maximum value of 52.3°C and the RH decrease from 51% to 17%. As a note drying process is effective in a condition with high temperature and low relative humidity. The figure clearly shows that the present of the heat exchanger increases the temperature and decreases the RH in the drying room. This is the reason for high drying rate for the case with heat exchanger.

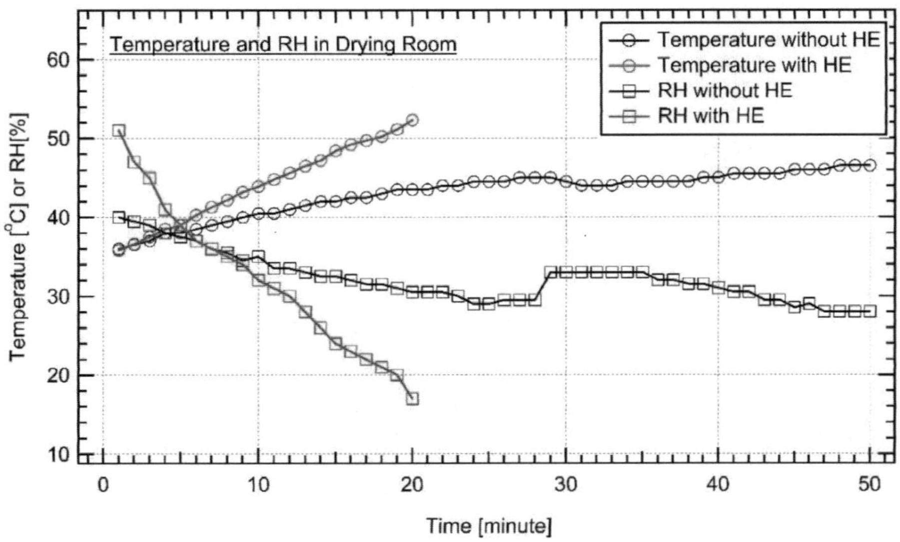

Figure 6. The history of temperature and relative humidity (RH) in the drying room

3.4. *Heat Exchanger effectiveness*

The effectiveness of the heat exchanger at different drying air speeds have been calculated using the equations shown in the method section. The results are shown in table 2. The table shows that the effectiveness of the present heat exchanger varies from 55.1% to 59.21% and higher at high speed. This suggests that the heat exchanger is very effective in the present application.

Table 2. The effectiveness of the heat exchanger at different speed

Parameter	Medium air speed		
	0.23 m/s	0.41 m/s	0.64 m/s
Effectiveness	55.1	59.06	59.21

The comparison of the performance parameter on the heat pump dryer with and without heat exchanger are shown in table 3. The data shows that the present of the heat exchanger in the system increase COP, TP and SMER in order of 15.11%, 4.81% and 58.62%, respectively.

Table 3. Performance comparison of the heat pump dryer with and without heat exchanger

Heat pump dryer	Performance parameter		
	COP	TP	SMER
Without Heat Exchanger	3.64	5.82	0.435
With Heat Exchanger	4.19	6.10	0.695
Increase	15.11%	4.81%	58.62%

4. Conclusions

The present of a heat exchanger in a heat pump dryer has been investigated experimentally. The drying load is wet clothes made of 100% cotton with initial weigh 245 gr. The results show the present of the heat exchanger increase the performance of the heat pump dryer. In the present experiment the COP, TP and SMER increase 15.11%, 4.81% and 58.62%, respectively. This is because the heat exchanger provides a better drying condition in the drying room with higher temperature and lower relative humidity in comparison with heat pump dryer without heat exchanger. The effectiveness of the heat exchanger is also high, it is above 50%. It is suggested to install a heat exchanger in a heat pump dryer.

Acknowledgement

The authors gratefully acknowledge that the present research is supported by Ministry of Research and Technology and Higher Education Republic of Indonesia. The support was under the research grant Hibah Penelitian Produk Terapan of Year 2017.

References

[1] Meyers S, Franco V H, Lekov A B, Thompson L and Sturges A 2010 *Proceedings of the ACEEE Summer Study on Energy Efficiency in Buildings* (California)
[2] Yadav V and Moon C G 2008 *Applied Energy* **85** pp 143–158
[3] Bansal P K, Braun J E and Groll E A, 2001 *Int. J. Energy Resour.* **25** pp 1315–1332
[4] Minea V 2013 *Int. J. Refrig.* **36** 643–658
[5] Braun J E, Bansal P K and Groll E A 2002 *Int. J. Refrig.* **25** pp 954–967
[6] Ameen A and Bari S 2004 *Energy Convers. Manag.* **45** pp 1397–1405
[7] Deng S and Han H 2004 *Appl. Therm. Eng.* **24** pp 2789–2800
[8] Mahlia T M I, Hor C G, Masjuki H H, Husnawan M, Varman M and Mekhilef S 2010 *Arab. J. Sci. Eng.* **35(1B)** 339–351
[9] Suntivarakorn P, Satmarong S, Benjapiyaporn C and Theerakulpisut S 2009 *Int. Jou. of Mech, Aerosp, Ind, Mechatron, and Manuf. Eng.* **3** pp 483-488
[10] Ambarita H, Nasution A H, Siahaan N M and Kawai H 2016 *Case Studies in Thermal Engineering* **8** pp 105-114
[11] Ambarita H, Nasution D M, Gunawan S and Nasution A H 2017 *IOP Conf. Series: Material Science and Engineering* **180** 012027

IOP Conf. Series: Materials Science and Engineering **308** (2018) 012028 doi:10.1088/1757-899X/308/1/012028

Numerical study on the effect of configuration of a simple box solar cooker for boiling water

H Ambarita[1]

Sustainable Energy and Biomaterial Centre of Excellent, University of Sumatera Utara, Jl. Almamater Kampus USU, Medan 20155 Indonesia

E-mail: himsar@usu.ac.id

Abstract. In this work, a numerical study is carried out to investigate the effect of configuration of a simple box solar cooker. In order to validate the numerical results, a simple a simple solar box cooker with absorber area of 0.835 m × 0.835 m is designed and fabricated. The solar box cooker is employed to boil water by exposing to the solar radiation in Medan city of Indonesia. In the numerical method, a set of transient governing equations are developed. The governing equations are solved using forward time step marching technique. The main objective is to explore the effect of double glasses cover, dimensions of the cooking vessel, and depth of the box cooker to the performance of the solar box cooker. The results show that the experimental and numerical results show good agreement. The performance of the solar box cooker strongly affected by the distance of the double glass cover, the solar cooker depth, and the solar collector length.

1. Introduction

Solar energy is a potential renewable energy resource and it will play an important role in fulfill future energy demand. The globe is yearly exposed by 3,400,000 Exa-Joule of solar energy. This energy can fill the present energy consumption of the world only with 1 hour 15 minutes of irradiation [1]. In the present time, only a tiny little portion of the human energy consumption filled by solar energy. However, according to REN21, since 2030 solar energy utilization will increase significantly. Recently, studies on solar energy have increased significantly [2]. Solar industry is developing constantly in all over the world because of increasing energy demand and also limitation on major energy source the fossil fuel. Indonesia has a big potency of solar energy and constant in every year [3]. Typically, solar energy is harvested in solar photovoltaic and solar thermal method. In the solar thermal method, the solar energy can be used as heat energy for solar drying process [4], solar adsorption refrigeration [5], solar desalination [6,7,8], solar water heater [9], solar cooker [10], etc.

There are many types of solar collector are found in literature. Flat-plate type solar collector is the simplest solar collector and it is very popular [11]. In the flat-plate type of solar collector a double glass cover is typically employed to reduce the heat loss from the top. Studies on the heat transfer characteristics and heat loss from a flat-plate type solar collector can be found in literature. Varol and Oztop [12] carried out a comparative numerical study on natural convection in inclined wavy and flat-plate solar collectors. It was revealed that flow field and thermal fields were affected by the shape of the air space (cavity) and heat transfer rate increases in the case of wavy cavity than that of flat cavity. The influence of inclination to the flat-plate and wavy solar collectors has also investigated [13]. Kumar [14] studied free convective heat transfer in trapezoidal profile of an enclosure of a box-type solar cooker. It was revealed that the values of convective heat transfer and top loss coefficient for rectangular enclosure are lower by 31-35% and 7 %, respectively. Those studies reveal that investigation on the flat-plate type solar collector has come under scrutiny. In this study, a flat-plate type solar collector that is used in a simple solar box cooker is studied numerically. The objective is to explore the effect of the configuration of the box cooker to the performance. The configuration in here includes dimensions of the solar cooker components. The results are expected to supply the necessary information in developing an effective and efficient solar box cooker.

2. Method

In this study, an experimental apparatus has been designed and fabricated to provide experimental data that can be used to perform numerical validation. The developed experimental apparatus is depicted in figure 1. It is a flat plate solar collector with double glasses cover in the top with absorber area of 0.835 m × 0.835 m. In order to decrease the heat loss, the envelop of the solar box cooker made of a series of insulation material rock wool, styrofoam, and wood. The figure shows the dimensions and configurations of these material. The load to the box solar cooker is water and it is placed in a cylindrical cooking container and located in the middle of the collector. The cocking container is made of aluminum with diameter and height of 30 cm and 12 cm, respectively.

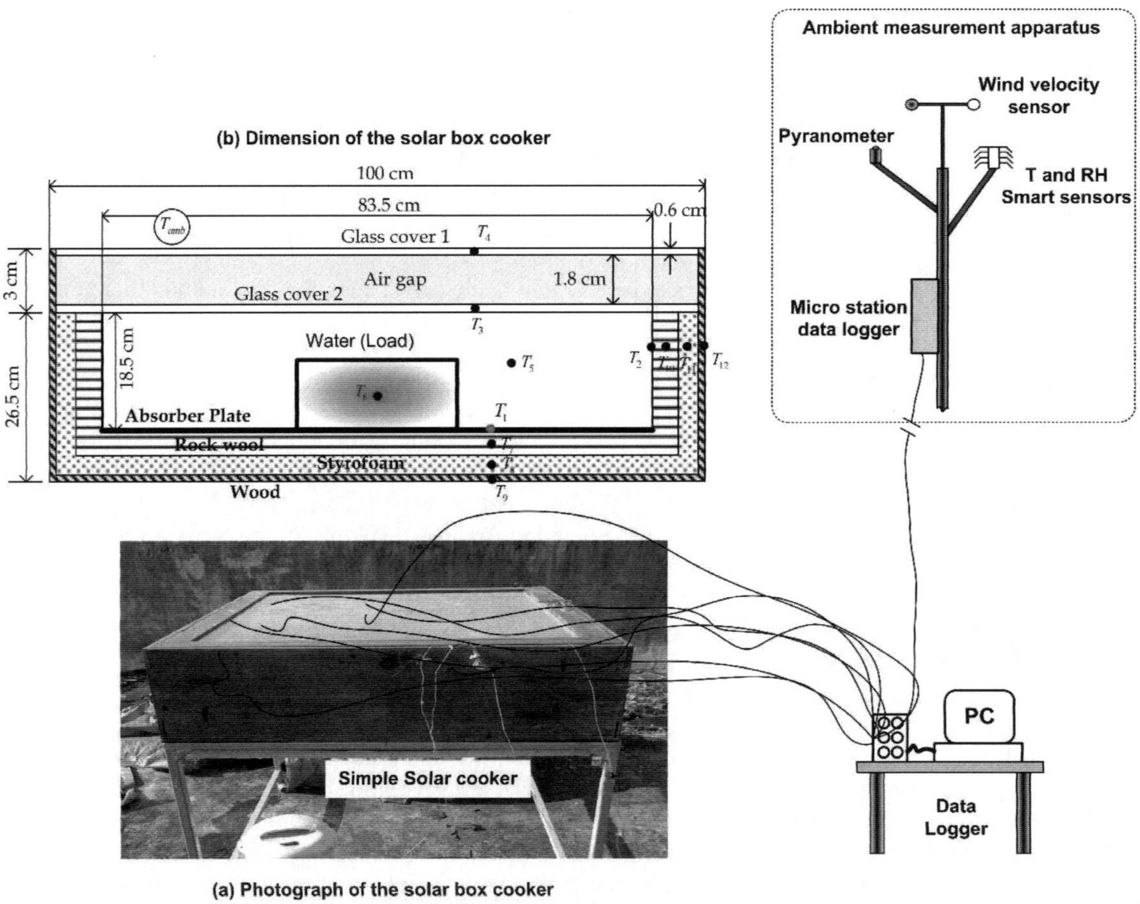

Figure 1. (a) Photograph of the solar cooker and (b) computational domain

2.1 Numerical Method

In this study, the solar box cooker is divided into 12 components. It is assumed that temperature and properties of each component are uniform. Figure 2(b) shows the dimensions and temperature notation of every component. To each component, the energy conservation is applied. The energy conservation for absorber plate gives:

$$m_p C_p \frac{\partial T_1}{\partial t} = \tau^2 \alpha I A_p - FC_{15}(T_1 - T_5) - FR_{13}(T_1 - T_3) - FR_{12}(T_1 - T_2) - \frac{1}{R_{17}}(T_1 - T_7) \quad (1)$$

where m_p [kg] and C_p [J/kg K] are mass and specific heat of absorber plate. The parameters of I [W/m^2], α, and τ are solar irradiance, absorptivity and transmission coefficients of the glass cover, respectively. The hot air inside the solar box cooker is assumed to be one substance with a homogeneous temperature of T_5. Thus, energy conservation to the air yield to:

$$(\rho V_{ol})_{air} \frac{\partial T_5}{\partial t} = FC_{15}(T_1 - T_5) - FC_{52}(T_5 - T_2) - FC_{53}(T_5 - T_3) - FC_{56t}(T_5 - T_6) - FC_{56w}(T_5 - T_6) \quad (2)$$

Where FC [W/K] is a factor to represent convective heat transfer from any surface. The subscript 1, 5, 2, 3, 6 refer to surfaces shown in Figure 1. And 6t and 6w refer to the top surface and the container wall surface of the container, respectively. The double glasses cover is made of glass with similar emissivity (ε) and thickness. The energy conservation gives:

$$m_g C_g \frac{\partial T_4}{\partial t} = (1-\tau) I A_g + FR_{34}(T_3 - T_4) + FC_{34}(T_3 - T_4) - FR_{4s}(T_4 - T_{amb}) - h_w A_g (T_4 - T_{amb}) \quad (3)$$

$$m_g C_g \frac{\partial T_3}{\partial t} = (1-\tau)\tau I A_g + FC_{53}(T_5 - T_3) + FR_{13}(T_1 - T_3) + FR_{23}(T_2 - T_3) - FC_{34}(T_3 - T_4) - FR_{34}(T_3 - T_4) \quad (4)$$

Where h_w [W/m^2 K], A_g and T_{amb} [°C] are convective heat transfer coefficient from the top glass to the ambient air and temperature of the ambient air, respectively.

As mentioned that the load for box solar cooker is water. In the process the cooking vessel and water receive the heat convectively from the air inside the solar box cooker through side wall and top wall of the vessel. The top surface of the cooking vessel also acts as a solar collector and it receives solar heat flux. Heat transfer from the absorber plate to the bottom and radiative heat transfer from the wall of the container are neglected. The temperatures of the container and the water are assumed homogenous. Thus, energy conservation in the container gives:

$$(m_f C_f + m_c C_c)\frac{\partial T_6}{\partial t} = FC_{56w}(T_5 - T_6) + FC_{56t}(T_5 - T_6) + \tau^2 \alpha I A_{ct} \quad (5)$$

Where the subscript f and c represent water and the container, respectively.

The temperature difference in each material of the wall of the solar box cooker is taken into account. The temperature of every material of the bottom wall of the solar cooker is also shown in Figure 1. Energy conservation in the rock wool, styrofoam, and wood show the following equation.

$$m_{rw} C_{rw} \frac{\partial T_7}{\partial t} = \frac{1}{R_{17}}(T_1 - T_7) - \frac{1}{R_{78}}(T_7 - T_8) \quad (6)$$

$$m_{st} C_{st} \frac{\partial T_8}{\partial t} = \frac{1}{R_{78}}(T_7 - T_8) - \frac{1}{R_{89}}(T_8 - T_9) \quad (7)$$

$$m_{wd} C_{wd} \frac{\partial T_9}{\partial t} = \frac{1}{R_{89}}(T_8 - T_9) - FC_{9a}(T_9 - T_a) \quad (8)$$

Where the subscript rw, st, wd, and a refer to rock wool, styrofoam, wood, and ambient, respectively. Furthermore, the energy conservation in every material of the wall of the solar box solar cooker (inner plate, rock wool, Styrofoam, and wood) will give the following equations:

$$m_p C_p \frac{\partial T_2}{\partial t} = FR_{12}(T_1 - T_2) + FC_{52}(T_5 - T_2) - FR_{23}(T_2 - T_3) - \frac{1}{R_{210}}(T_2 - T_{10}) \quad (9)$$

$$m_{rw}C_{rw}\frac{\partial T_{10}}{\partial t}=\frac{1}{R_{210}}\left(T_2-T_{10}\right)-\frac{1}{R_{1011}}\left(T_{10}-T_{11}\right) \tag{10}$$

$$m_{st}C_{st}\frac{\partial T_{11}}{\partial t}=\frac{1}{R_{1011}}\left(T_{10}-T_{11}\right)-\frac{1}{R_{1112}}\left(T_{11}-T_{12}\right) \tag{11}$$

$$m_{wd}C_{wd}\frac{\partial T_{12}}{\partial t}=\frac{1}{R_{1112}}\left(T_{11}-T_{12}\right)-\frac{1}{R_{12a}}\left(T_{12}-T_a\right) \tag{12}$$

All of the governing equations, equation (1) to equation (12), are converted from partial differential equation form into linier equation system by using discretization technique. Here, the used technique is forward time step marching. In this work, the value of $\Delta t = 1\sec$ is used due to stability consideration. This value shows stability.

3. Results and Discussions

3.1. Solar Irradiance and Numerical validation

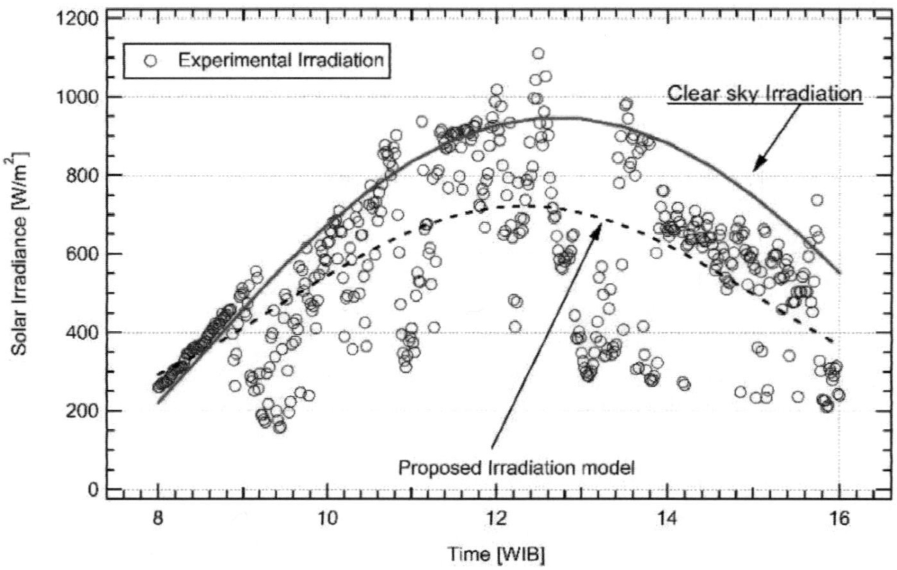

Figure 2. Solar Irradiance during experiment and modelled for numerical simulation

As a note, the source of energy for solar cooker is solar irradiation. The solar irradiation during experiment is measured. The measured data will be modelled and used as an energy input to the numerical simulation. The theoretical clear sky irradiance, measured solar irradiance and the modelled one are shown in figure 2. In the modelled one, the maximum solar irradiance is 722 W/m^2 and it occurred at 12.19 of local time. The total solar energy resulted by the solar radiation by the proposed model is 16.2 MJ/m^2. This shows that the solar energy from measurement and from the proposed model is the same. This model will be used in the numerical simulation as heat flux to the absorber plate of the solar box cooker.

A numerical validation test has been performed to the method before it is employed to analyse the problem. Three experiments have been carried out. The temperature history of cooking vessel at loads of 3 kg, 5 kg, and 6 kg water were selected for comparison. Here, numerical simulations using the developed method were carried out at the same load as experiments. Temperature history from numerical results and experimental results are shown in figure 3. The figure reveals that for all cases the temperature history from the experiment and numerical result do agree well. In other word, the numerical simulation accurately predicts the value of temperature and the boiling time of the water. Based on the above numerical validation, it can be said that the developed numerical method can now be employed to analyse the effect configuration of the solar box cooker for boiling water.

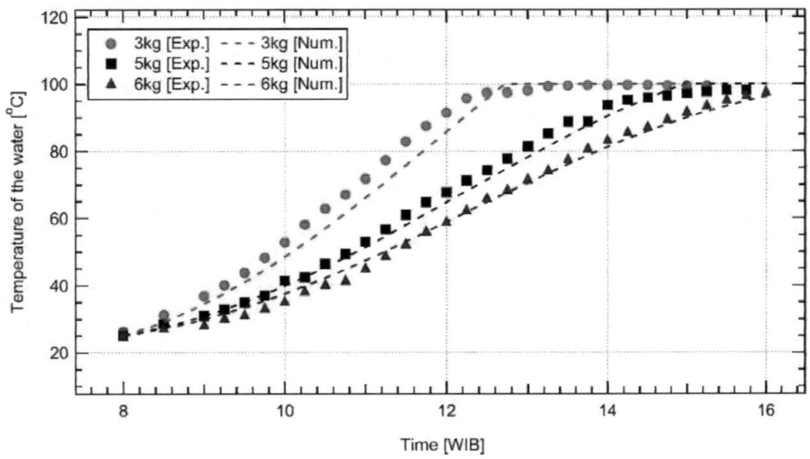

Figure 3. Numerical validation of the proposed model

3.2. Effect of the distance of the double glass cover

The objective of installing double glass cover is to reduce the heat loss from the solar collector. In this section, the effect of the configuration of the glass cover was investigated in this section. The distance of the glass cover was varied from 1 cm to 20 cm. The heat loss from the solar collector to the surrounding as a function a time at several distances of double glass cover are shown in Figure 4. In general, the heat loss increases as time increasing and reaching a maximum at around 14.00 local time. This is because the temperature inside the solar box cooker increases with increasing time. After reaching the maximum value, the heat loss decreases with increasing time. Here, the temperature is already decreased. It can be seen that the heat loss decreases as the distance of the double glass cover decreased. This suggests that increasing the distance of the double glass cover will decrease the heat loss to surrounding. This fact should be further analysed to search the optimum distance of double glass cover.

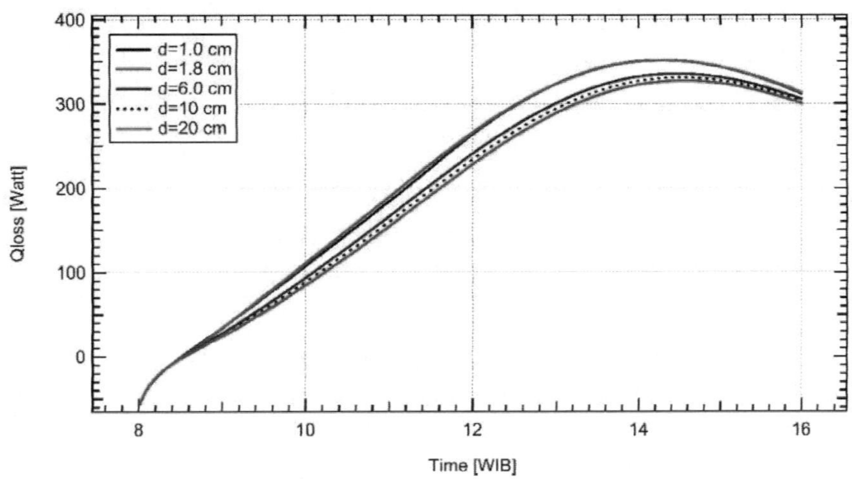

Figure 4. Effects of the distance of double glasses cover

IOP Conf. Series: Materials Science and Engineering **308** (2018) 012028 doi:10.1088/1757-899X/308/1/012028

3.3. *Effect of the Solar cooker depth*

The other parameter that is investigated here is the solar collector depth. The original depth of the solar box cooker is 18.5 cm. The simulations have been performed at different depths. The depth was varied from 18.5 cm to 50 cm. The effect of the solar cooker depth to the boiling time is shown in figure 5. The figure shows that boiling time increases with increasing solar cooker depth. This is because the higher solar cooker depth will increase the area of the wall and it will increase the heat loss. In other word, the energy to heat the water in the container will decrease and as a result the boiling time will increase. This suggests that increasing solar cooker depth will decrease the performance of the solar box cooker.

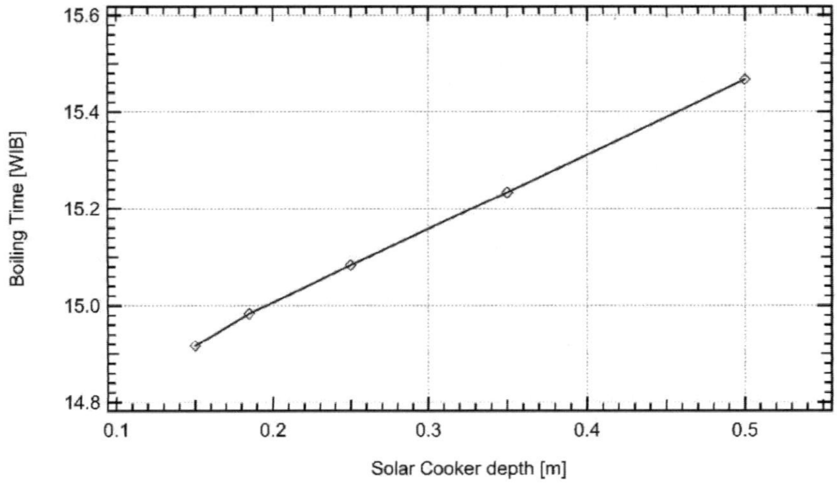

Figure 5. Effects of the solar box cooker depth to the boiling time

3.4. *Effect of the collector area*

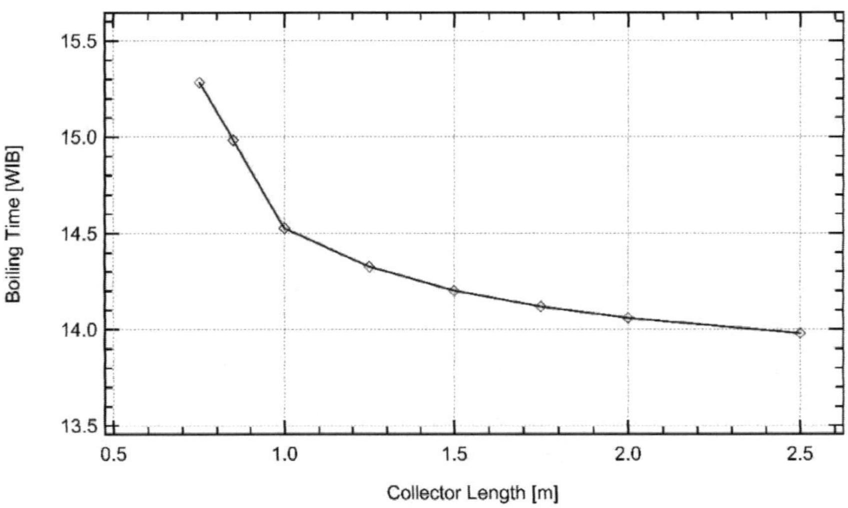

Figure 6. Effects of the collector length

The effect of the collector area to the performance of the solar box cooker is also investigated. In the present solar box cooker, the collector area is made of square plate with original dimension of 0.835 m × 0.835 m. The length of the solar box cooker was varied from 0.835 m to 2.5 m. The boiling time as a function of solar box cooker to boiling time is shown in figure 6. The figure shows that boiling time decrease with increasing solar collector length. This is because increasing collector area will increase the solar energy collected by the solar cooker. However, the rate of increasing collector area is decreased. This is because the increasing of collector area also increases the heat loss. Figure 6 shows this fact clearly. The boiling time at collector length 2.0 and 2.5 is almost similar. This fact suggests that there exists the optimum area for maximum performance of solar box cooker.

4. Conclusions

A numerical study has been carried out in order to study the effect of solar box cooker configuration. A numerical method to solve all of the governing equations has been developed. The conclusion of the study are as follows. The increasing of the distance of the double glass cover will decrease the heat loss to surrounding. The increasing solar cooker depth will decrease the performance of the solar box cooker. The boiling time decrease with increasing solar collector length. However, the rate of increasing collector area is decreased. This fact suggests that there exists the optimum area for maximum performance of solar box cooker.

Acknowledgments

The authors gratefully acknowledge that the present research is supported by Ministry of Research and Technology and Higher Education Republic of Indonesia. The support from my students Howard Li, Teguh Wirawan and Felix Onn.

References

[1] Thirugnanasambandam M, Iniyan S, and Goic R 2010 *Renewable and Sustainable Energy Reviews* **14** 312-322
[2] Kannan N and Vakeesan D 2016 *Renewable and Sustainable Energy Reviews* **62** 1092-1105
[3] Ambarita H 2017 *Journal of Physics: Conference Series* **801(1)** 012093
[4] Dina S F, Ambarita H, Napitupulu F H and Kawai H 2015 *Case Studies in Thermal Engineering* **5** 32-40
[5] Ambarita H and Kawai H 2016 *Case Studies in Thermal Engineering* **7** 36-46
[6] Ambarita H 2016 *Case Studies in Thermal Engineering* **8** 346-358
[7] Ambarita H 2016 *IOP Conference Series: Materials Science and Engineering* **180(1)** 012024
[8] Setyawan E Y, Napitupulu R A M, Siagian P and Ambarita H 2017 *IOP Conference Series: Materials Science and Engineering* **237** 012012
[9] Ambarita H and Sitepu T 2017 *IOP Conference Series: Materials Science and Engineering* **237** 012014
[10] Sitepu T, Gunawan S, Nasution D M, Ambarita H, Siregar R E T and Ronowikarto A D 2016 *IOP Conference Series: Materials Science and Engineering* **180(1)** 012032
[11] Pandey K M and Chaurasiya R 2017 *Renewable and Sustainable Energy Reviews* **67** 641-650
[12] Varol Y and Oztop H F 2008 *Building and Environment* **43** 1535-1544
[13] Varol Y and Oztop H F 2007 *Building and Environment* **42** 2062-2071
[14] Kumar S 2004 *Renewable Energy* **29** 211-222

IOP Conf. Series: Materials Science and Engineering **308** (2018) 012029 doi:10.1088/1757-899X/308/1/012029

Transient thermal stresses analysis and thermal fatigue damage evaluation for skirt attachment of coke drum

H Ambarita[*1,2], A S Siahaan[1*], H Kawai[1], and M Daimaruya[1]
[1]Department of Mechanical Systems of Engineering, Muroran Institute of Technology 27-1 Mizumoto-cho, Muroran 050-8585, Japan
[2]Sustainable Energy Research Centre, Faculty of Engineering, University of Sumatera Utara, Jl. Almamater Kampus USU Medan 20155, Indonesia

[*]E-mail: himsar@usu.ac.id

Abstract. In the last decade, the demand for delayed coking capacity has been steadily increasing. The trend in the past 15 to 20 years has been for operators to try to maximize the output of their units by reducing cycle times. This mode of operation can result in very large temperature gradients within the drums during preheating stage and even more so during the quench cycle. This research provide the optimization estimation of fatigue life due to each for the absence of preheating stage and cutting stage. In the absence of preheating stage the decreasing of fatigue life is around 19% and the increasing of maximum stress in point 5 of shell-to-skirt junction is around 97 MPa. However for the absence of cutting stage it was found that is more severe compare to normal cycle. In this adjustment fatigue life reduce around 39% and maximum stress is increased around 154 MPa. It can concluded that for cycle optimization, eliminating preheating stage possibly can become an option due to the increasing demand of delayed coking process.

1. Introduction

Coke drum are categorized into pressure vessel used in oil refinery to convert the heavy residual into lighter product by using thermal cracking due to delayed cooking process. In delayed cooking process, the heavy residual feed oil is heated into thermal cracking temperature in order to separate petroleum coke from the lighter. The remains coke then quenched to make possible to safely take out from the drum. Nowadays, because of the increasing demand of lighter oil is increasing affecting the increasing of coke drum instillation in whole world.

Coke drum undergoes cyclic high temperature and suddenly cooling. Thus its operational life is much shorter in comparison to other equipment in oil refinery. In fact, coke drum is designed based on ASME Boiler and Pressure Vessel Codes Section VIII Division I [1]. Concisely, most of coke drum is not designed for cyclic load. It is convinced that, the fracture mechanism in a coke drum is dominated by thermal fatigue.

Mainly coke drum is operated in four stages; preheating, filling or feeding, cooling and cutting process, with switching process between preheating and filling stage as a transition. In pre-heating stage the empty coke drum is heated with hot steam until temperature 350^{0}C. After the pre-heating stage the hot feed material enter the coke drum with temperature range between 440^{0}C to 500^{0}C, with internal pressure around 350 kPa. The thermal cracking process occurred in the filling stage, which separate the lighter and heavy parts. Typically, coke drum for this studies require 2800 minutes to complete one cycle; Preheating stage from 0 – 500 minutes, filling stage from 500 to about 2000 minutes, cooling stage from 2000 to 2400 minutes and switching stage from 2400 to about 2800 minutes [2].

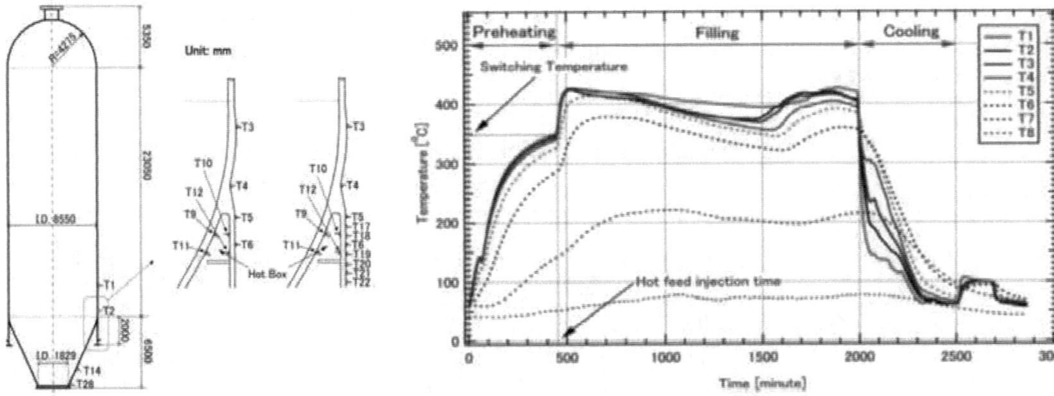

Figure 1. Coke Drum and Typical Coke Drum Cycle

In the present because the high demand of lighter oil, high rate production in oil refinery is extremely needed [3]. Nowadays, petroleum companies tends to cut the cycle time of delayed cooking process due to enhance the production rate, thus coke drum cycle will be as short as 2000 to 2400 minutes, which means more severe cycle and shorter fatigue life of the drum.

The main objective of this research is to observe the effect of shortening cycle time of coke drum, with consideration to mutual benefit of shorter fatigue life due to higher stress intensity compare to enhance production process of coke. Further analyses will conducted to compare fatigue life of coke drum on full cycle, preheating excluded, and cutting stage excluded. The result can be used to estimate fatigue life of coke drum with consideration of two stages exclusion.

2. Method

Thermocouples and strain gauge were inserted in to coke drum, located around shell-to skirt junction area. Operational strain and temperature history of coke drum were collected which in minute interval. Coke drums made by low alloy steel SA-387 Gr.11 CL.2. The cladding inside the coke drum is made of 3 mm thick SA 240TP. 410S material. Thickness of shells vary within the range 28.5 mm – 34 mm and the skirt thickness are 24 mm. Ratio between diameter and wall thickness of the coke drum varied from 125 to 150 mm/mm. The coke drum material density is 7850 kg/m3 and 40.6 W/mK for thermal conductivity. For the insulator we used ambient temperature model outside of the skirt and shell of the coke drum. In this analyses the wind, piping, and seismic overturning moments are excluded.

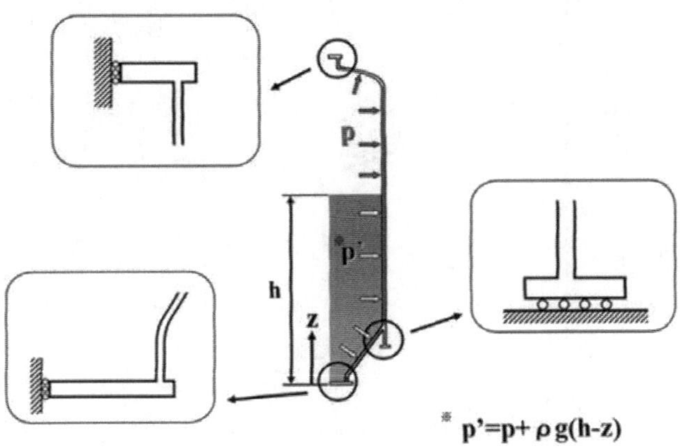

Figure 2. Model of Structural Boundary

Numerical analyses then performed on selected cycle using ANSYS code. FEM model and structural boundary is shown in figure 2 above. Using temperature dependent material properties (ASME Boiler and Pressure Vessel Code section II part D), shown in table 1. The model for numerical analyses is half segment of the coke drum. There are two mesh model were tested in this simulation which is 25675 and 41351 nodes. The strain result is almost the same but for the better result the model with 41351 nodes was selected for analyses.

Table 1. Temperature dependent material properties

Temperature (°C)	Properties		
	Young modulus (GPa)	Thermal Conductivity (W/mK)	Coeff.of thermal exp (/°C)
0	205.5	41.05	-
25	204	40.9	0.0000116
100	200	40.6	0.0000121
150	197	40.4	0.0000124
200	193	40.1	0.0000128
250	190	39.5	0.0000131
300	186	38.7	0.0000132
350	183	37.8	0.0000135
400	179	36.8	0.0000139
450	174	35.8	0.0000142
500	169	34.8	0.0000143
550	164	33.9	0.0000146
600	157	32.8	0.0000148

Analyses on estimation coke drum fatigue life were performed, which is treated as follows: strain comparison from selected cycle which gathered from strain gauge is compared to the numerical result. Selected cycle have total time of 2771 minutes. Similar with the temperature plot history, we can divided the strain result to preheating, filling, cooling and cutting stage.

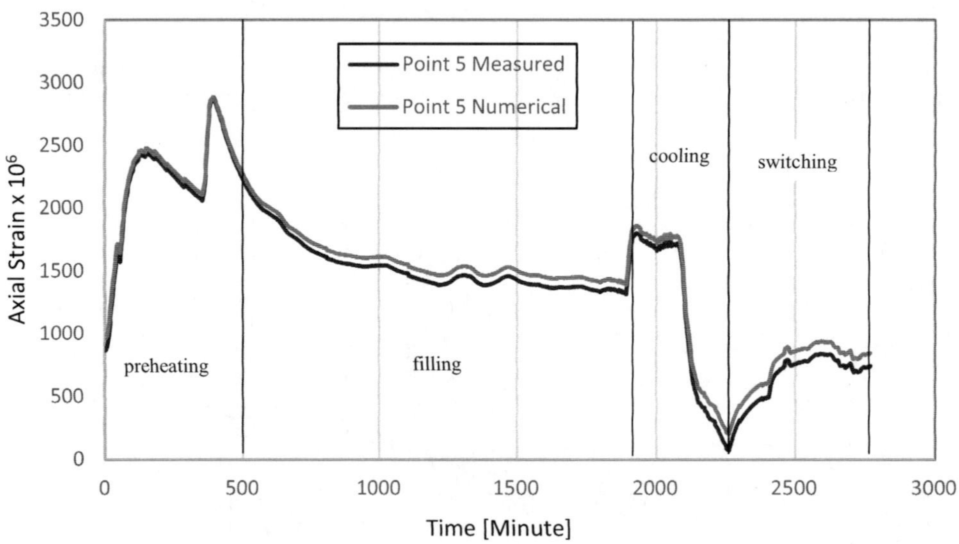

Figure 3. Comparison of Numerical and Measured axial strains

The comparison result show for full cycle were performed and show a good agreement between numerical and measured data shown in figure 3 above. After validating result between numerical and measured data, some adjustments for next simulation were implemented. For the first adjustment the cutting stage is excluded which total time of 2398 minutes. The second is when the preheating stage is excluded that cut total time cycle to 2416 minutes. Further analyses will performed to compare the result of the exclusion of these two stages.

3. Result and Discussion

The main objective of this research is to observe the effect of shortening the cycle time of coke drum. The thermal fatigue life of coke drum carried out in the most severe part of the coke drum at the shell-to-skirt junction area. Good agreement between numerical and measured strain result, allowed us to perform further analyses of fatigue life estimation of coke drum using numerical code. Because the data which we gather were strain data, fatigue life estimation which will we performed is strain life fatigue analyses.

The elastic modulus using for this equation is based on temperature was iterated from the input data from ASME Boiler and Pressure Vessel Code section II part D. From numerical analyses for full cycle which is shown in contour of stress intensity in fig. 4(a) we found that the shell-to-skirt junction is the most severe area especially in area around point 5. This result have the same result compare to previous research of the same coke drum by Oka et.al [2]. The result shows the highest axial strain distribution in junction area. For full cycle it was found that maximum stress intensity is 416 MPa.

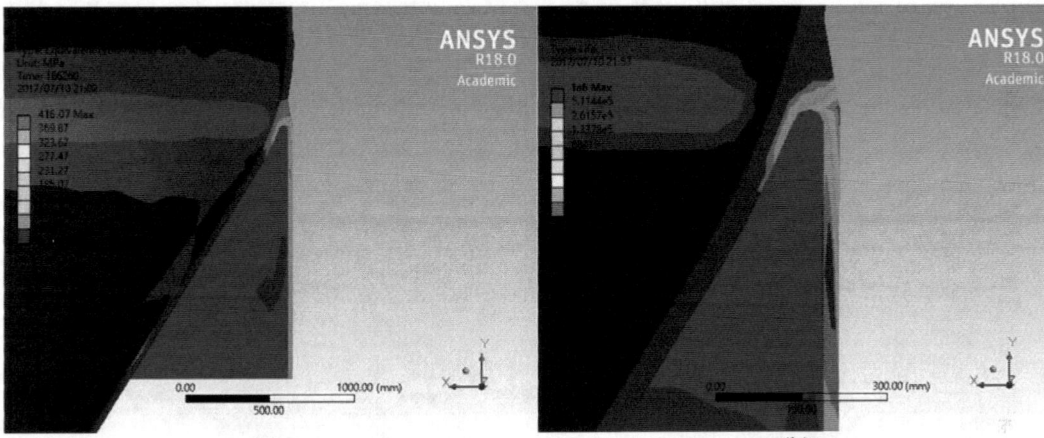

(a) (b)

Figure 4. (a) Contour of stress intensity and (b) contour of fatigue life of full cycle

From the contour of fatigue life in figure 4(b) above we found the estimation for fatigue life of full cycle is around 2394 cycles, for this particularly selected cycle. From contour of fatigue life in figure 4(b) it also shown shell-to-skirt junction is the area with the shortest fatigue life, comparing to another part in coke drum. Therefore for future research some focuses due to stress development and maintenance in this area need to be implemented.

As the aforementioned before, some adjustment was conducted. Using numerical analyses the estimation of coke drum's fatigue life with elimination for each preheating and cutting stages were performed. The first adjustment is full cycle without preheating stage. Contour of stress intensity in fig. 5(a) shows the same point of most severe part of coke drum also occurred in point 5. Although the stress intensity shows the same area of occurrence, but there are increasing of maximum stress around 97 MPa.

Fatigue life estimation for coke drum without preheating stage is around 1930 cycles, shown in figure 5(b). There is a decreasing of fatigue life of 19%, compare to full cycle. The deflation of fatigue life most likely occurred because of sudden heating due to the absence of preheating stage, which the objective of preheating stage is to avoid severe thermal shock when hot feed material is injected.

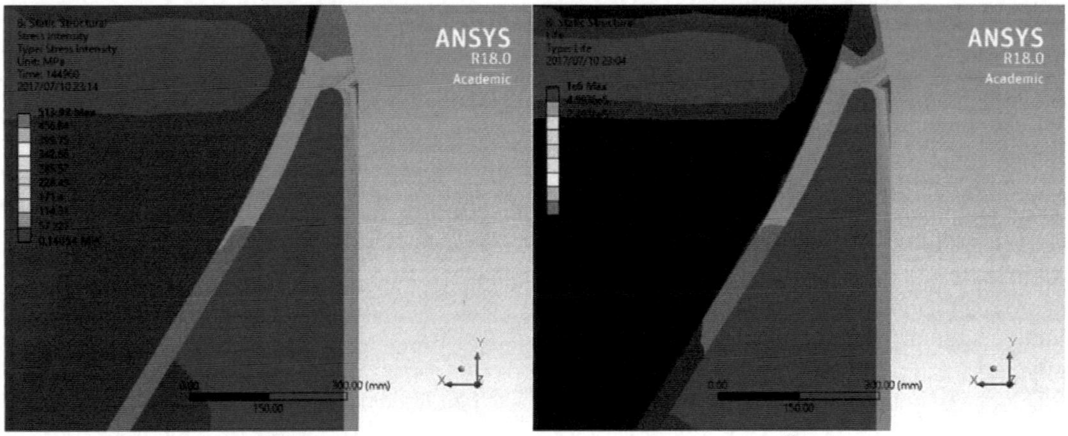

(a) (b)

Figure 5. (a) Contour of stress intensity and (b) contour of fatigue life of cycle without preheating stage

For the second adjustment is coke drum cycle without cutting stage. In this type of cycle, it found the increasing of maximum stress intensity of 154 MPa compared to normal cycle. From contour of fatigue life in figure 6(b), it was found that the estimation life of coke drum decreasing drastically to 1430 cycles. Around 39% compare to the normal cycle. This phenomena was not expected before, because in cutting stage there are no sudden change in stress and temperature. Therefore, further understanding on cutting stage's characteristics is necessary.

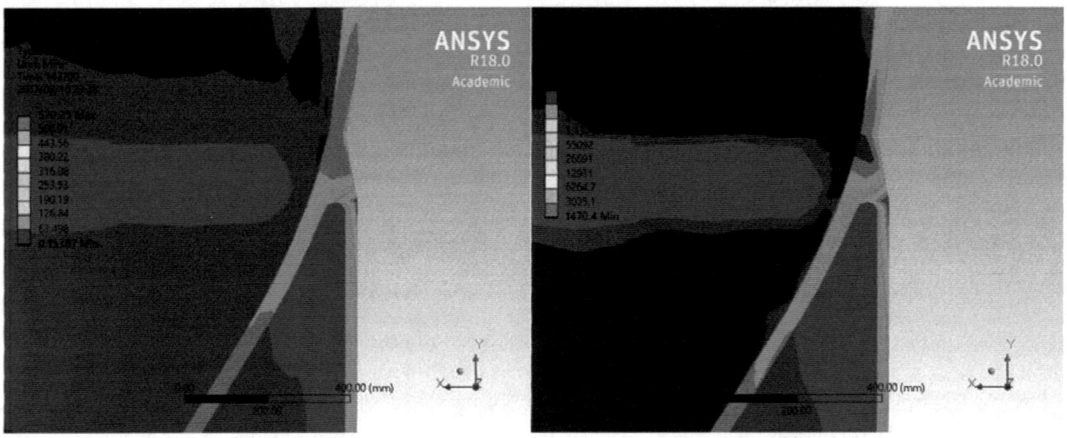

(a) (b)

Figure 6. (a) Contour of stress intensity and (b) contour of fatigue life of cycle without preheating stage

The main objective of cutting stage is to cut the coke bed, causing the coke bed to collapse in on itself and for the coke water slurry to drain through the pilot hole and out of the bottom the drum through the raised chute and into the coke pit or pad below the coke drum structure. When the coke have been cut from the drum, the chute is lowered, the bottom cover is re-attached and bolted to the coke drum and the feed line connected to its mating section on the bottom cover. At this time the drum is steam tested for tightness and then is heated by introducing steam into the feed line, which shown by the increasing temperature for some amount of time in cutting stage.

For some reason, cutting stage is also act as tempering in heat treatment. After rapid cooling stage, there occurred high residual thermal stress in material, thus cutting stage serve as intermediate stage between cooling and preheating of next cycle to removes internal pressure and residual thermal stress. Compared to the absence of preheating stage, fatigue life of material is much shorter when in the absence of cutting stage. In preheating stage, after rapid heating, filling stage might acted as heat treatment. As we can see in the in figure 3, strain rate in this filling stage is showing no significant fluctuation, which we can that this stage turn into steady state. It also shown in figure 1 for temperature plot, after preheating stage temperature is almost steady and it turn to steady heat transfer. Thus, even with the absence of preheating it turn to less damage compare to the absence of cutting stage.

Other things is, when some part of the plat is cooled, contraction of the outer layer causes tension. Failure when it occurs, is initiated somewhere at the surface or its near vicinity. When the parts is heated, however, expansion of the surface layers causes compression which is balanced by tensile stresses in the central region [4]. Although shear induced by the compression may cause surface spalling, the more critical type of failure occurs when a tensile crack develops in the inner region. Some research was conducted this phenomena [5, 6, 7, 8, 9], in which some uncertainties existed regarding the validity of the calculations of the stresses and in which the rate of the loading was much higher for cooled specimen. These data showed that alumina had much higher strength in the case of heating than in the case of cooling. Depending on specimen size, the strength in heating can be as great as two times the strength in cooling.

With the absence of cutting stage, the cycle will undergoes simultaneous rapid cooling and sudden rapid heating. Thus, fatigue life of material in case of the absence of cutting stage is more severe compare to the absence of preheating stage, not to mention cutting stage is the stage where water jets cut into the coke bed, cleaning the remaining coke bed which still adhere in inner wall of coke drum. Without cutting stage, the remaining coke bed in inner shell will aggravate to the next cycle, such as the disruption heat distribution from inside wall to outside part coke drum and the occurrence of hot spot and cold spot due to remaining coke' which is not taken into consideration in simulation. Thus fatigue life estimation without cutting stage in this simulation most likely was predicted to high compare to the real situation if cutting stage indeed will eliminated [10].

The option to shorten delayed cooking process will eventually lead to more severe stress and shorter fatigue life. Analyses optimization shows that comparing between the absence of preheating stage and cutting stage for enhance production, indicate that the cycle without preheating stage can become an alternative. With moderate rise in stress intensity and fatigue life reduction.

4. Conclusion

In this research, numerical analyses to estimate fatigue life of coke drum were performed. From the numerical analyses it was found that the most severe area is in shell-to-skirt junction, for every kind of adjustment, either it was full cycle, without preheating and without switching stage.

In the absence of preheating stage it turn that maximum stress intensity increase of 97 MPa compare to normal cycle as well as the decreasing of fatigue life around 19% from normal cycle. In the absence of switching stage it turn that maximum stress intensity increase of 154 MPa compare to normal cycle and the decreasing of fatigue life around 39%.

The main conclusion is the absence of cutting stage is more severe compare to the absence of preheating stage. As the aforementioned before, cutting stage acted like a heat treatment, and the absence of cutting stage resulting in unavailability of intermediate stage between cooling and preheating stage which means rapid cooling and rapid heating repeatedly. Thus for cycle optimization, eliminating preheating stage can be an option, due to the increasing demand of petroleum product.

References

[1] ASME 2007 ASME Boiler & Pressure Vessel Code, Section II, Part D, Properties-Material (ASME)
[2] Oka A, Ambarita H, Daimaruya M, Fujiki H 2011 *ASME Pressure Vessel Technology* **133 (6)** 061210
[3] Ramos A, Rios C, C Vargas, J Tahara, and T Hasagewa 1997 *ASME PVP* **359** pp 291-298
[4] S. S Manson 1966 *Thermal Stress and Low-Cycle Fatigue* (McGraw-Hill) p 303
[5] W B Crandall et.al 1955 *J Am Ceram Soc* **38** p 44
[6] Glenny et.al 1958 *Trans Brit Cearam Soc* **57** (10) p 645

[7] Oka M, Ambarita H, Kawashima K, Daimaruya M 2010 *ASME Pressure Vessels and Piping* **7** pp 37-43
[8] Ambarita H, Daimaruya M, Fujiki H, Kawai H, Kobayashi H and Oka M 2012 *Journal of Thermal Stresses* **35(12)** pp 1095-1107
[9] Oka M, Ambarita H, Daimaruya M, Fujiki H 2010 *Journal of Thermal Stresses* **33(10)** pp 964-976
[10] Daimaruya M, Oka M, Fujiki H, Ambarita H, Yamada T and Tomizawa K 2009 The 12th International Conference on Fracture 7 pp 5185-5194

IOP Conf. Series: Materials Science and Engineering **308** (2018) 012030 doi:10.1088/1757-899X/308/1/012030

Integrated biomass pyrolysis with organic Rankine cycle for power generation

T B Nur [*1,2], **A W Syahputra**[1]

[1]Department of Mechanical Engineering, Faculty of Engineering, Universitas Sumatera Utara, Padang Bulan, Medan 20155, Indonesia.

[2]Sustainable Energy and Biomaterial Center of Excellence, Universitas Sumatera Utara, Padang Bulan, Medan 20155, Indonesia.

[*]E-mail: taufiq.bin_nur@usu.ac.id

Abstract. The growing interest on Organic Rankine Cycle (ORC) application to produce electricity by utilizing biomass energy sources are increasingly due to its successfully used to generate power from waste heat available in industrial processes. Biomass pyrolysis is one of the thermochemical technologies for converting biomass into energy and chemical products consisting of liquid bio-oil, solid biochar, and pyrolytic gas. In the application, biomass pyrolysis can be divided into three main categories; slow, fast and flash pyrolysis mainly aiming at maximizing the products of bio-oil or biochar. The temperature of synthesis gas generated during processes can be used for Organic Rankine Cycle to generate power. The heat from synthesis gas during pyrolysis processes was transfer by thermal oil heater to evaporate ORC working fluid in the evaporator unit. In this study, the potential of the palm oil empty fruit bunch, palm oil shell, and tree bark have been used as fuel from biomass to generate electricity by integrated with ORC. The Syltherm-XLT thermal oil was used as the heat carrier from combustion burner, while R245fa was used as the working fluid for ORC system. Through Aspen Plus, this study analyses the influences on performance of main thermodynamic parameters, showing the possibilities of reaching an optimum performance for different working conditions that are characteristics of different design parameters.

1. Introduction

Fuel lately, is a hot topic to be discussed on various occasions. This is driven by increasing demand and fuel prices. In the meantime, the source of fuel oil and gas is decreasing. As a consequence, it is a necessity to seek other sources. One alternative is the utilization of renewable energy and used to replace the use of fuel oil and natural gas [1]. Biomass is the most widely used renewable energy source in the world today. It is used mostly in solid form and, to a lesser extent, in the form of liquid fuels and gas. The utilization of biomass for energy production has increased at only a modest rate in modern times [2].

Bioenergy can be produced from a variety of biomass feed stocks, including forest, agricultural and livestock residues; short-rotation forest plantations; energy crops; the organic component of municipal solid waste; and other organic waste streams. Through a variety of processes, these feed stocks can be directly used to produce electricity or heat, or can be used to create gaseous, liquid, or solid fuels [3-4].

Pyrolysis is a thermochemical decomposition process at elevated temperature in absence of oxygen. The pyrolysis gas contains mainly H_2, CO, CO_2, CH_4, and others light hydrocarbons such as C_2H_2,

IOP Conf. Series: Materials Science and Engineering **308** (2018) 012030 doi:10.1088/1757-899X/308/1/012030

C_2H_4, C_2H_6, C_3H_6, C_3H_8. Pyrolysis of biomass yields gases, liquids (so-called bio-oils), and a carbonaceous residue (so-called bio-char). The obtained yields depend on the feedstock composition and pyrolysis conditions. There are three types of pyrolysis: fast pyrolysis with a typical maximum bio-oil yield of around 60-80 wt. % at very short residence times, convectional pyrolysis with equal yields of product at middle residence time (minutes), and as the last, slow pyrolysis with a maximum solid yield of around 60-80 wt. % at very long residence time (hours) [5].

2. Technology Description

Figure 1 illustrates an ORC system that utilizes heat from biomass pyrolysis. Biomass is heated to the working temperature. Then the pyrolysis gas heats the organic working fluid, and expander to utilize energy contained in the organic working fluid for power generation.

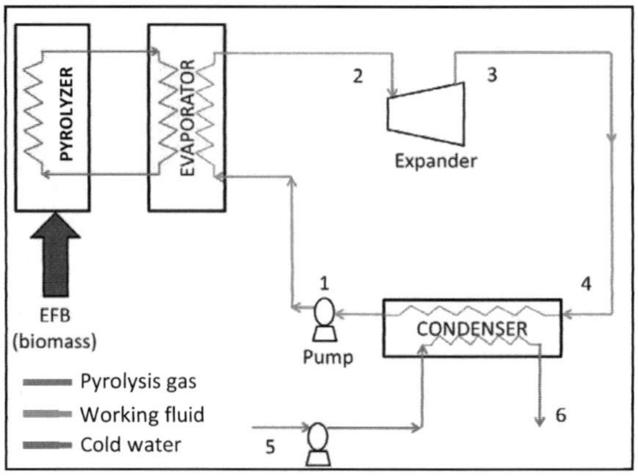

Figure 1. Schematic diagram of the Basic ORC system

The ORC operation principle is the same as the conventional Rankine cycle [6]. But, in this analysis, the working fluid is an organic compound of low boiling point instead of water, thus decreasing the temperature needed for evaporation. A basic ORC consists of four processes: increasing pressure of the working fluid through a pump, high temperature heat addition through an evaporator, expansion of high temperature as well as high pressure fluid through a turbine/expander, and low temperature heat rejection through a condenser.

3. Methodology

3.1. Aspen Plus flowsheet

Figure 2 depicts the Aspen Plus flowsheet of the ORC with empty fruit bunches pyrolysis system. Seven unit operation models (RYield, RGibbs, SSplit, Flash2 Pump, Compr, HeatX) have been used to simulate the system. The biomass is defined as an unconventional component (processes with solids). The biomass is first sent in a RYield reactor to decompose into its elemets (C, H, O, N, S, etc.). RGibbs reactor is used to simulate pyrolysis of biomass. The decomposed elements will react in the RGibbs reactor. RGibbs reactor models chemical equilibrium by minimizing Gibbs free energy.

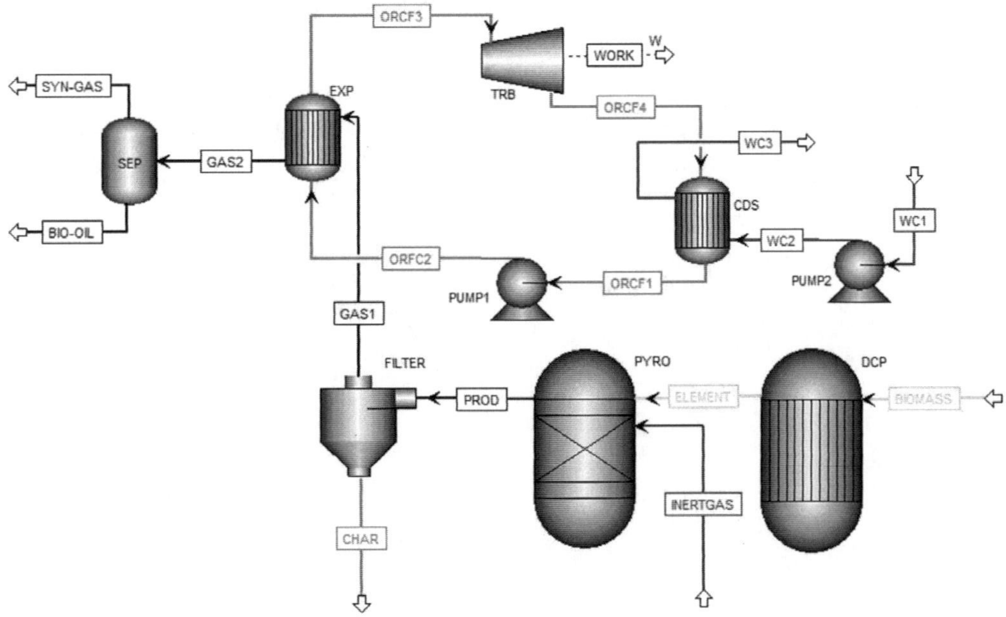

Figure 2. Aspen Plus simulation model for ORC plant with a regenerator unit.

Referring to Figure 2, flue gas leaving RGibbs reactor with temperature of 400 °C then passes through evaporator to raise the temperature of ORC working fluid up to 130 °C by means of the heat transfer medium. The hot working fluid of ORC is then expanded in the turbine/expander and at the same time producing power before entering regenerator. As the temperature of the vapor at the expander outlet is more than that of the liquid at the inlet of the evaporator, it is possible to enhance the efficiency of the system through a regenerator where the low pressure vapor from the expander (state 3) supplies heat to the stream of liquid working fluid entering regenerator at the other side (state 5). The heat of condensation is conducted away via cooling tower or used further as hot water for heating purposes. A pump re-increases the working fluid pressure before passing regenerator. After the working fluid was preheated in the regenerator, the working fluid is re-evaporated in the evaporator and superheated.

3.2. System performance

Based on the energy equilibrium system, the mathematical model of ORC developed are given in Table 1.

Table 1. Mathematical model of ORC [7].

Component	Mathematical Model
Pump	$W_p = \dot{m}\ (h_2 - h_1)$
Evaporator	$Q_{in} = \dot{m}(h_3 - h_2)$
Expander	$W_{exp} = \dot{m}\ (h_3 - h_4)$
Condenser	$Q_{out} = \dot{m}\ (h_4 - h_1)$

where W_p, Q_{in}, W_{exp}, Q_{out}, $\dot{m}$, and h are the power required by pump, heat transferred into system, the power produced by expander, heat transferred out from system, mass of biomass used per time, and enthalpy, respectively. The thermal efficiency of ORC power plant system can be calculated as follow:

IOP Conf. Series: Materials Science and Engineering **308** (2018) 012030 doi:10.1088/1757-899X/308/1/012030

$$\eta_{therm} = \frac{W_{expander} - W_{pump}}{Q_{in}} \quad (1)$$

3.3. *Input data*

The biomass utilized in this study is EFB as waste from palm oil mill which are available much at North Sumatera province. The composition of the EFB is given in Table 2. The operating parameters and other main assumptions for the plant calculation are summarized in Table 3.

Table 2. Composition of EBF [8].

EFB	
Proximate analysis (wt.%)	
Fixed carbon	18.84
Volatile matter	80
Moisture content	20
Ash content	5.89
Ultimate analysis (wt.%)	
Ash content	5.89
Carbon	48.63
Hydrogen	5.81
Sulphur	0.09
Nitrogen	0.63
Oxygen	38.95
Heating values (MJ kg^{-1})	
Lower heating value	18.5

Table 3. Main operational conditions and assumptions for plant calculation.

Temperature EFB (°C)	25
Type of working fluid for ORC unit	R245fa
Mass flow rate of working fluid (kg hr^{-1})	2500
Temperature of working fluid at the inlet expander (°C)	130
Outlet pressure of expander (bar)	2
Pressure losses during process (%)	0
Heat losses during process (%)	0
Isentropic efficiency of expander and pump (%)	70

4. Results and discussion

4.1. *Pyrolysis of Biomass*

The yield of pyrolysis product (char, bio-oil, and syngas) are shown in Figure 3. At 300 °C, the amount of char is about 28%. This amount decreases with increasing temperature pyrolysis. This also happens to syngas. On the contrary, the production of syngas increases with increasing temperature. At higher temperatures, the water gas shift reaction started, carbon monoxide and hydrogen were created, and the amount of carbon began to decrease. This part significantly differs from the experimental one. Water was evaporated below 200 °C and collected after the cooler in a condensate container, so it could not react with carbon in the reactor at higher temperature.

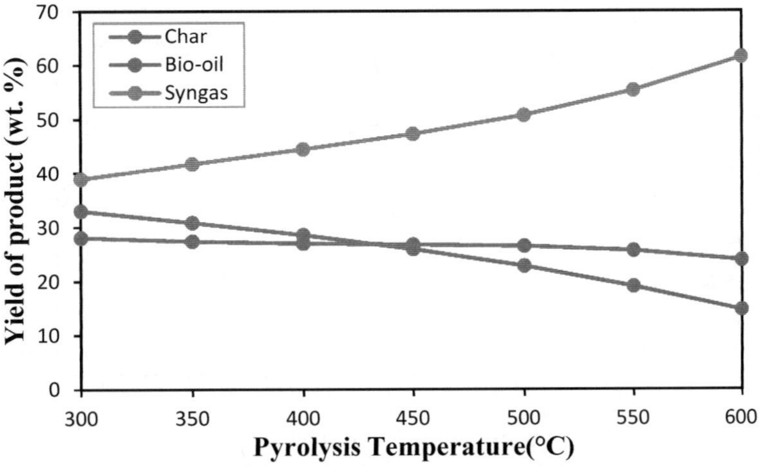

Figure 3. Yield of pyrolysis product.

The gas composition of pyrolysis product shown in Figure 4. H_2 and CO mole fractions increase with increasing temperature, whereas CO_2, and CH_4 show the opposite trend. The concentrations of light hydrocarbons were in a range on 400-500 °C, after that, light hydrocarbons were decomposing, which resulted in increases of H_2 concentrations. On the other hand, a half of the volume of produced gas was created below 400 °C that means without H_2.

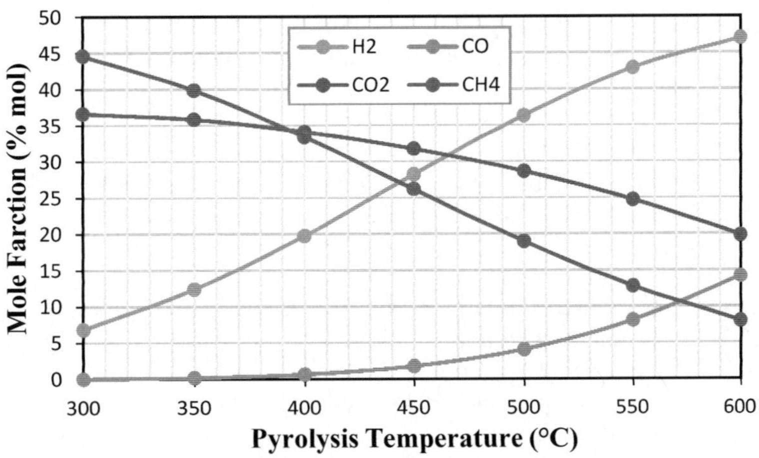

Figure 4. Mole fractions of H_2, CO, CO_2, CH_4 in the gas product as functions of pyrolysis temperature.

4.2. ORC Analysis

The ORC plant performance is estimated to improve with increase in working pressure of inlet expander as shown in Figure 5. The system generated 15.33 kW of net power at the 9 bar of inlet pressure of expander. By raising the inlet pressure of expander until it reaches 13 bar, the net power increased by 16%.

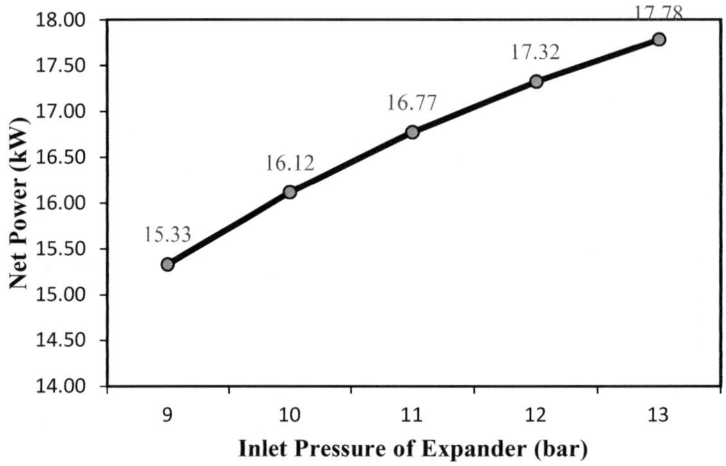

Figure 5. The effect of inlet pressure of expander on net power.

5. Conclusions

In this study, it has been analyzed an ORC integrated with pyrolysis biomass. The plant layout has been designed to utilize the energy from the EFB as fuel which are available from the waste of palm oil mill by using existing Aspen Plus unit operation block. From this preliminary design, 628 EFB per hour can generate 17.78 kW net electricity. Increasing inlet pressure of expander will increase the power output and total electrical efficiency. However, many tasks should be done to see the effect of changing design parameters on the net power output and total electrical efficiency. In addition, increasing the working temperature of pyrolysis resulted in increased amount of gas produced. Contrary to bio-oil and charcoal.

Acknowledgments

This work has been fully supported by TALENTA USU (Universitas Sumatera Utara) 2017, No: 43/UN5.2.3.1/PPM/KP-TALENTA USU/2017.

References

[1] Haryati T 2006 Biogas: waste from farm for alternative energy *Wartazoa* **16**:3 pp 160-169
[2] Toklu E 2017 Biomass energy potential and utilization in Turkey *Renew. Energ. Rev.* **107** pp 235-244.
[3] Ntona E, Arabatzis G, Kyriakopoulos G L 2015 Energy saving: views and attitudes of students in secondary education *Renew. Sustain. Energy Rev.* **46** 1-15
[4] Rosillo-Calle F, de Groot P, Hemstock S L, Woods J 2007 *The Biomass Assessment Handbook: Bioenergy for a Sustainable Environment* (Earthscan: London)
[5] Lestinsky P, and Palit A 2016 Wood pyrolysis using aspen plus simulation and industrially applicable model *GeoScience Engineering* **62**:1 pp 11-16
[6] Velez F, Segovia J J, Martin M C, Antolin G, Chejne F, Quijano A 2012 A technical, economical and market review of organic Rankine cycles for the conversion of low-grade heat for power generation *Renew. Sust. Energ. Rev.* **16** pp 4175-4189
[7] Cengel A Y and Boles M A 2006 *Thermodynamic: An Engineering Approach* Ed. 5 (USA: McGraw-Hill)
[8] Wijono A 2014 Feasibility study of biomass from empty fruit bunch palm oil tree for steam power plant application and its environment effect *Proceeding of National Symposium RAPI XIII-2014 FT UMS* (Indonesian Version)

IOP Conf. Series: Materials Science and Engineering **308** (2018) 012031 doi:10.1088/1757-899X/308/1/012031

Design of river height and speed monitoring system by using Arduino

T H Nasution*, E C Siagian, K Tanjung, Soeharwinto

Department of Electrical Engineering, Universitas Sumatera Utara, Medan, Indonesia

*E-mail: tigor.nasution@usu.ac.id

Abstract. River is one part of the hydrologic cycle. Water in rivers is generally collected from precipitation, such as rain, dew, springs, underground runoff, and in certain countries also comes from melt ice/snow. The height and speed of water in a river is always changing. Changes in altitude and speed of water can affect the surrounding environment. In this paper, we will design a system to measure the altitude and speed of the river. In this work we use Arduino Uno, ultrasonic sensors and flow rate sensors. Ultrasonic sensor HC-SR04 is used as a river height meter. Based on the test results, this sensor has an accuracy of 96.6%.

1. Introduction

Monitoring technology increasing along the development of information technology especially in the field of IoT [1-2]. Monitoring technology has been widely used in various fields. Direct monitoring helps us to conclude a problem or make decisions. It is also supported by advances in remote communications technology for control and monitoring [3-4]. In disaster management, this is very petrified. Therefore, in this research we develop a river monitoring technology.

River is a part of the hydrologic cycle. The height of the river surface is a condition that must be considered especially in the rainy season. If the river is not able to accommodate the water volume, the water will inundate the land and this condition is called the flood. The impact of flooding can be reduced by monitoring the altitude and the speed of the river flow so it could be an early warning before the flood happened.

There were several studies done before on monitoring technology especially for disaster management. Some researchers have designed monitoring technology to monitor air pollution [5-6]. In addition, telecommunication monitoring has been carried out to monitor water quality [7-8]. Research on river water monitoring has also been conducted primarily for monitoring the water quality of the river [9].

Different from previous studies, in this study we will conduct the design of speed monitoring technology and the water level in detects flooding. The design is done by using Arduino as the control center. Arduino commonly used for the design of monitoring technology [10].

2. Materials and Methods

2.1 System Requirements

In this system design, it uses Arduino Uno, G3 / 4 water velocity sensor and HC-SR04ultrasonic sensors. Arduino UNO is a microcontroller board based on ATmega328 [11]. The Arduino UNO has 14 digital

IOP Conf. Series: Materials Science and Engineering **308** (2018) 012031 doi:10.1088/1757-899X/308/1/012031

input / output pins, 6 analogue inputs, a 16 MHz Crystal oscillator, a USB connection, a power jack, an ICSP header, and a reset button.

The HC-SR04 ultrasonic sensor is an ultrasonic wave-based distortion sensor. This sensor working principle is similar with the ultrasonic radar [12]. The ultrasonic wave is transmitted and then received back by the ultrasonic receiver. The distance between transmit time and receive time is a representation of the distance of the object. This sensor is suitable for electronic applications that require distance detection [13]. It can be seen in figure 1.

Figure 1. HC-SR04 Ultrasonic Sensor

The G3 / 4 Water Velocity Sensor consists of a plastic body, rotor and hall-effect sensor [14]. When water flows through the rotor, the rotor spins. The rotation speed of the rotor changes along the speed of the water flow. The output hall sensor is a pulse signal. It can be seen in figure 2.

Figure 2. G3/4 Water Velocity Sensor

2.2 Design of Water Height Measurement

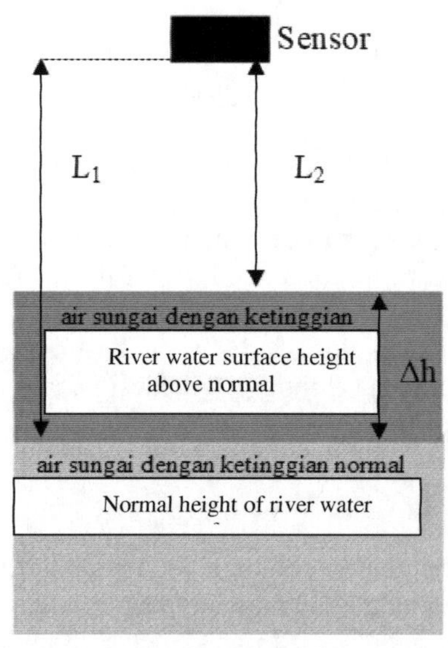

Figure 3. Ultrasonic Sensor Design Formula

IOP Conf. Series: Materials Science and Engineering **308** (2018) 012031 doi:10.1088/1757-899X/308/1/012031

Ultrasonic sensors are generally used as a distance meter. In a flood early warning system, ultrasonic sensors are used as a measure of river water level. The ultrasonic sensor is placed on top of the river water facing downward as shown in figure 3.

To calculate the height of river water then the formula used is:

$$\Delta h = L_1 - L_2$$

While:

Δh : Changes of water level

L_1 : Distance between the sensor with normal height river surface.

L_2 : Distance between the sensor with the river above the normal surface.

2.3 Hardware Design

In the phase of hardware design, the system is built for flood-prone river settlement areas. This system consists of several sensors, microcontroller and Arduino. The sensors used in this system are ultrasonic sensors and water velocity sensors. Ultrasonic sensor is used to measure the height of river water while the water velocity sensor serves the measurement of the river water speed. The sensors measurement data will be processed by Arduino microcontroller. The schematic circuit can be seen in figure 4.

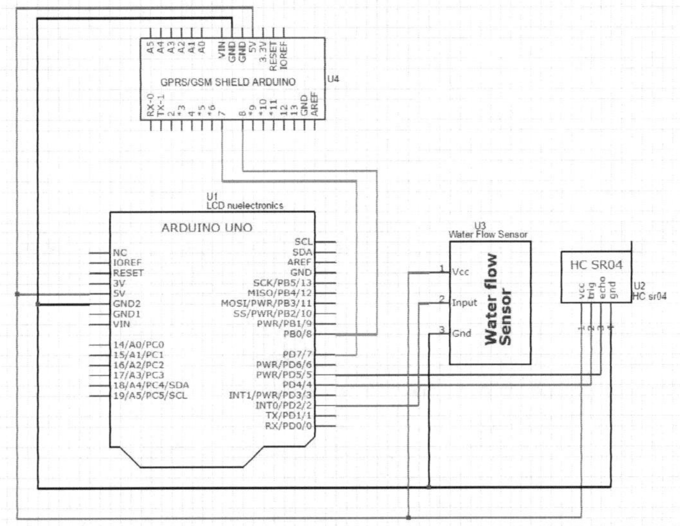

Figure 4. The schematic circuit of system

Here is an explanation of the schematic circuit drawing system:

1. Arduino Uno serves as a mainboard that controls all the components in the system. Arduino Uno uses a 11.1 volt Li-Po battery as its power supply.
2. The water velocity sensor has three pins namely vcc, signal and gnd. The signal pin is connected to the 2nd pin of the Arduino because the Arduino Uno reads numbers of pulse signals from the water velocity sensor using its interrupt function. Arduino has two pins that also have interrupt function which are 2nd and 3rd pins.
3. HC-SR04 Ultrasonic Sensor has four pins namely vcc, triger, echo and gnd. Triger pin is used to emit ultrasonic waves while the echo pin serves to receive ultrasonic waves. Triger pin on ultrasonic sensor is placed to 4th pin in Arduino and echo pin connected to 5th pin of Arduino.

3. Result and Discussion

3.1 Ultrasonic Sensor as Water Height Measurement Tool Test and Analysis

The test of ultrasonic sensor performed in a 50cm tub. The ultrasonic sensor is placed on top of the tub and connected to the Arduino. While the Arduino is connected to the laptop via a USB cable to see the

IOP Conf. Series: Materials Science and Engineering **308** (2018) 012031 doi:10.1088/1757-899X/308/1/012031

results of sensor measurements in the serial monitor. Then the tub is filled with water until the water level reaches a predetermined height. The sensor measurement results will be compared with the measurement manually (measuring the distance of the sensor to the water surface by using the meter). In this test, the distance between the ultrasonic sensor and the water surface is 136 cm.

Ultrasonic sensor test results as a height meter can be seen in table 1. From the table can be seen that the difference between the measurement of the sensor with manually measurement results. Water surface at an altitude of 0.08 m to 0.27 m has a measurement difference of 4 cm while the water level at an altitude of 0.31 m to 0.47 m has a measurement difference of 3 cm. This is because the average accuracy of distance measurements by using ultrasonic sensors at a distance of 0.8 meters to 1.3 meters is 96.6%. Although this sensor has an average accuracy of only 96.6%, it can still be tolerated to be implemented in this system.

Table 1. Testing Results Table

Water level (m)	Measured water height by sensor (m)	Distance from the sensor to water surface (m)	Distance by sensor (m)	Accuracy (%)
0.08	0.12	1.28	1.24	96,87
0.14	0.18	1.22	1.18	96.72
0.19	0.23	1.17	1.13	96.58
0.24	0.28	1.12	1.08	96.42
0.27	0.31	1.09	1.05	96.33
0.31	0.34	1.05	1.02	97.14
0.35	0.38	1.01	0.98	97.02
0.41	0.44	0.96	0.93	96.87
0.47	0.51	0.89	0.85	95.50

3.2 Waterflow Sensor Testing

During this section, the water velocity sensor testing has been done. This test was carried out by placing the water velocity sensor in the stream and monitor the measurement results in the monitor series. Sensor measurement results can be seen in figure 5.

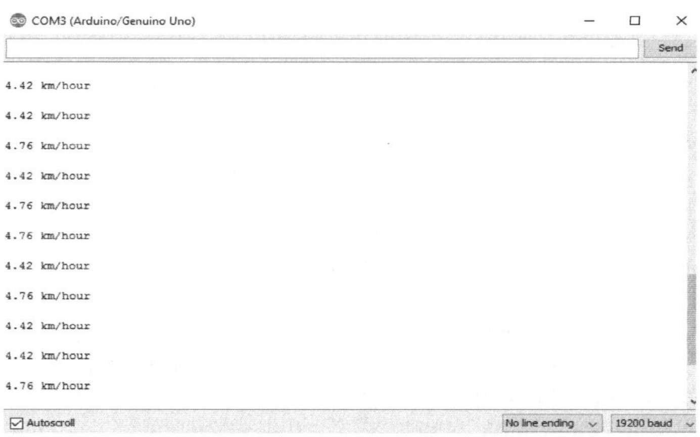

Figure 5. Waterflow Sensor Testing Result

3.3 Implementation on River Flow

System testing is done by combining the entire system device and a pre-designed testing program. The test was performed on March 1, 2017 where the test duration was approximately 27 minutes. This test is done in the Juanda River Deli. In this test, the distance of the ultrasonic sensor to the surface of the river at normal altitude is 1.65 m.

Figure 6. Assembling Devices Constructions Results

During this test, the first thing to do is the installation of hardware consisting of mainboard & sensor to its construction as shown in figure 6. Then the construction is placed at the edge of the river whose position has been determined before. The position of the apparatus shall be fixed so that the distance of the ultrasonic sensor to the normal surface of the river water remains unchanged. Then the tool is turned on to start sensor readings and perform data monitoring. Implementation of the system can be seen in figure 7.

Figure 7. Implemented Tools

River water level data graph can be seen in figure 8. In figure 8 can be seen that at 09:31 until 09:55 show the level of river water surface decreased from 65 cm to 60 cm. At 09:39, the water level of the river is 62 cm above the normal surface while at 09:42 the water level rises again to 64 cm above the normal surface. This indicates there was an error reading of the ultrasonic sensor that may be caused by noise. Noise in this case can be the floating objects flowing in the river that reduced the accuracy of sensor measurement.

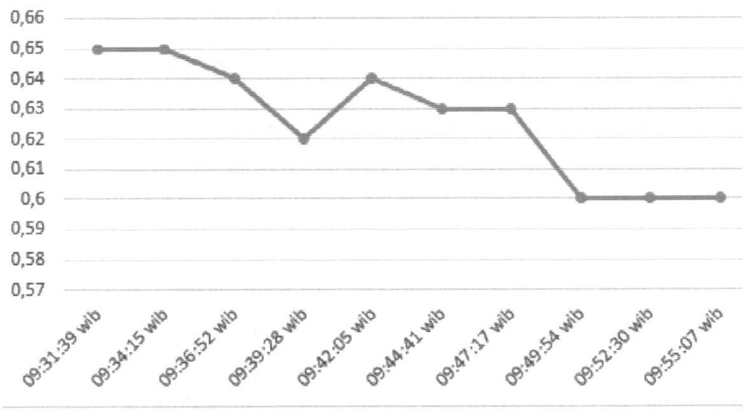

Figure 8. River Water Height Graph

4. Conclusion

From the test results above can be seen that the system can be realized in the river flow to monitor the height of river water and water velocity. From testing the water level is known that HC-SR04 sensor has an average accuracy about 96.6%. While implemented in the river flow there are several water level errors due to the floating objects flowing in the river through the ultrasonic sensor.

5. Future Works

For further research, this system will be developed by sending river water level data and stream flow rate data into the database server. Furthermore, web pages can be created to monitor the data remotely. The system is also can be applied to perform early detection of floods.

Acknowledgments

The authors gratefully acknowledge that the present research is supported by Ministry of Research and Technology and Higher Education Republic of Indonesia. We thank the Vice Rector for Research, Service, and Cooperation University of Sumatera Utara that have supported this research activity. All the faculty, staff members and lab technicians of Electrical Engineering Department, whose services turned my research a success. My Wife, Erna Kurnia and children, Zafran Ahmad Nasution and Zaura Aisya Nasution, family members and friends, without whom I was nothing; they not only assisted me financially but also extended their support morally and emotionally.

References

[1] Baker T, Asim M, Tawfik H, Aldawsari B, Buyya R 2017 An energy-aware service composition algorithm for multiple cloud-based IoT applications *J. Netw. Comput. Appl.* **89** pp 96–108

[2] Gubbi J, Buyya R, Marusic S, Palaniswami M 2013 Internet of Things (IoT): A vision, architectural elements, and future directions *Futur. Gener. Comput. Syst.* **29**:7 pp 1645–1660

[3] Soeharwinto, Sinulingga E, Siregar B 2017 Remote Monitoring of Post-eruption Volcano Environment Based-On Wireless Sensor Network (WSN): The Mount Sinabung Case *J. Phys. Conf. Ser.* **801**:1 p 12084

[4] Nasution T H, Muchtar M A, Siregar I, Andayani U, Christian E, Sinulingga E P 2017 Electrical appliances control prototype by using GSM module and Arduino *4th International Conference on Industrial Engineering and Applications, ICIEA 2017*

[5] Khedo K K, Perseedoss R, Mungur A 2010 A wireless sensor network air pollution monitoring system *Int. J. Wirel. Mob. Networks* **2**:2 pp 32–45

[6] Siregar B, Nasution A B A, Fahmi F 2016 Integrated pollution monitoring system for smart city in *2016 International Conference on ICT For Smart Society (ICISS)* pp 49–52

[7] Danve S and Barabde M 2015 Real Time Water Quality Monitoring System *Int. J. Innov. Res.*

Comput. Commun. Eng. **3:6** pp 5064–5069

[8] Rahmat R F, Syahputra M F, Lydia M S 2016 Real Time Monitoring System for Water Pollution in Lake Toba *2016 International Conference on Informatics and Computing (ICIC)* pp 383–388

[9] Halmi M I E, Jirangon H, Johari W L W, Rachman A A R, Shukor M Y, Syed M A 2014 Comparison of Microtox and Xenoassay light as a near real time river monitoring assay for heavy metals *Sci. World J.* **2014**

[10] Gertz E and Justo P D 2012 *Environmental Monitoring with Arduino*

[11] Drymonitis A 2015 *Introduction to Arduino*

[12] Indoware Ultrasonic Ranging Module HC - SR04 2013 *Datasheet* pp 1–4

[13] Mutiara G A, Agung A A G, Handayani R 2016 Sensor comparation for smart parking system *Proceeding of 2015 1st International Conference on Wireless and Telematics, ICWT 2015*

[14] Singh P and Saikia S 2017 Arduino-based smart irrigation using water flow sensor, soil moisture sensor, temperature sensor and ESP8266 WiFi module *Proceeding of IEEE Region 10 Humanitarian Technology Conference 2016 R10-HTC 2016*

Structural integrity of power generating speed bumps made of concrete foam composite

B Syam*[1], M Muttaqin[2], D Hastrino[2], A Sebayang[2], W S Basuki[3], M Sabri[1], S Abda[2]
[1]Impact and Fracture Research Center, Department of Mechanical Engineering, Faculty of Engineering, Universitas Sumatera Utara (USU)
[2]Department of Mechanical Engineering, Faculty of Engineering, Universitas Sumatera Utara
[3]Department of Chemistry, Faculty of Science, Universitas Sumatera Utara

*E-mail: bstsyam@gmail.com

Abstract. In this paper concrete foam composite speed bumps were designed to generate electrical power by utilizing the movements of commuting vehicles on highways, streets, parking gates, and drive-thru station of fast food restaurants. The speed bumps were subjected to loadings generated by vehicles pass over the power generating mechanical system. In this paper, we mainly focus our discussion on the structural integrity of the speed bumps and discuss the electrical power generating speed bumps in another paper. One aspect of structural integrity is its ability to support designed loads without breaking and includes the study of past structural failures in order to prevent failures in future designs. The concrete foam composites were used for the speed bumps; the reinforcement materials are selected from empty fruit bunch of oil palm. In this study, the speed bump materials and structure were subjected to various tests to obtain its physical and mechanical properties. To analyze the structure stability of the speed bumps some models were produced and tested in our speed bump test station. We also conduct a FEM-based computer simulation to analyze stress responses of the speed bump structures. It was found that speed bump type 1 significantly reduced the radial voltage. In addition, the speed bump is equipped with a steel casing is also suitable for use as a component component in generating electrical energy.

1. Introduction

Oil palm empty fruit bunches (EFB) is the waste produced by palm oil processing plant. Palm oil production in Indonesia reaches up to six million tons per year, as a matter of fact it also produces solid waste up to about 2.5 million tons per year [1]. In our laboratory the waste has been intensively used for several products, such as for several light structure products: tiles, parking bumper, and speed bumps [2-4].

In search of energy harvesting originated from waste materials in oil palm industry, empty fruit bunches (EFB) were used to produce EFB concrete composite to form a speed bump. As shown in Fig. 1, the structure of the speed bumps is equipped with casing and attached on top of the apparatus top plate. With this connection, the speed bump is subjected to both static (speed up to 10 km/hr) and impact load (speed more than 10 km/hr) resulted from vehicles passing on them.

In this paper, the structural integrity of the speed bump is checked using static load, the net weight of vehicle. The analysis was carried out using ANSYS 17.0. In addition, the speed bumps are also subjected to loadings generated by vehicles pass over the power generating mechanical system.

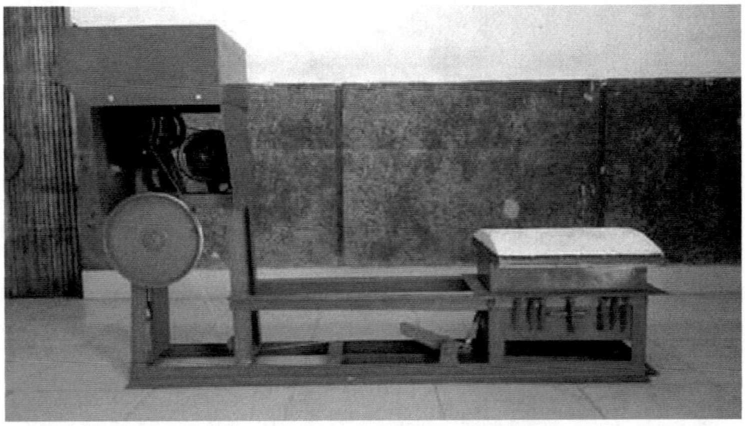

Figure 1. Energy harvesting (power generation) using speed bump

2. Materials and Method

2.1. Material

As shown in figure 1, the speed bump is placed on a bed equipped with steel casing attached to the top plate of the power generation set (genset). The speed bump receives load from wheel cars passing throuh the street (figure 2). To withstand both static and impact loads the speed bumps have to be designed and manufactured so that they meet engineering characteristics and other requirements.

Figure 2. Speed bump power generation station

There are several class of the confoam materials. In this research we choose type A5 [6] in which its physical and mechanical properties of the material are shown in table 1.

Table1. Confoam mechanical properties

Type spesimen	S_{uc}(MPa)	S_{ut}, (MPa)	ν	E (MPa)
A5	2.1	0.18	0.2	10.1

2.2. Geometry and Dimension

Geometry and dimensions of the speed bumps is shown in figure. 2, respectively.

IOP Conf. Series: Materials Science and Engineering **308** (2018) 012032 doi:10.1088/1757-899X/308/1/012032

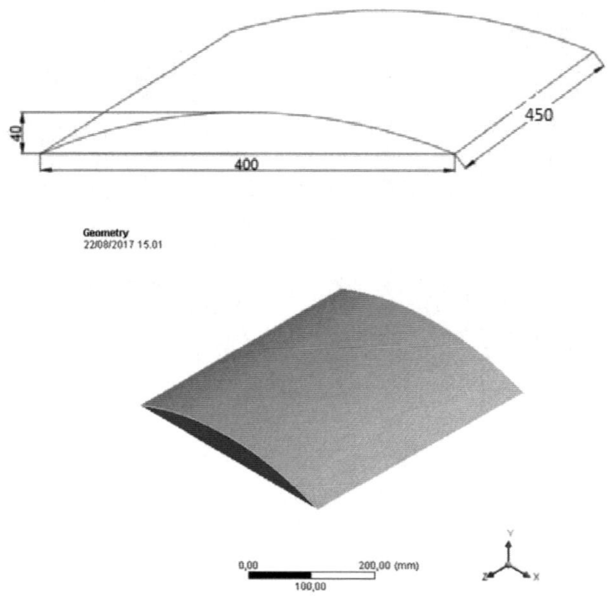

Figure 3. Geometry and dimension of speed bump

3. Numerical Simulation

3.1. Geometry and Load Model

The geometry and dimensions of the finite element models are shown in figure 3. For simulation, one model is fixed on the bed plate and the second model is enclosed with steel casing (not shown). The latest model is actually the one meets speed bump requirement in strength. We use static load of 3.29 tonf calculated from one fourth of vehicle weight permitted for class III road. We use ANSYS 17.0 to calculate the stress distribution in x, y, principle direction using 3-D elements. For steel case model we also calculate the von-Mises stress.

3.2. Stress contour

Figure 4 shows the stress distributions on x, y, and principle direction on the speed bump models. All calculations are presented in table 2. It can be shown that the normal stress σ_x is obtained that However, by comparing the static strength of speed bump material with stresses, and in accordance with failure theory the speed bumps will fail under static loading, except for bar diameter of three inches. We observed that it decreased the maximum stress significantly from 0.6 MPa to 0.0001 MPa. The result dictates that we may use this design for power generation speed bump. It is also interesting to observe the response of speed bumps when they are enclosed with steel or polymeric composite bars (reported in other paper) and shown in table 3.

Table 2. Max stress in x, y, and principle direction

Calibration	Speed bump concrete foam (without steel plate)	Speed bump using steel plate
σ_x	0,17152	25,514
σ_y	0,213	45,889
σ_1	2.2312	56,254

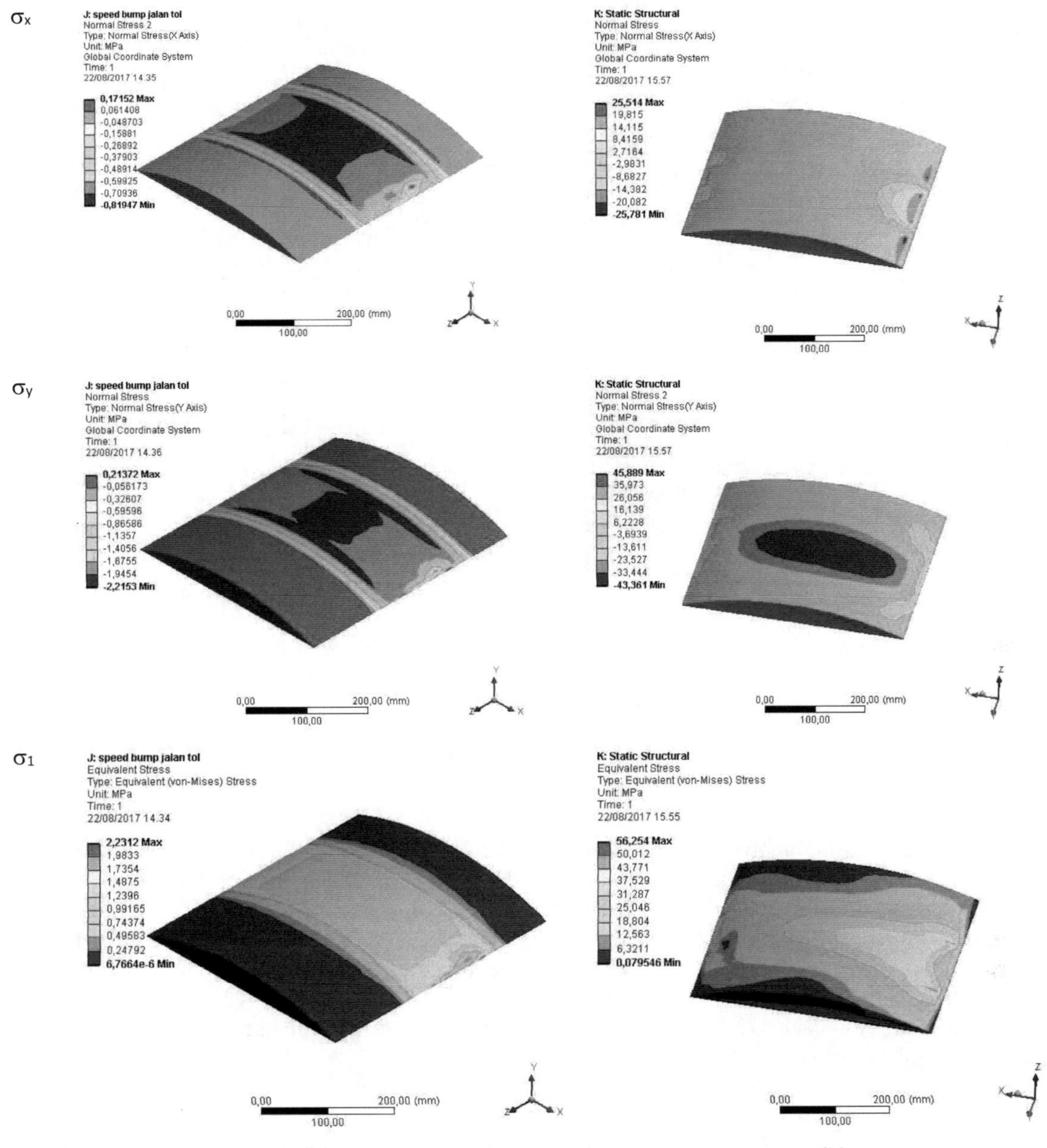

(a) (b)

Figure 4. Contour stress distribution speed bump
(a) without steel plate (b) Using Steel Plate

Table 3. Photographs of the tested materials

Type / Testing	Photo	Static Structural	Crack or not
A1			Crack
A2		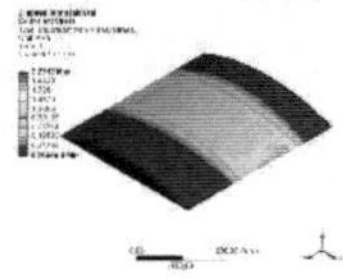	Not cracked
B1			Not Cracked
B2	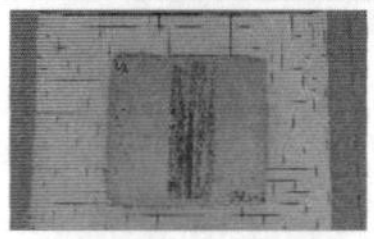		Not Cracked
C1	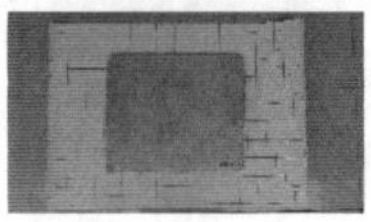	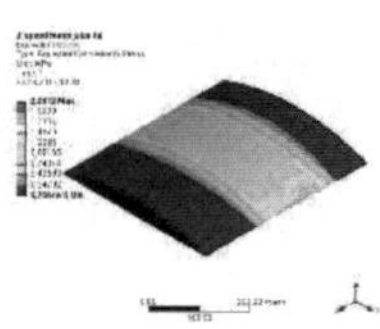	Crack
C2	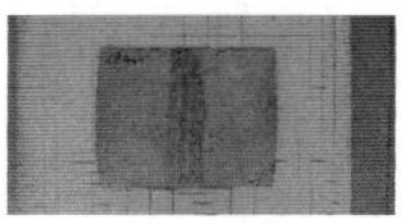		Crack

4. Conclusions

This paper discussed the analysis of speed bump made of concrete foam composite which isused to generate electrical power. Two models were proposed, they are: (a) speed bump without steel plate, (b) speed bump using with steel casing. To observe the structure integrity of the speed bumps the models were analyzed using a FEM-based numerical softwares. It was obtained that speed bump without steel plate reduces the stresses and may absorb axial loading resulted from vehicle tyres; however, it fails to sustain radial loading, except for the using steel plate. From the FEM analysis we also found that speed bumps enclosed with steel case (or polymeric case) are also suitable for use as part of a system in producing electrical energy.

ACKNOWLEDGEMENTS

The authors acknowledge the research fund obtained from Kemenristekdikti under PUPT scheme 2016.

5. References

[1]. Nuryanto 2000 The utilizing of empty fruit bunch for chemistry [in Indonesia: Pemanfaatan Tandan Kosong Kelapa Sawit Sebagai Sumber Bahan Kimia] *Warta PPKS* **8**(3) pp 137-144

[2]. Irwan A, Syam B, Basuki W S 2016 Analysis of Tile Mechanical Response from Confoam Materials Reinforced by EFB Fiber Due to Impact of Hail Rainfall [in Indonesia: Analisa Respon Mekanik Genteng dari Bahan Confoam Diperkuat Serat TKKS Akibat Beban Impak Hujan Es] *Proceeding of Seminar DIES USU*

[3]. Hastrino D, Pasi I, Muttaqin M, Bustami Syam B, Basuki W S, Sabri M, Abda S 2016 Structural Analysis of Speed Bump Alloys Concrete Foam and Polymeric Foam Materials with Various Cavity Using Ansys Software [in Indonesia: Analisa Struktur Speed Bump Paduan Bahan

Concrete Foam dan Polymeric Foam dengan Variasi Rongga Menggunakan Sofware Ansys 14.5] *Proceeding of Seminar DIES USU*

[4]. http://id.wikipedia.org/wiki/Polisi_tidur, acessed by : 5 mei 2017.

[5]. Syam B, Wirjosentono B, Rizal S 2013 Development of Polymeric Foam Composite Foam Composite Materials Reinforced Blanks of Palm Oil Empty Fruit Bunches (TKKS) for the Making of Various Industrial Products of Transportation and Sports [in Indonesia: Pengembangan Bahan Komposit Busa Polimer (Polymeric Foam Composite) Diperkuat Serat Tandan Kosong Kelapa Sawit (TKKS) untuk Pembuatan Berbagai Produk Industri Transportasi dan Olah Raga] *Reprot submitted to PENPRINAS MP3EI DIKTI*

[6]. Syam B, Fitriadi N 2015 *Concrete Foam Composite Production and Testing* Medan, Indonesia

IOP Conf. Series: Materials Science and Engineering **308** (2018) 012033 doi:10.1088/1757-899X/308/1/012033

Redefining the magic square on numerical characters

M K M Nasution[1*], **Sawaluddin**[2]

[1]Teknologi Informasi, Fasilkom-TI, Universitas Sumatera Utara, Padang Bulan 20155 USU Medan Indonesia

[2]Matematika, FMIPA, Universitas Sumatera Utara, Padang Bulan 20155 USU Medan Indonesia

E-mail: mahyuddin@usu.ac.id

Abstract. As a number system, the magic square is different from the others. Characteristic depends not only on size but also depends on numerical character in computation. This paper has redefined the term of magic square formally, by exposing the inductive general characteristics of cases to numerical ordering of numbers.

1. Introduction

The magic square is a numerical system that has long filled the human life [1], which has a special character [2]. The magic square although square, but computational nature is different from matrix and also other number systems [3]. There is a term of different understanding with formal formulation in mathematics about magic square [4, 5]. In addition, different types of magic square also give effect to the formalization of this system [6]. Therefore, redefinition needs to be done formally by first exploring some of the basic characteristics of numbering cases.

Redefining an object is not as easy as building a definition from scratch [7, 8]. Sometimes certain terms in language deny the truth of the statement to be built [9]. Therefore, some characteristics generated so far are only disclosed as axiom of definition support, and as assumptions. Characteristics excavated from the cases are inductively generalized. Thus, this paper aims to rediscover the formal definition of magic square by considering the odd size of magic square and some of numerical characters.

2. Problem Definition

Sometime the dictionary does not precisely define something: The term of magic square has been termed as

T *A square containing a number of integers arranged so that the sum numbers is the same in each row, column, and main diagonal and often in some or all of the other diagonals.*

T in Merriam-Websters online dictionary[1], thereby it provides improper understanding to who makes the study about the magic square.

The magic square is a two-dimensional array a_{ij} with size $n \times n$ or $i, j = 1, \ldots, n$, n can be odd or even. For n is odd, the magic square contains a number with the index $k \in I$, $k = 1, \ldots, n \times n$

[1] https://www.merriam-webster.com/dictionary/magic%20square

being entered in a way into the two-dimensional array with which n mod $2 = 1$ [2]. So each row, column, and two main diagonals have the same number. In other words, for first row we have the sum of numbers in first, second, ... rows

$$\Sigma r_1 = \sum_{i=1,j=1}^{1,n} a_{ij}, \quad \Sigma r_2 = \sum_{i=2,j=1}^{2,n} a_{ij}, \quad \sigma r_n = \sum_{i=n,j=1}^{n,n} a_{ij}. \tag{1}$$

For first column we obtain the first, second, ... columns

$$\Sigma c_1 = \sum_{i=1,j=1}^{n,1} a_{ij}, \quad \Sigma c_2 = \sum_{i=1,j=2}^{n,2} a_{ij}, \quad \Sigma c_n = \sum_{i=1,j=n}^{n,n} a_{ij}. \tag{2}$$

8	1	6
3	5	7
4	9	2

Figure 1. Positive integers in 3×3 magic-square.

For two main diagonals we have the sum of numbers are

$$\Sigma d_l = \sum_{j=i,i=1}^{n,n} a_{ij}, \quad \Sigma d_r = \sum_{j=n,i=1}^{1,n} a_{ij}, \tag{3}$$

where

$$\Sigma r_1 = \Sigma r_2 = \ldots = \Sigma r_n = \Sigma c_1 = \Sigma c_2 = \ldots = \Sigma c_n = \Sigma d_l = \Sigma d_r. \tag{4}$$

Suppose we take the simplest example of a magic square measuring 2×3 or $n = 3$ (odd). Entries involve integers $1, 2, 3, 4, 5, 6, 7$, and 9 and the result is like Fig. 1, but for the same size can also be submitted the numbers -1,-2,-3,-4,-5,-6,-7,-8 and -9 with which a magic square (see Fig. 2).

-8	-1	-6
-3	-5	-7
-4	-9	-2

Figure 2. Negative integers in 3×3 magic-square.

or by involving the numbers 2,4,6,8,10,12,14,16, and 18, and we get a magic square (see Fig. 3).

16	2	12
6	10	14
8	18	4

Figure 3. Integers with a differentiator 2 in 3×3 magic-square.

or with the real numbers 0.1,0.2,0.3,0.4,0.5,0.6,0.7,0.8 and 0.9, and a magic square of them is Fig. 4.

0.8	0.1	0.6
0.3	0.5	0.7
0.4	0.9	0.2

Figure 4. Decimal numbers in 3×3 magic-square.

IOP Conf. Series: Materials Science and Engineering **308** (2018) 012033 doi:10.1088/1757-899X/308/1/012033

or etc. Thus, numbers have a uniform pattern or a difference of up and down the same number for all entrain of a magic square,

$$e_k = k * b \tag{5}$$

for $k = 1, \ldots, n \times n$, where b is the difference between two adjacent numbers in a sequence.

Although the sum of entry numbers for each row, column, and main diagonal are the same. The arrangement of numbers on the first magic square reveals that for diagonal the numbers $1, 7$ and 4 have a sum is 12, whereas the diagonal of the numbers $3, 9$, and 6 have a sum is 18, on the other hand, the diagonal with the number $1, 3$ and 2 have a sum 6 while for the diagonal of numbers $9, 7$ and 8 have a sum is 24. Therefore, this sample example cancels the explication of the magic square T above.

Also, suppose $n = 5$, the size of the magic square is 5×5, for example by involving integers of $1, 2, 3, 4, 5, 6, 7, 8, 9, 10, 11, 12, 13, 14, 15, 16, 17, 18, 19, 20, 21, 22, 23, 24$, and 25 the result is like Fig. 5.

17	24	1	8	15
23	5	7	14	16
4	6	13	20	22
10	12	19	21	3
11	18	25	2	9

Figure 5. Positive integers in 5×5 magic-square.

The sum of numbers on each row, column, and main diagonal is 65. However, the sum of numbers on the diagonals parallel to one of the main diagonals is also 65, whereas the sum of numbers on diagonals parallel to the other diagonal is not equal to 65.

Therefore, each magic square has a common and special characteristics. The common characteristics generally hold as axioms in definition, whereas the special characteristics as the conditions in assumptions or statements for theories.

3. An Approach

To express the definition of magic squares formally, we will classify the magic square characteristics into two parts [2]. That is generally applicable characteristics, or characteristics that apply specifically. This classification is based on the properties and order of numbers that fill the square-shaped array based on numerical characteristics [10].

To reveal the general characteristics of magic squares, we do a study of some cases of odd-sized magic squares, ranging from $n = 3$ to $n = 15$. This general characteristics is inductively formed and becomes the axioms that support the definition about magic square. While the special characteristics are derived will form the theory specifically about the properties of magic square [11, 12]. Excavation of special characteristics is disclosed on the basis of cases that inductively also form specific formulations.

4. An Investigation of Axioms

To extract the general axioms of the magic squares we re-present a sequence of the magic squares of $n = 3, 5, 7, 9, 11, 13, 15$ or odd only, like Figs. 1., 5., 6., 7., 8., 9., and Fig. 10. From the entire magic square described above that all satisfy Eqs. (1) to (9) so that a general definition of magic squares can be formally expressed as follows.

30	39	48	1	10	19	28
38	47	7	9	18	27	29
46	6	8	17	26	35	37
5	14	16	25	34	36	45
13	15	24	33	42	44	4
21	23	32	41	43	3	12
22	31	40	49	2	11	20

Figure 6. Positive integers in 7×7 magic-square.

47	58	69	80	1	12	23	34	45
57	68	79	9	11	22	33	44	46
67	78	8	10	21	32	43	54	56
77	7	18	20	31	42	53	55	66
6	17	19	30	41	52	63	65	76
16	27	29	40	51	62	64	75	5
26	28	39	50	61	72	74	4	15
36	38	49	60	71	73	3	14	25
37	48	59	70	81	2	13	24	35

Figure 7. Positive integers in 9×9 magic-square.

68	81	94	107	120	1	14	27	40	53	66
80	93	106	119	11	13	26	39	52	65	67
92	105	118	10	12	25	38	51	64	77	79
104	117	9	22	24	37	50	63	76	78	91
116	8	21	23	36	49	62	75	88	90	103
7	20	33	35	48	61	74	87	89	102	115
19	32	34	47	60	73	86	99	101	114	6
31	44	46	59	72	85	98	100	113	5	18
43	45	58	71	84	97	110	112	4	17	30
55	57	70	83	96	109	111	3	16	29	42
56	69	82	95	108	121	2	15	28	41	54

Figure 8. Positive integers in 11×11 magic-square.

Definition 1. Let e_k is a sequance of numbers on pattern Eq. (4) $e_k = k * b$, b is difference between two adjacent numbers and $k = 1, \ldots, m$, and let a_{ij} is an array where $i, j = 1, \ldots, n$ such that $m = n \times n$, $n \bmod 2 = 1$. An odd magic square is a_{ij}, we notated by $[a_{ij}]$, with a unary-operation f

$$\Sigma a_{ij} = f(e_k), \tag{6}$$

and

$$\Sigma r_i = \sum_{i,j=1}^{i,n} a_{ij}, i = 1, \ldots, n \tag{7}$$

and

$$\Sigma c_i = \sum_{i=1,j}^{n,j} a_{ij}, j = 1, \ldots, n \tag{8}$$

IOP Conf. Series: Materials Science and Engineering **308** (2018) 012033 doi:10.1088/1757-899X/308/1/012033

and

$$\Sigma d_l = \sum_{i=1,j=i}^{n,n} a_{ij} \tag{9}$$

and

$$\Sigma d_r = \sum_{i=1,j=n}^{n,1} a_{ij} \tag{10}$$

so that $\Sigma r_1 = \ldots = \Sigma r_n = \Sigma c_1 = \ldots = \Sigma c_n = \Sigma d_l = \Sigma d_r$.

93	108	123	138	153	168	1	16	31	46	61	76	91
107	122	137	152	167	13	15	30	45	60	75	90	92
121	136	151	166	12	14	29	44	59	74	89	104	106
135	150	165	11	26	28	43	58	73	88	103	105	120
149	164	10	25	27	42	57	72	87	102	117	119	134
163	9	24	39	41	56	71	86	101	116	118	133	148
8	23	38	40	55	70	85	100	115	130	132	147	162
22	37	52	54	69	84	99	114	129	131	146	161	7
36	51	53	68	83	98	113	128	143	145	160	6	21
50	65	67	82	97	112	127	142	144	159	5	20	35
64	66	81	96	111	126	141	156	158	4	19	34	49
78	80	95	110	125	140	155	157	3	18	33	48	63
79	94	109	124	139	154	169	2	17	32	47	62	77

Figure 9. Positive integers in 13×13 magic-square.

122	139	156	173	190	207	224	1	18	35	52	69	86	103	120
138	155	172	189	206	223	15	17	34	51	68	85	102	119	121
154	171	188	205	222	14	16	33	50	67	84	101	118	135	137
170	187	204	221	13	30	32	49	66	83	100	117	134	136	153
186	203	220	12	29	31	48	65	82	99	116	133	150	152	169
202	219	11	28	45	47	64	81	89	115	132	149	151	168	185
218	10	27	44	46	63	80	97	114	131	148	165	167	184	201
9	26	43	60	62	79	96	113	130	147	164	166	183	200	217
25	42	59	61	78	95	112	129	146	163	180	182	199	216	8
41	58	75	77	94	111	128	145	162	179	181	198	215	7	24
57	74	76	93	110	127	144	161	178	195	197	214	6	23	40
73	90	92	109	126	143	160	177	194	196	213	5	22	39	56
89	91	108	125	142	159	176	193	210	212	4	21	38	55	72
105	107	124	141	158	175	192	209	211	3	20	37	54	71	88
106	123	140	157	174	191	208	225	2	19	36	53	70	87	104

Figure 10. Positive integers in 15×15 magic-square.

In particular it is found that there is one main diagonal having the same number of sequences: Fig. 1 revealing the numbers $4, 5, 6$ filling the main diagonal for the 3×3 magic square; Fig. 5 illustrates the order of numbers $11, 12, 14, 14$, and 15 in the main diagonal of the 5×5 magic square; Fig. 6 reveals that the order of numbers in the main diagonal of 7×7 magic square are $22, 23, 24, 25, 26, 27, 28$; and etc. This characteristic applies specifically, but cannot be taken as a general axiom. For example for 5×5 magic square (Fig. 5) because if two left columns are

IOP Conf. Series: Materials Science and Engineering **308** (2018) 012033 doi:10.1088/1757-899X/308/1/012033

removed and alternate with two in the right column and vice versa, a different arrangement will be obtained, whereas Eq. (1) to Eq. (4) remain valid, see Fig. 11.

8	15	1	17	24
14	16	7	23	5
20	22	13	4	6
21	3	19	10	12
2	9	25	11	18

Figure 11. Positive integers in other 5×5 magic-square.

Therefore, there is a general presumption of characteristic as follows

Assumption 1. If $[a_{ij}]$ is an odd magic square where $i, j = 1, \ldots, n$ and $n \bmod 2 = 1$, then there is one main diagonal d_r with a same difference b.

Under the Assumption 1, the other main diagonal d_l does not have the same difference b.

Another diagonal $d_{(l)}$ and $d_{(r)}$, or not main diagonal d_r and d_l, parallel to d_l and d_r have different characteristics, respectively. All diagonals $d_{(r)}$ not only have different results, but their characteristics depend on n. Thus there are characteristics that are not generally applicable, i.e.

Assumption 2. If $[a_{ij}]$ is an odd magic square where $i, j = 1, \ldots, n$, then all diagonals $d_{(r)}$ have the sum of n numbers $\Sigma d_{(r)}$ equal to Σd_r, $n = 5, 7, 11, 13, \ldots$, or

$$\Sigma d_{(r)_\ell} = \Sigma d_r; \ell = 1, 2, \ldots, (n-1); n = 5, 7, 11, 13, \ldots \tag{11}$$

Under the Assumption 2, for $n = 3, 9, 15, \ldots$ in general the $\Sigma d_{(r)_\ell}$, $\ell = 1, 2, \ldots, (n-1)$ do not equal to Σd_r. For example, the sum result of

$$\begin{array}{ccccccccccc} & & 6 & & & 1 & & & 8 & & \\ 3 & & & \neq & & & 7 & \neq & & 5 & \\ & 9 & & & 4 & & & & & & 2 \end{array}$$

$\Sigma d_{(r)_1} = 18 \neq \Sigma d_{(r)_2} = 12 \neq \Sigma d_r = 15$, but $(\Sigma d_{(r)_1} + \Sigma d_{(r)_2})/2 = \Sigma d_r$. While for $n = 9$ and $n = 15$, we have different results of sum about numbers in diagonals. For $n = 9$ from d_r to right-up (left-down) we have

$$\begin{aligned} \Sigma d_{(r)_1} &= 57 + 78 + 18 + 30 + 51 + 72 + 3 + 24 + 45 &= 378 \\ \Sigma d_{(r)_2} &= 67 + 7 + 19 + 40 + 61 + 73 + 13 + 34 + 46 &= 360 \\ \Sigma d_{(r)_3} &= 77 + 17 + 29 + 50 + 71 + 2 + 23 + 44 + 56 &= 369 \end{aligned}$$

so $\Sigma d_{(r)_4} = 378$, $\Sigma d_{(r)_5} = 360$, $\Sigma d_{(r)_6} = 369$, $\Sigma d_{(r)_7} = 378$, and $\Sigma d_{(r)_8} = 360$. We have one characteristic for $\ell = 3$ and $\ell = 6$ that is $\Sigma d_{(r)_j} = \Sigma d_r$. For $n = 15$ from d_r to right-up (left-down) we have

$$\begin{aligned} \Sigma d_{(r)_1} &= 138 + 171 + 204 + 12 + 45 + 63 + 96 + 129 + 162 + 195 + 213 + 21 + 54 + 87 + 120 \\ &= 1710 \\ \Sigma d_{(r)_2} &= 154 + 187 + 220 + 28 + 46 + 79 + 112 + 145 + 178 + 196 + 4 + 37 + 70 + 103 + 121 \\ &= 1680 \\ \Sigma d_{(r)_3} &= 170 + 203 + 11 + 44 + 62 + 95 + 128 + 161 + 194 + 212 + 20 + 53 + 86 + 119 + 137 \\ &= 1695 \end{aligned}$$

in pattern similar to them are $\Sigma d_{(r)_4} = \Sigma d_{(r)_7} = \Sigma d_{(r)_{10}} = \Sigma d_{(r)_{13}} = 1710$, $\Sigma d_{(r)_5} = \Sigma d_{(r)_8} = \Sigma d_{(r)_{11}} = \Sigma d_{(r)_{14}} = 1680$, whereas $\Sigma d_{(r)_6} = \Sigma d_{(r)_9} = \Sigma d_{(r)_{12}} = 1695$. Thus, for $\ell = 3, 6, 9, 12$ we have $\Sigma d_{(r)_\ell} = \Sigma d_r$. Or

Assumption 3. If $[a_{ij}]$ is an odd magic square where $i, j = 1, \ldots, n$, then all diagonals $d_{(r)_\ell}$, $\ell = 3, 6, 9, 12, \ldots$, have the sum of n numbers $\Sigma d_{(r)_\ell}$ equal to Σd_r, $n = 9, 15, \ldots$, or

$$\Sigma d_{(r)_\ell} = \Sigma d_r; \ell = 3, 6, 9, \ldots, (n-1); n = 9, 15, \ldots \tag{12}$$

Based on cases in Figs. 1, 5, 6, 7, 8, 9, all diagonals parallel to d_l have different sums of numbers in diagonals and does not equal to Σd_l (sum of numbers in main diagonal), i.e.

Assumption 4. If $[a_{ij}]$ is an odd magic square where $i, j = 1, \ldots, n$, then all diagonals $d_{(l)}$ have the sum of n numbers $\Sigma d_{(l)}$ does not equal to Σd_l, $n = 3, 5, \ldots$, or

$$\Sigma d_{(l)} = \Sigma d_l, n = 3, 5, 7, 11, 13, \ldots \tag{13}$$

In some cases, we have formulated some of the assumptions underlying the redefinition of the magic squares.

5. Conclusion

Some characteristics of the magic square have been inductively expressed, and redefined the intent of the term magic square formally, albeit based only on odd magic squares. In some cases some assumptions are derived inductively, but theoretically may be generally applicable, so there needs to be proof. Therefore, the future work will be to produce the characteristic of the even magic square, then to math the assumptions into a generally proven statement.

References

[1] W S Andres 1960 *Magic Squares and Cubes*, New York: Dover.
[2] M K M Nasution 2012 Algebraic on magic square of odd order n *Proceedings of the 1st IMT-GT Regional Conference on Mathematics, Statistics and Their Application.*
[3] J Chernick 1938 Solution of the general magic square *Mathematical Association of America* **45(3)**.
[4] T G Room 1955 A new type of magic square *The Mathematical Gazette* **39(330)**.
[5] B W Westbury 2003 *R*-matrices and the magic square *Journal of Physics A: Mathematical and General* **36(7)**.
[6] D H Pelletier 1987 Merlin's magic square *Americal Mathematical Montly* **94(2)**.
[7] M K M Nasution 2017 SumutSiana *TALENTA Conference on Engineering, Science and Technology*
[8] M K M Nasution 2017 Social network mining (SNM): A definition of relation between the resources and SNA *International Journal on Advanced Science, Engineering and Information Technology* **6(6)**.
[9] M K M Nasution 2017 Studies on behaviour of information to extact the meaning behind the behaviour *Journal of Physics: Conference Series* **801(1)**.
[10] C H Barton, and A Sudbery 2003 Magic squares and matrix models of Lie algebras *Advances in Mathematics* **180(2)**.
[11] M K M Nasution 2000 The number triangle of Laurent polynomial on knots *Jurnal Matematika dan Terapannya* **1(2)**.
[12] O W S Huang, H N H Cheng, and T-W Chan 2007 Number jigsaw puzzle: A mathematical puzzle game for facilitating players' problem-solving strategies *The First IEEE International Workshop on Digital Game and Intelligent Toy Enhanced Learning*, DIGITEL '07.

IOP Conf. Series: Materials Science and Engineering **308** (2018) 012034 doi:10.1088/1757-899X/308/1/012034

Effect of distributed generation installation on power loss using genetic algorithm method

A Hasibuan[*1], **S Masri**[2], **W A F W B Othman**[2]

[1)]Department of Electrical Engineering, Universitas Malikussaleh, Indonesia
[2)]School of Electric and Electronic Engineering – Universiti Sains Malaysia, Malaysia

[*]E-mail :arnawan@unimal.ac.id

Abstract. Injection of the generator distributed in the distribution network can affect the power system significantly. The effect that occurs depends on the allocation of DG on each part of the distribution network. Implementation of this approach has been made to the IEEE 30 bus standard and shows the optimum location and size of the DG which shows a decrease in power losses in the system. This paper aims to show the impact of distributed generation on the distribution system losses. The main purpose of installing DG on a distribution system is to reduce power losses on the power system.Some problems in power systems that can be solved with the installation of DG, one of which will be explored in the use of DG in this study is to reduce the power loss in the transmission line. Simulation results from case studies on the IEEE 30 bus standard system show that the system power loss decreased from 5.7781 MW to 1,5757 MW or just 27,27%. The simulated DG is injected to the bus with the lowest voltage drop on the bus number 8.

1. Introduction

The need for electric power is now increasing due to the increasing demand for load due to the rapid economic growth. This economic growth is driving the acceleration of the industrial world to use equipment that requires more electrical energy. Therefore, the supply and quality of electrical energy needs to be improved.Electrical energy supply that there is generally a conventional power plant. Conventional power plants are generally designed on a large scale, centralized, and built away from load centers requiring transmission and distribution networks to supply electricity. Conventional electric power system consists of three parts, ie generation, transmission and distribution associated with the load.

Long transmission and distribution networks from power plants to load centers result in greater power losses, so technical efforts are required to reduce such losses such as network shortening, reconductor, insertion of the substation, capacitor installation, automatic voltage regulator (AVR), Replacement of connectors and so on.

In recent decades, distributed generation began to be developed in the developed countries to support the electricity needs of the country. Distributed generation is considered as an appropriate solution to overcome the shortage of energy supply and to overcome the problem of power distribution systems such as power loss, the balance of the system, and also address the critical load that is experiencing drop voltage.

This scattered plant is a small-scale plant connected to a local distribution system often called "Distributed Generation" (DG). DG characteristic is small scale usually between 50 kW to 400 MW, distributed and close to load center (closed to load), interconnection

With distribution system. This distributed generation can be environmentally friendly plant, can limit the construction of new transmission lines, reliably in response to changes in load, reducing the use of fossil fuels, the deregulation of the electricity market and a number of other advantages [1].

CIGRE has defined the Distributed Generation as all generating units with a maximum capacity up to 50 MW range and coupled to the distribution network [2]. IEEE defines Distributed Generation as a generation that generates energy in a smaller capacity than conventional power stations and can be mounted almost at any point of the power system. IEA (2002) defines Distributed Generation as units that generate energy on the consumer side or in a local distribution network [3].

Distributed Generation is often referred to as the on-site generation, dispersed generation, embedded generation, decentralized generation, or distributed energy. Basically, DG generate electrical energy from some small-capacity energy source and connected directly to the distribution network.

Pepermans, J. et al defines DG as a generator connected to the transmission network [4], distribution or on the customer side. International Council on Large Electric Systems (1999) defines DG as:

- The plant is not centered
- Usually connected to network distribution system
- Capacity between 50 kW to 100 MW.

In the definition of distributed generating (DG) capacity there are different definitions and currently the definitions are used as follows:

1. The Electric Power Research Institute defines DG as a generator with a capacity of a few kilowatts to 50 MW.
2. Gas Research Institute defines DG as a generator with a capacity of between 25 kW and 25 MW.
3. Preston and Rastler determine DG sizes ranging from a few kilowatts to over 100 MW.
4. Cardell defines DG as a generator with a capacity of between 500 kW and 1 MW.
5. International Conference on Large High Voltage Electric Systems (CIGRE) defines DG as a generator with a capacity of between 50 and 100 MW.
6. However there was a presentation on the capacity of DG differentiated as follows [5] :
7. Micro (1W to 5 kW),
8. Small (5 kW up to 5 MW),
9. Medium (5 MW up to 50 MW),
10. Large (50 MW up to 300 MW).

DG placement can be done in substations and distribution can also be installed on the buses that directly supply the load. Before installing the need to analyze the impact of DG placement for buses which different needs to be done to determine where to place the best.

Due to the distributed power plants are located near load centers, the distributed generation in addition to directly serve a load requiring additional electrical energy can also be interconnected to the electricity distribution network to optimize the reliability of the distribution network. Based on the benefits and advantages of these distributed generators, the authors will simulate the location optimization and distributed generating capacity of a power distribution network in its effect on the minimization of the distribution network losses. The power distribution network used for this study isIEEE 30 bus test distribution system and the optimization method used genetic algorithm.

2. Methodology

2.1. Power Flow Calculation

Power flow calculations are paramount in planning, operating and developing systems, to obtain the operating characteristics of the generation or transmission system in steady state.In the power flow calculation, the operating system is assumed to be in the 3 phase equilibrium conditions, so that the calculation can be represented by a series of one phase that exists [6]. Components of the system

IOP Conf. Series: Materials Science and Engineering **308** (2018) 012034 doi:10.1088/1757-899X/308/1/012034

(generation, transmission, load, transformer, etc.) Are represented by the equivalent circuit and a bus in the system that are classified into 3 parts:

1. Reference bus / Slack bus / Swing bus that has voltage magnitude parameter | V | And the phase shrinkage voltage(θ),
2. Generator bus / control voltage bus / P-V bus which has active power parameter (P) and voltage magnitude | V |,
3. Load bus / P-Q bus having active power parameter (P) and reactive power (Q).

General conditions applicable to power flow studies:

1. Voltage on the bus generator or reference bus, set to the value V+j0 (V∠0°), While the voltage on the load bus, set to the value 1+j0 (1∠0°),
2. The load on the system is assumed to be a generator that supplies negative power.
3. One of the bus generators is set as a slack bus, generally chosen which has the largest power capacity, and
4. Testing convergence, to declare a power mismatch tolerance using 0:01 pu values and voltage magnitude and angles using a value of 0.001.

The power equation in electric power system is divided into 2 (two) ie:

1. Active / reactive power, as indicated by Equation (1) and (2);
2. Power active losses on the network, is formulated with Equations (3).

$$P_i^{(k)} = \sum_{j=1}^{n} \left|V_i^{(k)}\right|\left|Y_{ij}\right|\left|V_j^{(k)}\right| \cos\left(\theta_{ij} - \delta_i + \delta_j\right) \tag{1}$$

$$Q_i^{(k)} = -\sum_{j=1}^{n} \left|V_i^{(k)}\right|\left|Y_{ij}\right|\left|V_j^{(k)}\right| \sin\left(\theta_{ij} - \delta_i + \delta_j\right) \tag{2}$$

where k = 1, 2, …, n.

$$P_{loss} = \sum_{i=1}^{j} G_i\left[V_i^2 + V_j^2 - 2V_iV_j\cos\left(\delta_i - \delta_j\right)\right] \tag{3}$$

In the power flow study, many methods can be used, but the common and widely used method is Newton Raphson.

2.2. Newton Raphson method in Power Flow Calculation

The Newton Raphson method is used to solve nonlinear algebraic equations simultaneously from some unknown variables with a linear approximation.

Generally formulated with:

$$\begin{bmatrix} \Delta P_2^{(k)} \\ \vdots \\ \Delta P_n^{(k)} \\ \Delta Q_2^{(k)} \\ \vdots \\ \Delta Q_n^{(k)} \end{bmatrix} = \begin{bmatrix} \frac{\partial P_2^{(k)}}{\partial \delta_2} & \cdots & \frac{\partial P_2^{(k)}}{\partial \delta_n} & \frac{\partial P_2^{(k)}}{\partial |V_2|} & \cdots & \frac{\partial P_2^{(k)}}{\partial |V_n|} \\ \vdots & \vdots & \vdots & \vdots & \vdots & \vdots \\ \frac{\partial P_n^{(k)}}{\partial \delta_2} & \vdots & \frac{\partial P_n^{(k)}}{\partial \delta_n} & \frac{\partial P_n^{(k)}}{\partial |V_2|} & & \frac{\partial P_n^{(k)}}{\partial |V_n|} \\ \frac{\partial Q_2^{(k)}}{\partial \delta_2} & \cdots & \frac{\partial Q_2^{(k)}}{\partial \delta_n} & \frac{\partial Q_2^{(k)}}{\partial |V_2|} & \cdots & \frac{\partial_2^{(k)}}{\partial |V_n|} \\ \vdots & \vdots & \vdots & \vdots & \vdots & \vdots \\ \frac{\partial Q_n^{(k)}}{\partial \delta_2} & \cdots & \frac{\partial Q_n^{(k)}}{\partial \delta_n} & \frac{\partial Q_n^{(k)}}{\partial |V_2|} & \cdots & \frac{\partial Q_n^{(k)}}{\partial |V_n|} \end{bmatrix} \begin{bmatrix} \Delta\delta_2^{(k)} \\ \vdots \\ \Delta\delta_n^{(k)} \\ \Delta\left|V_2^{(k)}\right| \\ \vdots \\ \Delta\left|V_n^{(k)}\right| \end{bmatrix} \tag{4}$$

In equation above bus 1 as reference bus. The Jacobian matrix provides a linear comparison between the changes in the voltage angle $\Delta\left|V_k^{(i)}\right|$ and the voltage magnitude ($\Delta P_k^{(i)}$) with changes in active power ($\Delta P_k^{(i)}$). and reactive power ($\Delta Q_k^{(i)}$).

In the simplest form of equation (4) can be written as follows:

$$\begin{bmatrix} \Delta P \\ \Delta Q \end{bmatrix} = \begin{bmatrix} H & N \\ J & L \end{bmatrix} \begin{bmatrix} \Delta \delta \\ \Delta |V| \end{bmatrix} \quad (5)$$

The number of Jacobian matrix elements of equation (5) is determined by$(2n - 2 - m) \times (2n - 2 - m)$where n is the number of buses on the system, while m is the number of buses on the system voltage control.

The value of $\Delta P_k{}^{(i)}$ and $\Delta Q_k{}^{(i)}$ is the difference between the scheduled value and the calculated value of power residuals:

$$\Delta P_k{}^{(i)} = P_k{}^{\mathrm{sch}} - P_k{}^{(i)} \quad (6)$$

$$\Delta Q^{(i)} = Q_k{}^{\mathrm{sch}} - Q_k{}^{(i)} \quad (7)$$

Correction bus voltage magnitude and angle with the general iteration (i + 1) is:

$$\delta_k{}^{(i+1)} = \delta_k{}^{(i)} + \Delta\delta_k{}^{(i)} \quad (8)$$

$$|V_k{}^{(i+1)}| = |V_k{}^{(i)}| + \Delta|V_k{}^{(i)}| \quad (9)$$

This process stops until convergence:

$$|V_k{}^{(i+1)} - V_k{}^{(i)}| \leq \epsilon \quad (10)$$

2.3. Genetic Algorithm

In a genetic algorithm, a set of parameters (an individual) or in biology is called a chromosome for a problem in this case is an object function, formed or encoded in binary form. And in each generation, a number of individuals (population) are evaluated in parallel to their matching, as the price of the object to be minimized. New and improved populations are generated from the old through the application of genetic operators such as selection, crossover, and mutation.

Selection is a simple operator to get the new chromosome from the most powerful. Strings that have a stronger fitness value have a greater chance of being selected and can follow the operations of other operators.

Crossovers randomly select a pair of parent and form two offspring through the corresponding segment exchange from the parent. Implementation of crossover is done on the genes of the two parent chromosomes which result in the genes from the two different mains combined into their new offspring.

Mutation is a random change from the position of the string. In a binary string presentation, a simple change means 0 to 1 and vice versa.

Operations are repeated until the specified number of generations has been reached. Genetic structure of the algorithm in out line as shown in Figures 1-3.

2.4. Research Steps

The steps in this research were conducted in two steps, first to determine the location of DG placement and the second is to determine the optimum DG capacity.

1. The first step calculating the power flow to the system tested using Newton Raphson method to get the condition of the voltage profile on the system bus.
2. From this first step can be determined the location where DG will be paired on the system with the indicator refers to the voltage profile on the bus that decreased below 0.9500 pu.

From the second step of this optimization calculation can already determine the capacity of DG to be installed on the distribution system.

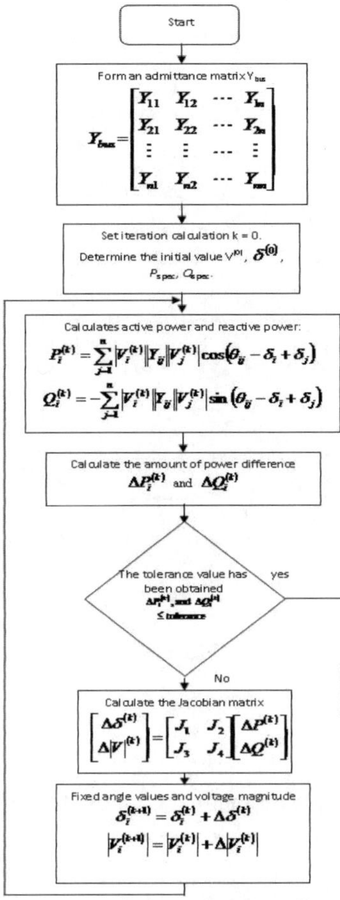

Figure 1. Flowchart calculation of power flow using Newton Raphson method

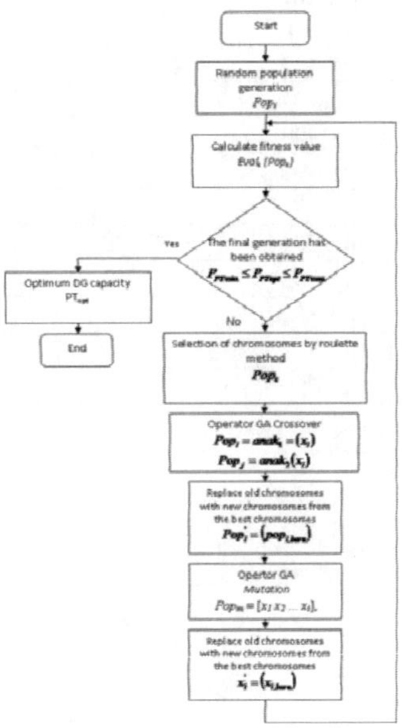

Figure 2. Flow chart of a genetic algorithm

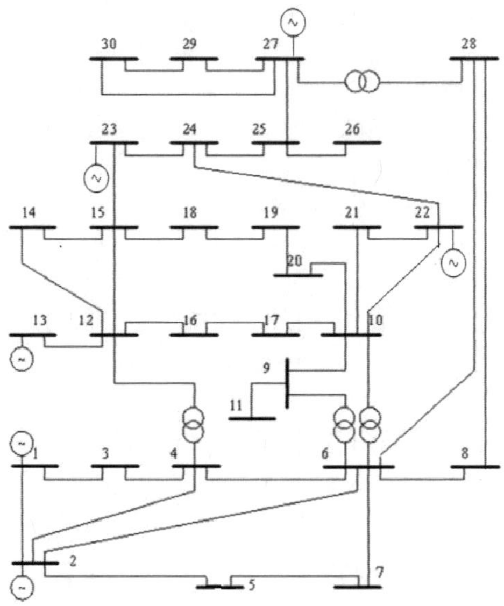

Figure 3. One line diagram of the IEEE 30 bus power system

3. Results and Discussion

3.1 Calculation Before Optimization

Calculation of power flow by using Newton Raphson method to show the voltage profile on each bus and line power loss obtained the Table 1.

Table 1. Profile of voltage after optimization

Bus Number	Voltage Profile (p.u.)	Bus Number	Voltage Profile (p.u.)	Bus Number	Voltage Profile (p.u.)
1	1.0000	11	0.9727	21	0.9916
2	1.0000	12	0.9804	22	1.0000
3	0.9731	13	1.0000	23	1.0000
4	0.9687	14	0.9693	24	0.9847
5	0.9736	15	0.9737	25	0.9870
6	0.9593	16	0.9704	26	0.9633
7	0.9522	17	0.9702	27	1.0000
8	0.9424	18	0.9587	28	0.9605
9	0.9727	19	0.9547	29	0.9730
10	0.9802	20	0.9599	30	0.9575

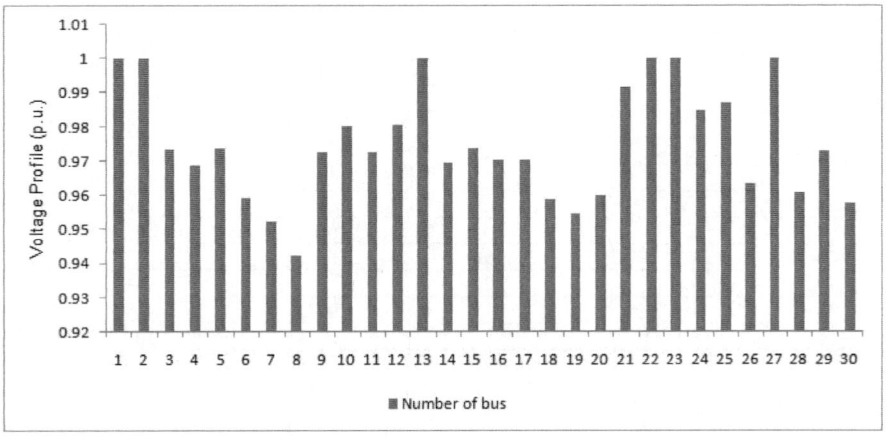

Figure 4. Graph of the voltage profile before optimization

Figure 4 shows graphically the voltage profile that occurred after the power flow analysis before the optimization, it is seen that the bus number 8 voltage profile below 0.9500 pu. It shows that under these conditions it is necessary to maintain the voltage profile of the bus going back well above 0.9500 pu. So that on bus 8 which is a load bus that has the lowest voltage magnitude is selected as the main location of DG installation. It is also possible to install DG on another bus which also has low voltage magnitude, but still on bus 8 also installed DG.

Table 2. Results of power flow analysis for power losses

Line	From bus	To bus	MW	MVAR
1	1	2	0,5849	1,7547
2	1	3	0,6134	2,3308
3	2	4	0,5074	1,4376
4	3	4	0,1009	0,4037
5	2	5	0,2672	1,0688
6	2	6	0,8029	2,4086
7	4	6	0,1697	0,6788
8	5	7	0,2806	0,6734
9	6	7	0,0333	0,0888
10	6	8	0,2375	0,9499
11	6	9	0,0000	0,5711
12	6	10	0,0000	0,4973
13	9	11	0,0000	0,0000
14	9	10	0,0000	0,2992
15	4	12	0,0000	0,3462
16	12	13	0,0000	2,2304
17	12	14	0,0707	0,1532
18	12	15	0,1548	0,2874
19	12	16	0,1122	0,2493
20	14	15	0,0057	0,0052
21	16	17	0,0388	0,0920
22	15	18	0,1348	0,2697
23	18	19	0,0267	0,0578
24	19	20	0,0184	0,0430
25	10	20	0,1057	0,2466
26	10	17	0,0421	0,1122
27	10	21	0,0923	0,2153
28	10	22	0,1099	0,2356
29	21	22	0,1441	0,2883

30	15	23	0,1476	0,2951
31	22	24	0,0939	0,1409
32	23	24	0,0724	0,1503
33	24	25	0,0550	0,0955
34	25	26	0,0799	0,1214
35	25	27	0,1035	0,1976
36	27	28	0,0000	0,3906
37	27	29	0,1553	0,2964
38	27	30	0,2956	0,5543
39	29	30	0,0603	0,1130
40	8	28	0,0513	0,1710
41	6	28	0,0096	0,0287
	Total		5,7781	20,5497

Power losses that occur before optimization can be shown in Table 2. Where it can be seen that the lowest loss of active power occurs in channel number 20 ie from bus 14 to bus 15 of 0.0057 MW. While the highest occurred in channel number 6 that is from bus 2 to bus 6 of 0.8029 MW. Total active power loss is 5.7781 MW.

3.2. Determination of Optimal Capacity DG

For this optimization step the maximum capacity referred is from DG type, ie from the intermediate type whose maximum capacity is 50 MW.

Table 3. Profile of voltage after optimization

Bus Number	**Voltage Profile (p.u.)**	**Bus Number**	**Voltage Profile (p.u.)**	**Bus Number**	**Voltage Profile (p.u.)**
1	1,0000	11	0,9951	21	0,9944
2	1,0000	12	0,9902	22	1,0000
3	0,9972	13	1,0000	23	1,0000
4	0,9972	14	0,9804	24	0,9847
5	1,0022	15	0,9880	25	0,9870
6	0,9996	16	0,9816	26	0,9633
7	1,0005	17	0,9826	27	1,0000
8	1,0029	18	0,9960	28	1,0012
9	0,9951	19	1,0048	29	0,9730
10	0,9929	20	1,0007	30	0,9575

Referring to Table 3 it can be seen that by injecting DG on three buses are buses 8, 7, and 19, indicates that DG increases the system voltage profile. We can see in table 3 that there is no voltage profile on any bus that is below the allowed standard limit of 0.9500 pu. We can also note on bus number 8 which previously experienced a decrease of 0.9424 pu, after the optimization using genetic algorithm to be 1.0029pu. All voltage profiles are within the permitted limits of between 0.9500 pu to 1.0500 p.u.

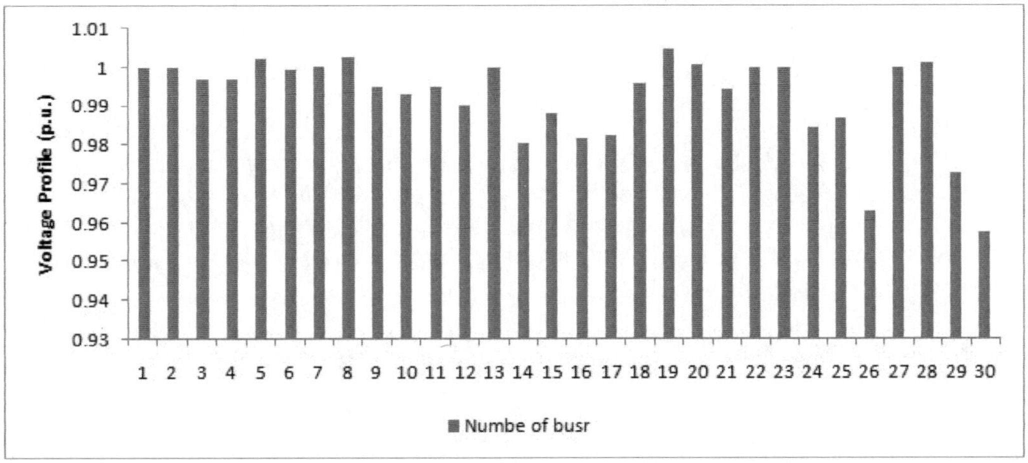

Figure 5. Graph of the voltage profile after optimization

In Figure 5 we can see that after the treatment with the installation of DG, seen voltage profile on bus number 8 is to rise above 1.0029. This shows that by adding DG to the voltage dropped bus and several other buses, this can improve the voltage profile of the bus, as well as fix some of the buses that have decreased as well.

Table 4. Results of the power flow analysis for total load and power generation

Bus	Load		Generation	
Number	MW	MVAR	MW	MVAR
1			1,0238	-0,8231
2	28,2100	16,5100	60,9700	2,3194
3	3,1200	1,5600		
4	9,8800	2,0800		
5				
6				
7	29,6400	14,1700		
8	-43,4939	-22,9205	50,0000	55,5514
9				
10	7,5400	2,6000		
11				
12	14,5600	9,7500		
13			37,0000	7,9983
14	8,0600	2,0800		
15	10,6600	3,2500		
15	4,5500	2,3400		
17	11,7000	7,5400		
18	4,1600	1,1700		
19	12,3500	4,4200		
20	2,8600	0,9100		
21	22,7500	14,5600		
22			21,5900	30,4908
23	4,1600	2,0800	19,2000	6,4892
24	11,3100	8,7100		
25				
26	4,5500	2,9900		
27			26,9100	6,6017
28				
29	3,1200	1,1700		
30	13,7800	2,4700		
		Generations Power	166.6938	53.0763

Demand Power	165.1181	63.5628
Power Line Losses	1.5757	5.7477
DG (bus 8)	28.3138	50.0000

From the results of this power flow analysis also found that the total generation, loading and loss of power (losses) on the network shown in Table 4 The total amount of power generation is P_{gen} = 166.6938 MW and Q_{gen} = 33.0763 MVAR. Total loading for is P_{load} = 165,1181 MW and Q_{load} = 63,5628 MVAR. While the total loss of power (losses) we need to see is the loss of active power only that is equal to P_{loss} = 1.5757 MW (27.27%). And it is seen that there is a decrease in power losses when compared with before the optimization of 5.7781 MW.

4. Conclusions And Suggestion

4.1. Conclusions

The results of the power flow analysis show that indicating the existence of a critical bus that is bus number 8. This indicates that the plant is in the system can not serve the load burden. So it should be understood that there should be an additional generator with DG installation. Installation of DG from the simulation result is bus 8. The improvement of voltage profile obtained is that all buses are within the permissible limits and nothing is below or above the standard value. From the installation of DG can also be known that the power losses on the channel on the system to decrease to 1.5757 MW or only 27.27% of the previous value of 5.7781 MW.

4.2. Suggestion

The suggestions that can be put forward for future research, use and planning are as follows:

1. This research may be used as a good basis for advanced research related to Distributed Generation, genetic algorithms, and other issues related to this study.
2. It is possible that the research will be used in the power system as a reference in planning the installation of DG to solve the problem of decreasing the voltage magnitude on the system and also can reduce the power losses.

References

[1] Viawan F A 2008 *Voltage Control and Voltage Stability of Power Distribution Systems in the Presence of Distributed Generation* (PhD Thesis: Chalmers University of Technology Gteborg Sweden)

[2] CIGRE 1998 *Impact of increasing contribution of dispersed generation on the power system* (CIGRE Study Committee no 37, Final Report)

[3] International Energy Agency IEA *2012 World Energy Statistical*

[4] Pepermans G, Driesen J, Haeseldonckx D, D'haeseleer, Belmans R 2003 *Distributed Generation: Definition, Benefits And Issues Working Paper Series 8,* (Faculty Of Economics And Applied Economic Sciences Center For Economic Studies Energy, Transport & Environment)

[5] Mondal A, Rani P 2017 Innovative Idea To Choose A Distribution Generation *International Journal of Industrial Electronics and Electrical Engineering* **5**:4

[6] Saadat H 2004Power System Analysis (Boston: McGraw-Hill)

[7] Adejumobi I A, Adepoju G A, Hamzat K A 2013 Iterative Techniques for Load Flow Study: A Comparative Study for Nigerian 330kv Grid System as a Case Study *International Journal of Engineering and Advanced Technology* (IJEAT) 3 : 1

[8] Gandomkar M, Vakilian M, Ehsan M 2005 A combination of genetic algorithm and simulated annealing for optimal DG allocation in distribution networks *Canadian Conference on Electrical and Computer Engineering*, Saskatoon, Sask.

[9] Bawan E K 2012 Dampak pemasangan Distributed Generation Terhadap Rugi-Rugi Daya *Jurnal Ilmiah Foristek* 2 : 21

[10] Budiman A, et all 2010 Penempatandan Penentuan Kapasitas Pembangkit Kecil Tersebar Menggunakan Algoritma Genetika Breeder Multiobjektif (*Tarakan Universitas Borneo*)

[11] Fahad S. Abu-Mouti, 2011 Optimal Distributed Generation Allocation and Sizing in Distribution Systems via Artificial Bee Colony Algorithm *IEEE Transactions On Power Delivery* **26** : 4

[12] Frías P, Gómez T, Cossent R, Rivier J 2009 Improvements in current European network regulation to facilitate the integration of distributed generation *Int J Electr Power Energy Syst* **31** : 9

[13] Gonzalez-Longatt F M 2008 Impacto de la Generación Distribuidaen el Comportamiento de los Sistemas de Potencia (*Universidad Central de Venezuela*)

[14] Kotb M F, Shebl K M, Khazendar M E 2010 Genetic Algorithm for Optimum Siting and Sizing of Distributed Generation *Proceedings of the 14th International Middle East Power Systems Conference* (MEPCON'10), Cairo University

[15] Krischonme B, Weerachai P 2013 Optimal Placement and Sizing of Distributed Generation for Power Loss Reduction using Particle Swarm Optimization *Energy Procedia* **34**

[16] Mwakabuta N S 2008 Study of the Application of Evolutionary Algorithms for the Solution of Capacitor Deployment Problem in Distribution Systems *40th Southeastern Symposium on System Theory* MC1.3 University of New Orleans New Orleans LA USA

[17] Ajay P. Vimal P, Raj D 2008 Optimization of Distributed Generation Capacity for Line Loss Reduction and Voltage Profile Improvement Using PSO *ELEKTRIKA* **10** : 2

[18] Philip P, Barker R W 2011 Determining the Impact of Distributed Generation on Power Systems: Part 1 - Radial Distribution Systems *12 IEEE* Retrieved 02 16

[19] Rau N S, Wan Y H 2014 Optimum location of resources in distributed planning *IEEE Trans Power Syst* **9**: 2014–20

[20] Katiyar S 2010 A Comparative Study of Genetic Algorithm and the Particle Swarm Optimization *Akgec International Journal Of Technology* **2** : 2

[21] Singh R K, Goswami S K 2010 Optimum allocation of distributed generations based on nodal pricing for profit, loss reduction, and voltage improvement including voltage rise issue *Int J Electr Power Energy Syst* **32** : 6

[22] Singh D, Singh D, Verma K S 2008 GA based Optimal Sizing & Placement of Distributed Generation for Loss Minimization *International Journal of Electrical and Computer Engineering* **2** : 8

[23] Srinivasa Rao R, Narasimham S V L, Raju M R, Srinivasa Rao A 2011 Optimal network reconfiguration of large-scale distribution system using harmony search algorithm *IEEE Trans. Power Syst* 26 : 3

[24] Srinivasa Rao R, Ravindra K, Satish K, Narasimham S V L 2013 Power Loss Minimization in Distribution System Using Network Reconfiguration in the Presence of Distributed Generation *IEEE Transactions On Power Systems* **28** : 1

[25] Alhajri M T, Abido MSEE M A 2009 Assessment of Genetic Algorithm Selection, Crossover and Mutation Techniques in Reactive Power Optimization *IEEE Congress on Evolutionary Computation* (CEC).

Harmonic reduction by using single-tuned passive filter in plastic processing industry

M I Fahmi*[1], U Baafai[1], A Hazmi[2], T H Nasution[1]

[1]Department of Electrical Engineering, Universitas Sumatera Utara, Medan, Indonesia
[2]Department of Electrical Engineering, Universitas Andalas, Padang, Indonesia

*E-mail: ikhwanfahmi@students.usu.ac.id

Abstract. The using of non-linear loads generated by industrial machines may result inconsistent harmonics that do not reach the IEEE 519 - 1992 standards. This study discusses the use of single-tuned passive filters in reducing harmonics in the plastics processing industry. The system modeling using matlab / simulink simulation resulted in total harmonic distortion (THD) of 15.55%, can be reduced to 4.77% harmonics in accordance with IEEE 519 - 1992 standards. From the simulation results also seen that single-tuned passive filter can reduce the harmonics of the current 82.23% harmonic that wants to be reduced and also can reduce other orders harmonics between 7% to 8%.

1. Introduction

Non-linear loads generated by industrial machinery may result the current harmonics that do not reach to IEEE 519 - 1992 standards[1]. The use of a control device composed of electronic components may cause the electrical load to be non-linear[2], [3]. This is due to the switching process of the components. The problem can be solved by doing a harmonic reduction by using a filter. Harmonics is a phenomenon caused by the operation of a non-linear electric load, where a basic frequency wave of 50 Hz or 60 Hz will occur and also cause the ideal current sinusoidal waveform and voltage wave to be not sinusoidal [4-5].

Previously there were several studies on harmonic reduction. We have conducted a literature study of research on single-tune passive filters. Several studies on the reduction of harmonics in the past researchers have been done in table 1.

In this research, we will do the harmonics reduction at a plastic processing factory. Processing machine used consists of electrical components classified as linear load category and non-linear load. Based on previous research, the harmonics reduction will be done by using a single-tuned passive filter.

Table 1. Passive Filter Researches

Name	Title of the Study	Research Field	Results
Young Sik Cho et all, 2011[6]	Single-tuned passive harmonioc filter design considering variances of tuning and quality factor	Single-tuned Passive Filter	THDi 5 decreased from 72.6% to 7.9%, the 7th harmonics from 61.8% to 11.2%, the 11th harmonics from 20.9% to 4.5%, the 13th harmonics from 13.2% To 2.9%

Sumit Kanwar et all, 2013[7]	Power Quality Improvement in Faulty Conditions using Tuned Harmonic Filters	Single-tuned Passive Filter	Installation of filters for three phase networks, can reduce THDi 6.40% to 1.46%
Zubair Ahmed Memon et all, 2012[8]	Harmonics Mitigation of Industrial Power System Using Passive Filters	Single-tuned Passive Filter	Passive filter reduces the total harmonic distortion from the source of 20.77 to 4.32%. The reduction results meets the IEEE 519 harmonic standards.

2. Methods

2.1 Measurement on PCC panel

To perform the harmonic measurements of the research object was done by using portable measuring instruments and PCC panel. Metering tool used is Metrel MI 2392 Power Q Plus Power Quality Analyzer. Measurements have been made on the object of research by means of direct measurement on PCC load. The results of the measurements obtained can be seen in table 2.

Table 2. Measurement Results by using Power Quality Analyzer Tool on PCC Panel

Phase Values						
Symbol	*Name*	L1	L2	L3	Total	Unit
V	*Voltage*	234.45	238.2	235.11	-	V
I	*Current*	365.94	383.27	396.54	-	A
THD V	*Voltage THD*	5.1187	5.4279	4.2477	-	V
THD V	*Voltage THD*	2.1836	2.2791	1.8068	-	%
THD I	*Current THD*	14.93	15.399	13.707	-	%
THD I	*Current THD*	54.004	58.332	53.819	-	A
P	*Active Power*	80.508	85.617	89.085	255.21	kW
Q	*Reactive Power*	29.65	31.694	27.498	88.842	kVAR
PF	*Power Factor*	0,94 ind	0,94 ind	0,96 ind	0,94 ind	
DPF	*Displacement Facto*	0,95 ind	0,95 ind	0,96 ind	-	
S	*Apparent Power*	85.794	91.295	93.233	270.23	kVA

In this study, it is assumed that a 3 phase load is considered to be balanced and the analysis is performed on only one phase. So the data used in the calculation and analysis is the result of measurement at phase L1.

From the table 2 measurement results above, it can be seen that the THDi current harmonic content in plastic processing industry load of 14.93%. This condition does not meet the IEEE 519-1992 standard ie the allowable THDi is less than 5%. As the resulting harmonics exceed the limits set by the IEEE 519-1992 standard, this research can be continued for harmonic reduction by using filters.

Figure 1 shows the harmonic wave spectrum of the measurement result. The results of the wave spectrum of harmonic current are randomly shaped so that they could not be identified as pure sinusoidal. This phenomenon is caused by a current disturbance known as harmonic current so it must be reduced so that the harmonic current content can be muffled under the standard set in IEEE 519-1992 standard. The measurement results show the total harmonic current distortion is 14.93% exceeds the IEEE 519-1992 standard limit of 7%.

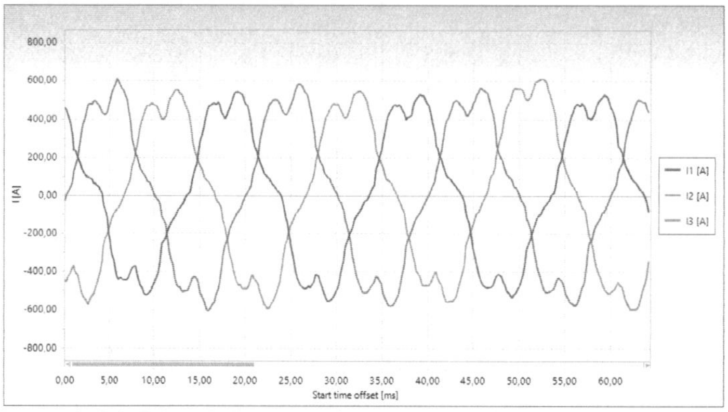

Figure 1. L1, L2, L3 Current Phase Harmonic Waves Spectrum

2.2 Single-tuned passive filter calculation

The calculation of the R, L, C filters used to reduce the current harmonics in this study by using the worst current harmonics exceeding the IEEE 519-1992 harmonic standard limits. The order of harmonics to be reduced is only the 5th order. The 3 phases loads are considered in a balanced state so that simulation and data analysis is done on one phase only, which is selected in phase L1 (phase R).

To determine the capacitor capacity (Qc) it is assumed that the power factor is improved from pf1 = 0.94 to pf2 = 0.99. To calculate the capacitor capacity required as a harmonic filter is:

$$Qc = P\{\tan(\cos^{-1} pf_1) - \tan(\cos^{-1} pf_2)\} \quad (1)$$

To determine the reactance of the capacitor (Xc) is as follows:

$$X_C = \frac{V^2}{Q_C} = X_C = \frac{(234{,}45)^2\ V}{17748{,}733\ VAR} = 3{,}0969423\ \Omega \quad (2)$$

To determine the capacitance of the capacitor (C) is as follows:

$$C = \frac{1}{2\pi f_0 X_C}$$

$$\boldsymbol{C = 1027{,}8\ \mu F}$$

Assuming the quality factor of single-tuned passive filter (Q) = 100, then determine the value of resistance (R) in the filter as follows:

$$R = \frac{X_n}{Q} \quad (3)$$

Based on the calculation of Single-tuned Passive Filters of R, L, C components which will be used in this study to reduce the current harmonics are shown in table 3.

Table 3. Single-tuned Passive Filter Calculation Results

No	Filter Components	Symbol	Unit	Value
1	Reactance	R	Ω	0,0061939 Ω
2	Inductance	L	H	0,3943 mH
3	Capacitance	C	F	1027,8 μF

2.3 Simulation without using filter

Figure 2 is a series of simulated object research system based on the measurement results. The simulation in figure 2 above is a circuit without using a filter. The circuit simulation results shows individual harmonic as shown in table 4.

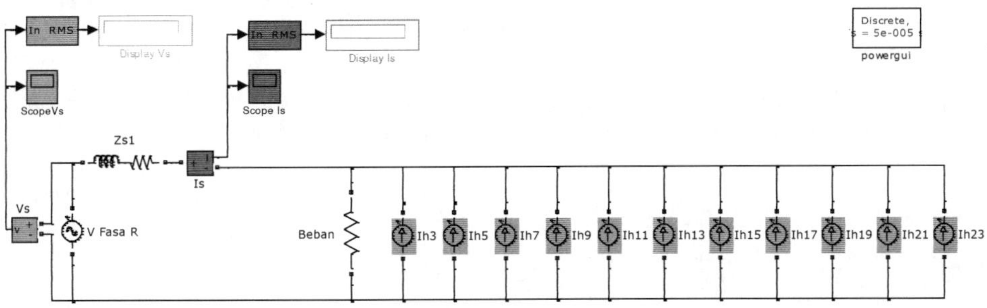

Figure 2. Matlab/Simulink Simulation without using Filter

Table 4. Simulation Results without Harmonic Filters (IHDi %)

IHD Order	Measurement	Fiterless Simulation	IEEE 519-1992 Standar
1	100,00	100,00	
3	2,20	2,31	7,0
5	14,22	14,97	7,0
7	2,35	2,47	7,0
9	0,84	0,88	7,0
11	1,21	1,27	3,5
13	1,03	1,08	3,5
15	0,49	0,51	3,5
17	0,84	0,88	2,5
19	1,02	1,06	2,5
21	0,24	0,25	2,5
23	0,81	0,84	1,0
Total THD	**14,95**	**15,55**	

Table 3 shows the result of simulation without using harmonic filter, total recorded current harmonic is 15.55% difference 0,6% from real measurement that is 14,95%. So that the simulation circuit in figure 2 can be illustrated as the condition of the harmonics in the plastics industry undertaken in this study. IHD value exceeding the IEEE 519-1992 standard limits that must be reduced by using a filter, in Table 3 it appears that only at the first five orders of the current IHD are well above the IEEE 519-1992 standard limit of 14.97% with the allowed limit being 7%.

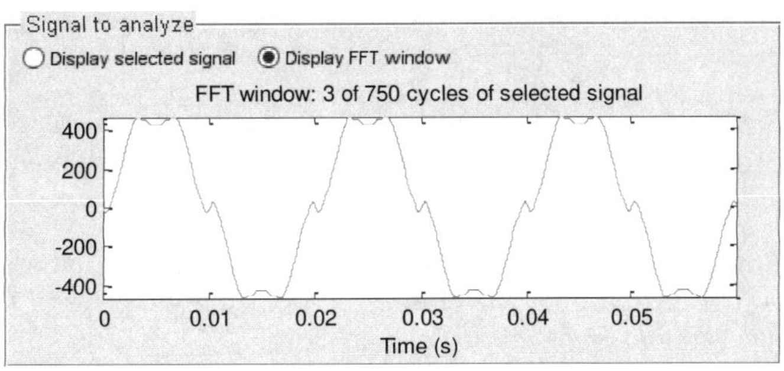

Figure 3. Filterless Simulation Test Results Waveform Graph

IOP Conf. Series: Materials Science and Engineering **308** (2018) 012035 doi:10.1088/1757-899X/308/1/012035

The current waveform generated by Matlab / Simulink simulating software without using a filter can be seen in figure 3. It shows that the current wave is not pure sinusoidal, there is a wave ripple and almost forming a square wave.

2.4 Simulation by using single-tuned passive filter

The circuit drawing of Matlab Simulation by using Single-tuned Passive Filter consisting of components R, L and C can be seen as figure 4. Single-tuned Passive filter simulation by using Matlab program is done in several stages same with the filterless simulation, then added with an RLC filter in parallel with load. RLC value has been calculated and obtained value R = 0,0061939 Ω, L = 0, 3943 mH, C = 1027,8 μF. This value became the block parameter value of the RLC filter.

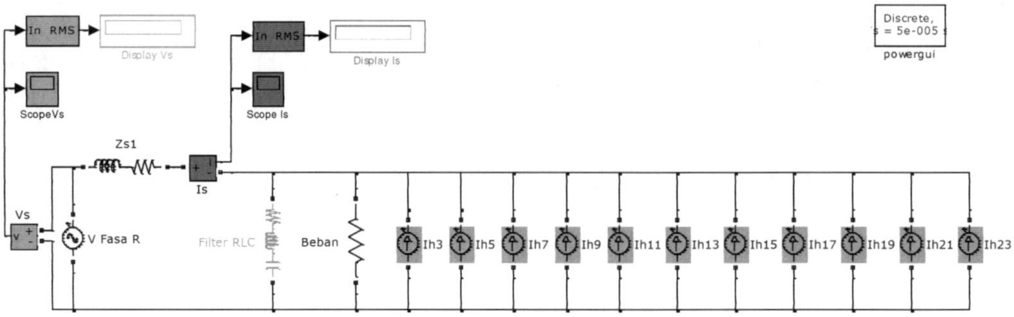

Figure 4. Matlab/Simulink Simulation by Using Single-Tuned Passive Filter

The result of current harmonic reduction by simulation using single-tuned passive filter shows the magnitude of individual harmonic flow as shown in table 5. Table 5 is the result of simulating current harmonic reduction by using single-tuned passive filter, this filter can reduce the total current harmonics from 15.55% to 4.77%. When it is viewed from individual harmonic currents, the simulation results show that the entire IHD Order of harmonics has been reach the IEEE 519-1992 standard.

Table 5. Single-Tuned Passive Filter Simulation Results (IHDi %)

IHD Order	Single-Tuned Passive Filter Simulation	IEE 519-1992 Standard	Description
1	100,00		
3	2,31	7,0	Match
5	2,66	7,0	Match
7	2,20	7,0	Match
9	0,80	7,0	Match
11	1,17	3,5	Match
13	1,00	3,5	Match
15	0,47	3,5	Match
17	0,81	2,5	Match
19	0,99	2,5	Match
21	0,23	2,5	Match
23	0,78	1,0	Match
Total THD	**4,77**		

The current waveform generated by using Matlab / Simulink with single-tuned passive filter can be seen in figure 5 shows that the resulting waveform has been sinusoidal shape that also indicating the system has been seperated from current harmonic interference.

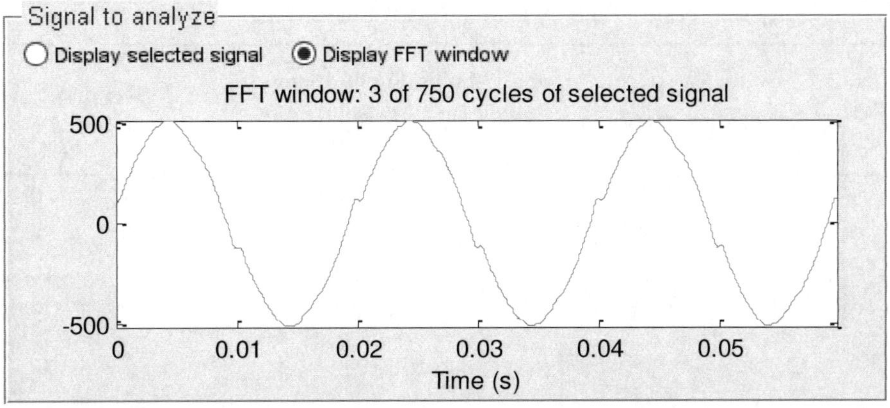

Figure 5. Single-Tuned Passive Filter Simulation Test Results Waveform Graph

3. Result and Discussion

Single-tuned Passive Filter can reduce harmonics to fulfill the IEEE 519-1992 standard in each orders. Single-tuned Passive Filter is able to reduce current harmonics from total THD by 15.55% to 4.77%, as shown in figure 6.

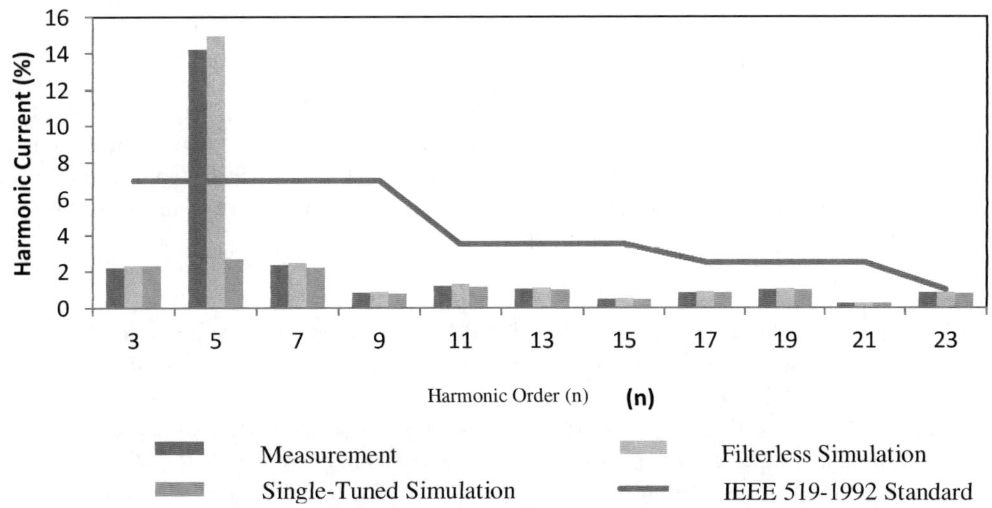

Figure 6. Graph of Harmonic Currenr with IEEE Standard 519-1992

Single-tuned passive filters can also reduce from current harmonics by 15.55% to 4.77%, indicating that the total THD current is reduced by 69.32% from filterless testing results. The filter reduction level in decreasing the harmonicity in each order can be seen in table 6. Table 6 shows that there is no change of harmonic current in 3rd order IHD ie 2.31%. In the 5th order IHD, which is the target of harmonic to be reduced, it can be seen that the single-tuned passive filter can decrease the current harmonics by 82.23% from the current harmonics of 14.97% can be reduced to 2.66%. The decrease of harmonic level on the 7th order IHD was 10.93%, the 9th order was 9.09%. Decrease in current harmonic after filter in 11th order and so on is relatively stable between 7% and 8% of current harmonics before using single-tuned passive filter.

Table 6. Harmonicity Decreasing Rate by Using Single-tuned Passive Filter

IHD Order	Filterless Simulation (IHDi %)	Simulation by Using Single-tuned Passive Filter (IHDi %)	Decrease Rate by Using Single-tuned Passive Filter
3	2,31	2,31	0,00%
5	14,97	2,66	82,23%
7	2,47	2,20	10,93%
9	0,88	0,80	9,09%
11	1,27	1,17	7,87%
13	1,08	1,00	7,41%
15	0,51	0,47	7,84%
17	0,88	0,81	7,95%
19	1,06	0,99	6,60%
21	0,25	0,23	8,00%
23	0,84	0,78	7,14%
Total THD	**15,55**	**4,77**	**69,32%**

4. Conclusion

Based on this research, then obtained the following conclusions:

1. Simulation modeling by using Single-tuned Passive Filter shows that THD (Total Harmonic Distortion) Current from 15.55% has been reduced to 4.77%. The Single-tuned Passive Filter can reduce the total THD of the system to fulfill the IEEE 519-1992 standard requirements.
2. Single-tuned passive filter has capability in decreasing one desired harmonic order and only slightly reducing other order harmonics.

References

[1] IEEE STANDARDS 2014 IEEE Std 519-2014; Recommended Practice and Requirements for Harmonic Control in Electric Power Systems *IEEE Std 519-2014 (Revision IEEE Std 519-1992)* **2014** pp 1–29

[2] Puchalapalli S and Pindoriya N M 2016 Harmonics assessment for modern domestic and commercial loads: A survey *International Conference on Emerging Trends in Electrical, Electronics and Sustainable Energy Systems ICETEESES 2016* pp 120–125

[3] Das J C 2015 *Power System Harmonics and Passive Filter Designs*

[4] Arrillaga J, Smith B C, Watson N R, Wood A R 2013 *Power system harmonic analysis*

[5] Mojiri M, Karimi-Ghartemani M, and Bakhshai A 2010 Processing of harmonics and interharmonics using an adaptive notch filter *IEEE Trans. Power Deliv.* **25**:2 pp 534–542

[6] Cho Y S and Cha H 2011 Single-tuned Passive Harmonic Filter Design Considering Variances of Tuning and Quality Factor *J. Int. Counc. Electr. Eng.* **1**:1 pp 7–13

[7] Kanwar S 2013 Power Quality Improvement in Faulty Conditions using Tuned Harmonic Filters *IOSR J. Electr. Electron. Eng.* **6** pp 1–9

[8] Memon Z, Uqaili M, Unar M Harmonics Mitigation of Industrial Power System Using Passive Filters *Mehran Univ. Res. J. Eng. Technol. Jamshoro Pakistan* **31** pp 355 – 360

IOP Conf. Series: Materials Science and Engineering **308** (2018) 012036 doi:10.1088/1757-899X/308/1/012036

Analysis of subsonic wind tunnel with variation shape rectangular and octagonal on test section

D Rhakasywi*[1], Ismail [1], A Suwandi [1], A Fadhli [2]

[1]Department of Mechanical Engineering, Faculty of Engineering, Universitas Pancasila, Srengseng Sawah – Jakarta 12640, Indonesia
[2]Student in Department of Mechanical Engineering, Faculty of Engineering, Universitas Pancasila, Srengseng Sawah – Jakarta 12640, Indonesia

E-mail: damora@univpancasila.ac.id

Abstract. The need for good design in the aerodynamics field required a wind tunnel design. The wind tunnel design required in this case is capable of generating laminar flow. In this research searched for wind tunnel models with rectangular and octagonal variations with objectives to generate laminar flow in the test section. The research method used numerical approach of CFD (Computational Fluid Dynamics) and manual analysis to analyze internal flow in test section. By CFD simulation results and manual analysis to generate laminar flow in the test section is a design that has an octagonal shape without filled for optimal design.

1. Introduction

The need for wind tunnels to solve complex aerodynamics problems is still needed today. The increasing need for fuel in vehicles requires an improvement in vehicle aerodynamics with the aim of more effectively and efficiently to save energy [1]. Upgrade force down on aerodynamics has grown rapidly especially on F1 racing cars. Aerodynamic forces are able to push the tire force against the road and as a result will increase the speed of the vehicle. To increase the downward force can be installing a spoiler. F1 car designers create their own wind tunnel to be able to conduct intensive research in improving vehicle aerodynamics [2]. Low open type wind tunnel is one easy way to solve aerodynamic problems. A CFD approach is also necessary to study the characteristics of aerodynamics in vehicles and wind turbines useful for power generation [3]. The correlation of clogging factors in wind tunnel experiments with Savonius wind turbine media in this study provides evidence of increased blockage correction in the range of 1-10% with an increase in the ratio of speed and the ratio of blockages, results show that for a deviation ratio approaching 10 and a speed ratio below 0.5 in the open-type test section [4]. CFD is a science that studies how to predict fluid flow, heat transfer, chemical reaction and other phenomena with solving mathematical equations. The fluid equation is built and analyzed with partial differential equation which explains the laws of mass conservation, momentum and energy [5].

Construction and testing of subsonic model wind tunnels with dimensions of 0.9 m x 0.9 m and having a test length section of 1.35 m obtained a maximum speed of 28 m/s [6]. The stable flow quality of a modified transonic wind tunnel for NACA 0012 produces the same surface pressure distribution and moment force when compared to other wind tunnel data [7].The effect of the strut shape on the diamond with the cylindrical cross section in subsonic wind tunnel result in a diamond shaped strut turns out to have a smaller lift force [8]. Experiments on turbulence intensity reduction in subsonic wind tunnels used a working air fluid, frequency and statistical analyzes were performed to obtain results in accordance with experimental findings [9]. The CFD simulation uses a combination of honeycomb for the management of turbulence intensity in subsonic wind tunnel using different length of honey comb, cell shape and screen with different open area ratio, obtained in accordance with the experimental and theoretical results [10]. Methods to improve the distortion of offshore wind flow using CFD simulations and wind tunnel tests concluded that the horizontal wind speed generated at an altitude of 80 m above sea level was 16% lower than the wind speed at 80 m above the island [11]. Studies on the performance of vertical axis wind turbines tested on a wind tunnel obtained similar results when installed on an electrical installation [12]. Mach number on wind tunnel tested which is affective by noise obtained at DNS (direct numerical simulation) that phase speed decreases at low frequency [13]. Modifications open circuit wind tunnels is conducted by varying some design contraction sections, test sections and diffuser section by using the simulation CFD [14]. CFD with rectangular model showed able to replicate wind tunnel measurements for velocity, turbulence intensity and pressure coefficient which have errors below 10% [15].

In this paper, tells what dimensions are most optimal for the design of rectangular or octagonal wind tunnel models to produce laminar flow and tells the maximum length of an optimal wind tunnel. The objectives of this study is to obtain a low and distributed turbulence current as well as produce winds that correspond to the design on the test section so that it can be used to analyze the specimens aerodynamically.

2. Methods

Methods the present computational and manual analysis is to validate result computation. The detail of each works is explained in detail as follows.

2.1. Computational work

The equation used in CFD is continuity [16].

$$\frac{\partial(\rho u)}{\partial x}+\frac{\partial(\rho v)}{\partial y}+\frac{\partial(\rho w)}{\partial z}; \qquad (1)$$

x-momentum.

$$\rho\left(u\frac{\partial u}{\partial x}+v\frac{\partial u}{\partial y}+w\frac{\partial u}{\partial z}\right)=\rho g_x-\frac{\partial P}{\partial x}+\frac{\partial}{\partial x}\left(2\mu\frac{\partial u}{\partial x}+\lambda\vec{V}\vec{V}\right)+\frac{\partial}{\partial y}\left[\mu\left(\frac{\partial u}{\partial y}+\frac{\partial v}{\partial x}\right)\right]+\frac{\partial}{\partial z}\left[\mu\left(\frac{\partial w}{\partial x}+\frac{\partial u}{\partial z}\right)\right]; \qquad (2)$$

y-momentum.

$$\rho\left(u\frac{\partial u}{\partial x}+v\frac{\partial u}{\partial y}+w\frac{\partial u}{\partial z}\right)=\rho g_y-\frac{\partial P}{\partial y}+\frac{\partial}{\partial x}\left[\mu\left(\frac{\partial v}{\partial x}+\frac{\partial u}{\partial y}\right)\right]+\frac{\partial}{\partial y}\left[2\mu\frac{\partial v}{\partial y}+\lambda\vec{V}.\vec{V}\right]+\frac{\partial}{\partial z}\left[\mu\left(\frac{\partial v}{\partial z}+\frac{\partial w}{\partial y}\right)\right]; \qquad (3)$$

z-momentum.

$$\rho\left(u\frac{\partial u}{\partial x}+v\frac{\partial u}{\partial y}+w\frac{\partial u}{\partial z}\right)=pg_z-\frac{\partial p}{\partial z}+\frac{\partial}{\partial x}\left[\mu\left(\frac{\partial w}{\partial x}+\frac{\partial u}{\partial z}\right)\right]+\frac{\partial}{\partial y}\left[\mu\left(\frac{\partial v}{\partial z}+\frac{\partial w}{\partial y}\right)\right]+\frac{\partial}{\partial z}\left(2\mu\frac{\partial w}{\partial z}+\lambda\vec{V}.\vec{V}\right); \qquad (4)$$

Equations 1 to 4 are used to calculate the fluid momentum value in the wind tunnel on the x axis, y-axis and then z-axis.

The condition of the air fluid properties used in the simulation is shown in table 1. Boundary conditions for wind tunnel air speed 5.8 m/s and then assume pressure on the outside of the wind tunnel 0 Pascal :

Table 1. Computation Condition

Computation Condition		
Model settings		3D, Steady
Fluid		Air
Fluid Properties	Density	1.225 kgm^{-3}
	Viscosity	1.7894 e-05 $kgm^{-1}s^{-1}$
	Cp	1006.43 $Jkg^{-1}K^{-1}$
	Thermal Conductivity	0.0242 $Wm^{-1}K^{-1}$
Boundary Condition	Velocity Inlet	5.8 ms^{-1}
	Pressure Outlet	0 Pascal

2.2. Manual Analysis

Table 2 describes the variation of wind tunnel design of rectangular and octagonal models. In these study variations in the test section with rectangular and octagonal forms to obtain a flow that produces a low turbulence value in the test section.

Table 2. Variations test section wind tunnel

Section	*All dimension are in (m)* *Length*	*Height*	*Width*	**Information**
VARIATIONS 1	Rectangular shaped test section and fillets throughout the wind tunnel			
settling chamber	0.613	1.226	1.226	+ fillet 0.2
Honeycomb	DH = 0.05	L = 0.3	Thk = 0.003	
Contraction (type 1)	0.9	1.226	1.226	+ fillet 0.2 ; AR = 6.073
Test section	1.0	0.5	0.5	+ fillet 0.2
Diffuser	4.143	1.226	1.226	+ fillet 0.2 ; AR = 6.073
VARIATIONS 2	Rectangular shaped test section and fillets throughout the wind tunnel			
settling chamber	0.613	1.226	1.226	+ fillet 0.25
Honeycomb	DH = 0.05	L = 0.3	Thk = 0.003	
Contraction (type 1)	0.9	1.226	1.226	+ fillet 0.25 ; AR = 4.174
Test section	1.2	0.6	0.6	+ fillet 0.25
Diffuser	3.572	1.226	1.226	+ fillet 0.25 ; AR = 4.174
VARIATIONS 3	Rectangular shaped test section and fillets throughout the wind tunnel			
settling chamber	0.613	1.226	1.226	+ fillet 0.25
Honeycomb	DH = 0.05	L = 0.3	Thk = 0.003	
Contraction (type 1)	0.9	1.226	1.226	+ fillet 0.25 ; AR = 3.036
Test section	1.4	0.7	0.7	+ fillet 0.25
Diffuser	3.0	1.226	1.226	+ fillet 0.25 ; AR = 3.036
VARIATIONS 4	Octagonal shaped test section throughout the wind tunnel			
settling chamber	0.613	1.225	1.225	
Honeycomb	DH = 0.05	L = 0.3	Thk = 0.003	
Contraction (type 2)	0.9	1.225	1.225	AR = 2.344
Test section	1.4	0.8	0.8	
Diffuser	2.43	1.225	1.225	AR = 2.344
VARIATIONS 5	(Cross section SC, Ct, TS = octagonal; DF = circle)			
settling chamber	0.543	1.086	1.086	
Honeycomb	DH = 0.05	L = 0.3	Thk = 0.003	
Contraction (type 2)	0.9	1.086	1.086	AR = 2.406
Test section	1.4	0.7	0.7	
Diffuser	3.0	Ø = 1.225		AR = 2.901

The continuity equation is a simple mathematical statement of the principle of mass conservation law [15]. In the continuity equation explains that the fluid flow velocity is inversely proportional to sectional area. In a pipe have a small sectional area then a high fluid flow rate.

$$A_1V_1 = A_2V_2 \tag{5}$$

3. Results and discussion

3.1. Computational work

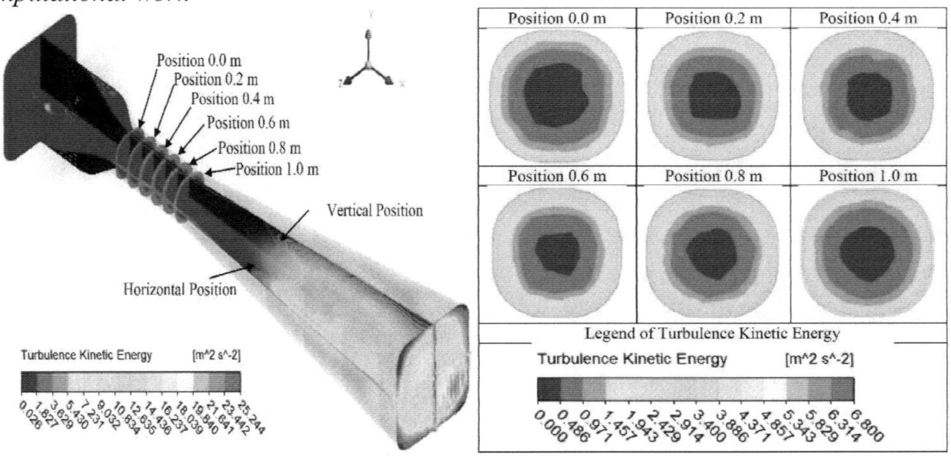

Figure 1. Turbulence kinetic energy for variation shape 1

Figure 1 shows profile turbulence kinetic energy variation 1 test section in rectangular shape of fillet at corner, the result of CFD simulation shows that the value of turbulence kinetic energy 0.486 m^2s^{-2} decreased at position 0.6 meter then maximum value at position 1 meter.

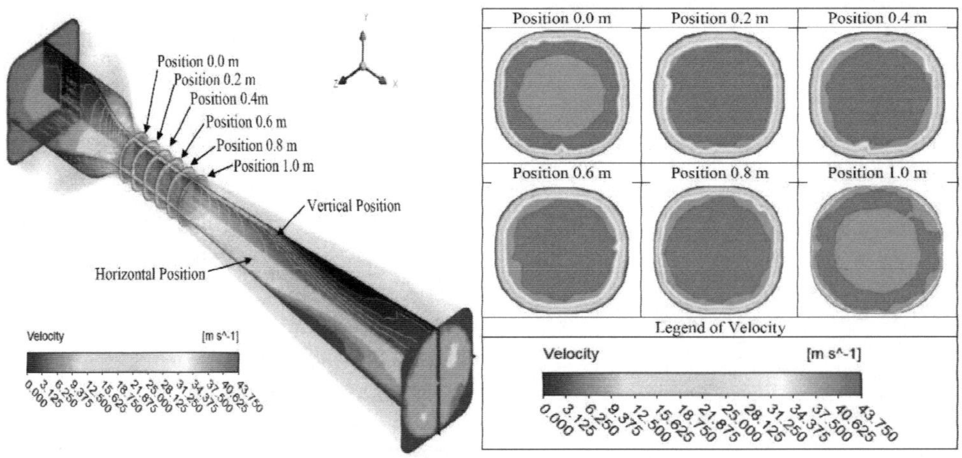

Figure 2. Velocity for variation shape 1

The velocity of the air fluid in the rectangular test section is shown in figure 2 the air velocity distribution is uniform at 0.2 m, 0.4 m, 0.6 m and 0.8 m positions in the inlet 0 m and the outlet 1 m occurs the not uniform of the fluid velocity.

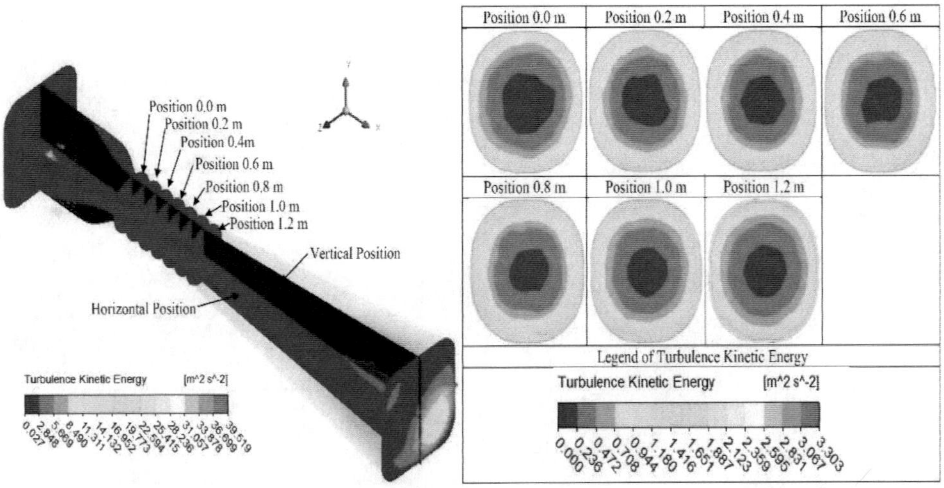

Figure 3. Turbulence kinetic energy for variation shape 2

Figure 3 shows a turbulent kinetic energy simulation that has a 0.25 meter fillet at the rectangular corner. The value of turbulent kinetic energy has increased in the position of 0.6 meters to 1.2 meters on the side near the wind tunnel wall.

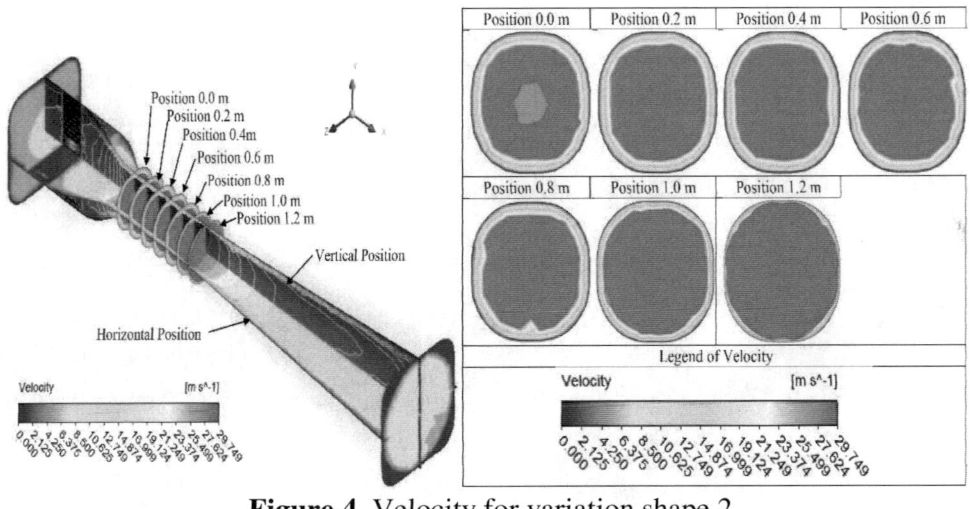

Figure 4. Velocity for variation shape 2

The fluid flow velocity that blows on the wind tunnel is shown in figure 4, for the position of 0 meter visible unstable speed then at the next position the fluid velocity uniformity.

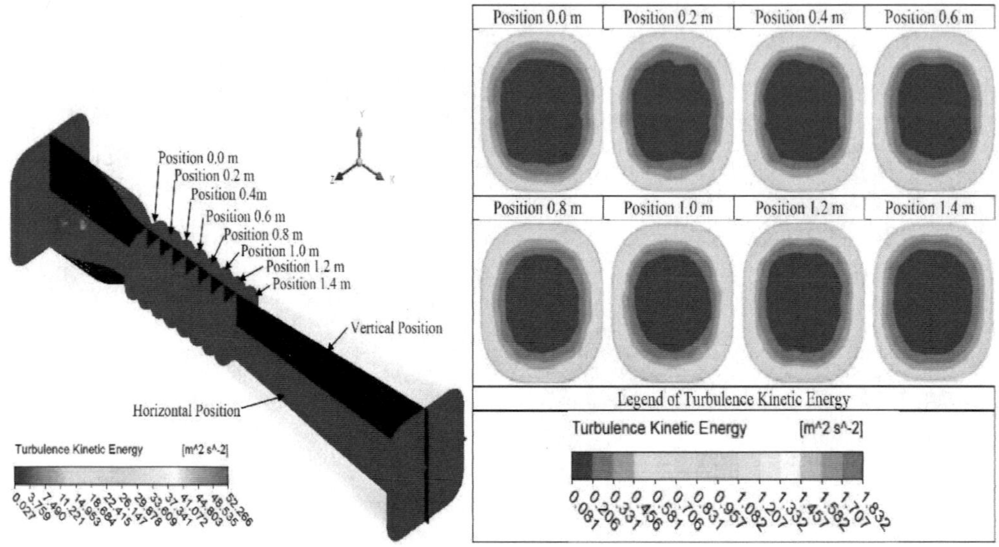

Figure 5. Turbulence Kinetic Energy for Variation Shape 3

At the 3 variations shown in figure 5 tells the turbulent value of kinetic energy has increased in the wall of the wind tunnel at the position of 1 m to 1.4 m.

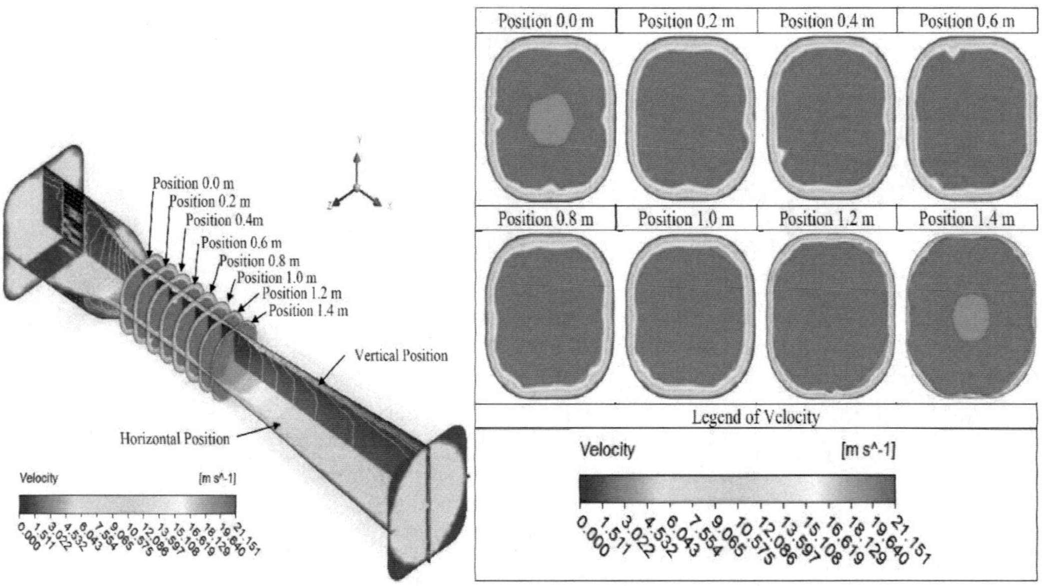

Figure 6. Velocity for variation shape 3

Figure 6 describes the results of the air fluid velocity contour flowing in the test section at positions 0 meters and 1.4 m does not provide uniform flow.

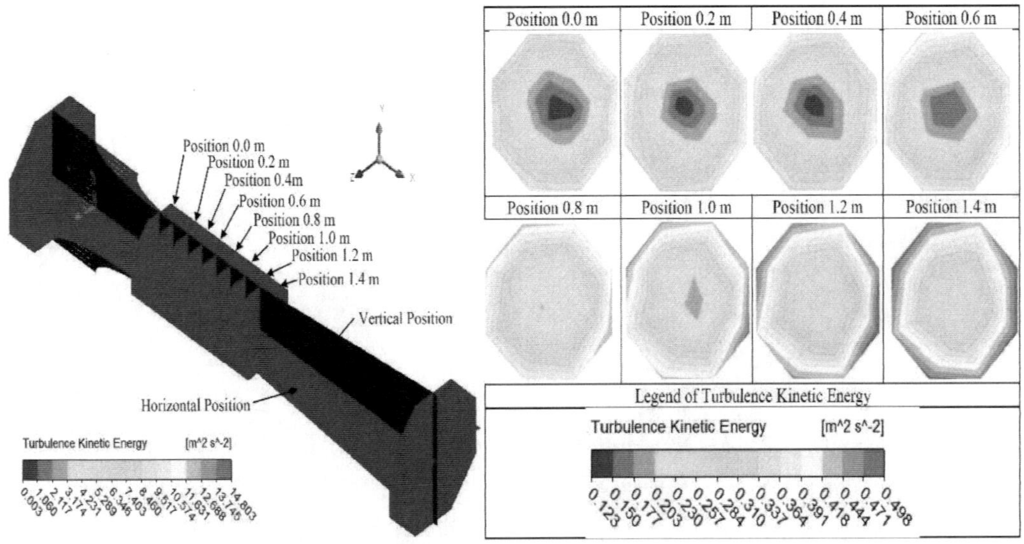

Figure 7. Turbulence kinetic energy for variation shape 4

Figure 7 describes the results of the CFD simulation for an octagonal section, the value of turbulence kinetic energy increased in the position of 1 meter to 1.4 m on the side of the wind tunnel. The turbulence kinetic energy value of 5 simulated variations was obtained in variation 4 which has the smallest kinetic energy turbulent value.

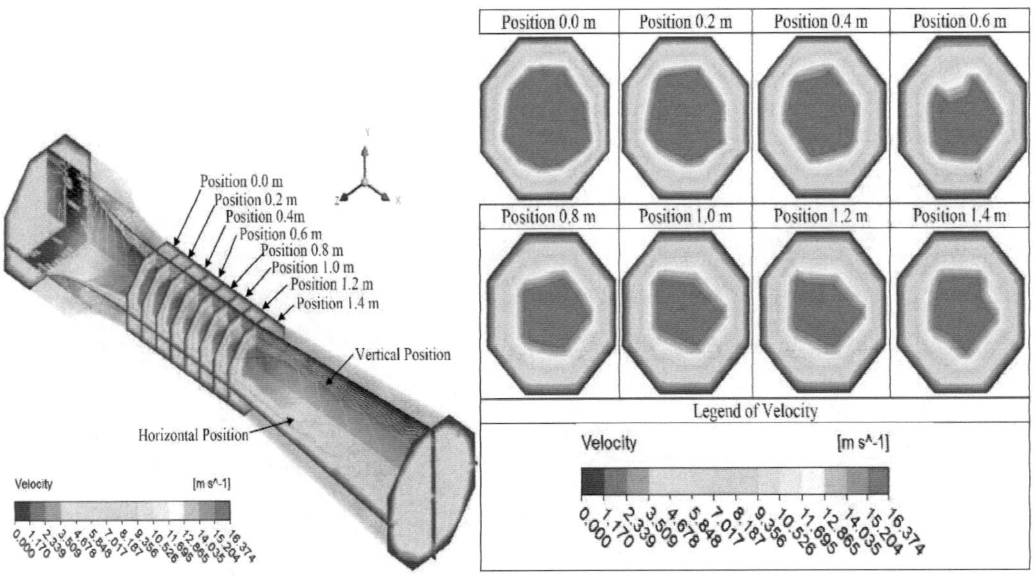

Figure 8. Velocity for variation shape 4

The fluid flow velocity in the octagonal sectional test section of figure 8 describes the maximum value of the fluid velocity flowing in the middle of the wind tunnel for all test section positions 16.374 ms^{-1}.

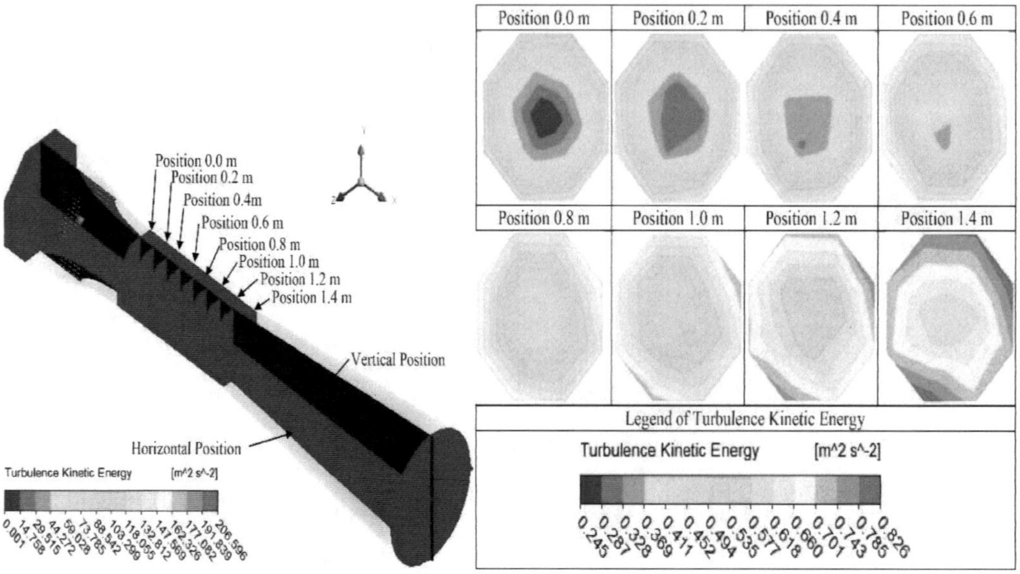

Figure 9. Turbulence kinetic energy for variation shape 5

Variation 5 in figure 9 explains the turbulence kinetic energy value for the octagonal test section. Turbulent kinetic energy value increased in the test section position 1.2 m to 1.4 m.

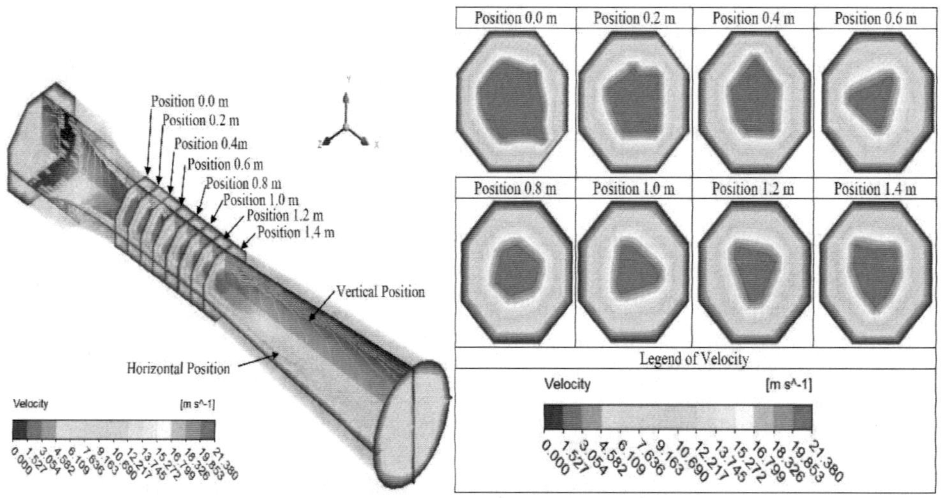

Figure 10. Velocity for variation shape 5

Figure 10 describes the fluid flow velocity in the octagonal test section and the circular section of the wind tunnel outlet. The maximum fluid flow rate is at the center of the test section.

3.2 Manual analysis

Table 3 describes the parameters that have been generated by the CFD simulation and comparison of manual analysis use the continuity equation. Parameters that have been presented from 5 variations of wind tunnel in table 3, the best variation are variation 4 because the variation has the greatest value.

Table 3. Parameters assessed from variations

Parameters assessed	**Var-1**	**Var-2**	**Var-3**	**Var-4**	**Var-5**
TKE (*test section*) (m^2s^{-2})	6.717	3.303	1.832	0.498	0.826
Velocity in *test section*	43.75	29.749	21.151	16.374	21.380
Angle *up flow* (°)	0.179	0.312	0.422	0.171	0.150
Angle *cross flow* (°)	0.167	0.256	0.241	0.202	0.533
Differential pressure (test section) (%)	0.23	0.047	0.019	0.008	0.015
Dimensions *test section* (m)	0.5	0.6	0.7	0.8	0.7
The total length of the wind tunnel (m)	6.656	6.285	5.913	5.343	5.913

4. Conclusion

From CFD simulations and manual analysis for optimal design resulted in laminar flow is an octagonal design for test section. The total length of the wind tunnel is 5.343 m which has a test section of 0.8 m with a velocity value at 16.374 ms^{-1} test section. The angle between the cross flow velocity and the speed along the wind tunnel in the test section is 0.171°. The lowest turbulence kinetic energy value produced by the wind tunnel is 0.498 m^2s^{-2}.

By modifying the design form on each wind tunnel module it will improve airflow quality. This is evident in the shape of the square shaped wind tunnel that the fillets provide in each corner, Produce the highest speed and low kinetic energy turbulent values.

Acknowledgment
This work was supported within the DRPM Kemenristek DIKTI - Universitas Pancasila (Project number: Nomor SP DIPA-042.06.1.401516/2017).

References

[1] Yuniarsih N & Rossbandrio W 2015 J. Integrasi **7** 19-22
[2] Kusumohadi C S,Yoga N G & Arrozi H 2015 *Low wind tunnel design for automotive testing* (Bahasa Indonesia) (UNJ Press)
[3] Muhammad B 2012 *Design, Construction, and Testing of Open-Loop Low-Speed Wind Tunnel* (Faculty Mechanical Engineering, Universiti Malaysia Pahang) p 67
[4] Sukanta R & Ujjwal K 2014 J Energy Convers Manag. **86** 418–427
[5] Miguel A G H, Ana I M L, Artur A J, Jose M P P, Yuliang W & Sun X 2013 Design Methodology for Quick and Low-Cost Wind Tunnel. (InTechOpen as the world's largest OA book publisher) p 28
[6] Arifuzzaman M & Mohammad M 2012 IOSR J. Eng **2** pp 83-92
[7] Amiria K, Soltani M R, Haghiri A 2013 J. Scientia Iranica B **3** pp 500-507
[8] Anwar U H, Waqar A, Ashraf A O & Erwin S 2017. J. Measurement **99 pp** 1-6
[9] Kaveh G, Mohammad R S, & Mojtaba D M 2011 J. Aero. Sci and Tech. **15** pp 137–147
[10] Vinayak K, Niranjan S, & Sandip D C 2011 J. Wind Eng. Ind Aerodyn **99** pp 37–45
[11] Joao S, F. M S, Antonio C, Ana E 2015 J. Wind Eng. Ind. Aerodyn **140** pp 87–94
[12] Qing'an L, Takao M, Yasunari K & Junsuke Murata 2016 *Renew Energy* **90** pp 291–300
[13] Stefan H, Jens B, Jorn S & Carsten S 2017 J. of Sound Vib. **402** pp 85–103
[14] Ismail, Samsul K, Purnomo, Sarjiya, Sulaiman T, Azhim A A, Inderanata 2015 J. Eng. Appl. Sci. **10** pp 8150-8156
[15] John K C, Hassam N C, Ben R G, Lik F S 2014 J. Wind Eng Ind Aerodyn. **125** pp 180–194
[16] Çengel Y A, Cimbala J M 2006 *Fluid mechanics: Fundamentals and Applications* (McGraw-Hill series in mechanical engineering)

IOP Conf. Series: Materials Science and Engineering **308** (2018) 012037 doi:10.1088/1757-899X/308/1/012037

Identification of tensile strength properties of abaca fiber by weakest-linkage approach-statistic property of fiber diameter

Suardi[*1]**, H Homma**[2]**, Abubakar**[1]

[1]Graduate School Departement of Mechanical Engineering, Universitas Sumatera Utara
[2]Visiting Professor, Department of Mechanical Engineering, Universitas Sumatera Utara

[*]E-mail: suardi@usu.ac.id

Abstract. Fiber reinforced plastics or metals (FRPor FRM) are usually ecological materials, because their specific strength defined as the strengthperunit mass is much larger than metal, and weight ofmachines and structuresfor transport made ofFRP can be significantly reduced so that the consumption of fossil fuel scan be saved to result in tremendous reduction of CO2emissions. However, when we consider life cycle assessment (LCA) of synthetic fibers like carbon fiber and glass fiber, we can recognize much CO2 emission in production of these fibers. Therefore, more ecological reinforcement fibers must be developed. For this end, we should utilization cellulose fibers derived from plant tissue structure as an alternative fibers for synthetic fibers, which are considered as carbon neutral materials, and natural degraded material. This study selectsabaca fiber, which is a natural fiber and is abundant in Indonesia, but its usagehas not been optimized for engineering material. The purpose of this study is to identify the mechanical strength of a single abaca fiber by statistical approach. First, weakest link theory and Weibull theory are used to discuss experimental data. 90 specimens of almost identical geometry and biological aspects are tested under tension. These data are analyzed by Weibull theory or other statistical theory. Final target is to look into optimal method to reduce scatter ratio, ratio of standard deviation to mean value, of less than 0.1, which is the level of metallic materials. If we can reduce scatter ration to such level, we can design machines and structures using abaca fiber in the same way as carbon fibers or glass fibers. Summary of Diameter Measurement the all mean value is 0.1 and standardeviasi. The t-Test showed that mean value of each part is estimated as sampling from group with the same mean value, at confidence level of 99%.

1. Introduction

Fiber reinforced plastics or metals (FRP or FRM) are usually considered as ecological materials, because their specific strength defined as strength per unit mass is much larger than metallic materials, and weight of structures and machines for transportation fabricated by FRP can be significantly reduced so that consumption of fossil fuels could be saved and as a result, CO2 emission into the atmosphere could be tremendously decreased [1].

However, carbon fires, grass fibers, Kevlar fibers such as man-made fiber are not always ecological materials if we consider their life cycle CO2 emission. For instance, The Japan Carbon Fiber Manufacturer Association (JCMA), estimated amount of CO2 emission at production stage of carbon fibers, assembly stage of CFRP car, usage (driving) stage of the car, and scrapped stage of the car after 10 years usage (driving) [2]. Fiber composite technology is progressing very rapidly. Basically fiber is divided into two type, namely natural fibers (organic fibers) and manmade fibers (organic and non-organic).Natural fibers are fibers obtained from fruits, stems, and leaves, and the fibers are highly

IOP Conf. Series: Materials Science and Engineering **308** (2018) 012037 doi:10.1088/1757-899X/308/1/012037

prospective commodities in the future as the commodity has advantages for a variety of industrial raw materials, and its significant contribution in saving the environment.Natural fibers have a single fiber which is the smallest unit of fiber that cannot be separated further mechanically, in general, a single bundle of macro fibrils. The macro fibril is composed of bundle of micro fibrils, which are cellulose chains, and are the main support of natural fiber materials [3].

Abaca (Musa Sapientum) is a plant that resembles a banana and originated in the Philippines which is a source of fiber spinning. In Indonesia, abaca banana also grows numerously in Kalimantan, North Sulawesi (especially on the island of Talaud in the village Essang) and Sumatra.

Abaca banana plant is a tropical plant that has advantages in terms of high fiber strength and wide usage for a raw material ofwrapping rope and cable in marine vessels, because of high resistance to salt water, a textile material, tea bag wrappers, wrapping tobacco, seat upholstery, crafts, cigarette paper and others.Natural fiber, especially, abaca fiber is produced in Sumatera as well as Kalimantan and Sulawesi in Indonesia as mentioned above, but its utilization for an engineering material is very limited. Main reason for this is wide scatter in mechanical properties of natural fibers. Integrity or reliability of structures and machines must be kept high enough to prevent users from fatal disaster by catastrophic failure.

According to the previous research[4], although the data on tensile strength of natural fiber is widely scattered, the scatter deceasesas the specific diameter variation rangedecreases. In order to use natural fibers for engineering materials, scatter of their tensile strength must be suppressed to the level of metallic materials. One way to suppress the scatter is to keep geometry, like a diameter, in a certain scatter band.

This research aims to evaluate diameter statistic properties of abaca fibers and to find a method to suppress the scatter of the tensile strength of abaca fibers.

2. Methodology

2.1. Abaca Fiber Preparation

Abaca fibers is used as a specimen obtained from the district of East Aceh, North Aceh and Aceh Jeumpa, Aceh Province. Then the fiber taken from the pseudo stem petals abaca has begun to bear fruit.

The following describes the of abaca fiber making process, starting from selecting, cutting and fiber-making positions on the stem or petals of abaca. Whereas the fiber starting from the position of the rod/bottom petals, because the rod/bottom petals are part of the oldest.

1. Then abaca fibers used in this study is the outer sheath item, namely strand 1 (one), 2 (two) and 3 (three).
2. Abaca petals cut to the same size of 120cm x 10cm,
3. Petals cut back to 3 (three) parts, namely the bottom, middle and top, each the size of 40cm.
4. Then the extraction is done using a wire comb. Extraction is done repeatedly, the extraction is done to separate the meat with fiber eyelids.
5. After extraction by using a wire comb, to get fiber is really clean, then soaking with clean water for 10 minutes.
6. Next step is cleaning back on the rest of the meat is still attached to the abaca fiber, using a blunt knife made repeated up to 15 times. This process also serves to straighten the fibers.
7. Once the fibers completely clean of its flesh, to get dry and strong fiber, then the fiber is dried (dried) using sunlight for one day (depending on weather).
8. Having obtained a dry fiber then conducted the election to be a single fiber tensile test specimens.

2.2. Specimen Geometry

Sample fibers are prepared as explained above. Then, an abaca single fiber specimen is prepared as follows. Each sample fiber is cut 80 mm long. Specimen gage length is 20 mm, and 30 mm both ends are molded into round grips as shown in figure 2. A grip is 30 mm long and made of plastic cylinder of

IOP Conf. Series: Materials Science and Engineering **308** (2018) 012037 doi:10.1088/1757-899X/308/1/012037

which one end is closed. Careful attention was paid so as that the fiber is aligned in the center of the grip.

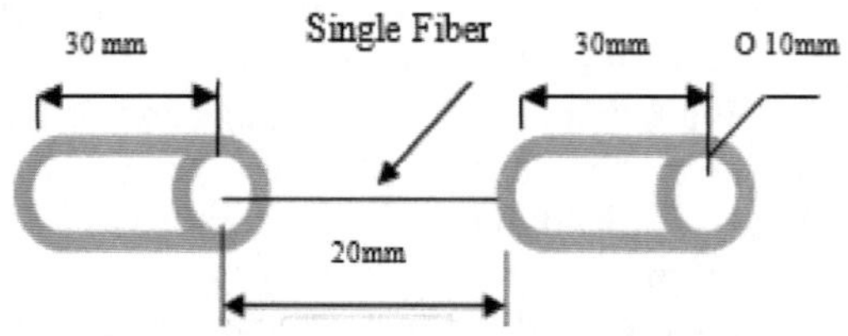

Figure 1. Specimen Geometry and Dimensions

2.3. Measurement of The Diameter

Diameter measurementis carried out using an optical digital microscope as shown in figure.3. Diameter measurement of single abaca fibers is done at three positions, near the grips, and the center of the specimen shown in figure 2. At each position, diameter is measured in two orthogonal directions as shown in figure 3. The six-measurement data for one specimen is averaged to define the specimen diameter.

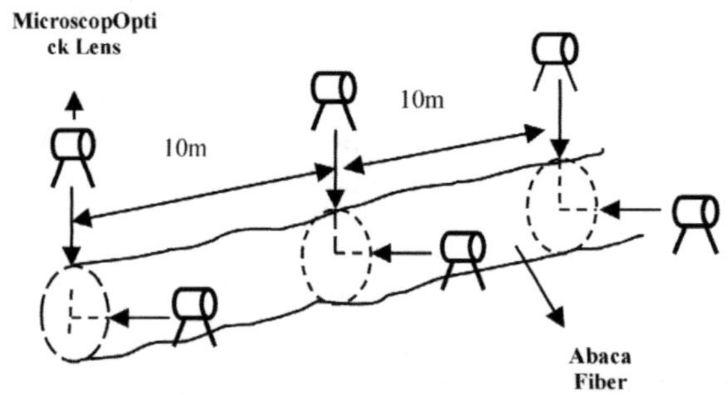

Figure 2. Schematic Diagram for Measuring The Average Diameter Of Fiber Fragment

3. Result and discussion

Diameter values measured for 30 specimens prepared from the bottom part of abaca sheathes are shown in table 1.

Table 1. Measured diameter of 30 specimens prepared from bottom part of abaca sheathes.

TP No.	Diameter	TP No.	Diameter
1	0.107	16	0.102
2	0.081	17	0.108
3	0.072	18	0.115
4	0.054	19	0.121
5	0.128	20	0.128
6	0.091	21	0.134
7	0.126	22	0.140
8	0.113	23	0.147
9	0.101	24	0.153
10	0.072	25	0.160
11	0.091	26	0.166
12	0.054	27	0.172
13	0.083	28	0.179

14	0.089	29	0.185
15	0.096	30	0.191
	Average		0.081

Using these raw data, a histogram is drawn in figure 6. According to an empirical law, 30 sample data are divided into six bands. Then, we count number of samples of which the diameter fall in each band to draw the histogram.

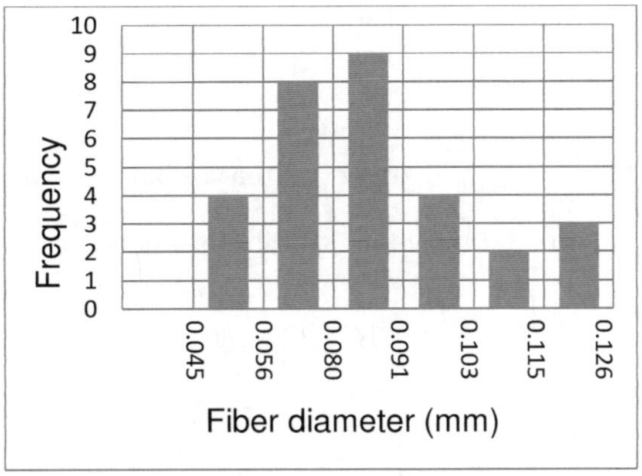

Figure 3. Histogram of fiber diameter in bottom part of abaca sheathes

From this histogram, we can calculate probability density for probability variable, fiber diameter, which is center value of each band. The probability density is plotted as function of diameter in figure 4. Similarly, probability densities for middle and top parts of abaca sheathes are plotted as a function of fiber diameter in figures 5 and 6.

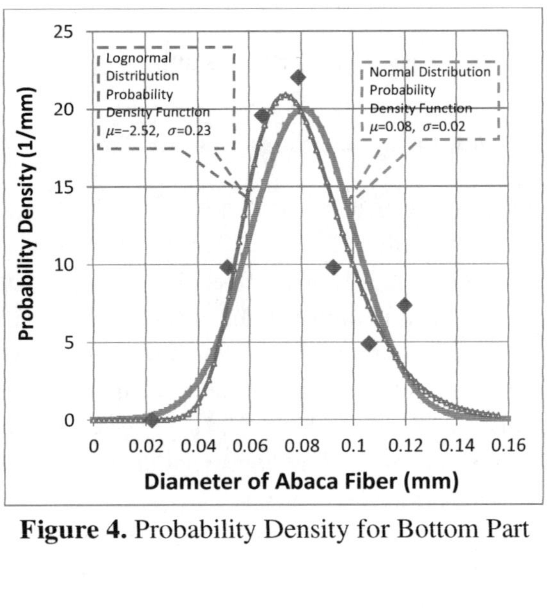

Figure 4. Probability Density for Bottom Part

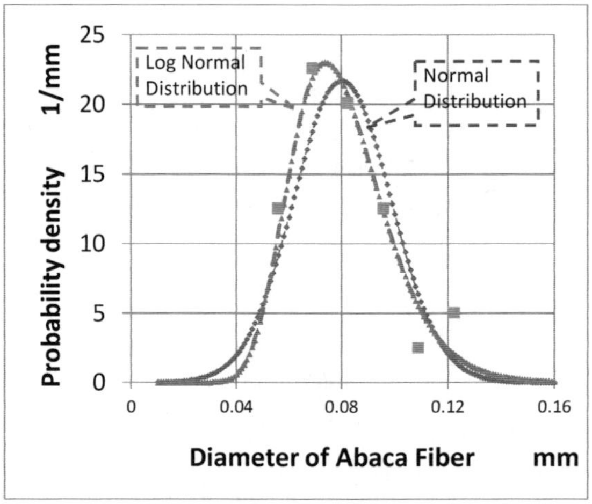

Figure 5. Probability Density for Middle part

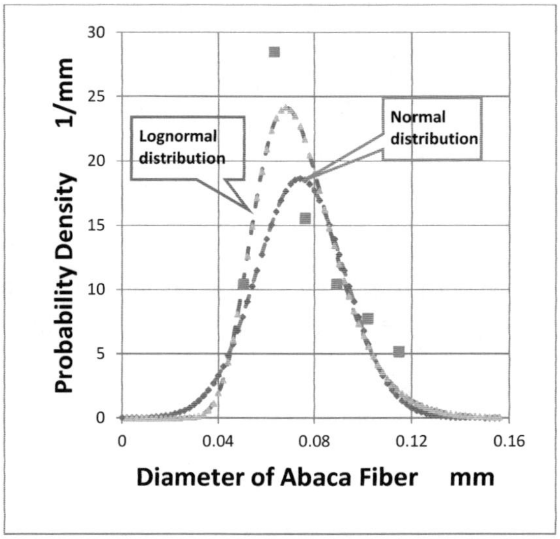

Figure 6. Probability Density for Top Part

As seen in the above figures, diameter data for all the parts fall on the normal distribution or log normal distribution line. Close examination shows that log normal distribution is more appropriate to express the scatter in the measured diameter of abaca fibers. Because measured dimensions are positive values, lognormal distribution that distribution variable must be positive is much more suitable to express the scatter rather than normal distribution of which distribution variable can be negative. E. Limpert, W. A. Stahel, and M. Abbt [4] summarize which parameters in various sience fields can be well expressed by lognormal distribution. A. S. Virk, W. Hall and J. Summer scales [5] show that cross section area of jute fiber can be well expressed by lognormal distribution. The results obtained by this research is consistent to the previous reserecher's results.

As shown in figure 7 to 9, diameter distributions of all the parts can be considered to follow lognormal distribution. Because random variable ln(D), natural logarithm of fiber diameter D, follows normal distribution, we can examine whether mean values of ln(D) for three parts are the same by t-Test. In table 2, Arithmetical mean values, D_{mean}, standard deviation values, s, geometric mean values, D^{*}_{mean} and standard deviation, s^{*}, of ln(D) for three parts are summarized. Then, we assume that natural logarithmic values of diameter for all the parts of abaca sheathes can follow the normal distribution and carry out t-Test to judge whether mean diameter values of three parts are statistically the same or not.

Table 2. Summary of Diameter Measurement

Part	D_{mean}	σ	D^*_{mean}	σ^*
Bottom	0.081	0.0196	-2.547	0.240
Middle	0.080	0.0184	-2.550	0.228
Top	0.074	0.0265	-2.613	0.236

D^*_{mean} *: mean value of lnD*
s^* *: standard deviation of lnD*

The t-Test result indicates at confidence level of 99% that the mean value of natural logarithm of diameter for three parts are not same and then, we can see that mean values of diameter for each part decreases as collecting position of fiber approaches top edge of abaca sheath.
Actually, t-Test suggests that mean diameter values of the three parts are significantly different, but it should be noted that the difference in mean values for the bottom and the middle part is very small.

4. Conclusion

1. Diameter distribution in the three parts is well expressed by logarithmic normal distribution as compared with normal distribution.
2. t-Test shows that mean values in the three partsaresignificant different and a mean value of diameter decreases as collection position approaches top end of abaca sheath. However, the mean values of the bottom and the middle part is quite small.
3. Diameter distribution for all the parts is well expressed by logarithmic normal, but it is necessary to examine statistical properties of abaca fiber produced in the other area.

Further experiments are necessary to identify statistical property of abaca fiber strength and to understand statistical properties in strength through statistical properties in geometry.

References

[1] Carbonfiber, Retrieved From http://carbonfiber.gr.jp/english/tech/pdf/lca.pdf
[2] Fukumoto Y 1987 Guidelines for stability design of steel structure, edited by Y. Fukumoto, Subcommittee on stability design committee on steel structures *Japan society of civil engineers* Tokyo Japan pp 37-76
[3] Mwaikambo L Y 2009 Tensile properties of alkalized jute fibres *BioResources* **4**:2 pp566-588
[4] Limpert E, Stahel W A, Abbt M 2001 Log-normal distributions across the Sciences: Keys and Clues *BioScience* **51**:5 pp 341-352
[5] Virk A S, Hall W, Summerscales J 2012 Modulus and strength prediction for natural fibre composites *Material Science and Technology* **28**:7 pp 864-871

Combined centralised and distributed mechanism for utilisation of node association in broadband wireless network

A Ulvan[*1], **M Ulvan**[1], **H Pranoto**[2]

[1]Department of Electrical Engineering, The University of Lampung Jl. Prof. S. Brodjonegoro No.1 Bandar Lampung 35145, Indonesia
[2]Department of Electrical Engineering, Medan State Polytechnique Jl. Almamater No.1 Kampus USU Medan 20155, Indonesia

[*]E-mail: ardian.ulvan@eng.unila.ac.id

Abstract. Mobile broadband wireless access system has the stations that might be fixed, nomadic or mobile. Regarding the mobility, the node association procedure is critical for network entry as well as network re-entry during handover. The flexibility and utilisation of MAC protocols scheduling have an important role. The standard provides the Partition Scheme as the scheduling mechanism which separates the allocation of minislots for scheduling. However, minislots cannot be flexibly reserved for centralised and distributed scheduling. In this paper we analysed the scheduling mechanism to improve the utilisation of minislots allocation during the exchange of MAC massages. The centralised and distributed scheduling is implemented in some topology scenarios. The result shows the proposed mechanism has better performance for node association than partition scheme.

1. Introduction

The MAC header of IEEE802.16 broadband wireless system is divided into two parts which are Generic MAC (GM) and Bandwidth Request (BR) [1]. GM is dealing with transfer data or MAC massages. In addition, BR has responsibility on sending bandwidth request packets to a base station (BS). In the legacy single connection of mobile station (MS) to BS, the connection management including node association and network entry seems have no critical issue. However, on the contrary, there are some issues when the extended connection and network such as relay network, multiple and mesh connection is present. What would happen when a BS is connected to multiple MSs? How can it manage the connection? These are some common issues that have to be answered and solved

In IEEE 802.16e standard, a MS are allowed to send data only at scheduled time which is decided by BS. The BS communicates to all MSs in the beginning of each frame in uplink map (UL-MAP). As we can found from the standard, the number of bandwidth allocation procedures and QoS are provided. However, particular scheduling method and reservation technique have not been standardised yet.

In Time Division Multiple Access (TDMA) mode, the data transfer has been scheduled by time. The time is divided between frames separated by guard time. Each frame is further divided between uplink (UL) sub-frame and downlink (DL) sub-frame. The UL sub-frame once again divided between three periods namely ranging period, bandwidth contention period and data uplink period. UL-MAP contains duration of these periods including time slots for each MS to send data to BS in uplink sub-frame. In uplink direction, MSs sent data to BS in corresponding allotted slots. Any MS is not allowed to use any other MS's uplink slots for data transfer. Therefore, to accomplish MS requirements the slots allocation is needed.

2. MAC scheduling and QoS Issues

In order to have a better support for QoS requirement the standard defines key parameters like delay, minimum reserved bandwidth, bandwidth request mechanism and flow type association. However, it is still not clear how the parameters would be employ, also when and where they should be used to obtain maximum QoS. It can be assumed that the implementation of an appropriate scheduling algorithm is the best way to support maximum application in parallel on this system.

Scheduling has to be done in both directions uplink and downlink. Downlink scheduling seems relatively easy because BS has all packets and their status. In contrast, the uplink scheduling seems more complex since it has various parameters such as numbers of client (MSs or RSs), delay bounds, associated flows and bandwidth requirements of each flow. In spite of different constraints between downlink and uplink direction, the different scheduling algorithm may or may not be implemented in both direction [1].

Additionally, it is necessary to investigate if the system require two scheduling components (at BS and at MS), or only single component of scheduling sitting at the BS can properly schedule all traffic. The IEEE 802.16 standard supports four different flow classes for QoS and the MAC supports a request-grant mechanism for data transmission in uplink direction. These flows are associated with packets at MAC level. Each connection has a unique flow type associated with it. The IEEE 802.16 standard does not define any slot allocation criterion or scheduling algorithm for any type of service. A scheduling module is necessary to design UL-MAP to provide QoS for each SS. Slots assignment for connections is done by BS and included the same in UL-MAP. The IEEE 802.16 standard support many traffic types (data, video, voice) with different QoS requirements. The 802.16 standard defines the following four types of service flow with distinct QoS requirement [1]:

- Unsolicited Grant Services (UGS)
- Real-Time Polling Services (rtPS)
- Non-Real-Time Polling Services (nrtPS)
- Best Effort (BE) Services

As described on IEEE 802.16e standard, each flow has its distinct QoS parameters. For UGS flow bandwidth request is not needed but for other flows queue length (incremental or aggregate) at MS is included in bandwidth request packet to represent the current demand. Bandwidth is always requested on a per connection basis. The IEEE 802.16e defines the following two ways for allocation of bandwidth grants [1]:

- Grant per Connection (GPC) Bandwidth is allocated to a connection and MS uses this grant only for this connection.
- Grant per Subscriber Station (GPSS) MS granted bandwidth aggregated into a single grant. This MS needs more intelligent to distribute this grant into various flows, running at this MS.

3. Scheduling Mechanism

This section point up the partition scheme mechanism which is described on the IEEE 802.16 standard, then we illustrate the centralised and distributed mechanisms as well as the node association in a network topology with relay capability.

3.1 Partition Scheme Mechanism

Obviously, in relay system the mesh mode topology is adopted. In IEEE 802.16 standard, the Time Division Multiplexing (TDM) radio access technology between clients (RSs and MSs) and between the clients and the mesh BS is take on. A radio channel is divided into physical slots (PSs) using time sharing, and multiple PSs are grouped as a frame. Figure 1 shows the frame structure.

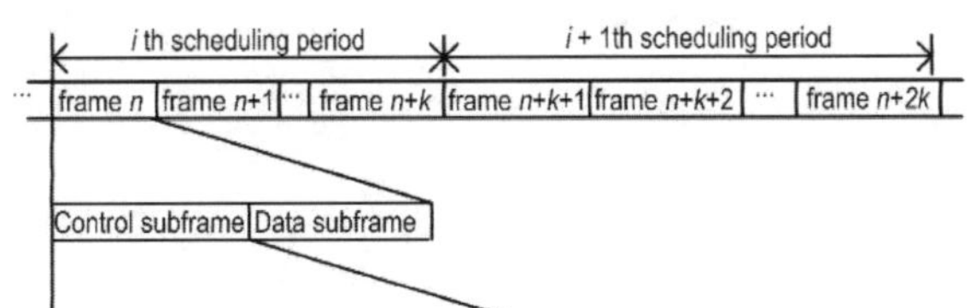

Figure 1. The frame structure of mesh network based on IEEE 802.16e

As described in [2], a frame is allocated for the purposes of control and management (control subframe) and information (data subframe). In the control subframe part, the signalling message is carried by the Transmission Opportunities (TOs). The TO for centralized scheduling is denoted as TOc, additionally the TOd is denoted for distributed scheduling.

The scheduling is to schedule minislots to serve different clients. The numbers of TOcs and TOds in a control subframe are configurable. On the other hand, the data subframe carries the user data and is further divided into 256 minislots. The transmission rate that the minislot can provide is assigned as r bits/sec. According to standard, the rate is not fixed but depends on several factors (e.g., channel coding, modulation, frequency band, etc) [1]. A minislot has a property called "minislot reuse" by which it can be reused by multiple clients as long as the clients are geographically separated or they do not interfere each other. In this paper we assume the multiple clients may be allowed to transmit in the same minislot.

The data traffics in the wireless mesh network can be divided into two categories [3]: intra-network traffic and inter-network traffic. Intra-network traffics are the traffics between two or more clients in the same wireless mesh network (i.e. MS to RS/BS, RS to BS, and vice versa). In addition, internetwork traffics are the traffics between the clients via different BSs, or the traffics from/to backhaul applications. A data subframe can simultaneously carry the intra and inter-network traffic data. The distributed scheduling mechanism is assigned for minislot allocation for the intranetwork traffic and the centralised scheduling mechanism is for the internetwork traffic. It is important to carefully schedule the minislots in a data subframe for both kinds of traffics to optimise minislot utilisation.

IEEE 802.16 recommends to partition a data subframe into two parts for the two schedulings mechanism which is known as the "partition" scheme. However, this may not be the best solution due to the fact that the partitioned boundary may not precisely capture the traffic pattern, and the minislots may not be fully utilized. Figure 2 shows the partition scheme mechanism. The first 128 minislots are assigned for inter-network traffic purposes. The rest 128 minislots are assigned for intra-network traffic.

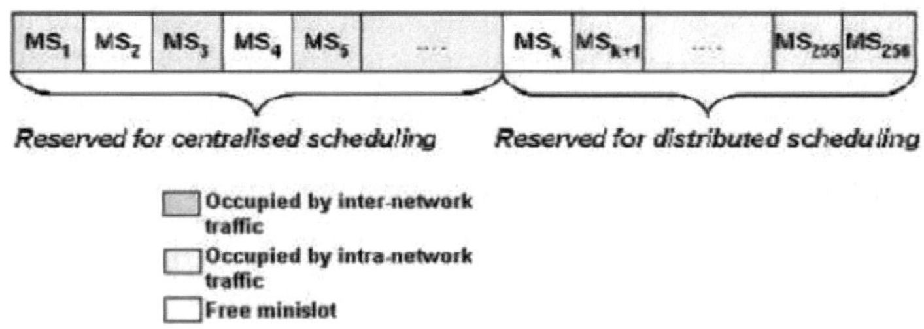

Figure 2. IEEE 802.16 Partition Scheme

3.2 Centralised Mechanism

The centralised scheduling mechanism for the communication between the BS to MSs has been specified on the IEEE 802.16e standard. The scheduling mechanism could also be adopted to make mobile possibility on relay station (mobile relay-MR). The BS acts as a scheduler and determines transmission and reception minislots for each client station (MS/MR). A scheduling tree rooted at the BS is established for the routing path between each client and the BS. A scheduling tree is denoted as follow and depicted on figure 3 below.

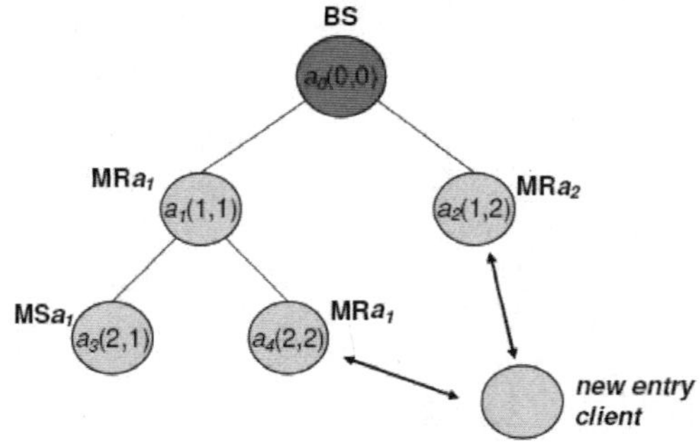

Figure 3. The scheduling tree of node association in centralised scheduling

The tree (T) can be written as follow:

$$T = \{a_0(k_{a_0}, n_{a_0}), a_1(k_{a_1}, n_{a_1}), a_2(k_{a_2}, n_{a_2}), \ldots a_i(k_{a_i}, n_{a_i}), \ldots\} \tag{1}$$

where k_{a_i} is the layer number, n_{a_i} is the position number in layer k_{a_i}, and (k_{a_i}, n_{a_i}) is the index number of the node. The index number of BS is (0,0).

Before a client executes the registration procedure to join the network, it selects a network node with smallest layer number from all neighbouring network nodes as the candidate node. Then the client sends the registration message (network entry procedure) to the mesh BS through the candidate node. Upon receipt of the registration message, the BS sets the new entry client as the child of the candidate node in the scheduling tree, updates the scheduling tree, and then broadcasts the scheduling tree to all Stations. By completing the procedure, the new entry client obtains a node ID.

The centralised scheduling consists of two stages. In the first stage, the BS collects bandwidth requests from all client stations. In the second stage, the BS calculates and then distributes the transmission and reception schedule to all Stations. The time period for exercising the two stages is called as *scheduling period*. The schedule assigned in the ith scheduling period is referenced in the $i+1^{st}$ scheduling period as shown in Figure. 1. Take Figure. 2 as an example, where MR_{a_i} (i = 1, 2, 3, 4) requests transmission rates, R_{u,a_i} and R_{d,a_i} for data transmission to and data receipt from the BS respectively. The MR_{a_i} first checks whether the number n_{f,a_i} of free minislots is smaller than $\left\lceil \frac{R_{u,a_i}}{r} \right\rceil + \left\lceil \frac{R_{d,a_i}}{r} \right\rceil$. If so, MR_{a_i} quits the centralised scheduling. Otherwise, the centralised scheduling exercises as follows: Based on the scheduling tree, the client station with larger k_{a_i} and smaller n_{a_i} transmits the UL signalling messages first, while the client station with smaller k_{a_i} and smaller n_{a_i} relays the DL signalling message first.

3.3 Distributed Mechanism

Additionally, the distributed scheduling will be assigned when client station CS_{a_s} has data to be transmitted with transmission rate R to its neighbour CS_{a_r}. The CS_{a_s} first checks whether n_{f,a_s}, is smaller than $\left\lceil \frac{R}{r} \right\rceil$. If so, CS_{a_s} quits the distributing scheduling. Otherwise, the scheduling is performed to reserve minislots for data transmission as shown in Figure. 6. Then, after the execution of distributed scheduling, CS_{a_s} can transmit data to CS_{a_r} in the reserved minislot without collision [4].

4. Flexibility and Utilised MAC Message

According to the scheme that have been described on previous section, the MAC massages flows for centralised and distributed scheduling mechanism can be seen on Figure. 4 and Figure. 5 respectively. The combination of centralised and distributed mechanism for node association MAC scheduling also have the positive effects since the minislot allocation can be more flexible, and the minislot utilisation is increased, shown in figure 6.

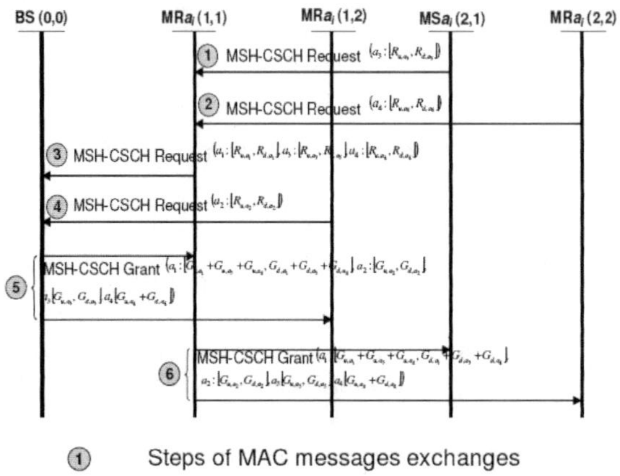

Figure 4. MAC massages exchanges for centralised scheduling

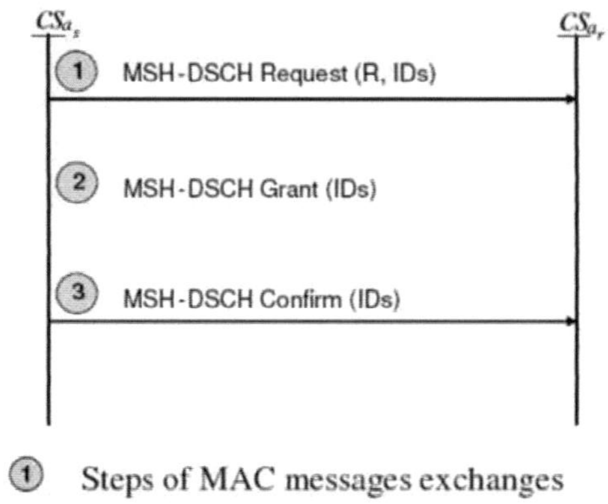

Figure 5. MAC massages exchanges for distributed scheduling

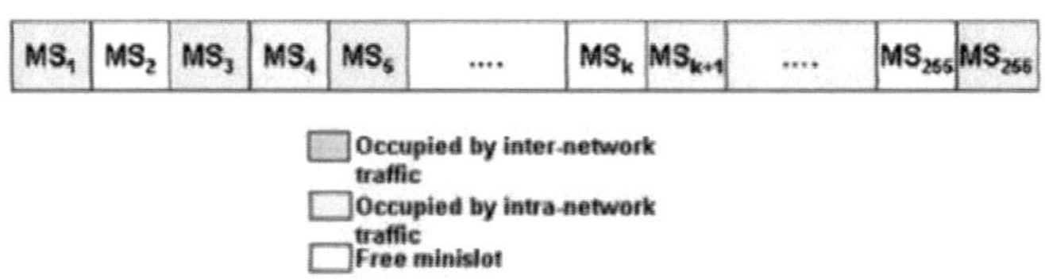

Figure 6. Flexible and utilised minislots by combine centralised and distributed mechanism.

5. Conclusion

The IEEE 802.16e standard has provided the partition scheme mechanism by dividing fixedly the 256 minislots for centralised and distributed scheduling. However, the partitioned boundary may not precisely capture the traffic pattern since the relay stations with multiple connections in mesh topology is obtainable, therefore the minislots may not be fully utilised. The use of combine centralised and distributed mechanism is more flexible and effective in utilisation of minislots.

Acknowledgment

This research work was supported by KLN Research Grant No. 160/UN26/8/LPPM/2015. The authors would also like to acknowledge the contributions of Directorate General of Higher Education, in supporting on the publication.

References

[1] Amendment 2: for Physical and Medium Access Control Layers for Combined Fixed and Mobile Operation in Licensed Bands 2006-02-28 (IEEE std 802.16e-2005/802.16cor1)

[2] Chen S M 2006 A Study on Distributed/Centralized Scheduling for Wireless Mesh Network *Proc. IWCMC* 2006 *ISBN:1-59593-306-9*
[3] Ulvan A, Bešták R 2007 Analysing the Node Association MAC Scheduling Mechanism Based on IEEE 802.16e *Proc. RTT 2007 ISBN 978-80-8070-735-4*
[4] Ulvan A, Bešták R 2007 *Transmission Performance of Flexible Relay-based Networks on the Purpose of Extending Network Coverage* (ser. Lecture Note in Personal Wireless Communications. New York: Springer)

IOP Conf. Series: Materials Science and Engineering **308** (2018) 012039 doi:10.1088/1757-899X/308/1/012039

Cancelation and its simulation using Matlab according to active noise control case study of automotive noise silencer

Alfisyahrin[1], **I Isranuri**[*2]

[1]Department of Electrical Engineering, University of Syiah Kuala Darussalam, Banda Aceh, Indonesia
[2]Department of Mechanical Engineering, University of Sumatera Utara. Jalan Almamater, Kampus USU, Padang Bulan, Medan 20155, Indonesia.

[*]E-mail: isranuri@yahoo.com

Abstract. Active Noise Control is a technique to overcome noisy with noise or sound countered with sound in scientific terminology i.e signal countered with signals. This technique can be used to dampen relevant noise in accordance with the wishes of the engineering task and reducing automotive muffler noise to a minimum. Objective of this study is to develop a Active Noise Control which should cancel the noise of automotive Exhaust (Silencer) through Signal Processing Simulation methods. Noise generator of Active Noise Control is to make the opponent signal amplitude and frequency of the automotive noise. The steps are: Firstly, the noise of automotive silencer was measured to characterize the automotive noise that its amplitude and frequency which intended to be expressed. The opposed sound which having similar character with the signal source should be generated by signal function. A comparison between the data which has been completed with simulation calculations Fourier transform field data is data that has been captured on the muffler (noise silencer) Toyota Kijang Capsule assembly 2009. MATLAB is used to simulate how the signal processing noise generated by exhaust (silencer) using FFT. This opponent is inverted phase signal from the signal source 180° conducted by Instruments of Signal Noise Generators. The process of noise cancelation examined through simulation using computer software simulation. The result is obtained that attenuation of sound (noise cancellation) has a difference of 33.7%. This value is obtained from the comparison of the value of the signal source and the signal value of the opponent. So it can be concluded that the noisy signal can be attenuated by 33.7%.

1. Introduction

Noise in this century is already a big enough problem to the whole society forward. Where the noise factor is increasing almost complete in various sectors, such as environmental and industrial sites. The cause of all this is a technological advancement that has increased since the start of the Industrial Revolution in the 17th century creation of industrial machines are on average almost cause noise.

Therefore, to overcome the noise of it needs to be studied and researched more about the noise factor by science or acoustic sound to a more developed science noise control. This science originated from the physical sciences Acoustics which later evolved to the various disciplines such as Mechanical Engineering, Architecture, Electrical, even chemistry. However, the discussion in this paper covers the scope of the Electrical and Mechanical Engineering is the science Active Noise Control.

We can look at some reference point of the research that has been studied by many people on Active Noise Control, such as research on noise cancellation using Active Noise Control System. Signals on the single channel Active Noise Control has reduced its best when low frequency noise at -

10.72 dB. This system is unstable when part of the signal is constantly changing. This section varies the displacement of an object in the room. Active Noise Control less than the maximum signal from the zone generated is also very relative silence to generate a large test area. We can see the set-up tool as follows [11].

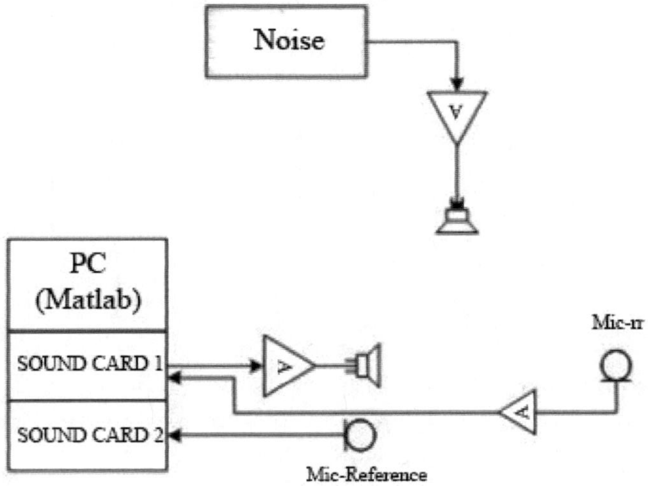

Figure 1. Hardware configuration on Active Noise Control

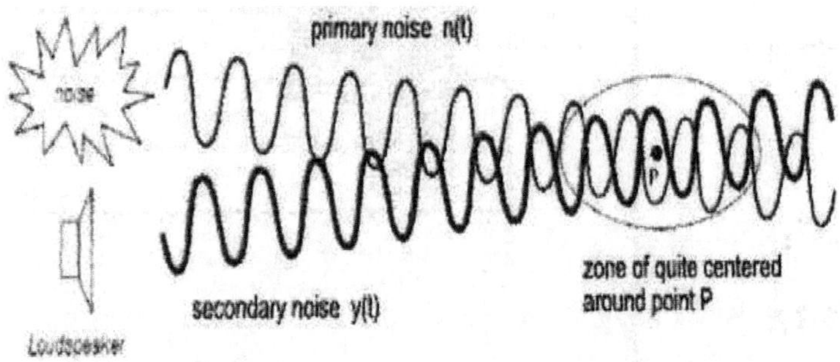

Figure 2. The two signal propagation

In the study of the Active Noise Control against the best vacuum cleaner in achieving the objectives of the Active Noise Control on signal generated simulation optimal under 6 dB. Radiation nosie on the vacuum cleaner is direct. This makes it difficult to study the setting up mikropone to obtain a good correlation with the noise signal [8].

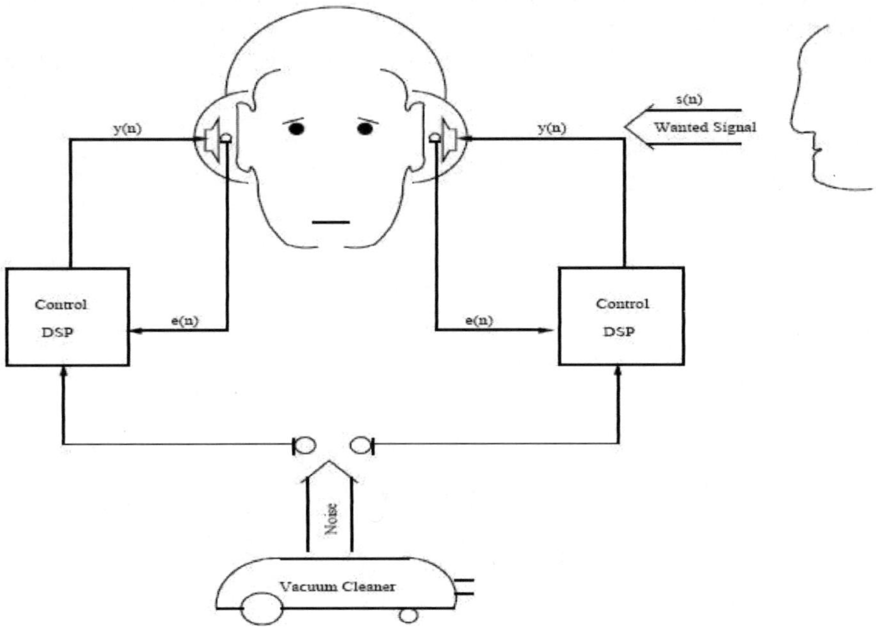

Figure 3. Active ear protectors of the second source

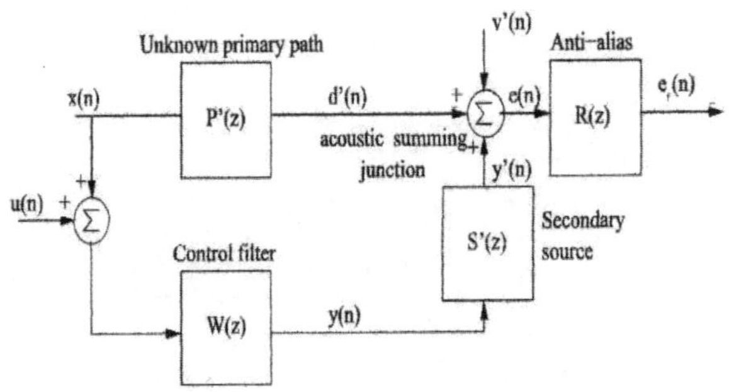

Figure 4. Block Diagram of the System channel Active Noise Control.

Research emphasis on low frequency noise or vibration passive has many disadvantages, largely because of concerns the volume silencer. With the use of signal processing becomes possible to use the Active Control. Noise secondary (artificial) should be produced, which weakens the noise that the primary (original) propagating in microphone] positioned [13].

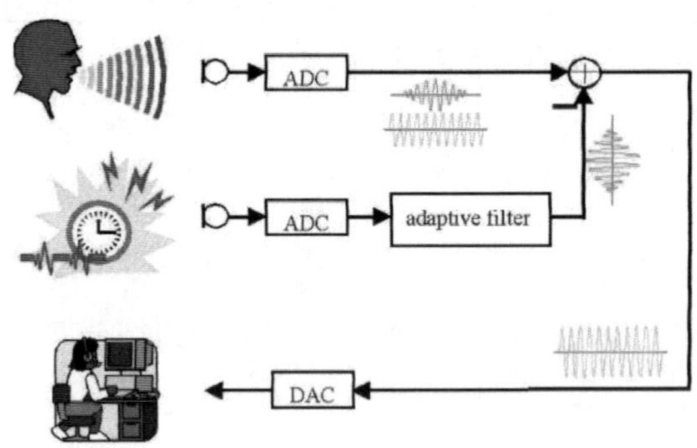

Figure 5. System Active Noise Control [2].

Modeling and controller design procedures for an FB-ANC for the sound system. Acoustic sound wave controlled modeled as a short. A feedback controller is designed then to attenuate the noise level around, involving a sensor fault with H8 control theory based on IIR models. The effectiveness of the proposed design procedure performed by test experiments. A sound field has been assumed in this study. A controller and systematic modeling procedures designed to examine a field of acoustic echo cancellation will become a staple further studies [12].

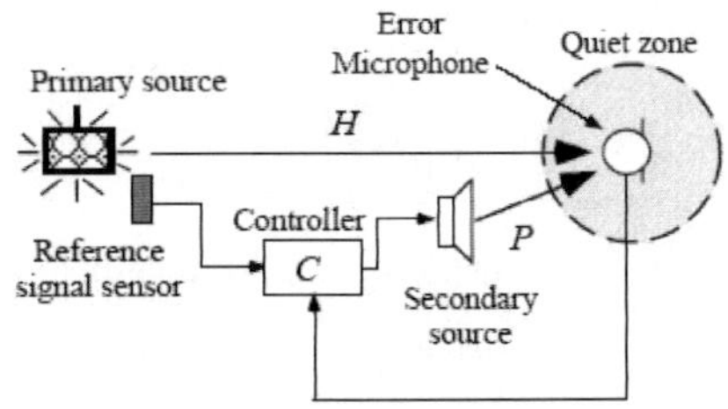

Fig. 1. Configuration of an SISOSE FF-ANC system.

Figure 6. ISOSE configuration FF- Active Noise Control System.

Active Control of Noise Using FXLMS

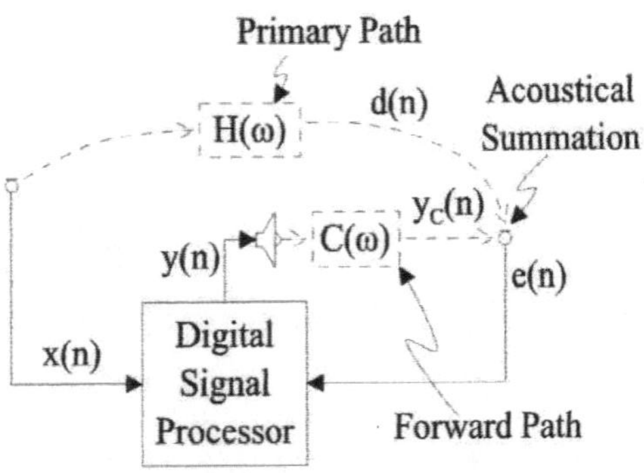

Figure7. Active Noise Control at the ear is placed on the primary path and the forward path.

In practical applications, methods of Active Noise Control salaing memepengaruhi with nonlinear. Nonlinear controller based on Volterra filter to apply in the form of multichannel filtering and useful data can be utilized on the environment. One of the aspects of the complex is the origin of efficient adaptation algorithms. In general, the so-called filtered-X LMS or NLMS algorithm. In this study we propose the use of affine projection technique, and we obtain in detail the so-called Filtered-X AP algorithm for homogeneous quadratic filter. According to the multichannel approach, the origin of this can be easily extended to general Volterra filter. Our extensive experiments to confirm the AP report compared to the classic technique LMS and NLMS algorithms with a limited increase concerns the computational complexity compared with the Active Noise Control acoustic. One application for the Active Noise Control is the active control of noise and vibration in a lathe. In the adaptive control of vibration at a fixed point behind this FXLMS control the display accordingly. It is also interesting for further describes the digital and analog [18].

Signal processing

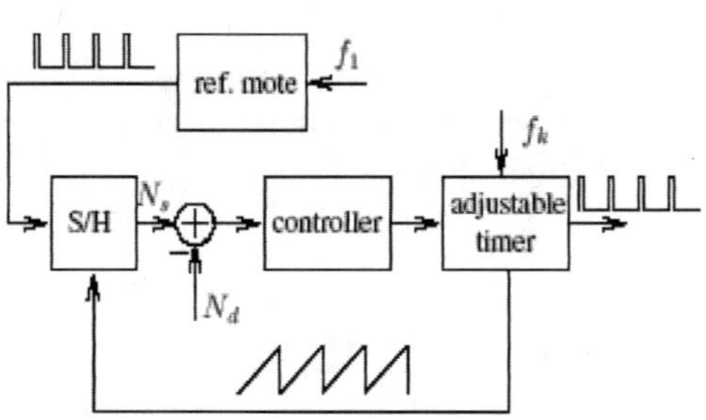

Figure 8. Synchronization signal processing block diagram of the Active Noise Control.

2. General

2.1 The concept of Active Noise Control

Active Noise Control sound plain language is countered by voice or a signal method that we get we must seek anti signal that is similar to the start signal but the opposite phase.

We can see an example in the description of Figure 9 which are described erect low amplitude signal with no sound condition.

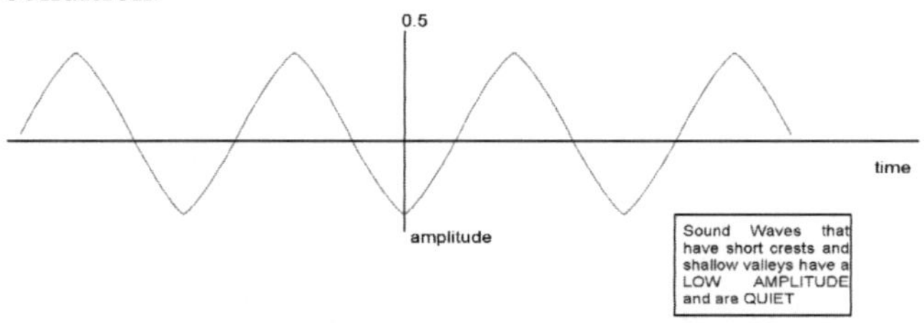

Figure 9. Low Upright stationary wave amplitude [7].

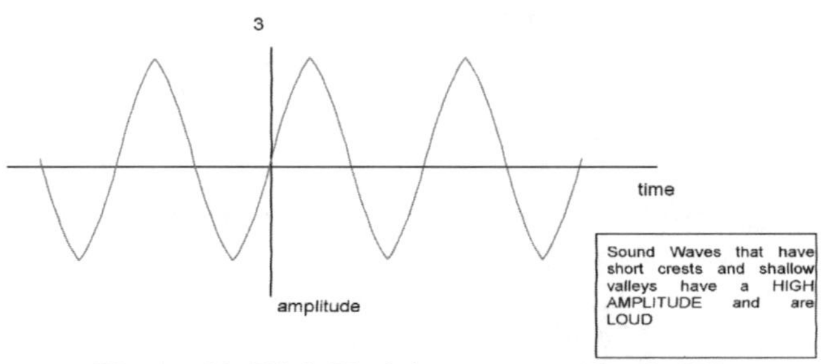

Figure 10. High Upright wave amplitude [7].

From the comparison Figure 8 and 9 can be seen that the amplitude at rest with the state of low and high amplitude. And looks at the low amplitude sound does not have to sound a loud high amplitude.

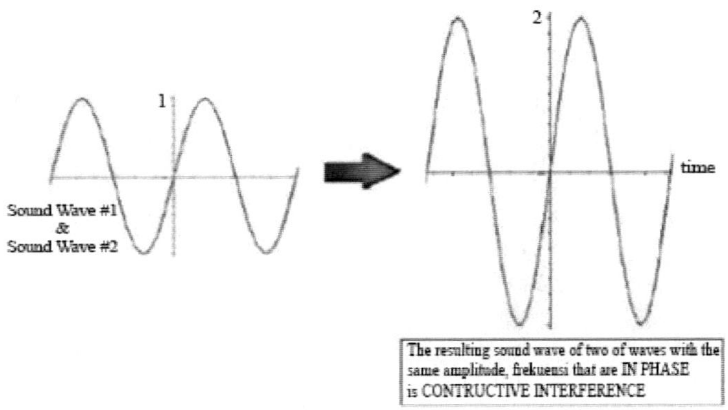

Figure 11. Upright waves on the phase shift [7].

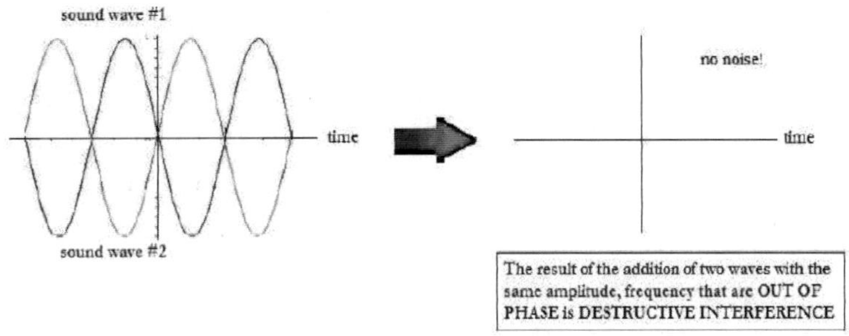

Figure 12. Two standing waves with a phase difference of 180° mutually exclusive [7].

Table 1. Data Sound Pressure Level (dB)

No	Sound sources	Intensity (dB)	Intensity (W/m^2)
1	Jet Plane at 30 m	140	100
2	Pain threshold	120	1
3	Strength of Rock concert in the open air.	120	1
4	Siren at 30 m	100	$1x10^{-2}$
5	Auto Interior, moving at 90 km / h	75	$3x10^{-5}$
6	Busyness / road congestion	70	$1x10^{-5}$
7	Casual conversations at 50 cm	65	$3x10^{-6}$
8	Radio calm	40	$1x10^{-8}$
9	Whisper	20	$1x10^{-10}$
10	Desir leaves	10	$1x10^{-11}$
11	Threshold of Hearing	0	$1x10^{-12}$

Source: Giancolli, Douglas. Physiscs Third Edition. New Jersey: Prentice Hall, Englewood Cliffs

In the Figure 12 is shown a signal source or signal 1 in an Active Noise Control system. Then the equation shown are:

$$y_1 = A \sin \varpi t \quad (1)$$

IOP Conf. Series: Materials Science and Engineering **308** (2018) 012039 doi:10.1088/1757-899X/308/1/012039

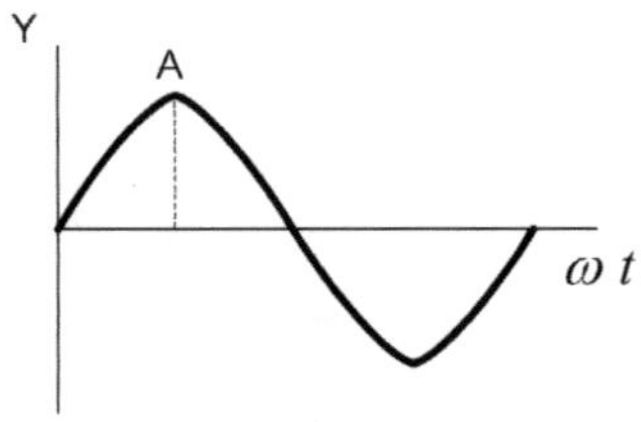

Figure 13. The source signals or signal 1

Figure 13 is shown opposing signals or signal 2 of an Active Noise Control system. The equation shown are:

$$y_2 = -A \sin \varpi t \quad (2)$$

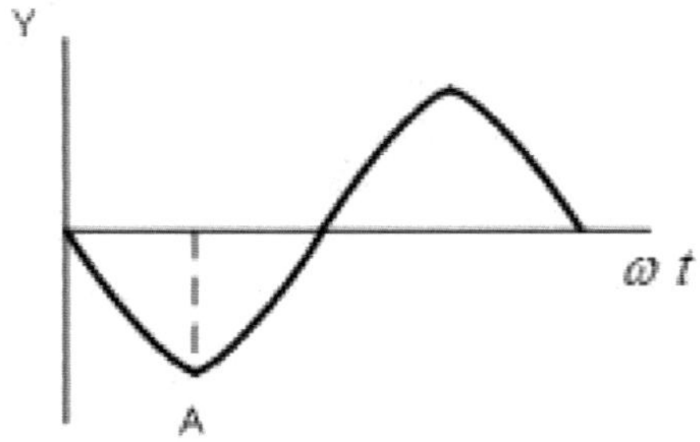

Figure 14. Opponents signal or signal 2

In the figure 14 is an active signal, wherein the signal 1 and signal 2 are members of the opposite phase signal 180° and cancel each other out.

Then the equation of incorporation Signals Start / Signal 1 to Signal Opponent / Signal 2 is:

$$Y_{\text{active noise}} = y_1 + y_2 \quad (3)$$

$$Y_{\text{activenois e}} = A \sin \varpi t - A \sin \varpi t$$

$$= 0$$

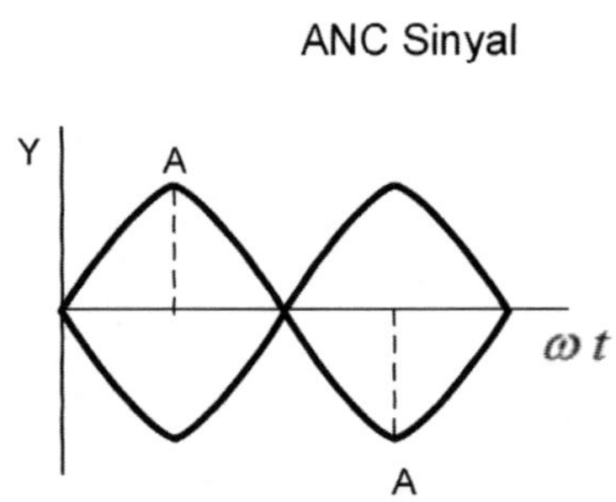

Figure 15. Active Noise Control Signal

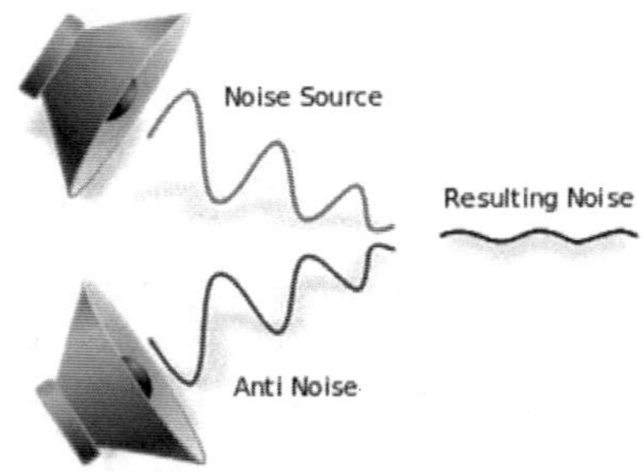

Figure 16. Experiments Active Noise Cancellation with Loudspeaker.

In this study, Active Noise Control is used to detect the signal attenuation in the muffler (silencer) car. Noisy signal to be generated is reduced to a minimum cultivated for comfort noise.

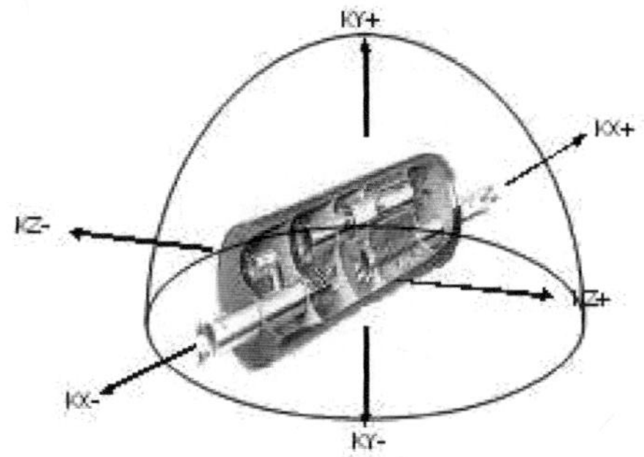

Figure 17. Exhaust (noise silencer) Cars

Figure 16 is a set of data collection to measure the noise that exist in the exhaust (noise silencer) then this setting is used to perform Active Noise Control Engineering is made thereof or signal virtual opponent is Noisy Signal Generators.

2.2 Noise Signal Generators

Noise generator is generating noise which serves to produce opposing signals, this signal can be regarded as a virtual signal from the signal source. The goal is the frequency and amplitude that are emitted by Instruments Signal Generators has similar characteristics to the signal source so as Active Noise Control can be realized. In making the character of many noise generator circuit configuration that can be applied to one of them is to use oscillator IC 555 is figure 17.

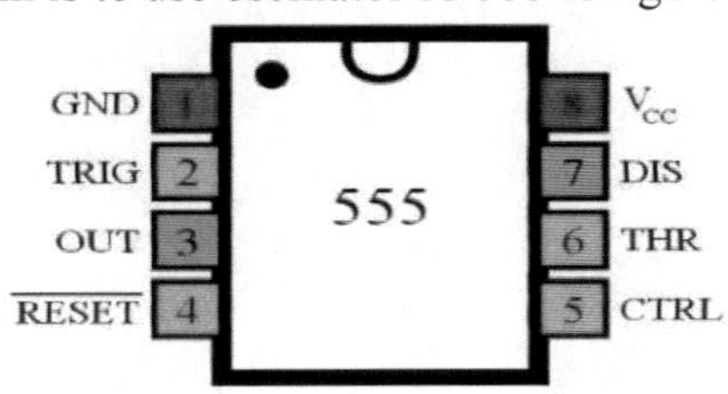

Figure 18. Oscillator IC 555.

IC 555 as a timer which is also a formula for the oscillator has a frequency and duty cycle:

$$f_r = \frac{1.44}{(R_1 + 2R_2)C_{ext}} \tag{4}$$

IOP Conf. Series: Materials Science and Engineering **308** (2018) 012039 doi:10.1088/1757-899X/308/1/012039

And for the duty cycle is:

$$\text{Duty cycle} = \left(\frac{R_1 + R_2}{R_1 + 2R_2} \right) 100\% \tag{5}$$

Before moving to the signal generator needs to know the value of the voltage to be generated by the IC 555 Oscillator. It needs to be taken into account for the voltage divider out of the value generated by a voltage source which is then inserted into the Oscillator. The voltage divider circuit as in Figure 18.

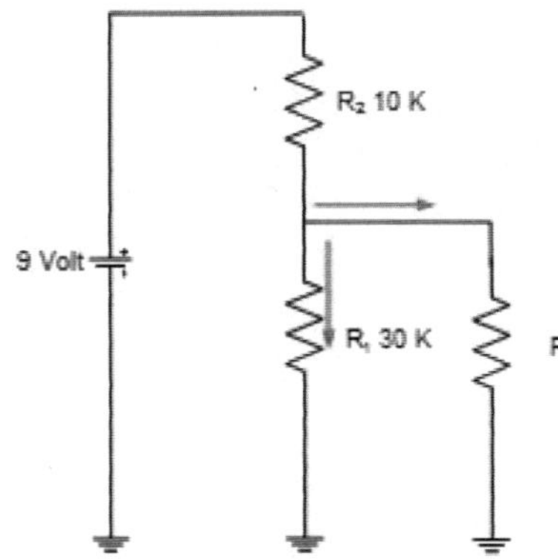

Figure 19. The voltage divider circuit

This voltage divider calculation:

$$Vout\,(unloaded) = \left(\frac{R_2}{R_1 + R_2} \right) V_S = \left(\frac{10K}{30K + 10K} \right) 9V \tag{6}$$

$$= \left(\frac{10K}{40K} \right) 9V = 2.25\ V$$

2.3 Phase Shifter

Phase shifter is a tool for moving phase on the signal. This technique is used for the purpose of reversing the phase 1800 the goal is to make the opponent signal (anti signal) at this research, to get the results of signal processing Active Noise Control.

Theory of the phase shifter by the input on transistors which then issued on foot of collector and emitter with a certain resistance so that the output signal can be adjusted phase shift can be seen in figure 19.

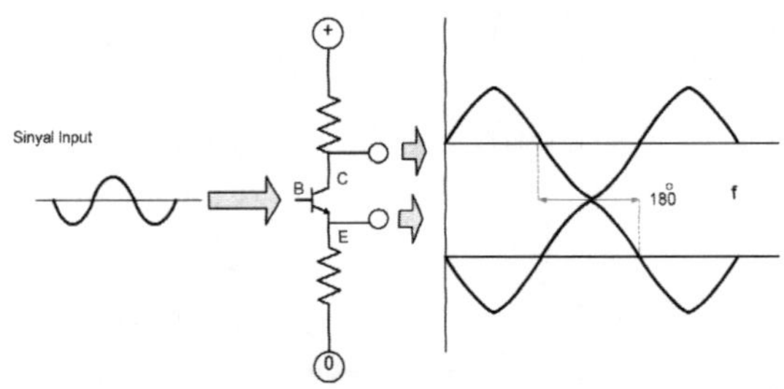

Figure 20. Process input and output of the phase shift.

180° collector signal is turned on and the input signal to the emitter signal. This analysis derived from Low Frequency Response amplifier can be seen in figure 20.

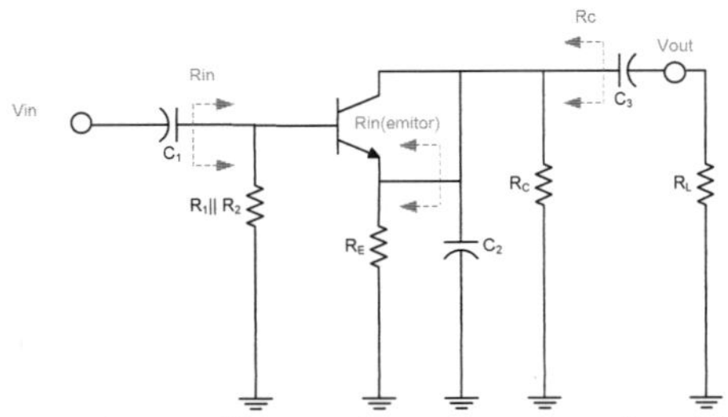

Figure 21. Low frequency AC coupling and bypass capacitor high-pass RC circuit.

The input current is the basis:
where $R_C = R_C \parallel R_L$
Circuit input RC

$$V_{base} = \left(\frac{R_{in}}{\sqrt{R^2 in + X^2 C_1}} \right) V_{in} \tag{7}$$

This condition occurs in the input RC circuit when $RC_1 = R_{in}$

$$\begin{aligned} V_{base} &= \left(\frac{R_{in}}{\sqrt{R^2 in + R^2 in}} \right) V_{in} = \left(\frac{R_{in}}{\sqrt{2R^2 in}} \right) Vin \\ &= \left(\frac{R_{in}}{\sqrt{2R^2 in}} \right) V_{in} = \left(\frac{1}{\sqrt{2R_{in}}} \right) V_{in} \\ &= 0.707 V_{in} \end{aligned} \tag{8}$$

low frequency can be calculated :

$$Xc_1 = \frac{1}{2\pi\pi fc_1} = R_{in} \tag{9}$$

$$fc = \frac{1}{2\pi\pi_{in} C_1} \tag{10}$$

then input source can be calculated :

$$fc = \frac{1}{2\pi(R_s + R_{in})C_1} \tag{11}$$

The process of achieving the 180° phase shift can be seen in figure 21.

IOP Conf. Series: Materials Science and Engineering **308** (2018) 012039 doi:10.1088/1757-899X/308/1/012039

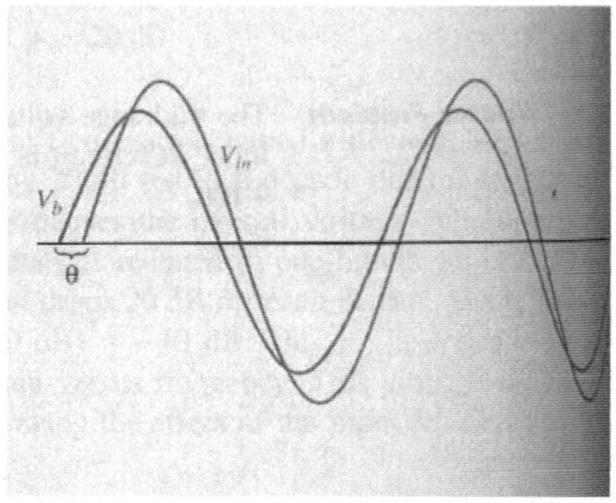

Figure 22. Input to the RC circuit is due to precede the base voltage of the input voltage at an angle below the intermediate frequency phase [18].

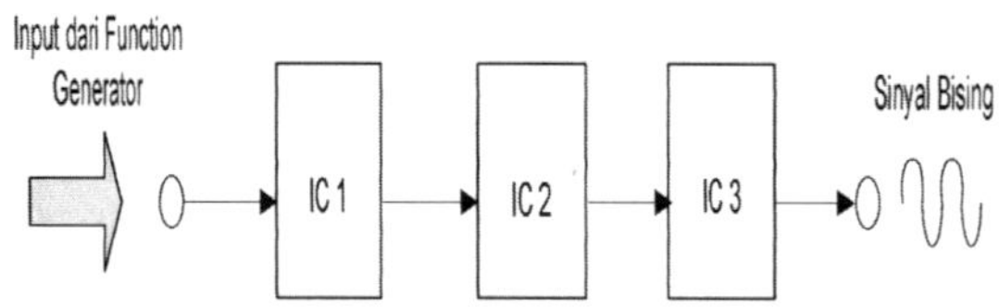

Figure 23. Block Diagram Noise Generator

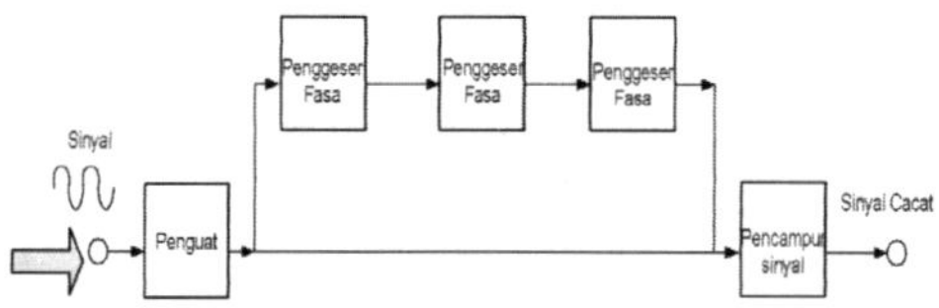

Figure 24. Block Diagram Phase Shifter.

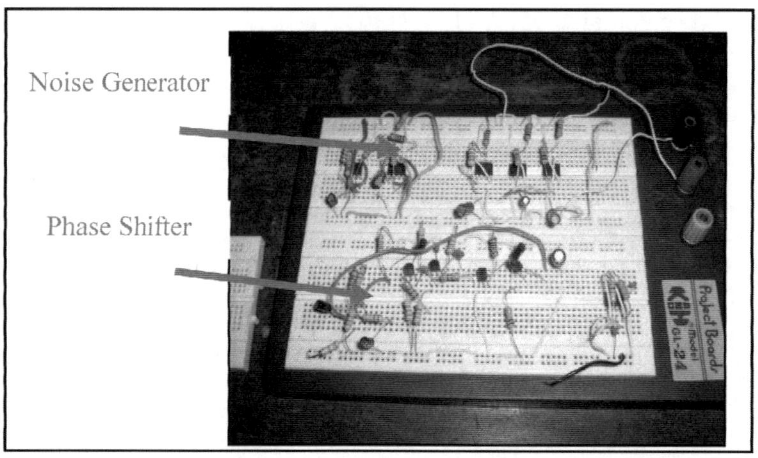

Figure 25. Setting Noise Signal Generators Instruments and Phase Shifter.

2.4 Fourier Transform

Theories about the Fourier transform signal that will be used equation is as follows:

$$X(\omega) = \int_{-\infty}^{\infty} x(t)e^{-j\omega t}dt: \tag{12}$$

The function f (x) is a periodic function if the function is repeated at a certain time with a single variable. This is a repetition of a certain time period in a vibration on the image can be seen in figure 25.

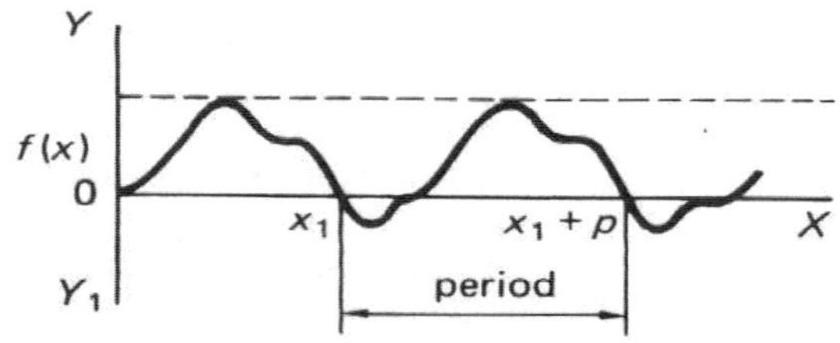

Figure 26. The function of a signal with a period of [22].

It is obvious that y = sin nx is a periodic function, where the distance value x increases from 0° to 360°. The period of 360° is said to 2π radians and maximum amplitude shift of position backrest, see figure 26.

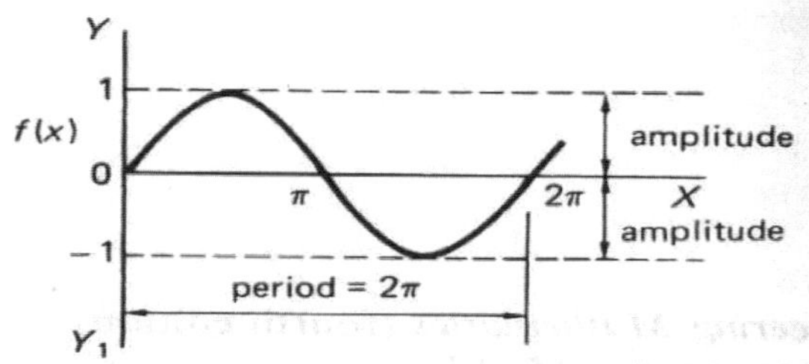

Figure 27. Amplitude [22].

Definition of Fourier series:The function f (x) which can be determined at the interval (-L, L) and outside interval and outside the specified interval pleh ii f (x + 2L), which is considered to be that f (x) has a period of 2L. Fourier series or Fourier expansion with respect to f (x) was defined as:

$$\frac{a_0}{2} + \sum_{n=1}^{\infty}\left(a_n \cos\frac{n\pi\pi}{L} + b_n \sin\frac{n\pi\pi}{L}\right) \tag{13}$$

With prices Fourier coefficients an and bn is determined by:

$$a_n = \frac{1}{L}\int_{-L}^{L} f(x)\cos\frac{n\pi\pi}{L}dx$$

$$b_n = \frac{1}{L}\int_{-L}^{L} f(x)\sin\frac{n\pi\pi}{L}dx \qquad n = 0,1,2,........ \tag{14}$$

With prices Fourier coefficients an and bn is determined by:

Every periodic function f (ωt) can be decomposed into an infinite trigonometric series and is called Fourier series. So that can be decomposed into a Fourier series, the function must meet the following conditions. That function is a periodic function and meet relation f (ωt) = f (ωt+2π) the period 2π.

Take a periodic function f (ωt) then the Fourier series for the function in the formula :

$$(\omega t) = A_0 + \sum_{n=1}^{\infty}[A_n \cos(n\omega t + B_n \sin(n\omega t)] \tag{15}$$

A_0, An, and Bn is called the Fourier coefficients and are determined by formula:

$$A_0 = \frac{1}{2}\pi\int_0^{2\pi} f(\omega t)\ d(\omega t)$$

$$A_n = \frac{1}{2}\pi\int_0^{2\pi} f(\omega t)\cos(n\,\omega t)\,d(\omega t)$$

$$B_n = \frac{1}{2}\pi\int_0^{2\pi} f(\omega t)\sin(n\,\omega t)\,d(\omega t)$$

where, n = 1, 2, 3, …

To simplify the calculation of Fourier series, is to use a Fourier transform tables. Table Fourier Transformation is a variety of forms, but the researchers took the simplest form of the signal to be processed.

Wavefrom is still a sinusoidal shape, by cutting in parts of randomness signals are very complicated.

Table 2. Fourier transformation.

$f(x)$	$f(y)$
$\left[f(x)=(1/2\pi)\int_{-\infty}^{+\infty}F(y)\,e^{+ixy}dy=FT^{+}\{F(y)\}\right]$	$\left[f(y)=(1/2\pi)\int_{-\infty}^{+\infty}F(x)\,e^{-ixy}dy=FT^{-}\{F(y)\}\right]$
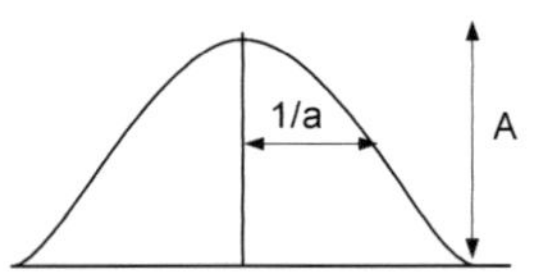 $A\ \exp(-a^2x^2)$ [*Gaussian*]	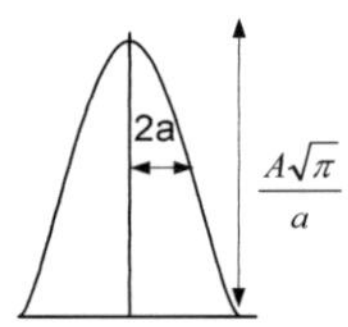 $\frac{A\sqrt{\pi}}{a}\ \exp(-y^2/4a^2)$ [*Gaussian*]

Source: Champeney, D.C. Fourier Transform and Their Physical Application: Academic Press, Inc

In this study, the use of Fourier calculations are Fourier transform using the above table is the equation of table 2.
Which we then use the equation calculates the signal for are:

$$\left[f(x)=\frac{1}{2}\pi\int_{-\infty}^{\infty}F(y)e^{+ixy}dy=FT^{+}\{F(y)\}\right] \tag{16}$$

$$A\exp(-a^2x^2) \tag{17}$$

Where,
A = Amplitude
a = Real constant
x = Function value

2.5 MatLab

MatLab is a software that is maximal for the execution and solution on methods and techniques of Noise Control. The software can provide solutions with signal processing methods using FFT (Fast Fourier Transform). the goal is to get Noise Control analysis, especially in this research is the Noise Control focused on exhaust (silencer).

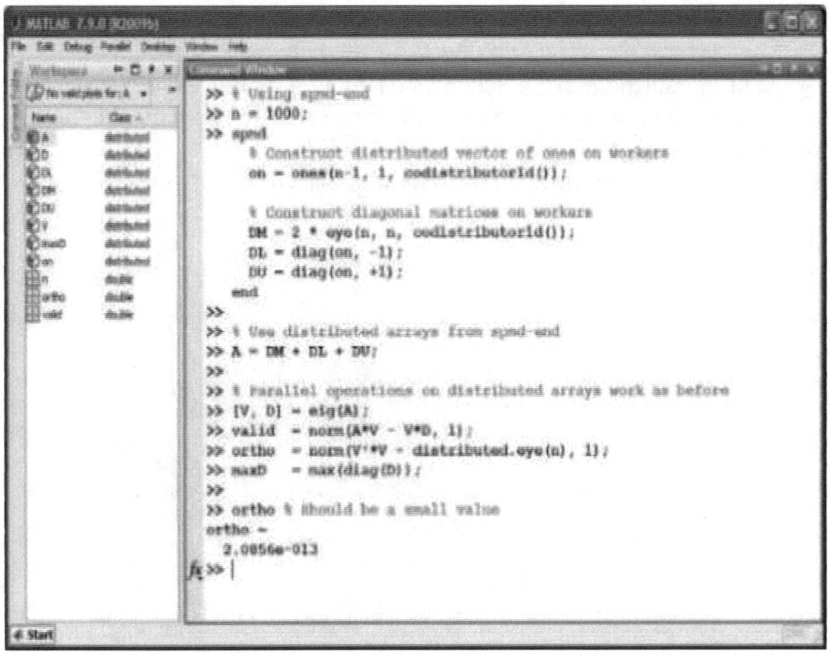

Figure 28. Menu on MATLAB in solving signal issues.

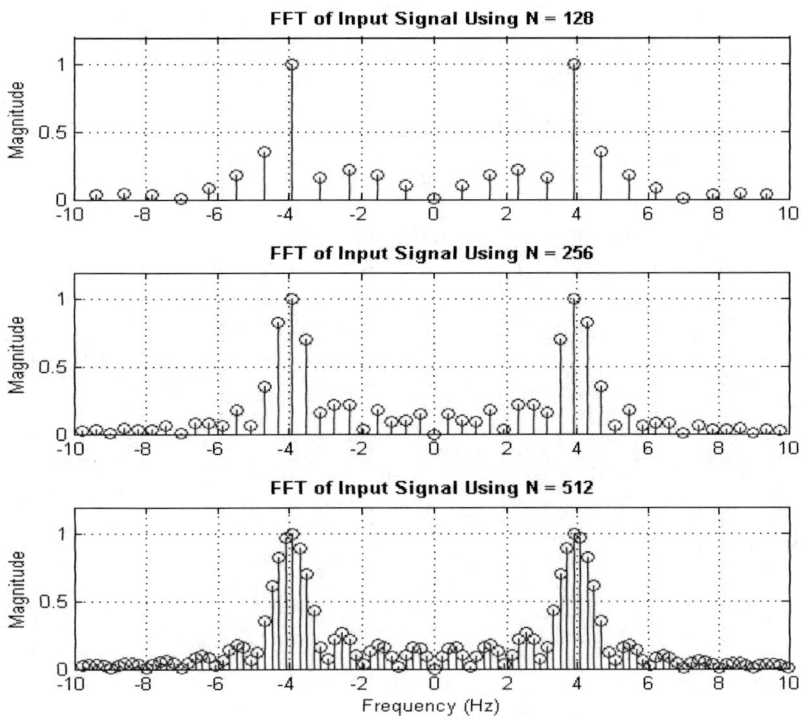

Figure 29. The signal processed by FFT.

IOP Conf. Series: Materials Science and Engineering **308** (2018) 012039 doi:10.1088/1757-899X/308/1/012039

3. Outline

The method in this study is a comparison between the results of simulation Noise Control derived from field data or data measuring muffler (silencer) Toyota Kijang Capsule assembled in 2009 with the data Instruments Signal Generators Noise with the results obtained from the simulation Noise Control signal processing using MatLab.

Comparison of results from research using data simulation calculations with the Noise Control Simulation Noise Control software that uses MatLab. Whether the results of calculations with the Signal Processing using Fourier better Transforms its reduction compared to simulations using MatLab.

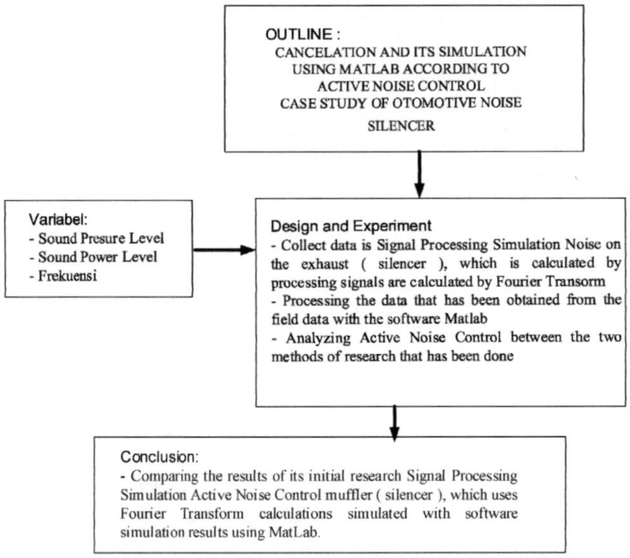

Figure 30. Framework concept

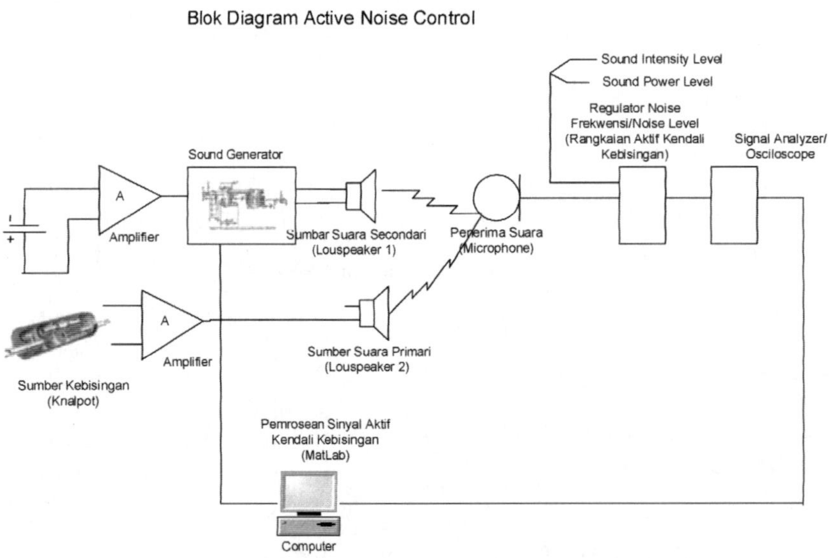

Gambar 3.3 Blok Diagram

Figure 31. Block diagram of Active Noise Control on the exhaust (silencer).

Figure 32. Instruments Noise Signal Generators Circuit

Table 3. Data Instrument Noise Signal Generators.

No	Frek	db X-	dB Z+	dBZ-
1	31.5	29.33	30.23	31.55
2	40	30.34	29.78	29.33
3	50	31.25	30.76	31.55
4	63	31.22	33.34	32.76
5	70	33.34	34.44	33.34
6	80	26.22	25.34	27.22
7	100	32.35	35.55	33.55
8	125	30.35	31.35	30.55
9	160	36.6	37.23	36.23

3.1 Results graph of the data Instrument Noise Signal Generators

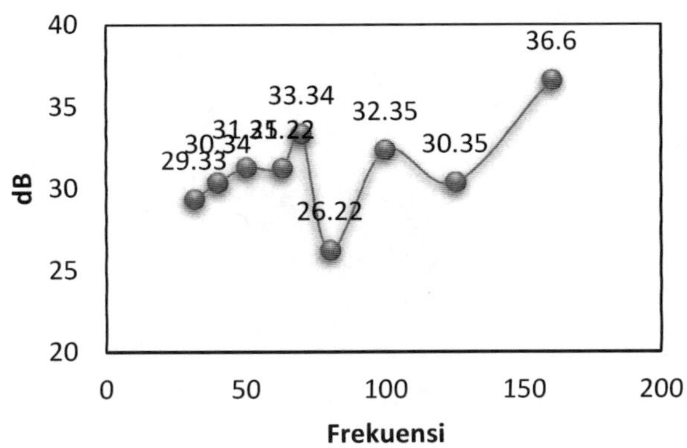

Figure 33. Value of Data Noise Signal Generator on the X- axis.

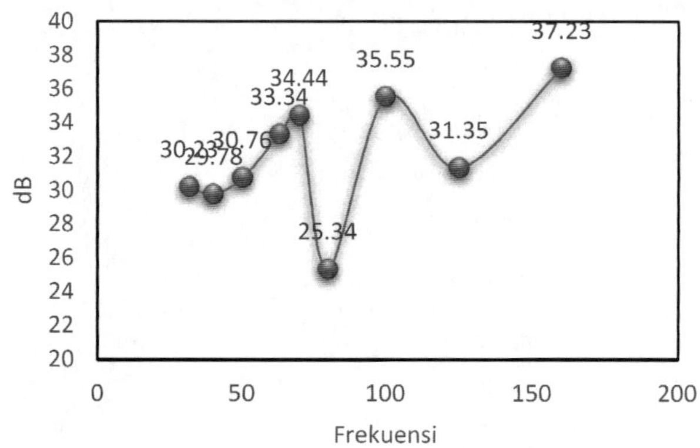

Figure 34. Value of Data Noise Signal Generator on the Z+ axis.

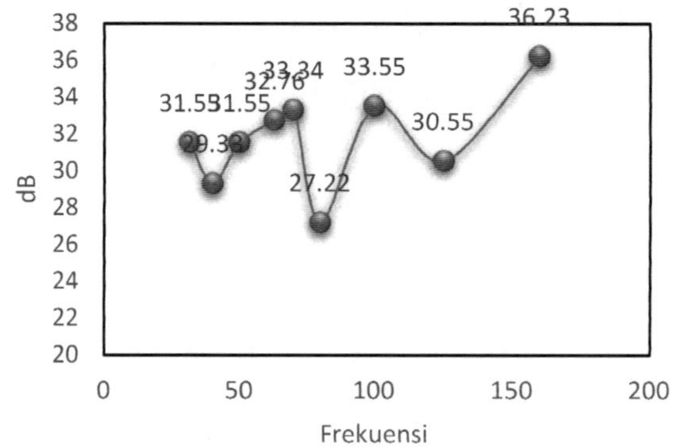

Figure 35. Value of Data Noise Signal Generator on the Z- axis.

3.2 Measurement Method and Data Collection noisy

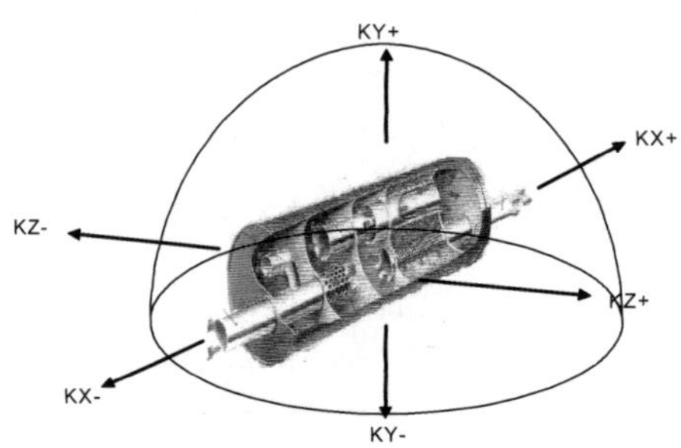

Figure 36. Measurement conditions Half Spherical.

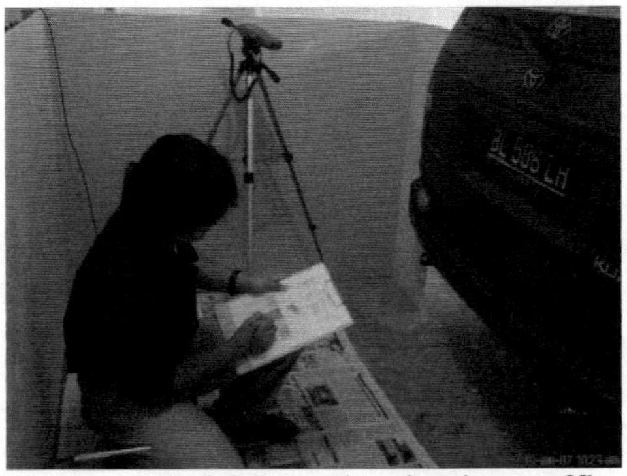

Figure 37. The process of data retrieval noisy muffler (silencer) Toyota Kijang Capsule assembly, 2009.

Table 4. Measurement data table muffler (silencer) Toyota Kijang Capsule assembly in 2009

No	1/3 Octave (Hz)	X-	Z+	Z-
1	31.5	26	25.9	26.3
2	40	26.7	27.8	26.7
3	50	35.7	35.5	35.5
4	63	26.8	26.9	26.3
5	70	27.2	26.3	26.5
6	80	28.3	28	27.9
7	100	35.2	35.2	34.9
8	125	27.3	27.4	27
9	160	25.6	26	26

3.3 The result of the measurement data graph muffler (silencer) Toyota Kijang Capsule assembly in 2009.

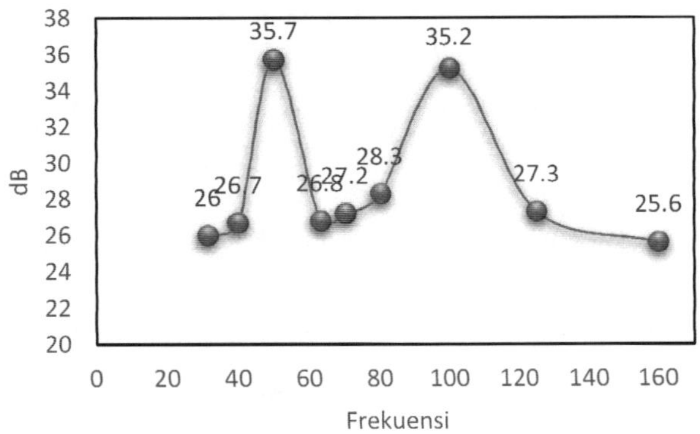

Figure 38. Value of Data Noise Signal Generator on the X- axis

IOP Conf. Series: Materials Science and Engineering **308** (2018) 012039 doi:10.1088/1757-899X/308/1/012039

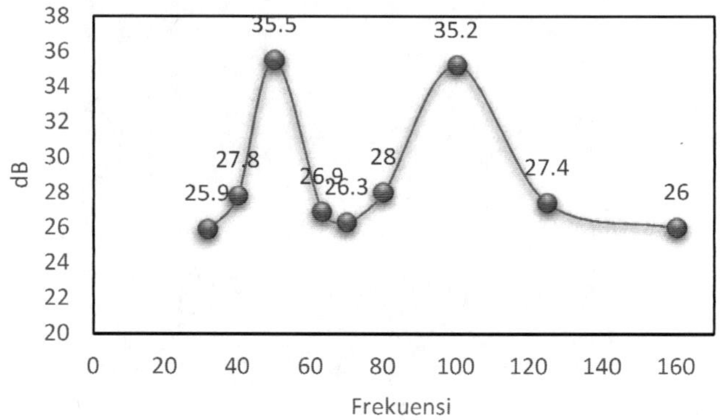

Figure 39. Value of Data Noise Signal Generator on the Z+ axis

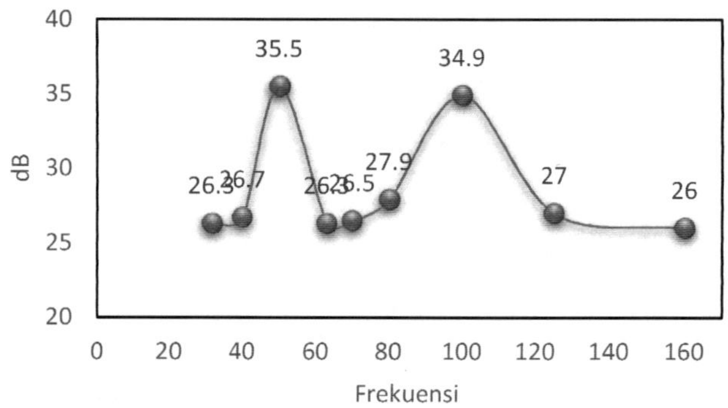

Figure 40. Value of Data Noise Signal Generator on the Z+ axis.

Discussion on the study is to conduct a simulation test signals processing from the exhaust (Silencer) using MatLab. The simulation results are then carried out the process of Noise Control signal after signal given opponents of the results of the data Noise Signal Generators. Then the results of this simulation compared with the results of previous research.

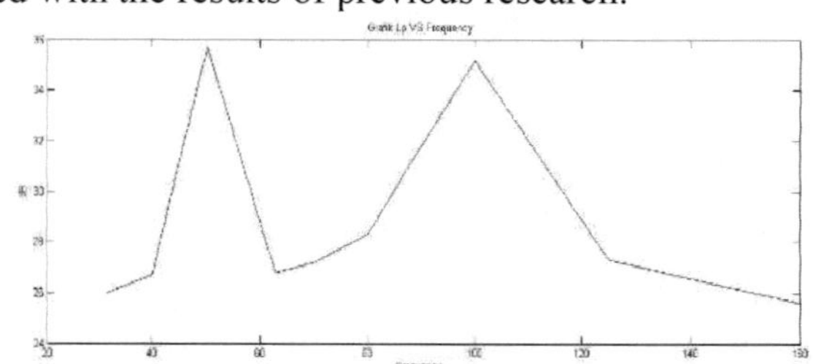

Figure 41. Graphic of data noise generator result on the X- axis.

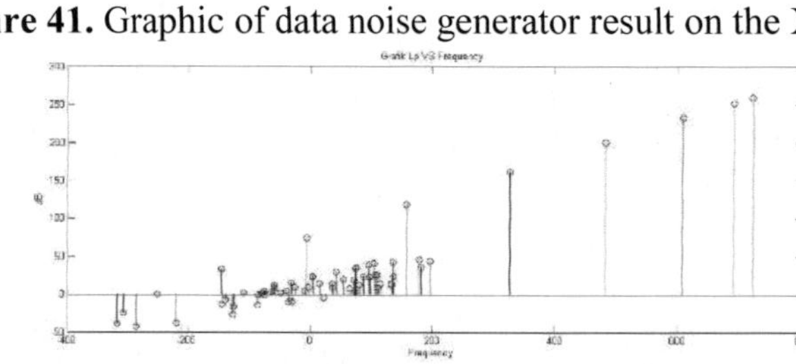

Figure 42. Graphic FFT of data noise generator result on the X- axis.

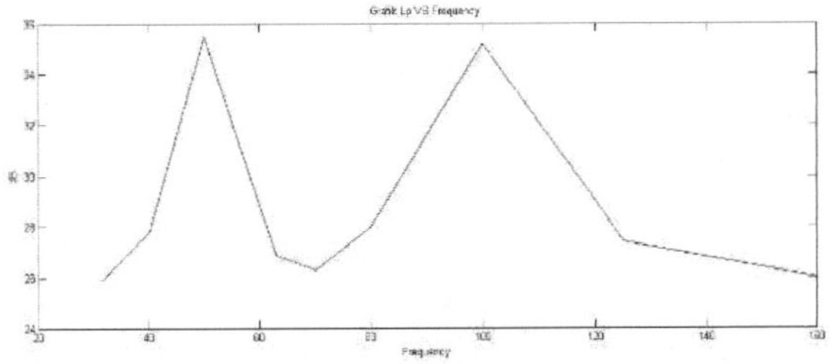

Figure 43. Graphic of data noise generator result on the Z+ axis.

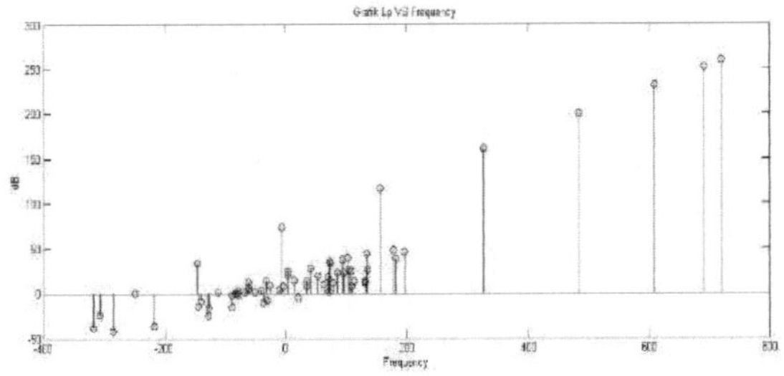

Figure 44. Graphic FFT of data noise generator result on the Z+ axis.

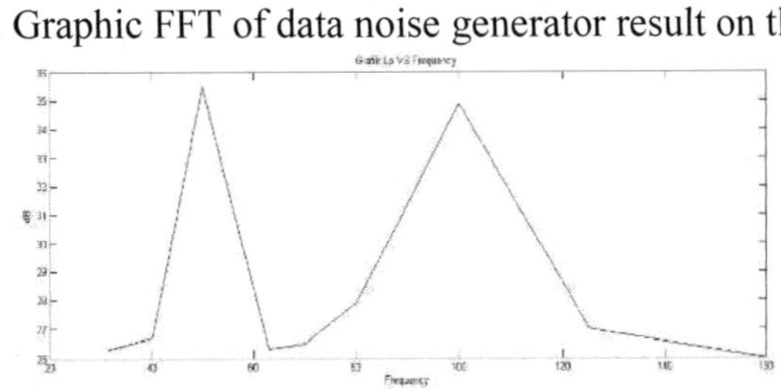

Figure 45. Graphic of data noise generator result on the Z- axis.

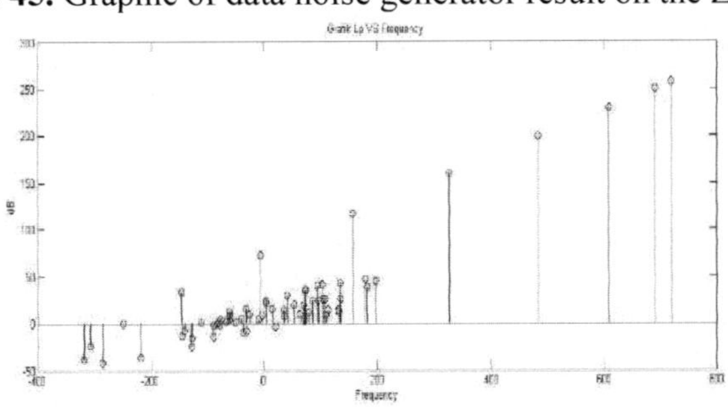

Figure 46. Graphic FFT of data noise generator result on the Z- axis.

4. Conclusion

Based on the conclusion of the study, the comparison of results of noise data in the can, a significant difference between the results obtained from data capture directly on the exhaust (silencer), compared with the results obtained from simulation using MatLab. By comparing the simulation results of Active Noise Control in the exhaust (silencer):

1. With Active Noise Control based simulation using MatLab pretty much closer to achievement. Spl produced quite closer to the data that has been obtained from the data field. So that the resulting error is in the range of 40% - 33.7% with 66.7% of phase shift.
2. While the achievements based on field data (measurements directly on the exhaust/ silencer). Active Noise Control Solutions achievements have 33.7% error by shifting the phase 66.7%.

IOP Conf. Series: Materials Science and Engineering **308** (2018) 012039 doi:10.1088/1757-899X/308/1/012039

Acknowledgements
Thanks to the professors and my promoter, fellow students, USU laboratory also Unsyiah, Who have helped me in doing this research.

References
[1] Diaz J, Egan J M, Vinolas J A Local Active Noise Control System Based on A Virtual-Microphone Technique For Railway Sleeping Vehicle Applications
[2] Doelle, Leslie L 1986*Acoustics for environment* [in Indonesia: *Akustik Lingkungan*](Montreal: Erlangga)
[3] Foremen, John E K 1990 *Sound Analysis and Noise Control*(New York: Van Nostrand Reinhold)
[4] Jadine T, Cheryl P, Elaina M, Meghan W, Kay T, Davies H*Active Noise Control Communication Headsets For The Entertainment Industry*
[5] *Animation and Acoustic Wave*ISVR (Institut of Sound and Vibration Research)27 Desember 2006
[6] Bana H, Fajar S2006*Noise Cancelletion Using ANC System***17**:1pp 9-16
[7] Peter R2003 *Active Noise Control in Home Environment*(Diss. Royal Institut of Technology)
[8] Benny S, Lars H, Ingvar C*Active Noise Control For Hearing Protection Using A Low Power Fixed Point Digital Signal Processor*
[9] Wilson, Charles E1989 *Noise Control Measurement, Analysis, and Control of Sound Vibration*(New York: Harper & Row Publisher)
[10] Alan V, Oppenheim, Alan S, Wilsky1997 *Sinyal & Sistem*(Jakarta: Erlangga)
[11] Beranek, Leo L 1959*Acoustic Measurement*(New York: John Wiley & Son, Inc)
[12] Harris, Cyril M 1979*Handbook of Noise Control*(New York: McGraw-Hill, Inc.)
[13] Zhang X H, Zhang L Y, Zou B, Du Z L. The Design of Virtual Low-Voltage Power Line Noise Generator
[14] Giancolli, Douglas *Physics Third Edition*(New Jersey: Prentice Hall Englewood Cliffs)

IOP Conf. Series: Materials Science and Engineering **308** (2018) 012040 doi:10.1088/1757-899X/308/1/012040

Crush analysis of the foam-filled bitubal circular tube under oblique impact

F Djamaluddin*, S Abdullah, A K Arrifin, Z M Nopiah
Departement of Mechanical and Material, Faculty of Engineering and Built Environment, Universiti Kebangsaan Malaysia, 43600 UKM Bangi, Selangor, Malaysia

*E-mail: fauzanman_77@yahoo.com

Abstract. This paper presents crashworthiness analysis of bitubal cylindrical tubes under different impact angular. The numerical solution of double cylindrical tubes are determined by finite element analysis (FEA). Moreover, the structure was impacted by mass block as impactor respect to longitudinal direction of the tubes. The model of structure was developed by non-linear ABAQUS sofware with variations of load angle and dimensions of tube. The outcome of this study is the respons parameters such as the peak crusing force (PCF), energy absorption (EA) and specific energy absorption (SEA), thus it can be expected this tube as the great energy absorber.

1. Introduction

To increase the crushing ability of thin walled and to decrease vehicle weight for fuel safety are the targets of automobile industry nowadays. These are some ways for these targets such as finding new materials and redesign play. In addition, the experiment test namely physical vehicle test more costly and for this reason, simulation is the best choice. With considering variables of tubes namely the geometry, size, and cross-section, some researchers study the behavior of thin-walled tubes to improve the energy absorption of material. Alexander et al. [1] was the pioneer studied the behavior of tubes under axial loading. Therefore, some scholars continued his work for instance, Reid et al. [2] studied the inversion and bending of tubes and Jones et al [3] compared the circular and square tubes subjected to axial loading of impact. The collapsible energy absorbers in different cross-sections, in particular circular and square tubes were studied by Alghamdi et al. [4]. To increase the energy absorption, the thin-walled tubes were filled by cellular material such as foams and honeycombs. Seitzberger et al. [5] was the first researcher studied empty and filled aluminum tubes. Meanwhile, the experimental and numerical solutions of tubes under axial and three-point bending were investigated by Guo et al. [6-7]. Then thin-walled aluminum subjected to oblique impact loading was analyzed by Reyes et al. [8-9]. The aluminum foam thin-walled tubes for beams had been investigated by several researchers. Li et al [10] performed experiment tests to study aluminum foam double tubes subjected to oblique loading.

2. Methodology

2.1. Crashworthiness indicator of structures

Several parameters were used to evaluate the crushing performance for instance, energy absorption (EA), specific energy absorption (SEA) and peak crushing force (PCF). EA is indicated as the total strain energy absorbed during deformation from the curve of crushing force-deflection. EA is defined in Eq. 1.

$$EA(d) = \int_0^d F(x)\,dx \tag{1}$$

where d is the effective stroke length and F(x) is the the instantaneous crushing load.

SEA is represented as the absorbed energy ratio (EA) to the structure mass (Mt) and is given by the formula in Eq. 2.

$$SEA = EA\,/\,M_t \tag{2}$$

2.2 Finite element models of tube under oblique impact

Figure 1 showed the models of the bitubal tube with dimensions, boundary and loading conditions. The tube was placed between two rigid wall that all of the tubes fixed at the bottom and at the other end impacted by a moving mass block with constant speed of 15 m/s. Aluminum foam filled thin – walled double circular tubes of the parameters design variable of tubes are the length (*L*) of 250 mm, the inner diameter tubes (*di*), outer diameter tubes (*do*), wall inner (*ti*) and thickness (*to*) can be seen in table 1. Seven angles of loading were considered ($\theta_1 = 0^\circ$, $\theta_2 = 5^\circ$, $\theta_3 = 10^\circ$, $\theta_4 = 15^\circ$ $\theta_5 = 20^\circ$, $\theta_6 = 25^\circ$, and $\theta_7 = 30^\circ$).

Table 1. Dimensions of bitubal circular tubes

do (mm)	*di* (mm)	*to* (mm)	*ti* (mm)
54	27 ; 25 ; 23	2 ; 1.8 ; 1.6	2 ; 1.8 ; 1.6

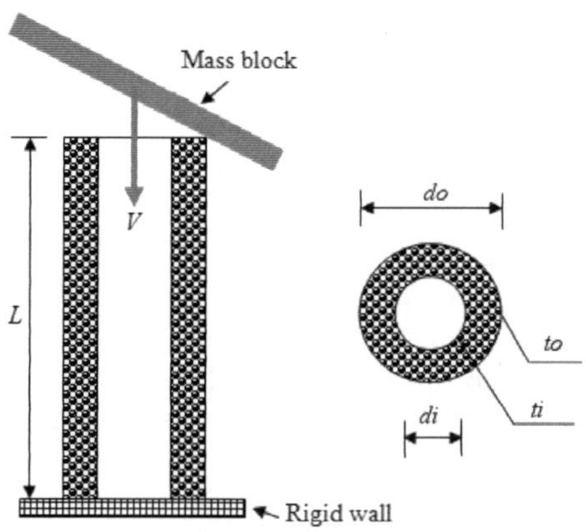

Figure 1. Schematic of aluminum foam filled double cylindrical tubes

Figure 2. Finite element model of cylindrical double

3. Result and Discussion

Figure 3 shows the crushing behaviour under angular loading. Energy absorption influenced the deformation pattern of the structure. It can be seen that tubes under pure axial loading caused progresive deformation behaviour, on the other side oblique impact switched from deformation mode to the bending mode.

IOP Conf. Series: Materials Science and Engineering **308** (2018) 012040 doi:10.1088/1757-899X/308/1/012040

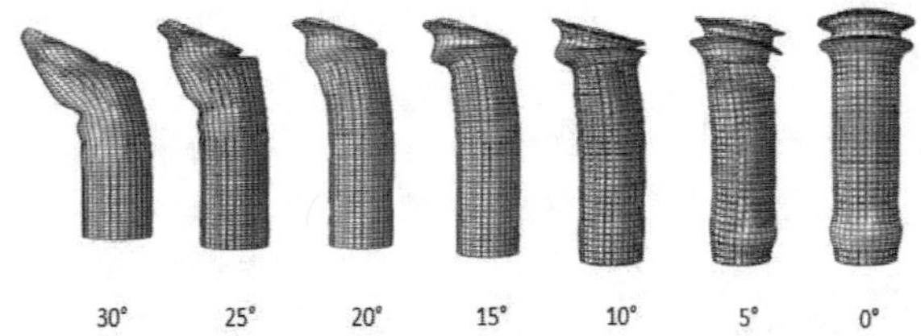

Figure 3. Crushing behaviour of aluminum foam tube under angular impact

The structures were simulated to observe the inner and outer thickness of foam-filled bitubal tubes on the impact responses. The energy (*E*), specific energy absorption (*SEA*) and peak crushing force (*PCF*) versus different variation of thickness outer wall and angle loads are plotted in F\figure 5. It shows that the effect of the outer wall thickness of tubes with different angle loads. Furthermore, the load carrying capacity decrease with the increase outer thickness of wall.

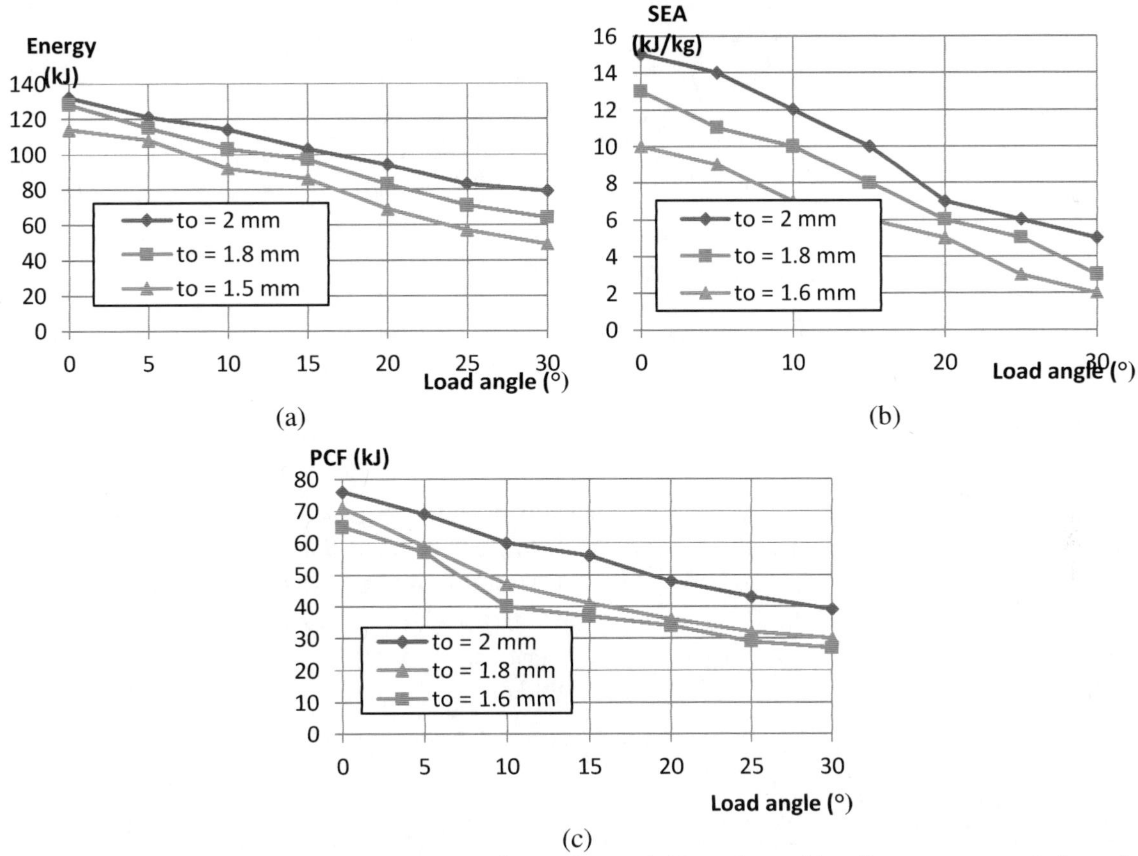

Figure 4. (a) Energy, (b) specific energy absorption and (c) peak crushing force versus variation of thickness outer wall and angle loads.

Figure 5 displays the influence of the different thickness of inner wall for double foam-filled tube. It can be seen that the energy absorption has not significant effect with three kinds of inner tube in different thicknesses. However, SEA and PCF increased value slightly with increasing inner thicknesses.

IOP Conf. Series: Materials Science and Engineering **308** (2018) 012040 doi:10.1088/1757-899X/308/1/012040

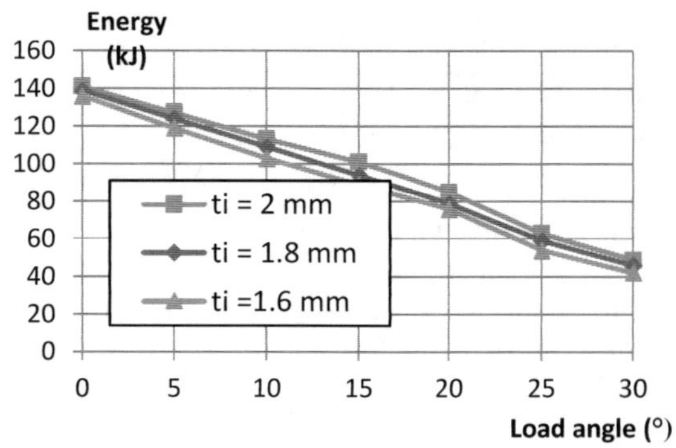

(a)

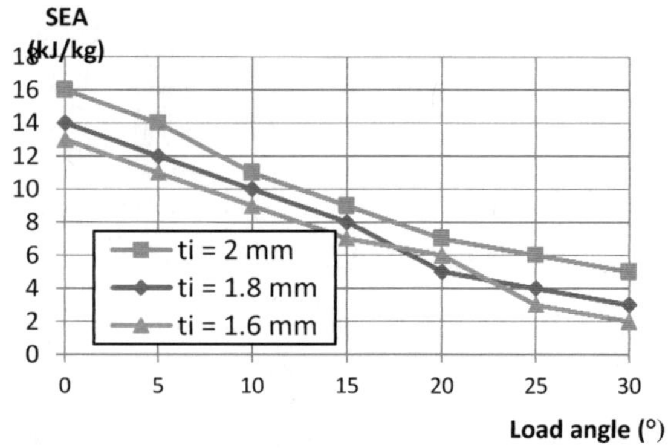

(b)

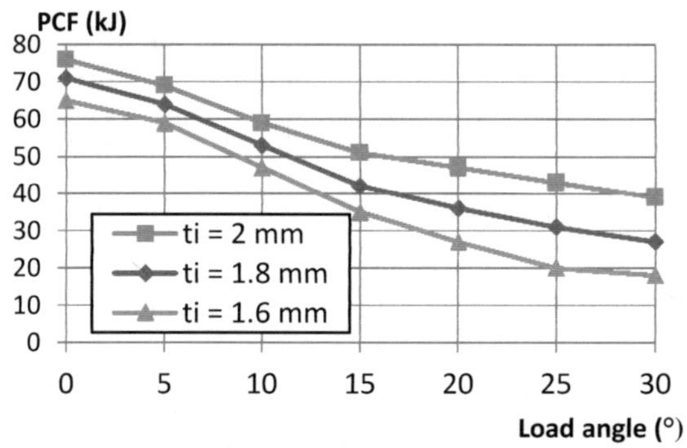

(c)

Figure 5. (a) energy, (b) specific energy absorption and (c) peak crushing force versus variation of thickness inner wall and angle loads.

The effect of the inner tube diameter of the foam-filled double tube was also studied on the energy absorption capacity and the results are shown in figure 6. It can be found that increasing diameter of the inner tube, therefore decreasing of the total energy and specific energy absorption and peak crushing force.

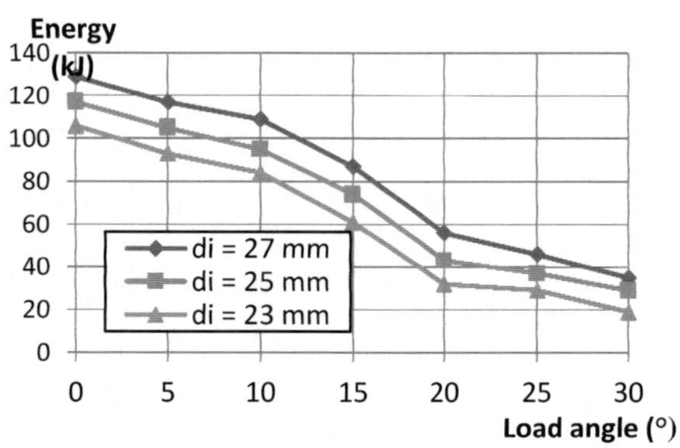

(a)

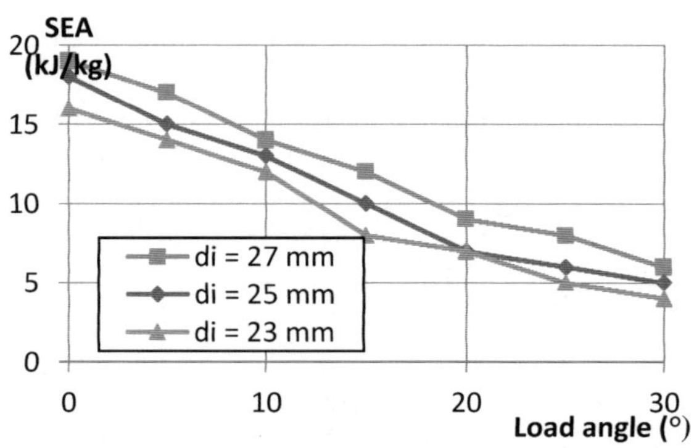

(b)

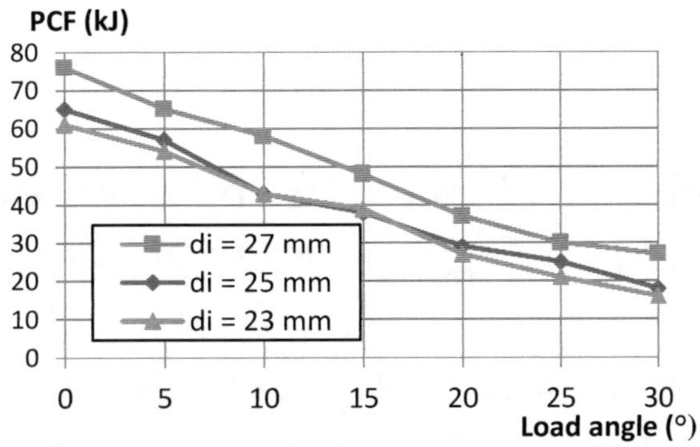

(c)

Figure 6. (a) The energy, (b) specific energy absorption (SEA) and peak (c) crushing force (PCF) versus variation of diameter inner wall and angle loads.

4. Conclusion

This paper presented the optimum conditions of aluminum foam cylindrical double tubes under oblique impact by considering various angles of loading and dimension tubes. Variation of geometrical dimensions such as inner and outer wall thickness, and inner diameters foam were observed to study the crash behaviour of bitubal circular tube. Finally, we found that the aluminum foam-filled double tubes are impacted by variation of angular loading. With increasing impact angle cause effects on the SEA and PCF which are decline of the value.

References

[1] Alexander J M 1960 An approximate analysis of the collapse of thin cylindrical shells under axial loading *Q J Mech Appl Math* **13** pp 10–5

[2] Reid S R 1993 Plastic deformation mechanisms in axial compressed metal tubes used as impact energy absorbers *Int J Mech Sci* **35**:12 pp 1035–52

[3] Jones N 1989 *Structural impact* (Cambridge University Press)

[4] Alghamdi A A A 2001 Collapsible impact energy absorbers: an overview *Thin-Walled Struct* **39**:2 pp 189–213

[5] Seitzberger M, Rammerstorfer R F, Degischer H P, Gradinger R 1997 Crushing of axially compressed steel tubes filled with aluminium foam *Acta Mech* **125** pp 93–105

[6] Guo L W, Yu J L, Li Z B 2010 Experimental studies on the quasi-static bending behavior of double square columns filled with aluminum foams *Acta Mech* **213** pp 349–58

[7] Guo L W, Yu J L 2011 Dynamic bending response of double cylindrical tubes filled with aluminum foam *Int J Impact Eng* **38**:2–3 pp 85–94

[8] Reyes A, Langseth M, Hopperstad O S 2002 Crashworthiness of aluminum extru- sions subjected to oblique loading: experiments and numerical analyses *Int J Mech Sci* **44**:9 pp 1965–84

[9] Reyes A, Langseth M, Hopperstad O S 2003 Square aluminum tubes subjected to oblique loading *Int J Impact Eng* **28**:10 pp 1077–106

[10] Li Z B, Yu J L, Guo L W 2012 Deformation and energy absorption of aluminum foam-filled tubes subjected to oblique loading *Int Journal of Mechanical Sciences* **54** pp 48–56

[11] Deshpande V S, Fleck N A 2000 Isotropic Constitutive Models for Metallic Foams *Journal of The Mechanics And Physics of Solids* **48**

IOP Conf. Series: Materials Science and Engineering **308** (2018) 012041 doi:10.1088/1757-899X/308/1/012041

Fatigue crack growth behaviour of semi-elliptical surface cracks for an API 5L X65 gas pipeline under tension

M S Shaari [*1], M R M Akramin [1], A K Ariffin[1], S Abdullah[1], M Kikuchi[2]

[1]Department of Mechanical and Materials Engineering, Faculty of Engineering & Built Environment, Universiti Kebangsaan Malaysia, 43600 Bangi, MALAYSIA
[2]Tokyo University of Scince, 2641 Yamazaki, Noda, Chiba 287-8510, JAPAN, Phone: +603-8921 6012

[*]E-mail: mohdshamil.shaari@yahoo.com

Abstract. The paper is presenting the fatigue crack growth (FCG) behavior of semi-elliptical surface cracks for API X65 gas pipeline using S-version FEM. A method known as global-local overlay technique was used in this study to predict the fatigue behavior that involve of two separate meshes each specifically for global (geometry) and local (crack). The pre-post program was used to model the global geometry (coarser mesh) known as FAST including the material and boundary conditions. Hence, the local crack (finer mesh) will be defined the exact location and the mesh control accordingly. The local mesh was overlaid along with the global before the numerical computation taken place to solve the engineering problem. The stress intensity factors were computed using the virtual crack closure-integral method (VCCM). The most important results is the behavior of the fatigue crack growth, which contains the crack depth (a), crack length (c) and stress intensity factors (SIF). The correlation between the fatigue crack growth and the SIF shows a good growth for the crack depth (a) and dissimilar for the crack length (c) where stunned behavior was resulted. The S-version FEM will benefiting the user due to the overlay technique where it will shorten the computation process.

1. Introduction

During the past 40 years, there have been many observations made on surface cracks under cyclic tensile loading. Since then, the continuity of research to evaluate the structural integrity in mechanical components vastly spread including the gas pipeline. It is very essential for oil and gas industry because the gas pipelines are viable to the fatigue loading that comes from the ground and vehicles [1]. The fact that the gas pipelines are buried under the ground is enough to include such environmental affect to the structure. The results of this environmental defect will cause the gas to leaks and lead to fire hazards and endangering human beings.

The API 5L X65 are one of the most commonly use type of material in oil and gas industry [2], thus it is considerable the importance of this study to the society generally and the industry specifically. The deterioration of an engineering structure happens when a cyclic loading presence on a period of time. The influence of other aspect such as corrosion, creep and material defect can add to the total cause. However, for a structure that happens to experience a direct interaction with the loading coming from energy generated from the oil and gas or even the environment aggressiveness (e.g. earthquakes, heavy vehicle on the ground) [3].

Numerous researches have conducted many studies for structural integrity for gas pipeline. However the involvement of surface cracks in gas pipeline is still expandable especially when it is related with the S-version FEM. This method was developed by Fish in year 1992 with the main objective, "to improve the quality of the finite element calculation in the regions of unacceptable errors" [4]. Since then, several researches were and still working in this area of specialty including Kikuchi and his team. It is widely accepted in fatigue and fracture mechanics field for example, since the last five years the number of publications increase progressively in varies topics from weld-heat-affected zone [5], internal cracks [6], multiple cracks [7], mixed mode cracks [8], heterogeneous material [9] and many more.

The objective of this study is to predict the fatigue crack growth (FCG) behavior of semi-elliptical surface cracks for API X65 gas pipeline using S-version FEM under tension loading.

2. S-version FEM Simulation Procedure

This section will be presenting the procedure to conduct the computational analysis using the S-version FEM. In this research study, the main material is the API 5L X65. However, the other materials are used such as SUS 304 and ASTM A533B to compare the fatigue behavior among the materials. Note that the load and geometry of the plates for all materials are similar. The API 5L X65 material used in this research study was acquired by a Malaysian owned oil and gas company. In order to attain the material properties, the monotonic test was conducted and consist of 223 GPa for the modulus Young's (E), 0.3 for the poisson ratio (υ), 472 MPa for yield and 648 MPa. The original concept of S-version FEM that was developed by Fish emphasized on the superimposed between local and global meshes.

Figure 1 shows a single crack referring to local mesh and it is then overlay together with a finite plate referring to the global mesh. Figure 2 shows the local and global meshes that were modeled in FAST using 20-node hexahedral elements. FAST is a pre-processing program written in C++ language. The geometry of the finite plate is 200mm by 200mm with a thickness of 20mm. The crack's length is 20mm and the depth is 4mm. The geometry of the local and global in this research study is using a specific ratio such as a/c of 0.4, c/b of 0.1, c/h of 0.1 and a/t of 0.2 where a is the crack depth, c is the crack length, b is the plate width, h is the plate length and t is the plate thickness.

Therefore, later on any computational can be performed by following the standard. In order to design a tension loading, the bottom of the plate is fixed in all direction while the upper side pulled with uniform load of 10MPa as shown by figure 3. The results of the displacement deviations in figure 4 validate the tension loading. Notice that the deviation is in y direction similar to the load applied.

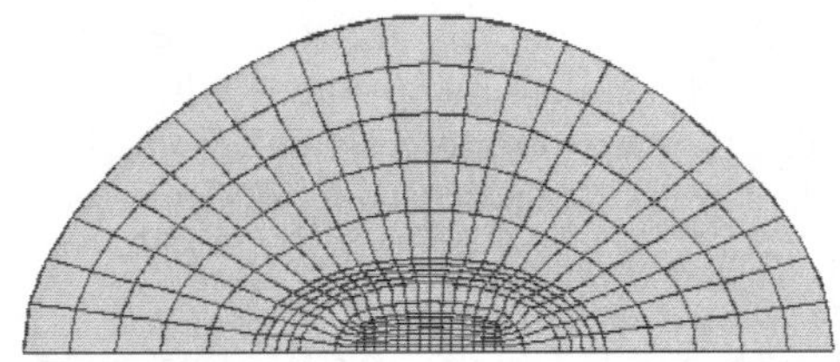

Figure 1. Local semi-elliptical surface crack

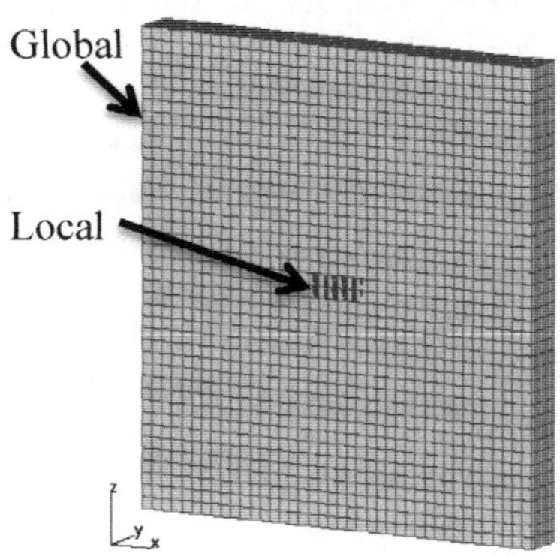

Figure2. Geometry of plate combines with the crack

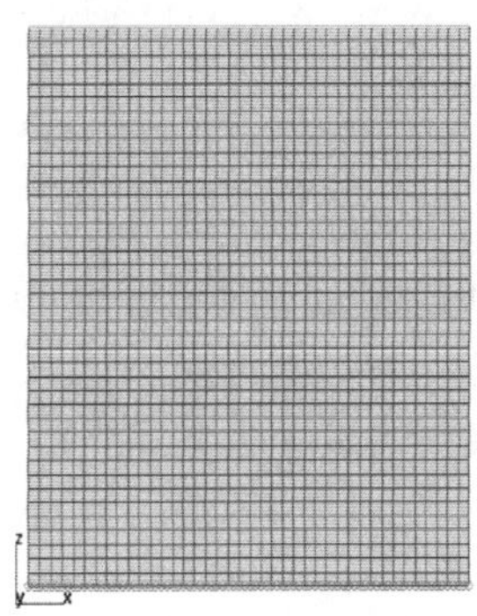

Figure 3. Global with boundary condition

IOP Conf. Series: Materials Science and Engineering **308** (2018) 012041 doi:10.1088/1757-899X/308/1/012041

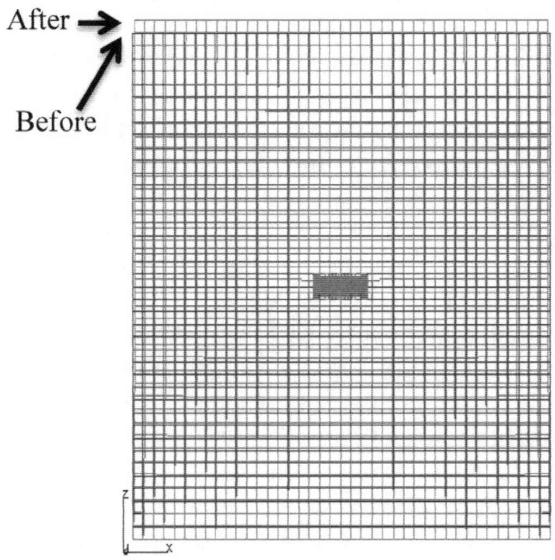

Figure 4. Displacement of global and local

3. S-version FEM Computational Results

The results of semi-elliptical surface cracks and the fatigue crack growth behavior will be discussed in this section. Figure 5 shows the semi-elliptical surface cracks according to the beach marks for all three types of materials. There are no notable differences between all types of materials accept the crack grows in plausible state. Figure 1 and figure 2 shows the axis for the crack and the propagation of the crack consists of two which are crack depth (a) and crack length (c). As for the center of the crack, it is noticeable at zero coordinate and the total length of the crack is 20mm.

The results produced by S-version FEM significantly showed vast crack propagation for the crack depth (a) and only small extent for the crack length (c). This is supported by [10] in discussing the shape of the surface cracks differ subjected to loads such as tension loading and bending loading. The effect of loading mode for the surface cracks influenced the evolution of the crack shapes prediction. It is because of the nature of the boundary and geometry of a rectangular plate most likely will be influencing the loading mode. The crack length (c) experiencing stunned development compare to the crack depth (a). To justify why the crack depth (a) propagates vastly than crack length (c), it can be reasoned by referring Figure 6. The equivalent stress intensity factors (Keq) is showing a better slope for the crack depth compare with the crack length. It is known in general that the SIF represent the crack growth movement and this explain the phenomenon.

It is norm to define the Paris exponents from the fatigue crack growth behavior especially from the experimental results. On the other hand, the S-version FEM already considers all the values before the simulation as inputs needed to compute the fatigue crack growth. Nonetheless, the purpose of the fatigue crack growth rate graph is still needed to observe the behavior especially when comparing the crack depth and crack length.

Figure 7 shows the cracks versus life cycles and as predicted by the fatigue crack growth rate graph, the crack length undergoing stunned growth unlike the crack depth. The increment for the crack growth is much smoother and proportional with the life cycles. Include in this study is the normalized SIF behaving along with the angular direction as shown in Figure 8. Notice the degree of angle (ϕ) representing the 90 degree of angular direction for the SIF. From the results, since the loading was applied uniformly, the SIF extend and propagate starting from 0 degree to 90 degree as can be seen from Figure 8. Note that the total angular direction is only 180 degree, hence the reason why the graph only consider it to be 90 degree. The results was normalized using Okada [11] and the calculations was based on the author concept.

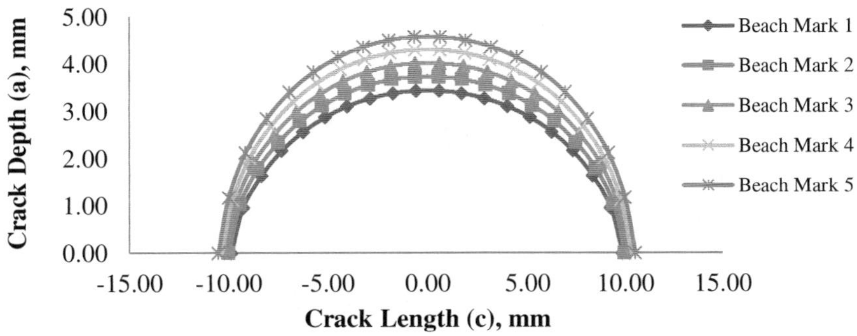

Figure 5. Fatigue crack growth for API X65 according to the beach marks.

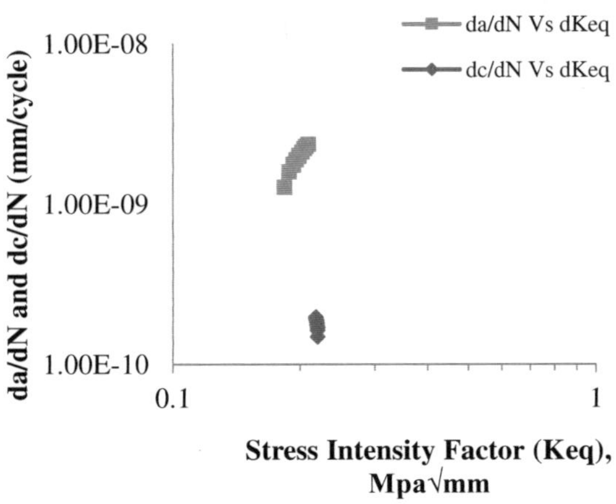

(a)

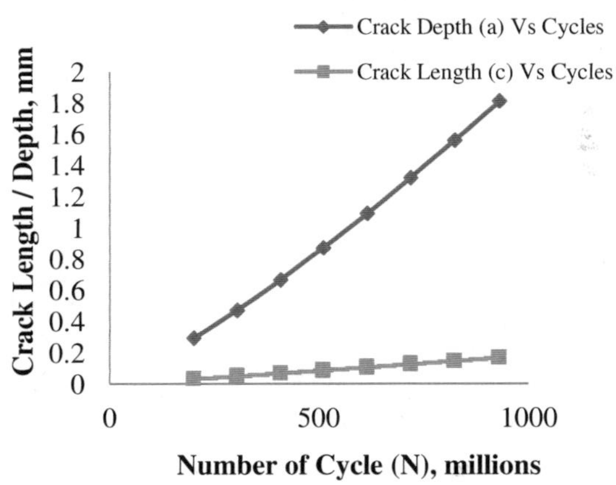

(a)

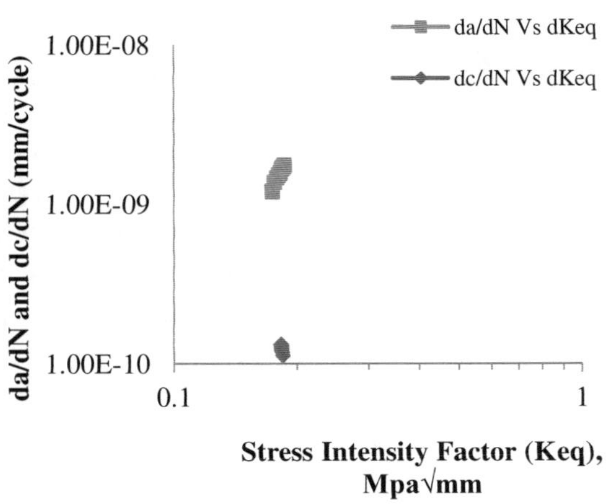

(b)

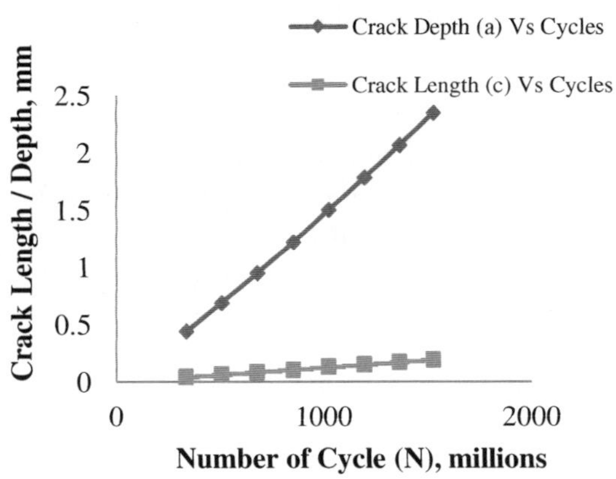

(b)

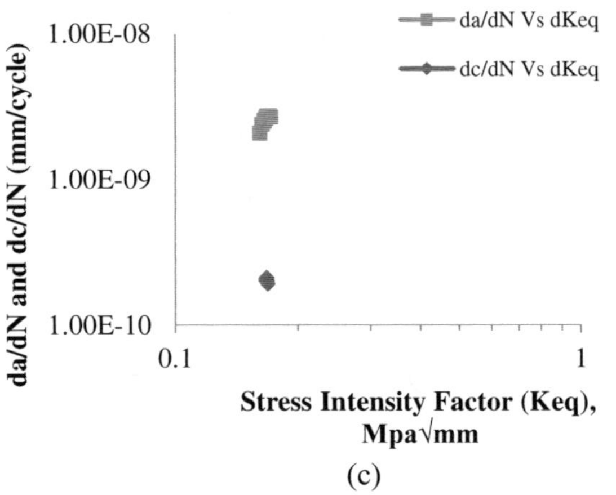

Figure 6. Comparison of FCGR between crack depth and crack length for (a) API X65, (b) SUS 304 and (c) ASTM A533B.

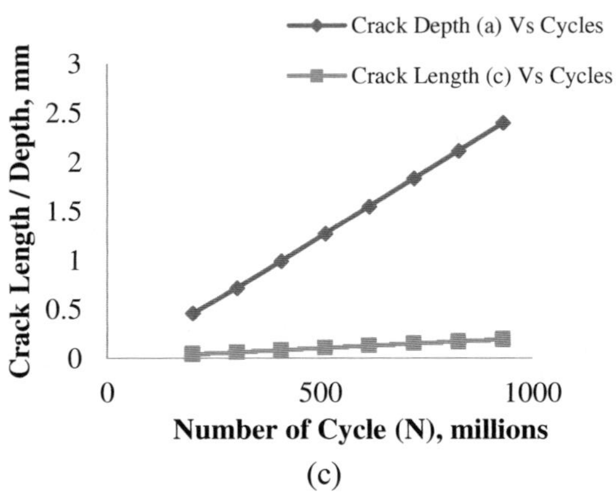

Figure 7. Comparison of crack length and crack depth for (a) API X65, (b) SUS 304 and (c) ASTM A533B according to the number of cycles.

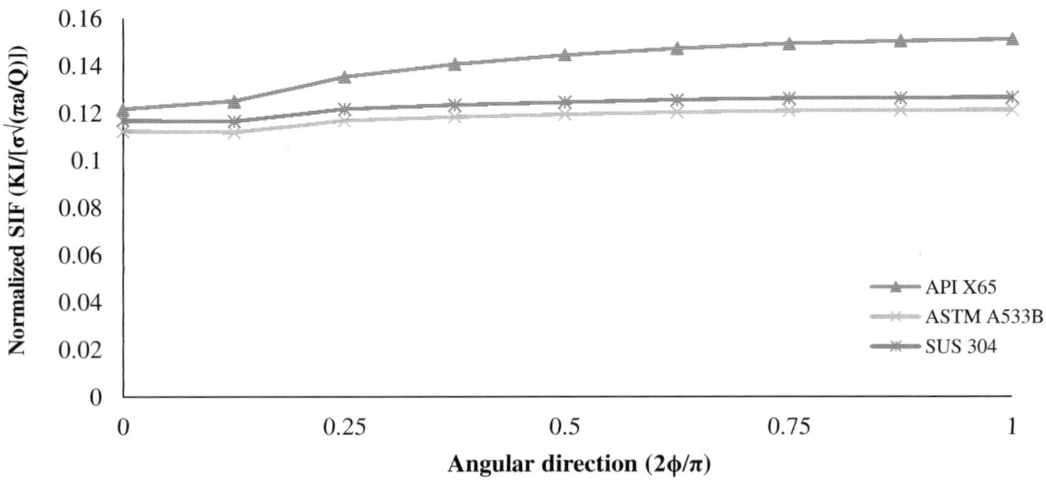

Figure 8. Comparison of normalized SIF according to the angular direction.

4. Conclusion

The prediction of the fatigue crack growth behavior of semi-elliptical surface cracks for API 5L X65 gas pipeline using S-version FEM under tension loading was performed successfully. The normalized SIF does very essentials especially if the research study related with the surface cracks and by comparing with different materials gave a significant value to the research conducted.

Acknowledgements

The author would like to acknowledge the Universiti Kebangsaan Malaysia (UKM), especially the laboratory of computational & experimental mechanics (CEM) for allowing the research to be conducted using the high performance computer (HPC).

References

[1] Hong S W, et al. 2015 Fatigue life prediction for an API 5L X42 natural gas pipeline. *Engineering Failure Analysis* In Press 2015(0)

[2] Fatoba O and Akid R 2014 Low Cycle Fatigue Behaviour of API 5L X65 Pipeline Steel at Room Temperature *Procedia Engineering* **74**:0 pp 279-286

[3] Mansor N I I, et al. 2014 A review of the fatigue failure mechanism of metallic materials under a corroded environment *Engineering Failure Analysis* **42** pp 353-365

[4] Fish J 1992 The s-version of the finite element method *Computers & Structures* **43**:3 pp 539-547

[5] Kikuchi M, et al. 2012 Crack growth analysis in a weld-heat-affected zone using S-version FEM *International Journal of Pressure Vessels and Piping* **90–91** pp 2-8

[6] Wada Y, et al. 2014 Fatigue growth of internal flaw: Simulation of subsurface crack penetration to the surface of the structure *Engineering Fracture Mechanics* **123** pp 100-115

[7] Kikuchi M, et al. 2010 *Closure Effect on Interaction of Two Surface Cracks Under Cyclic Bending* ASME 2010 Pressure Vessels and Piping Division/K-PVP (Bellevue, Washington, USA.: ASME)

[8] Kikuchi M, Wada Y, Suga K, 2011 Surface crack growth simulation under mixed mode cyclic loading condition *Procedia Engineering* **10** pp 427-432

[9] Kikuchi M, et al. 2014 Fatigue crack growth simulation in heterogeneous material using s-version FEM *International Journal of Fatigue* **58** pp 47-55

[10] Lin X B and Smith R A 1999 Shape evolution of surface cracks in fatigued round bars with a semicircular circumferential notch *International Journal of Fatigue* **21**:9 pp 965-973

[11] Okada H., et al. 2005 Three dimensional virtual crack closure-integral method (VCCM) with skewed and non-symmetric mesh arrangement at the crack front *Engineering Fracture Mechanics* **72**:11 pp 1717-1737

IOP Conf. Series: Materials Science and Engineering **308** (2018) 012042 doi:10.1088/1757-899X/308/1/012042

Modeling reliability measurement of interface on information system: Towards the forensic of rules

M. K. M. Nasution[1], Darwin Sitompul[2] and Marwan Harahap[3]

[1]Teknologi Informasi, Fasilkom-TI, Universitas Sumatera Utara, Padang Bulan 20155 USU, Medan, Indonesia
[2]Teknik Mesin, Fakultas Teknik, Universitas Sumatera Utara, Padang Bulan 20155 USU, Medan, Indonesia
[3]Matematika, Fakultas Matematika dan Ilmu Pengetahuan Alam, Universitas Sumatera Utara, Padang Bulan 20155 USU, Medan, Indonesia

E-mail: mahyuddin@usu.ac.id

Abstract. Today almost all machines depend on the software. As a software and hardware system depends also on the rules that are the procedures for its use. If the procedure or program can be reliably characterized by involving the concept of graph, logic, and probability, then regulatory strength can also be measured accordingly. Therefore, this paper initiates an enumeration model to measure the reliability of interfaces based on the case of information systems supported by the rules of use by the relevant agencies. An enumeration model is obtained based on software reliability calculation.

1. Introduction

Talking about the current machine, it cannot be separated from the software or the system that controls it [1]. Thus, an information system as the implementation of rules that apply in the management become an integral part of the rule [2]. If the information system is built on the principles of software engineering, then the information system can be measured its reliability in carrying out its duties, and this is also used as a measure of the strength of the rule. This is as a basis for its possible forensic form of applicable regulations [3].

The measurement of reliability of a program that represents the information system can be summarized by involving the principle of software reliability [4], by involving a probability calculation of the ongoing process flow by involving a flowchart or not [5]. The approach to the reliability model is based on the ease of enumeration [6]. This paper aims to reveal an enumeration model based on a single case, i.e. Kabupaten Road Management (KRM) system [7], in order to obtain a model to reveal the force of the applicable rules of something related.

2. Review and Motivation

An information system is built to support decision making quickly and precisely [8]. Such a system is built on the agreement and rules set that have been determined both by stakeholders and system designers [9]. The system is built from various technologies according to that provision [10]. The technology of translating those provisions into the system involves programming. In programming the size of the failure for a given time proportionately is the number of errors left and each error contributes equally to the size for the i-failure occurs is λ_i, and is declared with

$$\lambda_i = \pi_0(N_0 - (i-1)) \tag{1}$$

with where π_i is the correct failure size per error and N_i is the number of correct errors introduced early in the software [11].

On the different side of the software engineering, the program is composed of modules that are created procedurally and will take a specific place dynamically in memory if the program is executed. Each program or module can be describe as an execution with a flowchart, which is indirectly a process with a starting point and ending at a single point of destination [12]. Thus, the x_i event occurs on the edge $e_i \in E$ from a graph $G = (V, E)$ and $\overline{x}_i$ components circumstance that e_i fails, with which V is the set of vertices and E is the set of edges. The probability of operation p_i from edge e_i is probability x_i, and failure is represented by $q_i = 1 - p_i$, so for operational circumstance with the limited edges $e_1, e_2, \ldots, e_5$ [13]. Likewise, the process that takes place in the use of a system as a description of procedures that apply to reality.

3. An Adaptive Approach

The interface model can be represented by a node, and the data stream process is expressed as an edge, whereby the whole process can be represented by a flowchart [14]. The interface operation in the system is viewed as a flowchart as well, but in this section it can be judged based on the need for a large number of data inputs [13, 9].

By recognizing interface operations that flow data, by it also recognize of necessity in the data processing system. This operation is expressed by the flow from the source node s to the destination node t [15]. Operation is to record data or information from a human source to the destination database with the opportunity of false or true is balanced. Therefore, the probability graph G and the nodes s, t which are expressed as the reliability of two terminals, or so-called connections, are written $Hub_2(G)$, or $An_2(G)$ [13].

The second operation in the data flow is the dissemination of appropriate information. The reliability of all terminals, $Hub_2(G)$, against the probability that there is a path from s to every other node, or is the probability that the directed graph contains at least one spanning-tree that is rooted in s. Thus a path operation will simultaneously involve the pairwise connection of the k nodes as the stated terminal, $2 \leq k \leq n$ [16]. The k-terminal reliability, $Hub_k(G)$, is the probability that for the destination node, the graph contains a path between each pair of k nodes, or expressed as the connect-s, T for the directed graph G, with which s is the source node and T is the set $k-1$ destination nodes. In addition, other measurements may still be possible. However, the general mechanism is to state that reliability issues exist in the existing order. For any probabilistic graph $T = (V, E)$, it expresses a state from G to a subset, defined as all arcs in S operate and all arc $E - S$ fails. A path operation is defined by defining the set of operations, by which $OP(G)$ is the set of circumstances deemed to operate, or expressed in the $FA(G) = -OP(G)$ as failure set. $OP(G)$ is sometimes referred to as a stochastic binary system, with which its members in a set of paths [17].

If the provision has established a system, and the system can be measured its reliability, then the provision can be measured ability.

4. Enumeration Model

Let a matter of discussion about a case of using the KRM (Kabupaten Road Management) system, whose use is based on provision, with a size of file is 1.4MB. The system is run in semi-manual, whereby data validation between one admin with another admin is done by sending data by courier. Surely at this time, this system is no longer feasible to use because not accordance with the information technology development. However, if the interfaces of this system still

IOP Conf. Series: Materials Science and Engineering **308** (2018) 012042 doi:10.1088/1757-899X/308/1/012042

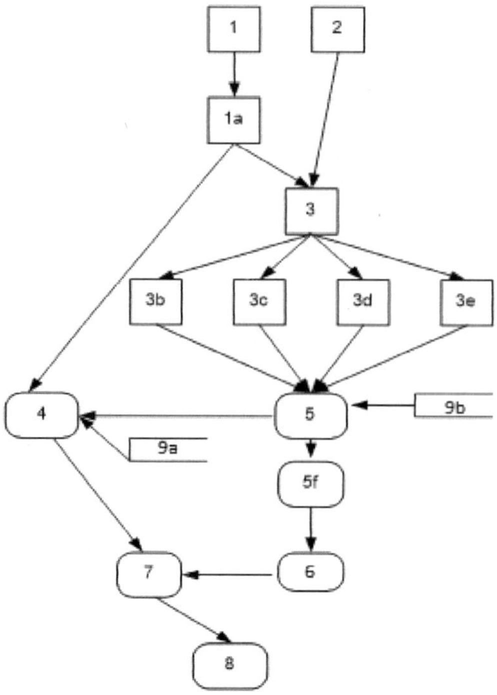

Figure 1. Data flow of the planning interface in KRM

keep the applicable rules of course the procedures in it can be transferred to another system. KRM has the following cycles: project approval, project preparation, project implementation, and project completion. This measurement model focuses on the planning module, which has the following interfaces: (a) Road Network Mastering Instructions (K1), (b) Bridge Inventory Log (K10), (c) Roads Analysis Form (A1), (d) Proposed Maintenance Form (P), (e) Large Work Study Form (P2), (f) Proposed Selected Weighing Work Form (P3), (g) Proposed Plan Form, (h) Definitive Plan Form, and (i) Maintenance Cost Table [7].

The sequence of data entries (numbers: $1, 2, \ldots, 9$) and the data analysis activities are carried out on/through whole interface of the planning part (a,b,...,f) see Fig. 1. According to the model, circumstance enumeration requires the production of all operating circumstances and determines a probability for each. The reliability of the terminal $Hub_{s,t}$ is obtained by summing all the 2^m probabilities of circumstance of G, which is sufficiently efficient in solving the problems inherent in measuring the reliability of the role of a provision directly [17]. Enumerate the perfect circumstance with G from the connect-s, t as Fig. 2, each edge corresponding to the number of activities: $e1 = 1$, $e2 = 2$, $e3 = 1a$ and so on, which can be tabulated as Table 1.

Furthermore, the minimum of the set of paths is called min-path, and it becomes a tree with root s with all arcs pointed from the root, $X = \{s\}$ is the set of nodes in the spanning-tree $T \subseteq E$, while $C \subseteq E$ is the set of edges which is left as a candidate for inclusion in the spanning-tree, up to $C = E$ [13]. Next, suppose min-path from G is registered. Suppose that the event that all edges are within min-path operates, and suppose $Pr[]$ denotes the probability of an event. Then reliability is none other than the probability that one or more events $\{E_i\}$ occur, with which $\{E_i\}$ is a jointed event, and therefore cannot simply summarize its probability. So in particular

$$Pr[E_1 \vee E_2] = Pr[E_1] + Pr[E_2] - Pr[E_1 \wedge E_2] \tag{2}$$

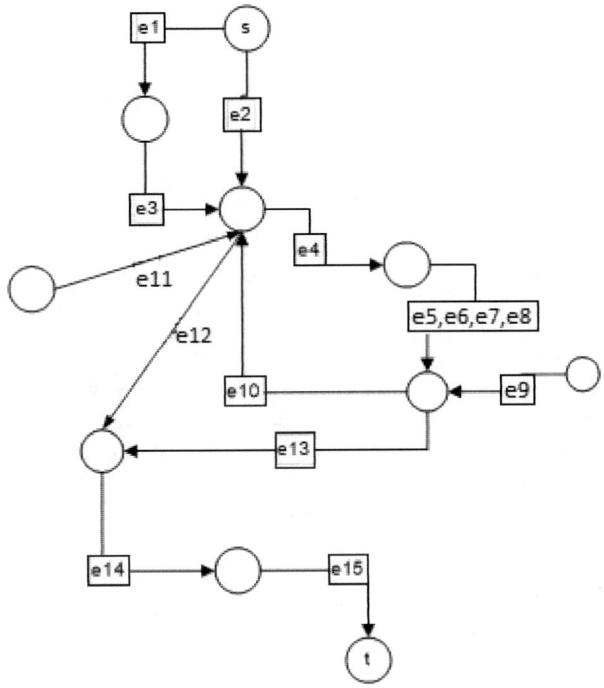

Figure 2. Connect-s, t graph of the planning interface in KRM

Table 1. Enumerate perfect circumstance

Code Number	Sequence of Edges
1	$e_1 e_3 \overline{e}_2$
2,3,4,5	$e_1 e_3 e_4 (e_5, e_6, e_7, e_8) \overline{e}_9$
6,7,8,9	$e_1 e_3 e_4 (e_5, e_6, e_7, e_8) \overline{e}_{10} \overline{e_2}$
10,11,12,13	$e_1 e_3 e_4 (e_5, e_6, e_7, e_8) \overline{e}_{11}$
14,15,16,17	$e_1 e_3 e_4 (e_5, e_6, e_7, e_8) e_{10} e_{12} e_{14} e_{15}$
...	and so on

whereas

$$An(G) = Pr[E_1 \vee E_2 \vee \ldots \vee E_n], \tag{3}$$

and then based on the concept of Eq. (1) we have

$$An(G) = \sum_{j=1}^{k} (-1)^{j+1} \sum_{I \subseteq \{1,\ldots,k\}, |I|=j} Pr[E_i] \tag{4}$$

with E_i being the event for all paths P_i with $i \in I$ operating. Therefore, interfaces with each other can be disclosed are not interdependent based on how the system works. Suppose $\overline{E}_i$ indicating the complement of events, the declared events, and generally written $D_i = \overline{E}_1 \cap \overline{E}_2 \cap \ldots \cap \overline{E}_n$ [17], and therefore we have

$$An(G) = \sum_{i=1}^{h} Pr[D_i] \tag{5}$$

IOP Conf. Series: Materials Science and Engineering **308** (2018) 012042 doi:10.1088/1757-899X/308/1/012042

The calculation of values for interface performance can be performed using Eq. (5) by substituting each interface probability value for error in data entry or process or by using edge based on Table 1. For example for interface 1 there is the data entry consists of 9 items, the probability of failure is 1/9 and the probability of success is 8/9, for interface number 1a there is a data entry is 20 items, the probability of failure is 1/20, and so on. So system reliability indicating regulatory toughness, the both can be measured at once.

5. Conclusion

The enumeration model of interface reliability of an information system using graph, logic, and probability concepts showed a good or not application interface. Using the case of the planning section interface of KRM system can be considered good because the failure factor is considered small according to the perfect enumeration that has been performed on each interface units in accordance with the data flow requirements in this system. It also illustrates that regulations set up to enforce KRM system are also considered good.

References

[1] L Harnefors, and H-P Nee 1998 Model-based current control of AC machines using the internal model control method *IEEE Transactions on Industry Applications* **14(1)**.
[2] M K M Nasution, and M S Lydia 2005 Metodologi pengelolaan proyek sistem informasi *al-Khawarizmi: Journal of Computer Science* **1(4)**.
[3] J Preece, and H D Rombach 1994 A taxonomy for combining software engineering and human-computer interaction measurement approaches: Towards a common framework *International Journal of Human-Computer Studies* **41(4)**.
[4] I Sommerville 1992 *Software Engineering*, Addison-Wesley Publishing Company, Wokingham.
[5] A A Abdel-Ghaly, P. Y Chan, and B Littlewood 1986 Evaluation of competing software reliability predictions *IEEE Transactions on Software Engineering* **SE-12(9)**.
[6] S N Weiss, and E J Weyuker 1988 An extended domain-based model of software reliability *IEEE Transactions on Software Engineering* **14(10)**.
[7] M Harahap, and M K M Nasution 2005 Model dan enumerasi pengukuran kinerja antarmuka aplikasi Kabupaten Road Management System *al-Khawarizmi: Journal of Computer Science* **1(2)**.
[8] P. G. W. Keen 1981 Information systems and organizational change *Communication of the ACM* **24(1)**.
[9] M Elfida, and M K M Nasution 2005 Perancangan antarmuka sistem informasi *al-Khawarizmi: Journal of Computer Science* **1(1)**.
[10] A R Dennis, J F George, L M Jessup, J F Nunamaker, and D R Vogel 1988 Information technology to support electronic meetings *Minneapolis* **12(4)**.
[11] J D Musa, A Iannino, and K Okumoto 1987 *Software Reliability: Measurement, prediction, application*, McGraw-Hill Book Company, New York.
[12] O Wing, and P Demetriou 1964 Analysis of probabilistic networks *IEEE Trans. Commun. Tech.* **COM-12**.
[13] L Fratta, U G Montanari 1973 A Boolean algebra method for computing the terminal reliability in a communication network *IEEE Trans. Circuit Theory* **CT-20**.
[14] D B Brown 1971 A computerized algorithm for determining the reliability of redundant configurations *IEEE Trans. Commun. Tech.* **R-20**.
[15] F Harary 1972 *Graph Theory* Addison-Wesley Publishing Company, Reading, Massachusetts.
[16] E Canale, H Cancela, F Robledo, G Rubino, and P Sartor 2012 On computing the 2-diameter-constrained K-reliability of networks *International Transactions in Operation Research* **20(1)**.
[17] M L Shooman 1968 *Probabilistic Reliability: An engineering approach* McGraw-Hill, New York.

IOP Conf. Series: Materials Science and Engineering **308** (2018) 012043 doi:10.1088/1757-899X/308/1/012043

Effective height of chimney for biomass cook stove simulated by computational fluid dynamics

Faisal[1)], A Setiawan[*1], Wusnah[2)], Khairil[3)], Luthfi[4)]

[1)] Mechanical Engineering Department of Malikussaleh University
[2)] Chemical Engineering Department of Malikussaleh University
[3)] Mechanical Engineering Department of Syiah Kuala University
[4)] Mechanical Engineering Department of Poly Technical of Lhokseumawe

[*]E-mail: adiseti@gmail.com

Abstract. This paper presents the results of numerical modelling of temperature distribution and flow pattern in a biomass cooking stove using CFD simulation. The biomass stove has been designed to suite the household cooking process. The stove consists of two pots. The first is the main pot located on the top of the combustion chamber where the heat from the combustion process is directly received. The second pot absorbs the heat from the exhaust gas. A chimney installed at the end of the stove releases the exhaust gas to the ambient air. During the tests, the height of chimney was varied to find the highest temperatures at both pots. Results showed that the height of the chimney at the highest temperatures of the pots is 1.65 m. This chimney height was validated by developing a model for computational fluid dynamics. Both experimental and simulations results show a good agreement and help in tune-fining the design of biomass cooking stove.

1. Introduction

Energy plays a very important role in people's lives. Almost all sectors of life can not be separated from the needs of energy consumption [1]. One of the major is household energy consumption, which is used for cooking [2]. Energy savings in this sector will have a significant impact on the welfare of society as well as the environment.

Biomass is one of energy source that is still widely used for cooking in households, especially those in rural areas in many countries around the world [3]. Estimated biomass energy use by 2030 is about 2.6 billion people still dependent on this fuel. Almost half of domestic energy use in many countries and 95% in poor countries used biomass fuel. [4]. Although developed countries tend to exchange the biomass uses to cleaner fuel sources, the reality shows that the use of biomass fuels in developing countries is increased [5], [6].

Biomass stove is a tool used to convert the energy contained in the fuel into heat through the combustion process. Fire and heat which is generated from the combustion process of biomass in the stove are used for cooking [7]. Stove is a term that refers to the physical structure that serves to retain and then direct the heat generated from the fuel burning process to the cooking pot as the target [8].

There are many disadvantages of traditional stove designs that are widely used by the community. These weaknesses are a challenge for engineers to design a better biomass stove [9]. There are two main things that are important to consider in the design to get a better biomass stove, i.e. energy saving and indoor air pollution problem [10].

The energy savings in a stove can be measured based on the value of its thermal efficiency. The thermal efficiency is influenced by the fuel combustion process factor and also the heat transfer process resulting from the combustion process of the fuel [11]. The stove with an open combustion process into the environmental air has good combustion process efficiency, but the heat transfer process is less effective. In contrast, the process of heat transfer in biomass stove with wall structures is usually more effective but has a low burning efficiency. The consideration of these two factors becomes an interesting thing in designing a better stove [12].

A good combustion process is when all the carbon elements present in the fuel completely react with the oxygen element [13]. There are many factors that influence a good combustion process. Airflow, turbulence and temperature are some of the factors that greatly affect the combustion process [14]. Control of the amount and rate of air supply to the combustion chamber is one of the things that can be done to create a better combustion process.

Traditional biomass stoves are generally single-pot stoves, which is pot as container is placed on the fire. Heat from the exhaust gas directly discharged into the environmental air. If there are two or more dishes, there will be a parallel cooking process, which is one burning for one cooking process. Biomass stove with multi pottery system is one of the ways advised by experts to improve the efficiency of the biomass stove. The main pot is usually located in a position that can receive direct heat from the combustion process in the combustion chamber. The next pot is positioned after the main pot which can absorb heat potency from the flue gas stream of combustion [15].

Cooking with traditional biomass stoves usually produces an amount of smoke pollutants that fills the kitchen room. Smoke pollutants caused discomfort to the cooker when working and may affect the taste of the food [16]. The most important thing is that the smoke pollution is harmful to human health, even if the smoke was generated from the perfect combustion process [10].

2. Chimney Uses in Biomass Stove

Chimney uses in a biomass stove will generates a continuous flow of draught through a stove that will be able to supply air for combustion needs in the combustion chamber. Sufficient air supply will create a better combustion process. The real combustion requires excess air to obtain sufficient amount of oxygen for better combustion process [13]. However, air contains another element which is inert and absorbs heat such as nitrogen. Too much excess air will produce useless excess heat, which will be absorbed by the inert gases then flow and discharged through the chimney into the environment. The stove with multi pot system also requires a chimney [17]. The natural draught of the chimney serves to draw the flue gas and flow it through each pot. The potential heat contained in the flue gas will be transfered to the pots when it contacted the surface of the pot. The natural draught of the chimney must be adequate to overcome the flow of frictional obstacles as it flows in the stove channel. In addition, the chimney also serves to overcome the smoke pollution in the kitchen room where the stove is operated. The chimney will direct the flue gas out of the kitchen to the environmental air [15].

Winiarski has developed the concept of chimney uses in the early 1980s by creating a stove that create clean combustion and increase heat transfer efficiency. Placing a short chimney on the stove can increase the draught and reduce emissions. This concept has been used in many stove models such as, Stove Z, Vesto Stove, Wood Gas Camp Stove, Rocket stove, Tso-Tso stove, etc [18]. The Eindhoven Group has also used a chimney on a stove that produces a clearer combustion process [9].

Chimney is a channel with a certain height that serves to generate natural air draught. The height of the chimney is the vertical distance measured from the bottom of the combustion chamber to the end of the chimney. The tests conducted by Stone and Shelton had shown the characteristics of the exhaust gas temperature at each chimney height [19]. The results of this study showed a general formula for the relation of temperature profile to chimney height variation.

Height of the chimney ($hchim$) is one of the main geometric parameters of the stove which refers to the vertical distance from the source of hot flue gas to the ambient air.

$$\Delta P_{induced} = gh_{chim}(\rho_{ambient} - \rho_{hot}) = gh_{chim}\rho_{ambient}\left[1 - \left(\frac{T_{ambient}}{T_{hot}}\right)\right] \quad (2.1)$$

The natural draught value of a chimney is determined by the height of the chimney and the temperature gradient of the flue gas in the chimney and ambient air. The pressure gradient described by equation (2.1) is translated into the mass flow rate ($\dot{m}$) flowing through the stove which depends on the height of the chimney, cross sectional area of the chimney (A), the loss coefficient (LC), the temperature of the combustion gas ($Thot$) and Ambient temperature ($Tambient$). The relationship is described as equation (2.2). The loss coefficient (LC) in this equation represents the losses associated with friction and viscosity of the flue gas due to stove geometry and the heat lost through the chimney wall. The results showed that $LC \approx 0.5$ without pots and $LC \approx 0.35$ with the pots on the stove [20].

$$\dot{m} = LC * A \left(\frac{\Delta P_{induced}}{R_s T_{hot}}\right) \sqrt{2gh_{chim}\left(\frac{T_{hot} - T_{ambient}}{T_{ambient}}\right)} \tag{2.2}$$

The consideration of the chimney uses in a biomass stove is a good thing. Chimney uses in biomass stoves function to create a better combustion and heat transfer process and also to overcome the pollution problems in the kitchen room. Based on these reasons, the determination of chimney height that affects the three aspects is very necessary to be done.

3. Biomass Stove Improvement Researches

Biomass stove improvement research was initially conducted in developing countries around 1970 by researchers incorporated in NGOs. Winarski is the one who had introduced the basic principles of biomass stove improvement [9]. The improvement of the stove at that time was intended for fuel savings due to concerns about deforestation for household energy demand [21], [22]. Biomass stove improvement efforts are then also intended to improve health by reducing the impact of air pollution and the safety of biomass fuel uses that are in line with climate change prevention programs [23], [24].

Early stove research was done experimentally in several stages. The research stage was carried out using several standard methods that have been developed, either for laboratoric test, such as Water Boilling Test (WBT), Combustion and Exhaust gas Test, or field test such as Control Cook Test (CCT), Kitchen Performance Test (KPT) and others [25], [26], [27]. Several field researches have also been conducted to understand how a stove can be received and used continuously over a long period of time [28], [29], [30]. Hundreds of stove implimentation programs had been done in many country of the world and showed different sizes, scope and type of the stoves, as well as its technological design improvement, distribution and financial benefits. So far the research has been aimed at developing a new stove design form with priority to improve the production process, marketing techniques and financial incentives of the stove applied [31].

Biomass stove research then also done in simulation along with the development of computer technology. Computational fluid dynamic (CFD) has been widely used to simulate fluid flow, heat transfer, pyrolisis and chemical reactions to the design process of several biomass stoves. CFDs have provided more information that helps designers to develop their design stoves. There are many academic papers containing general information relating to the simulation of CFDs in cooking stoves. However, only a few clearly explain the reasons for choosing a model specification and display the actual scanned documentation of the use of CFD apps.

There are specific examples of simulations of small-scale biomass burning equipment using CFDs that help to understanding the applied approaches. Menghini et al. [32], [33], describes the use of CFD applications to simulate the equipment used to burn wood. The purpose of the simulation is to optimize the heat transfer process in the heat exchanger in the equipment. No modeled combustion reactions and no discussion of exhaust emissions are produced. The simulation result is a heat of advection without reaction to estimate the heat from the combustion process.

Ravi et al. [34], [35] has developed a simulation method for a sawdust fuel stove. The process of simulating CFDs is done iteratively by assuming that there is a heat flux from the flame returning to the fuel. The fluid flow is modeled on steady state and uses the steady flow formula and assumes that there is no transient flow occurs. The completion of long time scale fuel consumption is preferred to be focused over shorter transient flow timing scales.

Bojko and Branc [36] describe the simulation process of CFDs in wood-fired stoves with detailed geometry models. The mass flow of pyrolysis gas occurring on the fuel surface is regarded as a constant fuel source. The heterogeneous combustion phase is ignored and is considered to be combustion occurring as a homogeneous phase. Burning diffusion flame is modeled by eddy-disipation method. Radiation transfer is modeled through the discrete ordinate method. The air entry boundary conditions are defined as mass flow that explicitly defined based on the naturally induced excess air ratio due to buoyant effects through inlet pressure conditions.

Bryden et al. [37], [38], and Urban et al. [39], have used a graphical evolution algorithm to optimize geometric parameters of heat transfer surfaces in a plancha biomass stove. The geometric boundary is applied to the channel boundary between the gas stream and the inside surface of the plancha stove. Burnham-Slipper [40], [41], has developed a process for using genetic algorithms, which automatically optimize the overall combustion chamber of the biomass stove.

Miller-Lionberg [8] has simulated a wood-fired biomass stove developed. The biomass stove is only a single pot and without a special chimney with a meaningful height. Provide results of flow conditions and temperature distribution. Testing was also conducted experimentally by installing a number of temperature measuring devices in the stove to validate the simulation results.

4. Reseach Method

The research is conducted on biomass stove that has been designed to fit the needs of cooking in the household. This biomass stove is a stove with multi pot system (double pot). The first pot is placed above the combustion chamber. It receives heat directly from the fire generated from the combustion process of the fuel. The second pot is expected to be able to receive heat from the flue gas before it is flowed through the chimney. A variable height chimney is placed at the end of the stove to produce the natural air draught that is required for stove operation. In addition to solve the smoke pollution problem, the function of the chimney primarily intended to improve the combustion process and heat transfer in the stove.

The test consists of two stages, e.i. the stage to determine the effect of chimney height on the biomass stove conducted by CFD simulation and experimental testing with the water boiling test method.

4.1 Simulation Method

The simulation process is initialized by drawing geometry and mesh of the biomass stove by using Gambit software as a preposessor. The geometry of the stove is drawn for a variety of chimney height i.e. 1.1 m, 1.45 m, 1.55 m, 1.65 m and 1.85 m. Furthermore, the meshing process is done using the automated method in Gambit.

The heat flow that occurs in the system is assumed only in the gas domain. Therefore the volume control only covers the gas domain volume whereas the solid area domain volume on the stove is ignored. The volume control domain is divided into a mesh or grid cell, which allows the represention by linear mathematical matrices where the fundamental equations constructed are solved by using boundary volume methods. Several different approaches have been applied to achieve the appropriate mesh. The mesh should be smooth enough to complete a small limit feature, such as biomass fuel mesh and the bottom of the pot mesh. The mesh of fluid domain must fit along a solid boundary to overcome the effects of the thermal boundary layer and momentum. The problem is very complicated actually because the boundary layer can be interconnected with one and another [42].

Mesh geometry significantly influences the simulation of unsteady flow and turbulences. Mesh evolves rapidly from the basic mesh size. The minimum scale of unsteady motion and turbulences completed in the simulation is a function of the local mesh size. The mesh specification should be good enough to complete the flow feature, but the number of cells should be as low as possible to maintain the practical speed of the calculation. Therefore the mesh density becomes increasingly

IOP Conf. Series: Materials Science and Engineering **308** (2018) 012043 doi:10.1088/1757-899X/308/1/012043

reduced to the part that is farther away from the basic mesh zone that requires a good resolution. Effort in meeting the requirements between adequate resolution and minimum number of cells are often a tremendous challenge.

The created mesh needs to be re-examined to ensure that the meshing process is done properly. The border layer feature checking scheme starts with the small elemental thickness continuing to the next larger elements with standard growth rates. Mesh checks are performed to ensure the mesh is well established so that the definition of geometry can be performed.

Figure 1. Colour view of boundarry layer on Gambit of the stove

After the process of making geometry and meshing was complete then simulation and analysis with CFD application was proceed. The selected system is 3D. The first step is to import the mesh model from Gambit to the CFD application. All mesh results on specimens that have been built on the Gambit program are imported in this process. Mesh checks need to be done to make sure the mesh and grid are good so that the analysis process can proceed. Furthermore, it is necessary to select the formula solver. The choice of solver formula is applied equally to all specimens and settings in the CFD program. The next step is to determine the basic model and equation so that the test results can convergen.

The determination of the boundary conditions in the stove uses many assumptions that reflect the overall system applied to the simulation. The fuel reaction conditions are assumed to be steady and without feedback to reduce the thermal response time of the whole system being modeled. The phenomenon of steady flow field motion is assumed to occur over a short time scale. The walls of the combustion chamber are assumed to achieve rapid thermal equilibrium with improved flow conditions and new reactions. The simulation does not attempt to explain the heat flow in the stove structure which is assumed to be isolated. The heat leaves the domain across the wall surface along the inside of the combustion chamber without conduction of heat along the domain boundary. The heat conduction along the boundary surface on all solid structure has been ignored except at the bottom of both pots. Operating pressure is set at 1 atm. Characteristics of the materials used are the characteristics of air at a constant state.

After determine the solution control parameters and initialization of the flow field is done. Furthermore iteration is run. When iteration begins then all defined conservation equations are solved simultaneously and parallel. After all processing steps are completed then post-processing steps can be performed to show the simulation results.

4.2 Experimentaly Method

Experimental tests were performed to get the best performance from the stove for each chimney variation. Performance parameters concerned are the water boiling time in pot 1 and thermal efficiency of the stove. Experimental chimney height variations have the same height as simulations.

IOP Conf. Series: Materials Science and Engineering **308** (2018) 012043 doi:10.1088/1757-899X/308/1/012043

So the result can be compared with the thermal effect obtained by simulation test. Figure 2 shown the experimental testing process performed on the stove.

Figure 2. Experimental stove testing

Testing is done by using water boiling test method. The test is carried out for stove operating conditions consisting of high power and low power. High power consists of cold start and hot start while low power is called simmer. Cold start is a process of operation that starts with the condition of the stove is still cold. Cold start operating conditions are common when the stove is operated in the morning when the stove has not absorbed the heat from the combustion process yet and the temperature is close to ambient temperature. Hot start is a process operation with the stove already in a hot condition because the stove has absorbed heat from the previous process and its temperature higher than the environmental temperature. While simmer is the process of operating stove with low heat to maintain the water boiling point for a certain time interval. Differences in temperature fluctuations in simmer conditions are not greater than 5 ^{o}C. Simmer operating conditions in the cooking process with the stove is usually done to make meals more cooked but not burned. The test time for high power starts when the fire lit and ends up when the water in the first pot boiled. While the time used for the simmer operation process in this test starts when the water in the first pot boiled and ends up in 45 minutes [25].

The amount of heat absorbed by water for each test operating condition can be calculated as the amount of useful energy. While the amount of energy supplied to the stove system can be calculated based on the amount of burned-out fuel mass multiplied by the calorific value held by the type of fuel used in the test. This calorific value calculation has considered the wood humidity and air humidity factors, charcoal and dust masses generated from the combustion process. A comparison of useful heat to the energy given for each operating condition is the thermal efficiency value of the stove under test.

In this test, three repetitions were performed for each chimney height variation. The data obtained then tabulated and calculated to get the average value. The results are then depicted in graphical form to show the relationship between water boiling time in pot 1 and thermal efficiency of the stove to the height variation of the chimney. So it can be compared with the thermal effect of the simulation result. Figure 3 showed the experimental testing process performed.

5. Testing Results and Discussion

The following sections report the CFD simulation results and experimental tests conducted to validate the simulation. Run time simulation is only about one third of the duration needed for cooking. While the thermal balance of the stove naturally takes more than an hour. Thermal aspects greatly affect the behavior of the stove so that temperature is one of the important and interesting properties in the simulation to be discussed. Although the parameters considered different from the simulation and experimental results but both of them are the thermal aspects of the stove system that closely relate

IOP Conf. Series: Materials Science and Engineering **308** (2018) 012043 doi:10.1088/1757-899X/308/1/012043

each other. So the thermal aspects obtained by simulation will be comparable with the experimental results.

5.1 Results of simulation test

After all processing steps are completed then post processing stage can be done to show the simulation result. Because it has a three-dimensional control volume, the CFD simulation results can be assigned to any location in the stove. The CFD simulation results have a clear visual where the domain retains the shape of the stove. Some actual results are evaluated and displayed in 3D space, including fluid flow pathways and various iso surfaces.

The main simulation results are reported in two dimensional forms. The forms provide a view on longitudinal plane of the stove. The simulation results are also given in cross section view in longitudinal plane image to show the conditions inside the stove. The temperature contours shown are the convergen results of the iterations performed. Each chimney height variation has different temperature contour. The temperature contours for each of the varied flue stove elevations are shown in Figures 3 to 7. Image shows visual orientation of the resulting contour plot graphics.

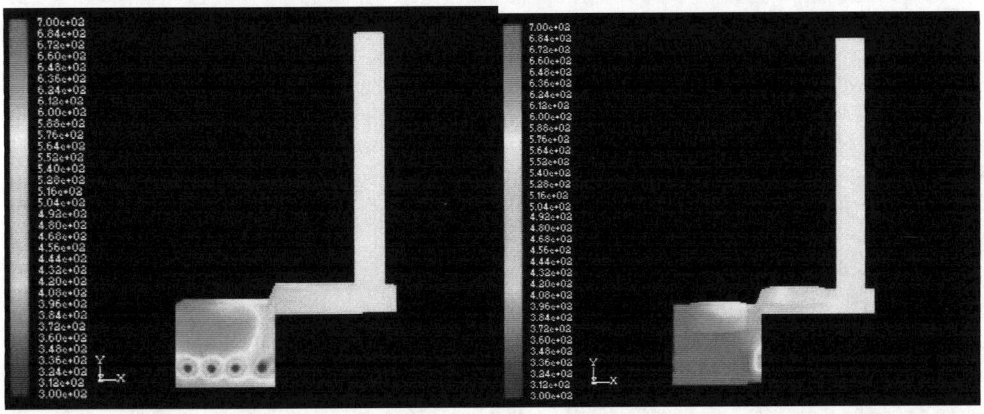

(a) (b)

Figure 3. Temperatures contour of fluid flow in stove with 110 cm height of the chimney (a) View on longitudinal plane, (b) Cross section view in longitudinal plane.

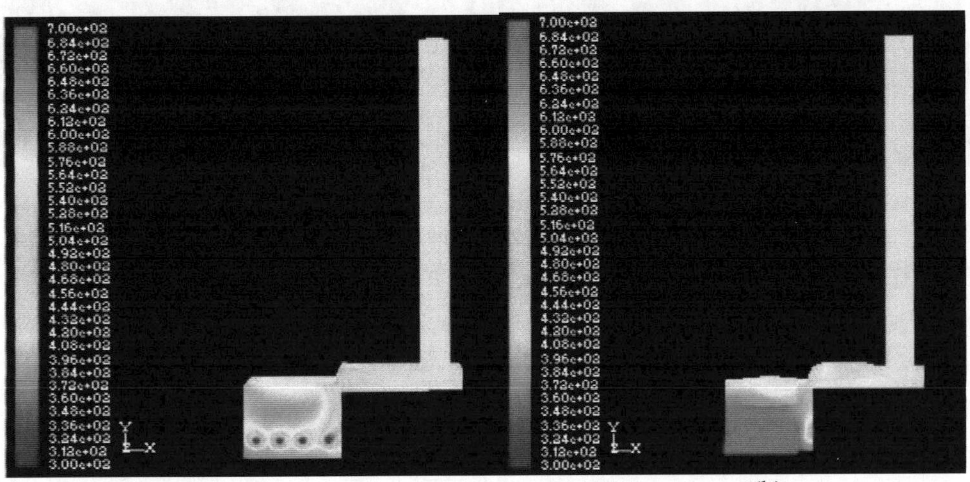

(a) (b)

Figure 4. Temperatures contour of fluid flow in stove with 145 cm height of the chimney (a) View on longitudinal plane, (b) Cross section view in longitudinal plane.

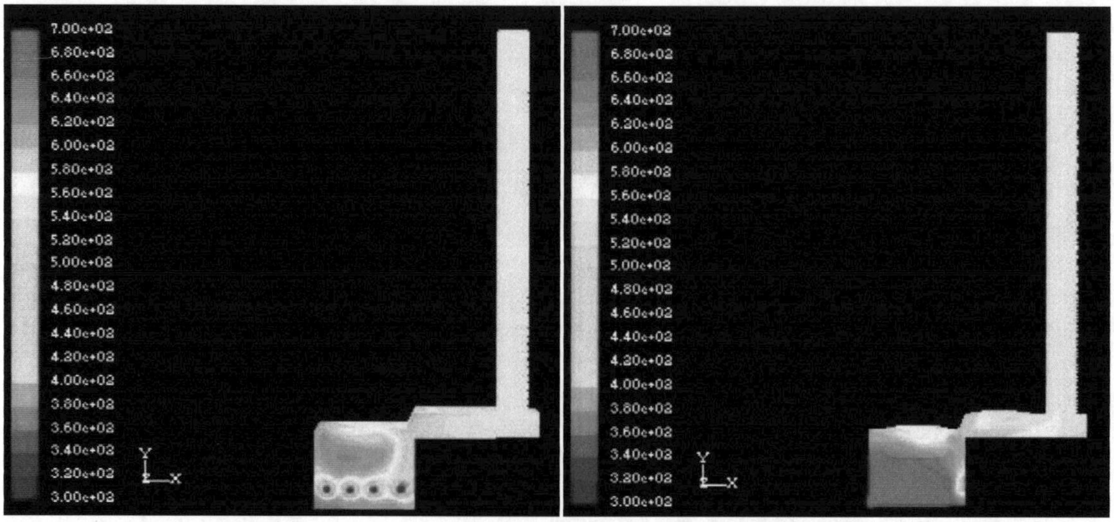

(a) (b)

Figure 5. Temperatures contour of fluid flow in stove with 155 cm height of the chimney (a) View on longitudinal plane, (b) Cross section view in longitudinal plane.

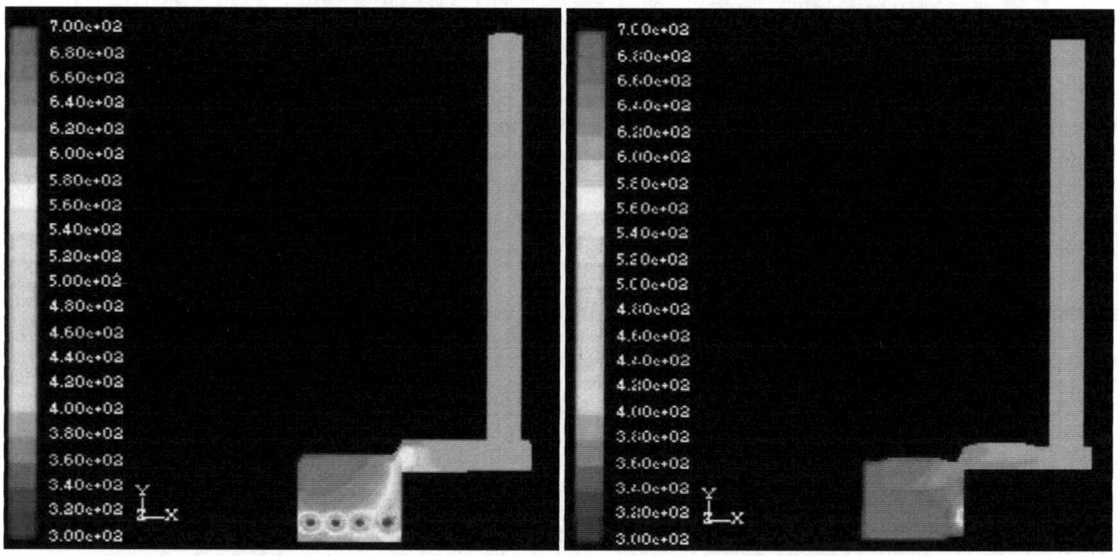

(a) (b)

Figure 6. Temperatures contour of fluid flow in stove with 165 cm height of the chimney (a) View on longitudinal plane, (b) Cross section view in longitudinal plane.

IOP Conf. Series: Materials Science and Engineering **308** (2018) 012043 doi:10.1088/1757-899X/308/1/012043

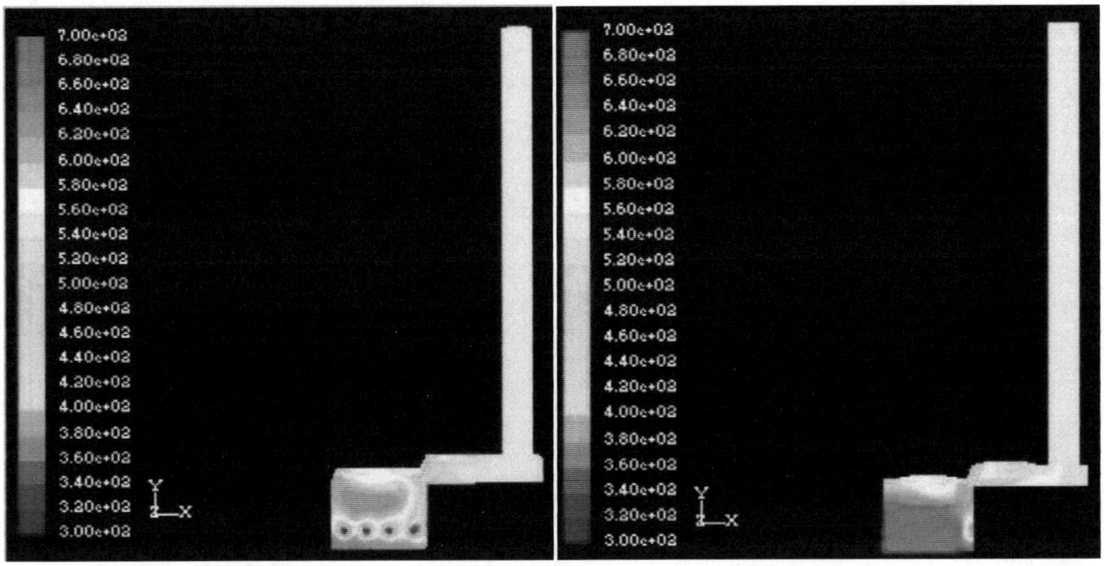

(a) (b)

Figure 7. Temperatures contour of fluid flow in stove with 185 cm height of the chimney (a) View on longitudinal plane, (b) Cross section view in longitudinal plane.

The results of the temperature distribution and flow pattern are useful for understanding the general effect of the chimney to the flame position, flow path and stretching. The temperature distribution combined with the heat flux on the solid surface is useful for optimizing heat transfer in the stove. Useful heat can be maximized by controlling and directing a better gas flow to the pot. More detailed statistical analysis of the temperature distribution values will also provide additional information for combustion scientists and stove designers such as information on fire position stability and possible emissions generation.

There is a lot of data obtained through simulation results that can be discussed. However, only a few data are selected and shown here to help provide information that can be used as a basis for determining the effective chimney height of the designed stove. The temperatures in each pot and on the out-let chimney are the data that is considered important that can show the thermal effect of the stove. Based on these considerations, the simulated results data of those sections are selected to show the effect of the chimney height variation.

Table 1. Simulation result data of temperature at pot 1, pot 2 and outlet to the chimney height variation

Chimney Height (m)	1.1	1.45	1.55	1.65	1.85
Paths	Temperature (K)				
Pot 1	645.26175	625.14104	628.47114	665.67156	623.29761
Pot 2	580.25653	569.51124	570.48503	619.02604	568.96524
Outlet	566.49196	555.13441	555.81833	609.92467	554.58899

The simulation results data of the temperature at pot 1, pot 2 and outlet of the stove to the height variation of the chimney can be seen in Table 1. Based on the temperature data, a comparison graph of temperature change at pot 1, pot 2 and the stove outlet for each chimney height variation is created as shown in Figure 8. Based on the graph, it can be seen that the height variation of the chimney results a difference thermal effect on the stove. Noted that the temperature at pot 1, pot 2 and outlet have different values for each chimney height variation. The stove simulation results with 1.65 m chimney height give the highest thermal effect at pot 1, pot 2 and out-let which is indicated with the highest temperature value at each part compared with the other height.

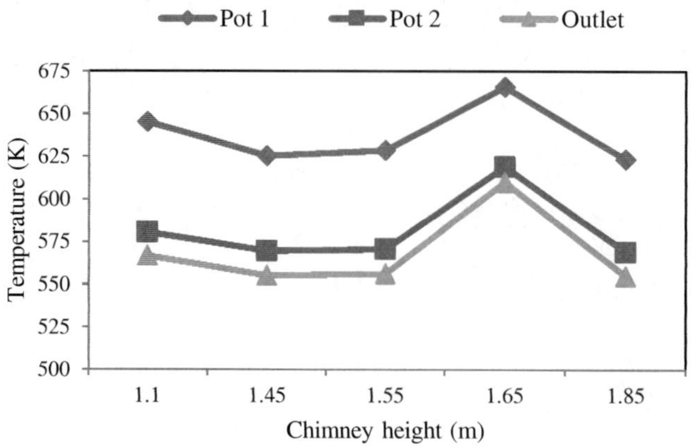

Figure 8. Graph of temperature at pot 1, pot 2 & outlet of the stove to the chimney height variation

5.2 Experimental Test Results

The parameters of experimental test results shown below were only the relation between water boiling time in pot 1 and the thermal efficiency of the stove. The values that described the thermal effect were the average value of the test results. Both values for each variation of the chimney are given in graphical form.

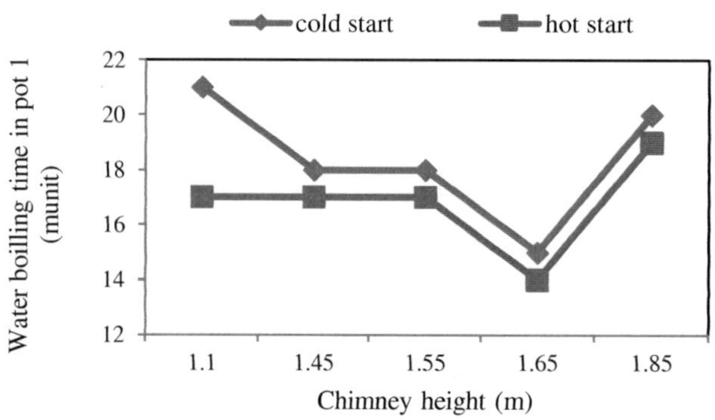

Figure 9. Graph of water boiling time in pot 1 of stove to the chimney height variation.

Figure 9 shows the graph of water boiling time in pot 1 of the stove for hight power operation. It can be seen that water boiling time in pot 1 for hot start is shorther than cold start. It is caused by the heat generated in the combustion chamber on the cold start does not immediately transfer to the water in the pot maximally, due to part of it was absorbed by the stove and other part transfered to the environment as heat losses. The stove will continue to absorb the heat from the combustion to the maximum until the stove reaches its thermal balance when the cold start operation. While the hot start operation is carried out when the stove condition has absorbed heat from the previous operating process and is already in thermal equilibrium and only part of it is transfered to the environment as heat losses. However, both graphs of both operations show similar tendencies. Both graphs showed that the shortest time is obtained at 1.65 m of chimney height.

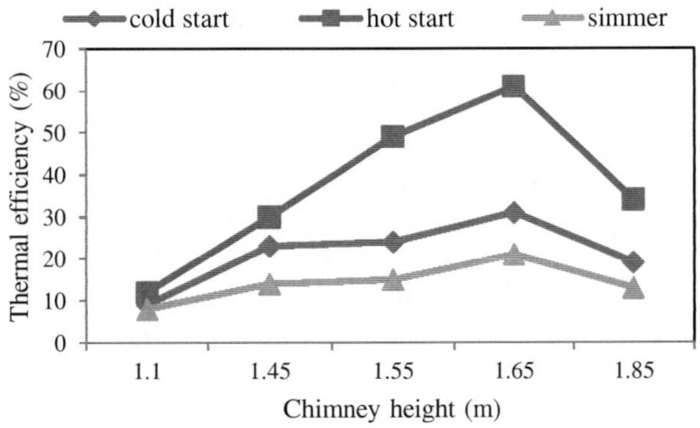

Figure 10. Graph of thermal efficiency of stove to the chimney height variation.

Figure 10 showed the thermal efficiency of the stove at cold start, hot start and simmer for each of the chimney height variations of the stove. The thermal efficiency for the simmer operating conditions is always below the cold start and the hot start. This is because simmer is a low power operating condition. The flame in the combustion chamber of the stove during simmer operation is attempted in small sizes only to retain water in pot 1 being at a constant temperature boiling point. While cold start and hot start is a high-energy stove operation condition. Figure 6 also showed that the thermal efficiency of the stove for hot start was always above the cold start. As explained earlier this is because the stove in hot-start operational condition has reached its thermal equilibrium so that more useful energy transfered to the water in the pot compared to the cold start condition. Overall, each graph shows a similar trend of efficiency value to the chimney height variation. It appears that the highest thermal efficiency of the stove is achieved at 1.65 m chimney height for all operating conditions, either for low power or high power.

6. Conclusions

The CFD simulations developed to evaluate the behavior of double pots biomass stove with the chimney have been achieved a fairly good process. The control volume assumed to be on the gas domain only by ignoring the solid domain of the stove has yielded satisfactory thermal effects. Accurate and well-finished stove simulation results have also been obtained with a relatively shorter time than required for cooking. Boundary layer resolution and accurate thermal effect values obtained through the simulation results are useful for selecting the effective height of the chimney on the stove.

The result of double pot system stove simulation can be correlated with experimental test result. Both test results showed that the best thermal aspect value is obtained for 1.65 m chimney height. This has been proven either through simulation or experimental tests. Based on the results obtained through both tests, it has provided data and information that assists in designing the biomass stoves. The biomass stove had shown a good performance so that it was believed to provide advancements in addressing efficiency and pollution problems compared to traditional biomass stoves generally.

References

[1] Siregar E *et al* 2013 *Indonesia Energy Outlook* BPPT p 104
[2] Nuryanti et al 2007 *The characteristics analysis of energy consumption at household sector* Yogyakarta p 21–2

[3] Modi V et al 2006 *Energy and the Millenium Development Goals*. New York: Energy Sector Management Assistance Programme, UNDP

[4] Ezzati M and Kammen DM 2002 *The health impacts of exposure to indoor air pollution from solid fuels in developing countries: Knowledge, gaps, and data needs*. Environ Health Perspect. 110(11):1057–68.

[5] Bruce N *et al* 2000 *Indoor air pollution in developing countries*: A major environmental and public health challenge. Bulletin WHO 78(9):1078–92

[6] Arnold M *et al* 2003 *Fuelwood revisited: what has changed in the last decade?* Center for International Forestry Research (CIFOR)

[7] Barnes DF *et al* 1993 *What Makes People Cook with Improved Stoves? A Comparative Review*. Wood Energy Dev. Planning, Policies Strategy Vol III III(37):137–68.

[8] Miller-Lionberg DD 2011 *A fine resolution CFD simulation approach for biomass cook stove development*. Colorado State University

[9] Bryden M *et al* 2006 *Design Principles for Wood Burning Cook Stoves* p38

[10] Ma R Lou 1982 *Cookstove Construction by the Terra-CETA Method*. Guatemala: Centro de Experimentacion en Technologia Apropiada p 1-14

[11] Zube DJ 2010 *Heat Transfer Efficiency of Biomass Cookstoves*. Colorado State University p. 156

[12] Vaccari M *et al* 2012 *Improved cookstove as an appropriate technology for the Logone Valley (Chad–Cameroon)*: *Analysis of fuel and cost savings*. Renew Energy 47 p 45–54.

[13] Hanby VI 2012 *Combustion and Pollution Control in Heating Systems* Springer London

[14] Surjosatyo A and Ani FN 2011 Study of Enhancing the Swirl Burner Performance On a Small scale Biomass Gasification *Int J Eng Technol* **11** 04 p21–38.

[15] Faisal 2015 *Advanced Improvement of Gahegi Biomass Stove for Salt Manufacture* Proceedings of In: The 1st Almuslim International Conference on Science, Technology and Society (AICSTS) Bireuen Indonesia p. 85

[16] De Koning HW *et al* 1985 *Biomass fuel combustion and health*. Bull World Health Organ. 63(1):11–26.

[17] Panwar N L *et al* 2009 *Mitigation of greenhouse gases by adoption of improved biomass cookstoves*. Mitig Adapt Strateg Glob Chang. 14(11027):569–78

[18] Ballard-Tremeer G and Jawurek HH 1996 Comparison of five rural, wood-burning cooking devices: Efficiencies and emissions *Biomass and Bioenergy* **11**(5) PP 419–30

[19] Brewer JP *et al* 2007 *Chimney Fires : Causes, Effects & Evaluation*. Chimney Fire Education and Research Task Force, Chimney Safety Institute of America 46168(317)

[20] Agenbroad J *et al* 2011 A simplified model for understanding natural convection driven biomass cooking stoves—Part 1: Setup and baseline validation *Energy Sustain Dev* **15**(2) PP 160–8

[21] Ruiz-Mercado I *et al* 2011 Adoption and sustained use of improved cookstoves *Energy Policy* **39**(12):7557–66

[22] Arnold M and Persson R 2003 Reassessing the fuelwood situation in developing countries *International Forestry Review* **5** (4) p379-383

[23] Smith KR 1994 Health, energy, and greenhouse-gas impacts of biomass combustion in household stoves *Energy Sustain Dev* **1**(4) p23–9

[24] Johnson M *et al* 2009 Supporting Information Quantification of carbon savings from improved biomass cookstove projects *Environmental Science and Technology* **43**, 2456–2462

[25] Bailis R *et al* 2007 *The Water Boiling Test (WBT) Version 3.0*. p1–38

[26] Bailis R *et al* 2004 *Controlled Cooking Test (CCT)* Household Energy and Health Programme, Shell Foundation, p1–8

[27] Bailis R and Edwards R 2007 *Kitchen Performance Test (KPT) p*1–32

[28] Agarwal B 1983 Diffusion of rural innovations: Some analytical issues and the case of wood-burning stoves *World Dev* **11**(4) p359–376

[29] Pandey S and Yadama GN 1992 Community Development Programs in Nepal: A Test of Diffusion of Innovation Theory *Soc Serv Rev*. **66**(4) p582–597
[30] Hessen JO *et al* 2001 Motivational Factors Related to Improving Indoor Air Quality in Rural Nepal *Mt Res Dev* 21 p148–153
[31] Gifford ML 2010 *A Global Review of Cook stove Programs*. CA: Energy and Resources Group UC Berkeley p. 2–31
[32] Menghini D *et al* 2007 Numerical and experimental investigations to lower environmental impact of an open fireplace *Exp Therm Fluid Sci* **31**(5) p477–482
[33] Menghini D *et al* 2008 Effect of excess air on the optimization of heating appliances for biomass combustion *Exp Therm Fluid Sci* **32**(7) p1371–1380.
[34] Ravi MR *et al* 2004 Development of a semi-empirical model for pyrolysis of an annular sawdust bed *J Anal Appl Pyrolysis* **71**(1) p353–374
[35] Ravi MR *et al* 2002 Use of CFD simulation as a design tool for biomass stoves *Energy Sustain Dev* **6**(2) p20–27
[36] Bojko M and Branc M 2009 A definition of the mathematical model of combustion process in the stove *J Appl Sci Thermodyn fluid Mech* **3**
[37] Bryden KM and McCorkle DS 2004 Evolutionary optimization of energy systems using population graphing and neural networks *Adv Eng Softw* **35**(5) p289–299
[38] Bryden KM *et al* 2003 Optimization of heat transfer utilizing graph based evolutionary algorithms *Int J Heat Fluid Flow* **24**(2) p267–277.
[39] Urban GL *et al* 2002 Engineering optimization of an improved plancha stove *Energy Sustain Dev* **6**(2) p9–19
[40] Burnham-Slipper H 2005 *Eritrean Stove Optimization: Literature Review* University of Nottingham p1–31
[41] Slipper B *et al* 2009 *Breeding a better stove : the use of computational fluid dynamics and genetic algorithms to optimise a wood burning stove for Eritrea* PhD thesis University of Nottingham
[42] Yuen GHY and Yuen KK 2009 *Computational Fluid Dynamics in Fire Engineering*. Butterworth-Heinemann p544

IOP Conf. Series: Materials Science and Engineering **308** (2018) 012044 doi:10.1088/1757-899X/308/1/012044

Car painting process scheduling with harmony search algorithm

M F Syahputra*[1], A Maiyasya[2], S Purnamawati[3], D Abdullah[4], W Albra[5], M Heikal[6], A Abdurrahman[7] , M Khaddafi[8]

[1,2,3] Department of Information Technology, Faculty of Computer Science and Information Technology, University of Sumatera Utara, Medan, Indonesia
[4]Department of Informatics, Universitas Malikussaleh, Aceh, Indonesia
[5,6,7] Department of Management, Universitas Malikussaleh, Aceh, Indonesia
[8]Department of Accounting, Universitas Malikussaleh, Aceh, Indonesia

*E-mail: nca.fadly@usu.ac.id

Abstract. Automotive painting program in the process of painting the car body by using robot power, making efficiency in the production system. Production system will be more efficient if pay attention to scheduling of car order which will be done by considering painting body shape of car. Flow shop scheduling is a scheduling model in which the job-job to be processed entirely flows in the same product direction / path. Scheduling problems often arise if there are n jobs to be processed on the machine, which must be specified which must be done first and how to allocate jobs on the machine to obtain a scheduled production process. Harmony Search Algorithm is a metaheuristic optimization algorithm based on music. The algorithm is inspired by observations that lead to music in search of perfect harmony. This musical harmony is in line to find optimal in the optimization process. Based on the tests that have been done, obtained the optimal car sequence with minimum makespan value.

1. Introduction

Developments in the industrial world increasingly experienced an increase in terms of quality and quantity of production. Starting from using skilled human labor to use the machine in conducting production activities so that the possibility of errors caused by human error (human error). In developed countries almost all industries do production activities using the power of the engine, so that humans only operate the machine in its production activities.

Although already using the power of machinery in conducting its production activities, every industrial company keep trying to make its production activities produce an optimum result. Scheduling is one of the efforts of industrial companies to produce optimum results. Production scheduling is attempted to obtain an effective job assignment at each work station, so that no job stacking occurs so as to reduce idle time or waiting time for subsequent processing [1].

Car industry factories in developed countries already use the machine in the process of painting the car body. In the painting process has not done scheduling optimally, where in this case there is still waiting time in the process of painting the next stage, if the painting process is done more than one variation of the form of the car in the same painting machine.

Flow shop scheduling is a scheduling model where the job to be processed entirely flows in the same product direction / path. In other words, jobs have routing work together. Scheduling problems

IOP Conf. Series: Materials Science and Engineering **308** (2018) 012044 doi:10.1088/1757-899X/308/1/012044

often arise if there are n jobs that will be processed on the machine m, which must be specified which must be done first and how to allocate jobs on the machine so that a scheduled production process is obtained [2].

2. Problem identification

In the process of painting cars with different car shape variations in the same painting machine, there are still a problem in waiting time. Therefore, an optimal solution is required to achieve the effective and efficient scheduling of car sequences in the painting process. How to get effective and efficient car order in the painting process, thus obtaining minimum makepan value in solving flow shop scheduling problems.

3. Previous Research

Scheduling flow shop in the search for the best optimization, has been a lot to solve various kinds of problems in everyday life. From small problems to complex problems with various methods in the settlement.

In solving the flow shop scheduling problem about the time of movement of a job that sometimes still has a free time when moving from one machine to another machine [3]. In this case [3] uses ant colony algorithm to solve the problem in finding the smallest makespan time.

Soukhal [4] in his research solve the problem of flow shop scheduling by using polynomial-time algorithm. Problems examined by Soukhal is about the truck carrying goods that will be delivered to the customer by considering the capacity of truck and transport time.

Boukef [5] conducts research on the flow shop scheduling problem in terms of minimizing the amount of prices for the production process and the delivery process undertaken on pharmaceutical drugs and the food industry. Boukef uses the genetic algorithm method to solve the problem.

Aulia [6] researches the problem of flow shop scheduling with permutation constraints on harmony search algorithm. The result of this algorithm can solve the problem well compared with the lower bound.

4. Methodology

HSA is a metaheuristic optimization algorithm based on music. The algorithm is inspired by observations that lead to music in search of perfect harmony. This musical harmony is in line to find opitmal in the optimization process. The search optimization process can be compared to the process of improvising jazz music. On the one hand, the perfection of harmony is determined by aesthetic sound standards. A musician always expects to produce a song with perfect harmony. On the other hand, an optimal solution for an optimization problem will be the best solution for objective and limited issues. Both of these processes will produce the best or optimum solution [7].

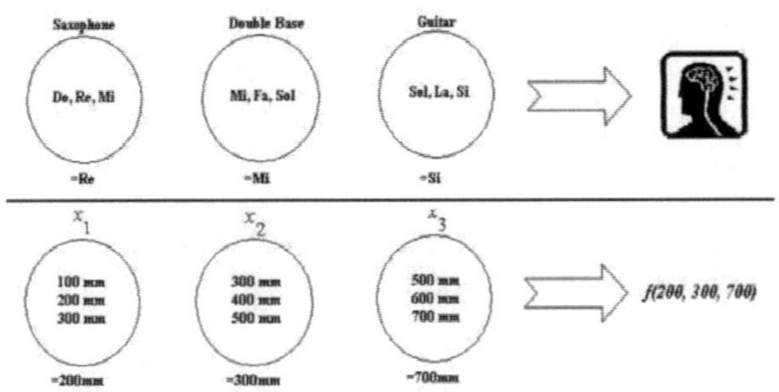

Figure 1. Improvisasion of Music Analogy

The analogy between musical improvisation and optimization techniques can be illustrated in figure 1:

Each music player (saxophonist, double bassist and guitarist) can be analogous to the variables (x1, x2, x3) and the tone levels of each musical instrument (saxophone = {Do, Re, Mi}; double bass = {Mi, Fa, Sol}; And guitar = {Sol, La, Si}) are analogous to the variables (x1 = {100, 200, 300}; x2 = {300, 400, 500}; and x3 = {500, 600, 700}). If the saxophonist produces a Re sound, the double bassist emits a sound of the Mi and the guitarist emits a Si sound, so all three will simultaneously create a new harmony (Re, Mi, Si). If this new harmony is better than the previous harmony, then the new harmony will be used. Likewise a new vector solution (200mm, 300mm, 700mm) will be used as well if it is better than the previous one for objective function value conditions [8].

In accordance with the above concept, HSA consists of five stages, namely:

A) Initialization of optimization problems and algorithm parameters.
B) Initialize harmony memory.
C) Improvisation of new harmony.
D) Harmony memory update.
E) Repeat steps c and d until the stop criteria are met.

implementation of harmony search algorithm in flow shop scheduling process can be described as below:

A. Initialization of Optimization Problems and Algorithm Parameters

In the first stage, the optimization problem is determined as follows:
Minimize (or Maximize) f (x)
With x i € Xi, i = 1,2, ..., N where f (x): an objective function,
Xi: decision variable i,
Xi: the set of decision variables, and
N: the number of decision variables

Next, determine the parameters needed in the HSA, namely:

a) Harmony Memory Size (HMS)

HMS is the number of vector solutions in harmony memory.

b) HMMM Memory Considering Rate (HMCR)

HMCR is a continuous value used as a harmonized improvisation parameter in HSA. The value of this parameter is $0 \leq \text{HMCR} \leq 1$.

c) Pitch Adjustment Rate (PAR)

PAR is a continuous value used as a harmonized improvisation parameter after the HMCR criteria are met. The value of this parameter is $0 \leq \text{PAR} \leq 1$.

d) Stop Criteria

Stop Criteria is the value used to stop the repetition of the new harmony improvisation.

B. Harmony Memory Initialization

In the initialization phase of Harmony Memory, solution vectors will fill the memory harmony according to the number of HMS. The solution vector is generated from randomly generated decision variables. The solution vector that is randomly generated is a sequence of jobs in the flow shop. Then each vector solution will be calculated its mask value. So it will produce the solution vector with the value makespannya respectively.

An example of data that will be solved based on flow shop problems is shown in table 1.

Table 1 Example of Time Flow Shop Scheduling

Machine/Car	1	2	3	4	5	6	7
1	1	3	5	2	4	5	3
2	2	4	1	3	5	4	2

From the data in table 1, the number of machines is 2 and the number of cars is 7. So the decision variable that is formed is

$$Xi = \{1,2,3,4,5,6,7\}$$

After the Xi decision variable is generated, then determine the number of HMS to be formed to accommodate the number of solution vectors in HM. Then the solution vector generation is generated randomly and calculated makespan from each solution vector.

C. Improvisation of New Harmony
At this stage, the new harmony vector x '= (x'1, x'2, x'3, x'4, ..., x'N) has three basic parts, namely:

A) Memory Consideration
The use of memory consideration is very important, this is the same as choosing the best individual in the genetic algorithms. Memory consideration will ensure the best harmony that will be brought to the last new harmony memory.

B) Pitch Adjustment
The pitch adjustment is fixed by a pitch band-width brange and a pitch adjusting rate rpa. Although in music, pitch adjustment is a tool to change the frequency, it is suitable for connecting a slight difference solution inside the HSA. In theory, patterns can be arranged linearly or nonlinear, but in practice linear adjustment is used. So it can be concluded xnew = xold + brange * € where xold is the lifestyle or solution of harmony memory and xnew is a new pattern after pitch adjusting action.

C) Random Selection
Random selection is useful for expanding the diversity of solutions. Although pitch adjustment has the same role, but pitch adjustment is limited to certain local pitch adjustments and so is local search. Random selection can run on a more advanced system to examine a variety of solutions to find the optimal global.
In this research, the improvisation process is described as follows:

Iteration 1
The random number generated is a1 = 0,432. Then the random number is adjusted to HMCR = 0.9. Because a1 <HMCR, then the decision variable is chosen randomly. Suppose the selected x1 '= 2. Next generate a random number to be adjusted with PAR = 0.3. Suppose that the random number generated is a2 = 0.512. Because a2> PAR, then the decision variable x1 '= 2 is retained.

Iteration 2
The random number generated is a1 = 0.925. Then the random number is adjusted to HMCR = 0.9. Because a1> HMCR, then the decision variable is chosen randomly from X1. Suppose the selected x1 '= 5.

Iteration 3
The randomized number generated from the previous iteration, e.g. a1 = 0.276, will be compared with the value of HMCR = 0.9 . Because the condition a1 < HMCR is fulfilled, the decision variable will be selected randomly, e.g. x1' = 3. The next step is generating random value to be adjusted by comparing the value with PAR = 0.3 . For example, if the value, defined by variable a2, is 0.114, and the rule a2 < PAR is fulfilled, then the decision variable x1' = 3 will be adjusted to the next variable, which is x2' = 4.

Iteration 4
The random number generated is a1 = 0.456. Then the random number is adjusted to HMCR = 0.9. Because a1 < HMCR, then the decision variable is chosen randomly. Suppose the selected x1 '= 6. Next generate a random number to be adjusted with PAR = 0.3. Suppose the random number generated is a2 = 0.212. Since a2 <PAR, then the decision variable x1 '= 6 is adjusted to the adjacent variable, ie x2' = 7.

Iteration 5
The random number generated is a1 = 0.941. Then the random number is adjusted to HMCR = 0.9. Because a1> HMCR, then the decision variable is chosen randomly from X1. Suppose the selected x1 '= 6.

Iteration 6
The random number generated is a1 = 0.672. Then the random number is adjusted to HMCR = 0.9. Because a1 <HMCR, then the decision variable is chosen randomly. Suppose the selected x1 '= 1. Next generate a random number to be adjusted with PAR = 0.3. Suppose the random number generated is a2 = 0.322. Because a2> PAR, then the decision variable x1 '= 1 is retained.

Iteration 7
The random number generated is a1 = 0,731. Then the random number is adjusted to HMCR = 0.9. Because a1 <HMCR, then the decision variable is chosen randomly. Suppose the selected x1 '= 3. Next generate a random number which will be adjusted with PAR = 0.3. Suppose that the random number generated is a2 = 0.843. Because a2> PAR, then the decision variable x1 '= 3 is retained.

From the above improvisation process obtained a new vector solution, namely:

$$X = [2\ 5\ 4\ 7\ 6\ 1\ 3]$$

Furthermore from the sequence of the car, it will be calculated its objective function to know the time.

D. Harmony Memory Update
After the improvisation process is completed it will get a new car sequence. Then from the sequence the car is calculated makespan. Having obtained the value of makespan, then the value of makespan is compared with the value of makespan contained in HM.
If the makespan obtained at the improvisation stage is smaller than the makespan contained in HM, then the sequence of car and the makespan value in HM will be replaced by the sequence of the car and the new makespan. However, if the makespan obtained at the improvisation stage is not smaller than the makespan contained in HM, the sequence of cars and makespan in HM will be maintained.

E. Stop Criteria
Criteria for quitting is a condition in which the improvisation process stops if it meets the stipulation of the stop criteria. However, if the determination of the stop criteria has not been met, then the improvisation process continues to be done repeatedly until it meets the stop criteria.

F. Objective Functions
The objective function of this research is using makespan, which is the time period of completion of a job which is the sum of all the processing time of a machine. The smallest makespan is the best use of machine and job combinations and ensures the job from start to finish is completed. Makespan is calculated using the following equation:

The calculation method of make span value is described in table 2.

Table 2. Makespan Calculation

Machine	Car	Duration	Start time	End time
1	2	3	0	3
1	5	4	3	7
1	4	2	7	9
1	7	3	9	12
1	6	5	12	17
1	1	1	17	18
1	3	5	18	23
2	2	4	3	7
2	5	5	7	12
2	4	3	12	15
2	7	2	15	17
2	6	4	17	21
2	1	2	21	23
2	3	1	23	24

From table 2, the car column has been arranged in the order of cars to be processed on each of the 1st and 2nd engines. In the sequence of the 1st car (car 2), the start time starts 0. Since each car is in the order of- 1 must start from minute 0. To know the start time on the 2nd car sequence (car 5), that is the duration of time and start time on the sequence of car 1 (car 2) summed. So the result is the amount of time starting from the order of the 2nd car (car 5). And so on until the last car. To fill in the end time, the duration of time and start time is added to each car.
For machine 2, the start time of the 1st car sequence (car 2) is taken from the engine end time 1 of the 1st car sequence (car 2). Then for the start time of the next car sequence, it is obtained from the comparison between the end time of the 1st car sequence (car 2) on the 2nd (j) machine with the end time of the 2nd car sequence (car 5) (i + 1). What is taken is the greatest end time. So that get the value makespan is the last sequence (car 3) on the order of the 7th car in machine 2, which is 24.

5. Result and Discussion
In this study, the data to be used is a combination of engines and the number of cars 2 x 20 based on the time that has been calculated. Testing the data using parameters, namely HMS = 5, HMCR = 0.9 and PAR = 0.3. The data will be tested 10 times with the number of improvisation 100, 500, 1000, 2000 and 2500.

Table 3. Test result

Replication	Iterations				
	100	500	1000	2000	5000
1	241.49	241.41	241.86	241.49	240.85
2	242.59	241.44	241.74	241.51	241.37
3	241.16	241.49	241.65	241.36	240.52
4	242.76	240.16	241.3	241.3	241.34
5	242.79	242.13	241.63	241.48	241.26
6	242.75	241.7	241.65	240.61	241.06
7	241.76	241.89	241.8	241.09	240.93
8	241.71	241.92	241.77	241.16	241.15
9	241.28	241.89	241.85	241.72	241
10	242.28	241.76	241.49	240.91	241.25
Makespan	241.16	240.16	241.3	240.61	240.52

From the test with the combination of engine and the number of cars 2 x 20, obtained results on each iteration with the value makespan vary. So obtained the best makespan value is 240.16 obtained on iteration to 500, with the order of cars 17-16-12-0-5-10-9-13-8-11-4-15-7-6-1- 18-19-2-14-3.

6. Conclusion

From the research that has been done can be concluded that, flow shop scheduling on car body painting at basecoat and clearcoat stage by using harmony search algorithm obtained minimum makepan value, that is 240.16 at iteration to 500 with sequence of car, that is 17-16-12-0- 5-10-9-13-8-11-4-15-7-6-1-18-19-2-14-3.

In the next study it is suggested to add the painting process from the beginning of the painting to the top coat (basecoat and clearcoat) process. And consider the number of machines in each painting process.

References

[1] Masruroh N 2008 Analysis of production scheduling using ampbell dudeck smith, palmer and dannenbring method at PT. Loka refractories surabaya *Journal of Industrial Engineering and Management* 158-171

[2] Ginting R 2006 Study the application of artificial immune system method in flow shop scheduling *Jurnal Sistem Teknik Industri* **7** pp 5-21

[3] Lamoudan T, Khoukhi F E, & Alaoui A E H 2011 Flow shop scheduling problem with transportation times, two-robots and input/output with limited capacity *International Journal of Intelligent Computing Research* **2** pp 244-253

[4] Soukhal A, Oulamara A, & Martineau P 2005 Complexity of flow shop scheduling problems with transportation constraints *European Journal of Operational Research* **161** pp 32-41

[5] Boukef H, Benrejeb M, & Borne P 2007 A proposed genetic algorithm coding for flow-shop scheduling problems *International Journal of Computers, Communications and Control* **2** pp 229-240

[6] Aulia I, Nababan E B, & Muchtar M A 2012 Application of harmony search algorithm in flow shop scheduling problem *Jurnal Dunia Teknologi Informasi* **1** pp 1-7

[7] Zong Woo Geem (ed.) 2009 *Music-Inspired Harmony Search Algorithm Theory and Applications* Springer: New York

[8] Chakraborty P, Roy G G, Das S, & Jain D 2009 An improved harmony serach algorithm with differential mutation operator *Fundamenta Informaticae* **95** pp 1-26

IOP Conf. Series: Materials Science and Engineering **308** (2018) 012045 doi:10.1088/1757-899X/308/1/012045

The effectiveness of stone ash and volcanic ash of mount Sinabung as a filler on the initial strength of self-compacting concrete

R Karolina*, **W Muhammad, M D S M Saragih, T Mustaqa**

Civil Engineering Department, Universitas Sumatera Utara (USU), Jl. Perpustakaan, Kampus USU Medan 20155 INDONESIA

*E-mail: rachmie_caroline@yahoo.co.id

Abstract. Self Compacting Concrete is a concrete variant that has a high degree of workability and also has great initial strength, but low water cement factor. It is also self-flowable that can be molded on formwork with a very little or no compacted use of compactors. This concrete, using a variety of aggregate sizes, aggregate portions and superplasticizer admixture to achieve a special viscosity that allows it to flow on its own without the aid of a compactor. Lightweight concrete brick is a type of brick made from cement, sand, water, and developers. Lightweight concrete bricks are divided into 2 based on the developed materials used are AAC (Autoclave Aerated Concrete) using aluminum paste and CLC (Cellular Lightweight Concrete) that use Foaming Agent from BASF as a developer material. In this experiment, the lightweight bricks that will be made are CLC type which uses Foaming Agent as the developer material by mixing the Ash Stone produced by Stone Crusher machine which has the density of 2666 kg / m³ as Partial Pair Substitution. In this study the variation of Ash Stone used is 10%, 15%, and 20% of the planned amount of sand. After doing the tasting the result is obtained for 10% variation. Compressive Strength and Absorption Increase will decrease by 25.07% and 39.005% and Variation of 15% compressive strength will decrease by 65,8% and decrease of absorbtion equal to 17,441% and variation of 20% compressive strength will decreased by 67,4 and absorption increase equal to 17,956%.

1. Introduction

1.1 Background

Present day construction is progressing very rapidly. The increase of development takes place in various fields, such as the construction of buildings, bridges, towers, and so on. Material in the selection of construction depends on several factors such as the relatively cheap price, has good strength and quality, raw materials constituent are easy to find and durable.

Due to the high number of development in the field of construction, the need for construction materials is also increased. One of the most used construction materials in construction is concrete. According to SNI-032847-2002, the concrete notion is a mixture of portland cement or other hydraulic cement, fine aggregate, coarse aggregate, and water, with or without additive that form a solid mass.

Since 1983 in Japan, the problems about the durability of concrete. To get a durable concrete, good quality control is required with the foundry that the experts do. The concrete problem is the need for adequate solidification to produce solid concrete. Air cavities are often trapped inside the concrete thus

IOP Conf. Series: Materials Science and Engineering **308** (2018) 012045 doi:10.1088/1757-899X/308/1/012045

affecting strength and endurance. with a self-compacted concrete mix, it can reduce the work of the experts and obtain a highquality concrete.

1.2 Problem Formulation

a) What's the definition of Self Compacting Concrete?
b) How's the *workability* of SCC (*Self Compacting Concrete)* with the substitution of the volcanic ash of mount Sinabung and stone ash?
c) How much is the compressive strength of SCC (Self Compacting Concrete) by using volcanic ash of mount Sinabung and stone ash?
d) How is the price comparison of SCC (Self Compacting Concrete) using volcanic ash of mount Sinabung and stone ash with normal SCC (Self Compacting Concrete)?

1.3 Purposes and Benefits

Purposes of this paper are:

1. To obtain a concrete mixture of SCC (Self Compacting Concrete) with the substitution of volcanic ash of mount Sinabung and stone ash.
2. Knowing the comparative price of normal SCC and SCC with stone ash substitution, and volcanic ash of Mount Sinabung.

2. Method

The method used in this study is an experimental study conducted at the Concrete Laboratory Faculty of Engineering Department of Civil Engineering University of Sumatra Utara. In general, the sequence of the study phase includes: a. Provision of porous concrete materials, b. Examination of materials, c. Mix Design, d. Creation of test samples with the addition of superplasticizer, e. Examination of slump flow value, f. Treatment of test samples (Curing), g. 24 hours of compressive strength test.

2.1 Mix Design

Calculation of the mix design can be seen in the attachment fully. From the calculation results of the mix design, 1m³ proportion of the concrete mixture obtained are as follows:

Table 1. Each variation of the concrete mixture proportion per m^3

Annotation	Cement (kg)	Sand (kg)	Water (kg)	Grevel 1-2 (kg)	Grevel 1-1 (kg)	Stone Ash (kg)	Volcanic Ash (kg)	Siperplasticzer (kg)
Final Innovation	510.66	715.28	155.00	291.28	416.107	416.107	71.52	6127.01

Figure 1 : The Materials

IOP Conf. Series: Materials Science and Engineering **308** (2018) 012045 doi:10.1088/1757-899X/308/1/012045

2.2 Concrete Compressive Strength Test

The test is done on 24 hours old concrete for each concrete day variation as many as 3 pieces. The day before the test according to the age plan, the concrete cylinder is removed from the soaking tub. Before a compressive strength test is performed, the test object is weighed. The concrete compressive strength test is done by using electrical compress machine with 2000 KN capacity.

3. Result and discussion

3.1 Compressive Strength of Cylinder Concrete

The concrete compressive strength test is done at 1 days old, test result can be seen in following graph:

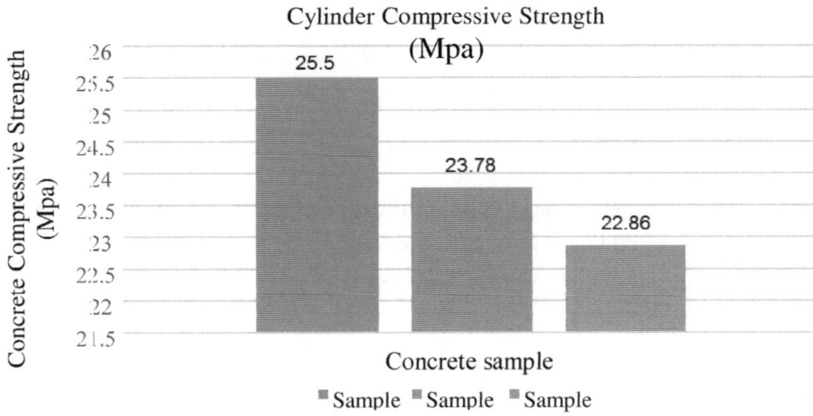

Figure 2: Concrete Compressive Strength

The highest compressive strength is 25.50 MPa.

3.2 Test Documentation

(a) (b) (c)

Figure 3: Test Documentation (a) Sample I (b) Sample II (c) Sample III

3.3 Analysis of the Impacts Inflicted by Concrete Innovation

SCC Concrete is significantly gives positive impacts to the environment by reducing energy consumption and reducing noise pollution generated by the vibrator engine. While the impact on the social is making the job easier and does not require a lot of reliable human resources in in the field processing. The last isthe impact of SCC concrete on the economic side is to directly squeeze the cost of the project on the vibratory and concrete pump processes, due to the nature of the Filling Ability of

IOP Conf. Series: Materials Science and Engineering **308** (2018) 012045 doi:10.1088/1757-899X/308/1/012045

SCC concrete that flows by itself throughout the mold surface and solidifies itself, thus reducing the number of workers.

The benefits of added ingredients on SCC Concrete Innovation, one of them is the use of volcanic ash of mount Sinabung:

- Provide a very good impact on the environment because it utilizes waste into something useful and reduce
  environmental pollution caused by the spread of volcanic ash of Mount Sinabung.
- While the impact on Social and Economy is able to involve the people arouand Sinabung mountain to build
  the spirit of gotong royong and of course to support the income of the people because the volcanic ash of
  mount Sinabung can be utilized as a cement substitution in concrete mixture.

4. Conclusion
From the results of research, analysis, and discussion that has been implemented, it can be concluded that:

1. This research is using Volcanic Ash of Mount Sinabung as a cement substitution of 15%, stone ash as a substitution of coarse aggregate equal to 15%, and superplastisizer MasteGlenium SKY 8614 to improve workability of concrete quality with dose of 1,2%.
2. 1 day old of compressive strength of 25.50 MPa is obtained

5. Suggestion
After seeing the results of the test and being aware of the possible deficiencies in this SCC concrete test, the author can provide suggestions as followst:

1. Need to do an additional test to complete the requirements and characteristics of SCC concrete including:
   - V funnel test = Filling Ability
   - V funnel at 5 minute test = Segregation Test
   - J – ring test = Passing Ability
   - L Shape box = Passing Ability
2. Further research is needed on SCC concrete, especially on waste materials, so that environmental wastes such as volcanic ash of mount Sinabung can be utilized well.

6. References

[1] Edward A dan Joseph I 2005 *Basic of building construction and method* [Buku terjemahan Dasar-Dasar Konstruksi Bangunan Bahan Bahan Dan Metodenya Jilid 2] Jakarta : Erlangga

[2] Efnarc Association, Specification and Guideliness for Self Compacting Concrete, http://www.efnarc.org/efnarc/SandGforSCC.PDF, 2002

[3] Herbudiman, Bernardinus, Sofyan Ependi Siregar, Kajian Interval Rasio Air Powder Beton Self-Compacting Terkait Kinerja Kekuatan Dan Flow (009m), Institut Teknologi Nasional (Itenas) Bandung

[4] Karolina, Rahmi., dkk. 2014. The Use of Volcanic Ash of Mount Sinabung Eruption as the Substitution of Fine Aggregate in Makin Batako (Mass-Produced Brick). Journal of Civil Engineering Research 2014, 4(3A): 178 – 186

[5] Karolina, Rahmi., dkk. 2014. Optimization of the use of volcanic Ash of Mount Sinabung eruption as the Substitution for fine Aggregate. ScienceDirect

[6] Okamura, H. and Ouchi, M., 'Self-compacting concrete-development, present, and future', Proc. 1st International RILEM Symposium on Self-compacting concrete,(1999) 3-14, RILEM Publication S.A.R.L.

[7] Okamura, H., Self Compacting High-Performance Concrete, Ferguson Lecture, New Orleans, 1996.

[8] Daniel W, Mungok C D, Supriyadi A Studi Perancangan Self Compacting Concrete (SCC) Untuk Beton Berkekuatan Tinggi (High Performance Concrete) Dengan Metode Aci, Universitas Tanjung Pura.

IOP Conf. Series: Materials Science and Engineering **308** (2018) 012046 doi:10.1088/1757-899X/308/1/012046

Genetic algorithm to solve the problems of lectures and practicums scheduling

M F Syahputra[1], R Apriani[2], Sawaluddin[3], D Abdullah[4], W Albra[5], M Heikal[6], A Abdurrahman[7] , M Khaddafi[8]

[1,2,3] Department of Information Technology, Faculty of Computer Science and Information Technology, University of Sumatera Utara, Medan, Indonesia
[4]Department of Informatics, Universitas Malikussaleh, Aceh, Indonesia
[5,6,7] Department of Management, Universitas Malikussaleh, Aceh, Indonesia
[8]Department of Accounting, Universitas Malikussaleh, Aceh, Indonesia

Email : nca_fadly@usu.ac.id | dahlan@unimal.ac.id | wahyuddin@unimal.ac.id | mohdheikal@unimal.ac.id | apridar@unimal.ac.id | khaddafi@unimal.ac.id

Abstract. Generally, the scheduling process is done manually. However, this method has a low accuracy level, along with possibilities that a scheduled process collides with another scheduled process. When doing theory class and practicum timetable scheduling process, there are numerous problems, such as lecturer teaching schedule collision, schedule collision with another schedule, practicum lesson schedules that collides with theory class, and the number of classrooms available. In this research, genetic algorithm is implemented to perform theory class and practicum timetable scheduling process. The algorithm will be used to process the data containing lists of lecturers, courses, and class rooms, obtained from information technology department at University of Sumatera Utara. The result of scheduling process using genetic algorithm is the most optimal timetable that conforms to available time slots, class rooms, courses, and lecturer schedules.

1. Introduction

Generally, the theory class and practicum class scheduling process in university is done manually. This method of scheduling has a low accuracy rate, depending on number of courses scheduled and class rooms available. Lee [1] stated that scheduling process correlates with time and resources allocation, which will ensure the entire task to be run smoothly. Problems found in timetable scheduling process correlates with the number of classes and the number of students enrolling for a course. The schedule collision can occur when there are students re-enroll for the previous course, or enroll for the more advanced course.

A scheduling process should fulfill each requirement in order to ensure that every scheduled task can be run smoothly [2]. In this case, class timetable scheduling process should meet the schedule of lecturers and students, the availability of class rooms, and the number of courses available, to ensure that there are no “collision” between a schedule to another schedule. Therefore, a method has to be implemented to perform theory class and practicum class scheduling process.

In this research, genetic algorithm will be implemented to perform theory class and practicum class scheduling process. Genetic algorithm is a heuristic search algorithm based on natural selection process mechanism, which is known as biological evolution process [3]. The algorithm has an optimal search method, and has an ability to perform search for the best solution on numerous solutions available. The implementation of genetic algorithm in this process should result in the more optimal theory class and practicum class timetable, which fulfills the requirements of zero schedule collisions. Previously we developed application to schedule diet for diabetes mellitus patient [4], where in other area of scheduling, used on container scheduling [5] and manufacturing [6] .

2. Problem identification

Generally, the manual scheduling process on theory class and practicum class timetable in the university has a low accuracy rate, caused by possibilities of schedule collisions between number of classrooms and courses available, along with lecturers' and students' schedule. Therefore, a method has to be implemented for performing the scheduling process on theory class and practicum class timetable.

3. Methodology

The general architecture of the system developed for this research is shown by Figure 1. The processes involved in this research include initialization of population, evaluate fitness function, selection, crossover, and mutation. The best result obtained from this process will be shown as final timetable schedule.

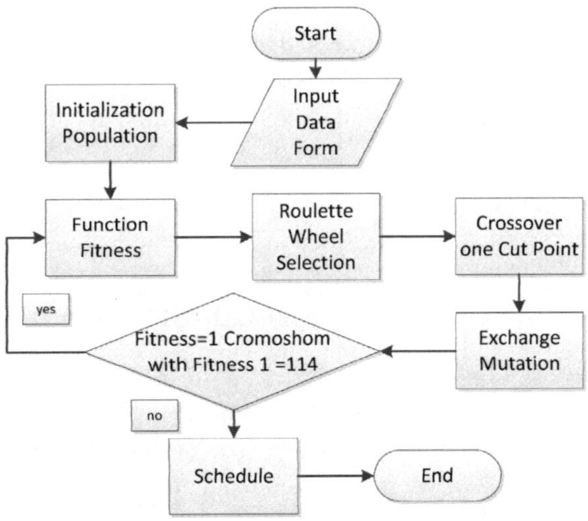

Figure 1. General architecture

When using genetic algorithm, several constraints have to be defined in order to enable the algorithm to work. In this research, the constraints are divided to two types, namely hard constraints and soft constraints. Hard constraints defined in this research are described as below:

a. Every course can only be held in a class room;
b. A lecturer can only teach for a course in one schedule;
c. A student can only attend a course for one schedule;
d. Theory class and practicum class should be placed in adjacent schedule slot;
e. The number of available class rooms has to meet the number of students attending for the course;
f. Each chromosome has to contain course name, lecturer, time slot, class name, and day.

IOP Conf. Series: Materials Science and Engineering **308** (2018) 012046 doi:10.1088/1757-899X/308/1/012046

Meanwhile, soft constraints defined in this research are described as below:
a. The lecturer can select the desired time slot for a course;
b. Each schedule has to be distributed evenly for each day;
c. Each schedule should not collide with another schedule.

3.1. Population initialization
The chromosome will be represented by $m\,x\,n$ = [114][8], where m represents the number of available courses, and n represents the number of available chromosomes, which combines course, lecturer, credit value, class room, day, and time slot into several pairs. The example of chromosome representation is shown by Table 1.

Table 1. Chromosome representation example

Chromosome	Course ID	Lecturer ID	Class ID	Credits	Room ID	Day ID	Time ID
1	1	26	1	2	6	4	5
2	2	26	2	2	5	3	4
3	3	6	1	2	3	2	2
114	114	15	1	3	2	4	1

From the chromosome representation expressed above, the initial population can be generated by providing the number of initial population. The population will be generated randomly by combining several individual from the population.

3.2. Evaluate fitness function
Fitness function value will be calculated by using (1):

$$fitness = \frac{1}{1 - (penalty1 + penalty2 + penalty3 + penalty4)} \quad (1)$$

where the penalty value is calculated from the aspects described as below:
a. There is a certain number of lecturer with multiple schedule within a time slot in one day;
b. There are pairs of theory class and practicum class scheduled within the same time slot in one day;
c. There are some courses within a semester scheduled within the same time slot; and
d. There are time slots being assigned by multiple courses within the same time slot.

The fitness function value can be calculated by checking each chromosome in an individual for penalty value, and dividing it with the number of chromosomes, which is 114 as described in population initialization process.

3.3. Selection
In this research, the roulette wheel selection method is implemented to perform selection between individuals. Each individual will be sorted based on the biggest fitness value, and the first two individual with highest fitness value will be used to perform crossover process.

3.4. Crossover
Crossover is the method to combine each gene in multiple chromosomes to generate a new individual. The process is done by switching each similar gene in chromosomes.

3.5. Mutation
Mutation is the next step to generate new individual after crossover. In this research, exchange mutation method is implemented to perform the mutation process. By using exchange mutation method, the value of several genes in each individual will be exchanged in order to generate a new individual.

4. Experiments and results
The system built for this research consists of three menus, namely File, which contains forms needed for scheduling process, namely: courses, lecturers and lab assistants, time slots, class rooms, and class groups; "Penjadwalan" (Scheduling), which contains parameter form to start scheduling process by using genetic algorithm; and "Informasi", which contains information about the built system. The appearance of home menu in this system is shown by Figure 2.

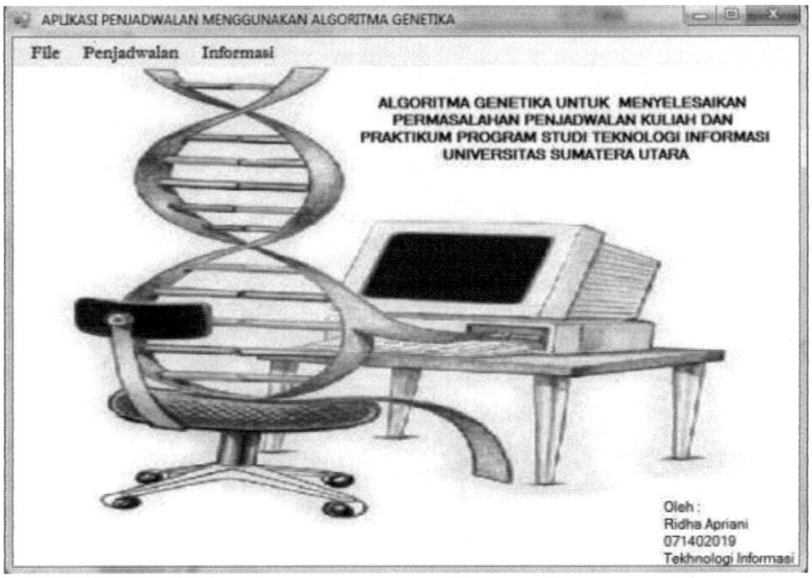

Figure 2. The appearance of home menu

After scheduling process is done, the result is shown by timetable format, as shown by Figure 3. The schedule timetable can be converted to Excel spreadsheet format. Each schedule is represented by course code, along with day and time slot.

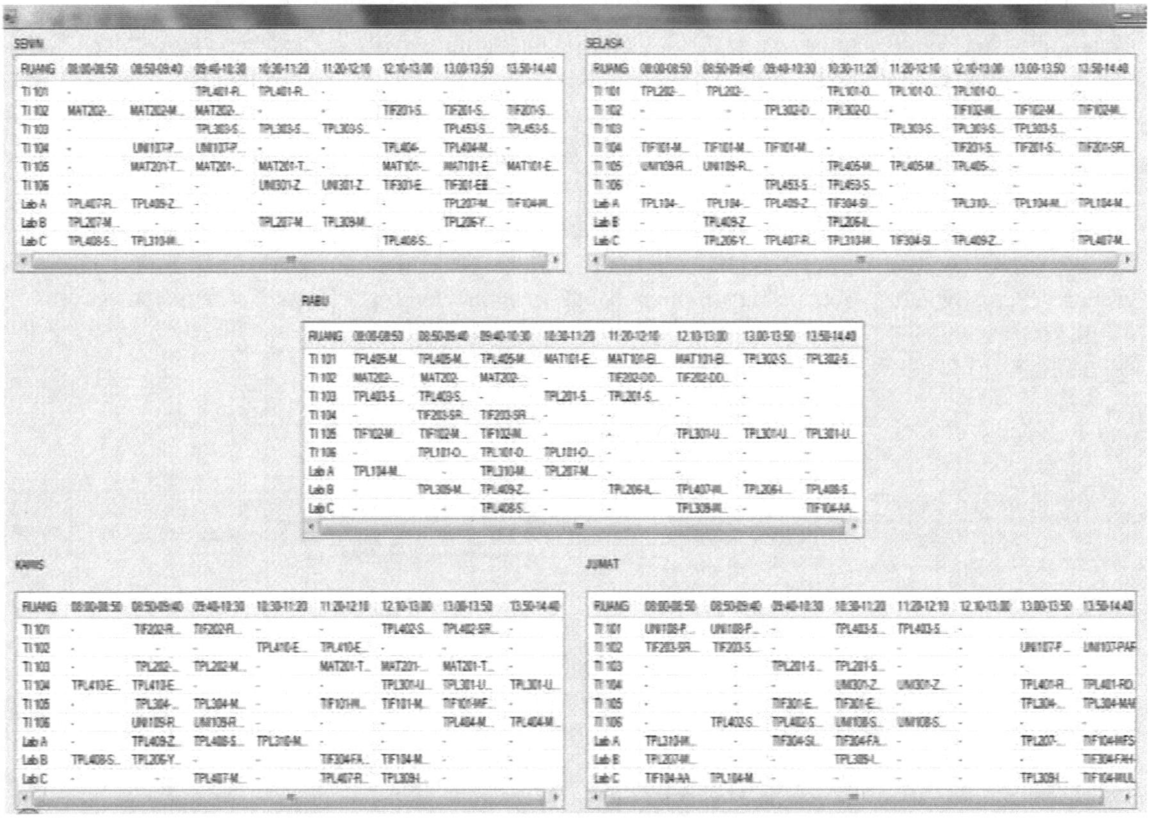

Figure 3. Result view

The result of experiment using genetic algorithm is shown by Table 2. The final fitness value of each experiment has to reach the value of 1 for obtaining best optimal result. The experiment result shows that the experiment utilizes fewer value of final population, can result in fewer final fitness value, which indicates that the result is not optimal. Meanwhile, the random function used in this research results in different result obtained when utilizing the same amount of initial and final population, compared with different amount of initial and final population.

Table 2. Experiment result

No.	Initial population	Final population	Max. Fitness	Min. Fitness	Average Fitness	Final fitness	Generation result	Iteration
1	100	500	1	0.10361	0.26806	1	458	228
2	100	500	1	0.10546	0.29074	1	514	256
3	20	200	1	0.22880	0.61459	1	866	432
4	50	200	1	0.22786	0.48364	1	272	135
5	50	200	1	0.21430	0.58106	1	548	273
6	20	120	1	0.26306	0.90015	0.8789	2000	999
7	50	120	1	0.27653	0.90349	0.8663	2002	1000
8	100	120	1	0.54776	0.92468	0.8864	2000	999
9	50	300	1	0.14537	0.36282	1	380	188
10	100	300	1	0.19552	0.38352	1	250	124
11	50	100	1	0.37414	0.96623	0.9764	2002	1000
12	75	200	1	0.25072	0.50389	1	234	116

IOP Conf. Series: Materials Science and Engineering **308** (2018) 012046 doi:10.1088/1757-899X/308/1/012046

5. Conclusion

In this research, genetic algorithm is used for arranging timetable schedule of theory classes and practicum classes in university. The result of timetable arrangement using genetic algorithm correlates on random function, which will affect the best result obtained from the experiment. The number of final population utilized in each experiment correlates to the final fitness value, which affects the result of the schedule generated by the algorithm.

For the future research, numerous variables can be used in the scheduling process. Also, parameter addition to the scheduling process is recommended to provide a better result.

References

[1] Lee, H.S.C. 2000. Timetabling Highly Constrained System Via Genetic Algorithm. Department of Mathematics. College of Science,University of the Philippines. Diliman. Quezon City

[2] M. Kumar, M. Husian, N. Upreti, and D. Gupta, "Genetic algorithm: Review and application," International Journal of Information Technology and Knowledge Management, vol. 2, pp. 451-454, 2010.

[3] Kusumadewi, Sri. 2003. Artificial Intelligence (Teknik dan Aplikasinya). Yogyakarta: Graha Ilmu.

[4] M F Syahputra, V Felicia, R F Rahmat and R Budiarto 2017. Scheduling Diet for Diabetes Mellitus Patients using Genetic Algorithm. Journal of Physics: Conference Series, Volume 801, Number 1.

[5] B. Skinner, S. Yuan, S. Huang, D. Liu, B. Cai, G. Dissanayake, ... and D. Pagac, "Optimisation for job scheduling at automated container terminals using genetic algorithm," *Computers & Industrial Engineering*, vol. 64, pp. 511-523, 2013.

[6] M. G. Filho, C. F. Barco, and R. F. T. Neto, "Using Genetic Algorithms to solve scheduling problems on flexible manufacturing systems (FMS): a literature survey, classification and analysis," *Flexible Services and Manufacturing Journal*, vol. 26, pp. 408-431, 2014.

[7] Y. Xu, K. Li, J. Hu, and K. Li, "A genetic algorithm for task scheduling on heterogeneous computing systems using multiple priority queues," *Information Sciences*, vol. 270, pp. 255-287, 2014.

IOP Conf. Series: Materials Science and Engineering **308** (2018) 012047 doi:10.1088/1757-899X/308/1/012047

Development of secondary chamber for tar cracking–improvement of wood pyrolysis performance in pre-vacuum chamber

Siahaan S[*1], Homma H[2,3], Homma H[3]
[1]Department of Mechanical Engineering, University of Sumatera Utara, Indonesia. Jl. Almamater, Medan, Indonesia.
[2]National Institute of Technology, Matsue College, Matsue, Japan.
[3]Visiting Professor, Department of Mechanical Engineering, University of Sumatera Utara, Jl. Almamater, Medan, Indonesia.

[*]E-mail: sihar_mechanical@yahoo.co.id

Abstract. Energy crisis and global warming, in other words, climate change are critical topics discussed in various parts of the world. Global warming primarily result from too much emission of carbon dioxide (CO2) in the atmosphere. To mitigate global warming, or climate change and improve electrification in rural areas, wood pyrolysis technology is developed in a laboratory scale, of which gases are directly applicable to the gas engine generator. Our laboratory has developed a prototype of wood pyrolysis plant with a pre-vacuum chamber. However, tar yield was around 40 wt% of feedstock. This research aims to reduce tar yield by secondary tar cracking. For the secondary tar cracking, a secondary pre-vacuum chamber is installed after primary pre-vacuum chamber. Gases generated in the primary pre-vacuum chamber are lead into the secondary chamber that is heated up to 1000 K. This paper reports performance of the secondary chamber for secondary tar cracking in homogeneous mode and heterogeneous mode with char.

1. Introduction

Today the energy crisis and global warming which lead to climate change are hot topics discussed in various parts of the world. The energy crisis has other two aspects, limited energy resources, and secure access to energy resouces. Global warming is a phenomenon of the increase in the average temperature of the atmosphere, ocean, and land. Temperature of the earth is getting warmer at this time. Average global temperatures have risen by more than 1 degree Fahrenheit over the last century, with average warming of as much as 4 degrees Fahrenheit in some regions [1]. The largest contributing source of greenhouse gas is burning of fossil fuels emitting carbon dioxide [2]. Since the Industrial Revolution, annual CO_2 emissions from fuel combustion dramatically increased from near zero to almost 32 $GtCO_2$ in 2012 [3]. On the other hand, recently biomass is increasingly receiving large attention, because the biomass is one of renewable energy resouces, and significant reduction of greenhouse gas emission is expected [4]. Primary energy consumption in the world will increase from 13,371 million tonnes of oil equivalent (Mtoe) in 2012 to 19,276 Mtoe in 2040 [5]. The energy crisis in South East Asia where the economic development and growth are stably high. According to South East Asia Outlook 2013 that 22% of ASEAN population 600 million, cannot access to electricity [6].

Total Indonesia's fossil energy resources was equivalen to 137000 Million tonnes of oil equivalent (Mtoe) in 2012. Renewable energy was accounted for 155.3 Gwe in 2012 [7]. To reduce CO_2 emission into atmosphere, and mitigate global energy crisis, many trials, and research and developments are carried out for alternative and renewable energy resources such as wind power, solar energy, micro-hydraulic power, geothermal energy and biomass energy. Biomass energy has been estimated that 20 kW power plant could replace 125 t CO_2 per year as compared to electricity generating unit through grid [8].

Pyrolysis is a thermochemical process during which biomass feedstocks are broken down using heat in the absence of oxygen. The pyrolysis process may be represented by a generic reaction such as [9]:

$$C_nH_mO_p \text{ (Biomass)} \xrightarrow{\text{heat}} \sum_{liquid} C_xH_yO_z + \sum_{gas} C_aH_bO_c + H_2O + C \text{ (char)}$$

The condensable gas may break down further into noncondensable gases (CO, CO_2, H_2, and CH_4), liquid, and char [9]. Jieheng Guo [10] carried out pyrolysis experiments at the heating rates of 20 $K.min^{-1}$, 40 $K.min^{-1}$ and 60 $K.min^{-1}$, respectively. Kinetics model wood pyrolysis as:

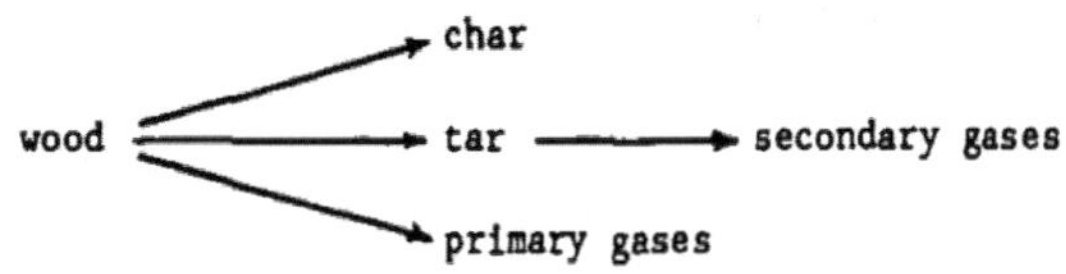

Figure 1. Kinetics Model wood pyrolysis [11]

In the previous experiments [12], the maximum temperature in the pre-vacuum chamber was 600 °C at the chamber bottom and 350 °C at the top of the chamber.

The tar yield from the pilot plant is significantly large, aroud 40 wt% of feedstok. This tar was produced under 600 °C, which may be primary tar according to the other researcher's experiment [13]. To increase the plant efficiency, the primary tar must be decomposed into secondary one.

Tar is a complex mixture of organics that is produced during thermochemical biomass conversion processes. The chemical formula of tar In the previous numerical analysis [14] was hypothetically defined as $C_9H_{18}O_4$ for convenience sake. Phuphuakrat Thana, Namioka Tomoaki, Yoshikawa Kunio [15] carried out thermal tar decomposition at temperature of 800°C. The pyrolysis tar was effetively decomposed, down to approximately 22% of the inlet level at the exit of the reformer. To improve the efficiency of tar reduction, either steam or air was introduced into the reformer, as a reforming agent [15]. Paethanom, et al [16] indicated that at the temperature 800 °C char had the optimum tar removal performance in which tar concentration could be reduced from 36.9 to 4.6 g/m3, which corresponds to 87.5% tar removal, whereas 600 and 1000 °C char possessed the performances of 82.9% and 81.6% tar removal, respectively. The adsorption capacity of char is related to the specific surface area and the fixed carbon content of the char. When temperature reaches about 850°C, secondary tar gets its peak yield and starts to cracking afterwards; however, primary tar is almost vanished [17]. M. L. Boroson [18] indicates that at temperature 600 C, primary tar yield decreases from 52.6 wt% of wood to 36.6 wt% of wood and at temperature 700 C, the decrease is significant, namely from 52.6% to 16.6%. Linear interpolation suggests that primary tar yield decreases by around 50% at temperature 650 C. P. O. Morf indicates that at 650 C, primary tar yield decreases from 31 wt% to 23 wt%. [19]. T. A. Milne, R. J. Evans and N. Abatzoglou research that at temperature 650 C, decrease of around 40 % in primary tar yield was obtained [20].

This research aims to develop simple and practical pyrolysis technology with enhanced plant performance, namely higher gas yields and lower tar, and CO_2 yields from previous research (Homma, et al) [12].

IOP Conf. Series: Materials Science and Engineering **308** (2018) 012047 doi:10.1088/1757-899X/308/1/012047

2. Experiment

2.1. Wood Feedstock and Apparatus

In this research, rubber wood is used for pyrolysis feedstock. In North Sumatera region, there are so many rubber plantations [21]. In figure 2, a pyrolysis system constructed for this research is schematically illustrated. All components of the system are connected with stainless steel pipe of ½ inch in diameter, and the combustion gas inside furnace is exhausted into the chimney and flows into the secondary chamber so that the pyrolysis gases from pre-vacuum chamber can be heated up to decompose existing tar. Collecting data in this experiment is done in two ways, namely without charcoal in the secondary chamber, and by inserting charcoal pyrolysis results of prevacuum chamber into the secondary chamber.

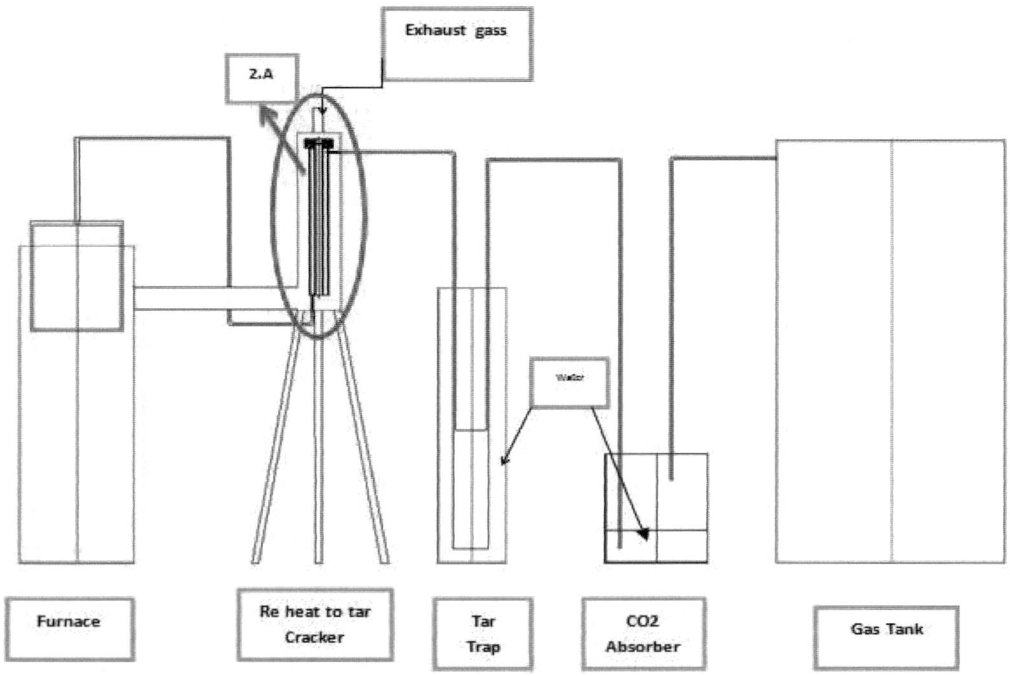

Figure 2. Schematic view of constructed pyrolysis system.

2.2. Experimental Procedures

The procedure is as follows:

a. In the pre-vacuum chamber filled with 5 kg rubber wood.
b. Water flows in the tar trap at the bottom and flows out from the top, and CO_2 absorber is not filled with water to measure total pyrolysis gaes.
c. Two K-type of thermocouples are installed properly, namely at pre-vacuum chamber bottom, and at the center of chimney just before the secondary pre-vacuum chamber.
d. The equipment is vacuumed.
e. Fuels (rubber wood pieces) in the furnace are ignited.
f. Temperature, and pressure in the pre-vacuum chamber and the secondary chamber are monitered.
g. Pyrolysis gases from the pre-vacuum chamber flow slowly into the seconday pre-vacuum chamber through stainless steel pipes after pyrolysis starts. After reaching 0.1 Mpa, the pressure is kept at 0.1 MPa by opening the valve and releasing pyrolysis gases into the tar trap, gas tank until the pyrolysis process is terminated.

IOP Conf. Series: Materials Science and Engineering **308** (2018) 012047 doi:10.1088/1757-899X/308/1/012047

h. So that the temperature in the secondary chamber can reach 800oC - 1000oC, outside wall of the secondary chamber is wrapped with isulation belt to minimize heat loss.
i. Record the temperature change every one minute during the pyrolysis process, and the pressure is maintained at 0.1 MPa until the pyrolysis process is complete.
j. After a pyrolysis experiment has been terminated, the pressure of the gas tank is measured to obtain the gas product volume. Weight of tar trapped in the tar trap is measured.

3. Experimental Result

3.1. Temperature

In the experiment, the exhaust gas temperature was measured just before inlet of the secondary chamber. The result is shown in Figure 3. The exhaust (combustion) gas was moderately heated up as a function of time. In the figure, an arrow mark indicates the moment when the secondary chamber was opened and pyrolysis gases flew out from the chamber. Before the valve opens, the temperature in the primary pre-vacuum chamber is almost same as the date obtained by the previous experiment. However, after 20 minute, the temperature measured by the current experiment is rather lower than that in the previous experiment. The reason for this phenomenon is not clear yet, but it may result from effect of the secondary chamber.

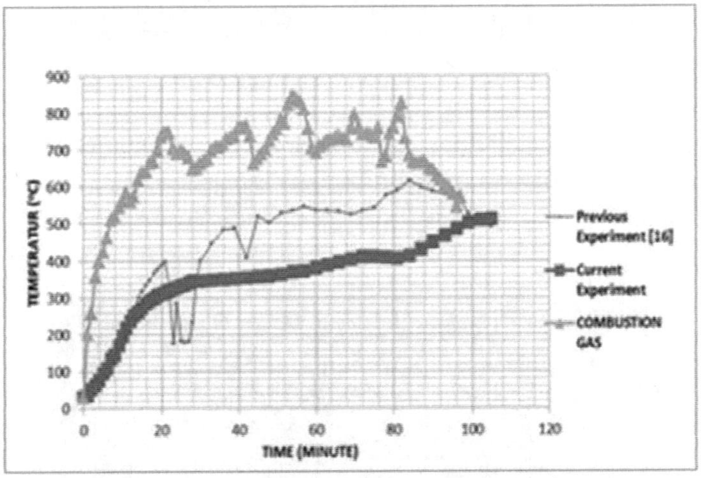

Figure 3. Temperature measure by experiment

The exhaust gas temperature reached 727 C (1000 K) when the valve was opened and then fluctuates around 750 C. For numerical analysis of heat and mass transfer in the secondary chamber, the exhaust gas temperature is approximated by two kinked line. After 20 minutes, the exhaust gas temperature was kept constant at 1000 K.

3.2 Heat and Mass Transfer Analysis Based On Measured Exhaust Gas Temperature

To examine performance of secondary tar cracking in the secondary chamber, heat and mass transfer analysis was carried out by use of Fluent based on measured exhaust gas temperature.

Numerical analysis was carried out for two heating scenarios as follows:

1. Exhaust gas temperature is constant, 1000 K
2. Exhaust gas temperature increases as a bi-linear function of time simulating measured results at the experiment.

In Scenario I, exhaust gas heated up to 1000 K flows into the chimneys inside and outside the secondary chamber, and the result is shown in Figure 4.

IOP Conf. Series: Materials Science and Engineering **308** (2018) 012047 doi:10.1088/1757-899X/308/1/012047

From the identification results shown in Figure 8, we present 5 samples from 15 testing images that we conduct

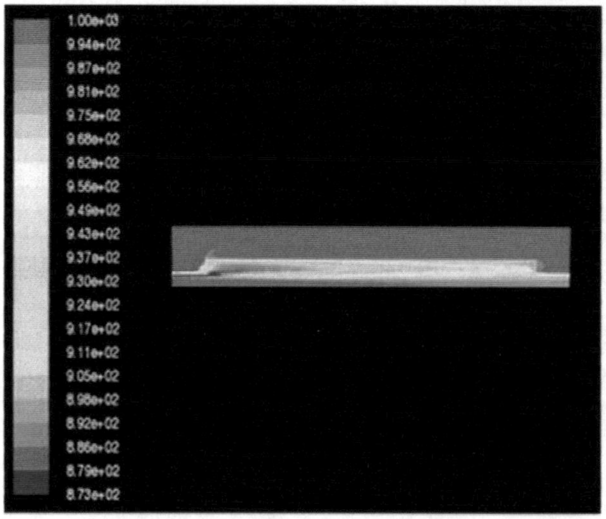

Figure 4. Temperature contours at 40 minutes after heating

In Scenario 2, exhaust gas heated up to the temperature as a function of time as measured in the experiment flows into the secondary chamber. At the same time, pyrolysis gases heated at 500 K flow in the secondary chamber according to the pressure measured in the experiment (Figure 5).

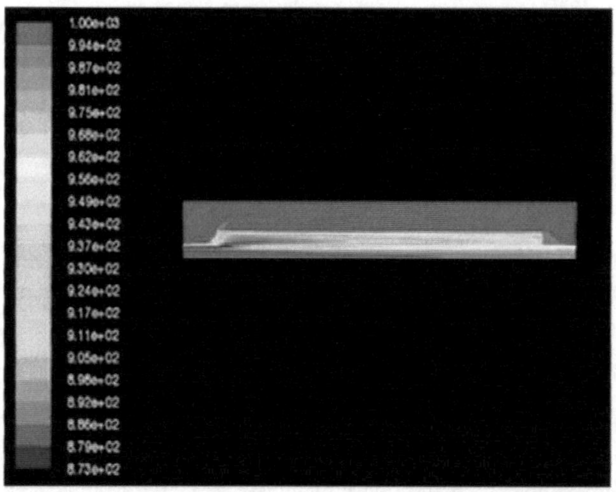

Figure 5. Temperature contours at 40 minutes after heating

As shown in figures 4 and 5, temperature distribution in the secondary chamber is almost the same for both Scenarios. At the upper part of the secondary chamber, pyrolysis gas temperature is above 950 K (677 C). This result can suggest that secondary tar cracking may take place in the secondary chamber.

IOP Conf. Series: Materials Science and Engineering **308** (2018) 012047 doi:10.1088/1757-899X/308/1/012047

3.3 Pyrolysis Yields
After the experiment is terminated, pyrolysis yields of tar, char, and gases were measured and shows in Table 1 and 2.

Table 1. Previous pyrolysis yield (mass) for 5 kg rubber wood and pressure 0.1 MPa

PREVIOUS DATA	Experiment			Average	%
	1	**2**	**3**		
Tar (kg)	2.09	1.73	2.07	1.96	39.29
Char (kg)	1.55	1.25	1.5	1.43	28.67
Gas (kg)	1.36	2.02	1.43	1.60	32.07

Table 2. Current pyrolysis yield (mass) for 5 kg rubber wood and pressure 0.1 MPa

CURRENT DATA	Experiment			Average	%
	1	**2**	**3**		
Tar (kg)	2.02	1.81	1.653	1.83	36.55
Char (kg)	1.399	1.443	1.764	1.54	30.71
Gas (kg)	1.581	1.747	1.583	1.64	32.74

3.4. Discussion
The experimental results shows that the tar yield is not significantly reduced from the result obtained by the previous experiment in which the secondary chamber was not utilized to promote the secondary tar cracking. To examine performance of the secondary chamber developed by this research, heat and mass transfer analysis was carried out by use of numerical fluid dynamics code Fluent.

The numerical analysis results show that:

1. Axisymmetric heat and mass transfer analysis was carried out for simplicity, although the geometry is not completely axisymmetric, especially, exhaust gas and pyrolysis gas inlets and outlets. However, other geometry and dimensions are the same as those of the secondary chamber used in the experiment.
2. Two scenarios are considered for the numerical analysis.
   a. The first one is an ideal case. The exhaust gas temperature is kept constant 1000 K
   b. The second is a case to simulate the experiment, namely, the exhaust gas is gradually heated up to 1000 K at 20 minutes after heating.

Numerical result obtained by Scenario 2 could suggest that the secondary tar cracking may surely take place in the secondary chamber, based on the temperature of more than 650o C and the residence time of around 50 seconds.

4. Conclusion
The pyrolysis technology developed by this research is rather simple and does not needs special technical skill and knowledge. Therefore, the pyrolysis plant can run without less maintenance, and it can be easily installed in rural areas where woods waste is available.

The experimental results showed that the tar yield was not significantly reduced from the previous experimental result. However, in order to enhance efficiency and performance of this pilot plant,

IOP Conf. Series: Materials Science and Engineering **308** (2018) 012047 doi:10.1088/1757-899X/308/1/012047

there are many parameters affecting performance of this pilot plant to be clarified. The experimental result and numerical analysis of tempearature distribution in the secondary chamber show that secondary tar cracking surely takes place in the secondary chamber, but the primary tar reduction is not apparent. Futher experiments are necessary to make sure the secondary tar cracking.

Acknowledgement
The authors would like to express their deepest appreciation to Mr. Kosasih Matseh who supplied rubber wood pieces for feedstock used in this research.

References
[1] 101 Climate change *The Science and impacts* Published by the Pew Center on Global Climat Change and the Pew Center on the States
[2] Global warming, climate change, Retrieved From http://timeforchange.org/cause-and-effect-for-global-warming
[3] Boden T and Blasing T J Record High 2010 Global Carbon Dioxide Emissions from Fossil-Fuel Combustion and Cement Manufacture Posted on CDIAC Site *Retrieved From http://cdiac.ornl.gov/trends/emis/prelim_2009_2010_estimates.html*
[4] Turner L A, et al. 2011 Bioenergy Review *Committee on Climate Change UK*
[5] Executive Summary (October 2014) *Analysis of low-growth scenarios for China and India and the climate change issue.* Japan: The Institute of Energy Economics.
[6] Agency I E (September 2013) *South East Asia Energy Outlook* (Paris, France. Retrieved From www.worldenergyoutlook.org)
[7] Indonesia Energy Outlook 2013 *Study of Indonesia Energy Outlook Ministry of Energy and Mineral Resources, Republic of Indonesia, Secretary General, Center for Data and Information, Budget Year 2013,* Jakarta
[8] Lata, Choubey T A K, Dubey A K, Singh P L, Gangil S 2014 *Significance of Biomass in Reduction of Global Warming* AISECT University Journal Vol. III/Issue V
[9] Prabir B 2010 *Biomass Gasification and Pyrolysis: Practical Design and Theory*, ISBN 978-0-12- 374988-8 (alk. paper), TP339.B355
[10] Jieheng G 2004 *Pyrolysis Of Wood Powder And Gasification Of Wood Derived Char* (Eindhoven: Technische Universiteit Eindhoven)
[11] Michael L B, Jack B H, John P L, William A P *Effect of Extra-particle Secondary Reactions of Fresh Tars on Liquids Yields in Hardwood Pyrolysis* (Department of Chemical Engineering and Energy Laboratory Massachusetts Institute of Technology Cambridge, MA 02139)
[12] Hiroki H, Hiroomi H, Yusrizal, Muhammad I 2013 Wood Pyrolysis in Pre-Vacuum Chamber *Journal of Sustainable Bioenergy Systems* **3** pp 243-249
[13] Oliver M P 2001 *Secondary Reactions of Tar during Thermochemical Biomass Conversion* (A dissertation submitted to the SWISS FEDERAL INSTITUTE OF T ECHNOLOGY ZURICH for the degree of Doctor of Technical Sciences. Diss. ETH No 14341)
[14] Hiroki H, Hiroomi H , Muhammad I 2014 Numerical Analysis on Wood Pyrolysis in Pre-Vacuum Chamber *Journal of Sustainable Bioenergy Systems* **4** pp 149 -160
[15] Thana P, Tomoaki N, Kunio Y 2010 Tar removal from biomass pyrolysis gas in two-step function of decomposition and adsorption *Applied Energy* **87** pp 2203–2211
[16] Anchan P, Kunio Y 2012 Influence of Pyrolysis Temperature on Rice Husk Char Characteristics and Its Tar Adsorption Capability *Energies* **5** pp 4941- 4951

[17] Ashraf E 2012 Modeling Of Secondary Reactions Of Tar (Srt) Using AFunctional Group Model *International Journal of Mechanical Engineering and Technology (IJMET)* ***3****:3* pp 123-136
[18] Boroson M L 1987 *Secondary reactions of tars from pyrolysis of sweet gum hardwood, thesis for doctor of philosophy* (MIT)
[19] Morf P O (1070) Secondary reactions of tar during thermochemical biomass conversion, thesis for doctor of technical science (Swiss Federal Institute of Technology Zurich)
[20] Milne T A, Evans R J, Abatzoglou N 1998 *Biomass gasifier "tar": Their nature, formation, and conversion National Renewable Energy Laboratory*/TP-570-25357
[21] *Directorate General of Plantation of the Ministry of Agriculture* 2014 Increased Production, Productivity And Quality of Annual Plants: Technical Guidelines for Developing Rubber Crops

IOP Conf. Series: Materials Science and Engineering **308** (2018) 012048 doi:10.1088/1757-899X/308/1/012048

Shear rate analysis of water dynamic in the continuous stirred tank

Tulus*[1], Mardiningsih[1], Sawaluddin[1], O S Sitompul[2], A K A M Ihsan[3]

[1]Department of Mathematics, University of Sumatera Utara (USU) Jl. Bioteknologi No.1, Kampus USU Medan 20155 INDONESIA
[2]Department of Information Technology, University of Sumatera Utara (USU) Jl. Universitas No.9, Kampus USU Medan 20155 INDONESIA
[3]Department of Mechanical Engineering, National University of Malaysia (UKM) 4600 Bangi, Selangor, MALAYSIA

*E-mail: tulus@usu.ac.id or tulus_jp@yahoo.com

Abstract. Analysis of mixture in a continuous stirred tank reactor (CSTR) is an important part in some process of biogas production. This paper is a preliminary study of fluid dynamic phenomenon in a continuous stirred tank numerically. The tank is designed in the form of cylindrical tank equipped with a stirrer. In this study, it is considered that the tank is filled with water. Stirring is done with a stirring speed of 10rpm, 15rpm, 20rpm, and 25rpm. Mathematical modeling of stirred tank is derived. The model is calculated by using the finite element method that are calculated using CFD software. The result shows that the shear rate is high on the front end portion of the stirrer. The maximum shear rate tend to a stable behaviour after the stirring time of 2 second. The relation between the speed and the maximum shear rate is in the form of linear equation.

1. Introduction

Continuous stirred-tank reactors (CSTR) are widely applied in the chemical, food, and pharmaceutical engineering fields. Zhang *et. al.* [1] did 3-dimensional Computational Fluid Dynamics (CFD) simulations that carried out to portray the flow characters in the CSTR of the mini-plant using speed of 400rpm. Studies about shear rate has been performed by some researchers. Kumar [2] analysed the effect of geometric parameters on the energy dissipation and shear rate. Pérez *et. al.* [3] had studied the relationship the rotational speed and the average shear rate in a stirred vessel by using a rigorous theoretical analysis. Safrit *et. al.* [4] noted that the effects of monomer content, molecular weight, temperature, and shear rate had to be investigated in the modeling of the reaction mass viscosity.

Johnsen [5] looked at particles in different rheology to develop a methodology for measuring settling velocities in static systems. In the study the variation of shear rates it has been used to get the comparison of experimental and theoretical velocities for Power-law values. Vlaev has simulated the shear rate distribution in the flow fields of two basic impellers comprising flat and fluid-foil blades and compared with measurement data obtained by electro-diffusion. Pérez *et. al.* [3] also showed that for both Newtonian and non-Newtonian power law fluids agitated in stirred vessels, the average shear rate γ in the fluid is a function of the rotational speed N of the impeller, i.e. $\gamma = \text{constant} \cdot N$ for the case of laminar flow, and $\gamma = \text{constant} \cdot N^{3/2}$ for the case of turbulent flow

This paper will show the distribution of shear rate in the slice of domain at the level of baffles. The distributions are considered in the speed variation.

2. Shear rate problem

Shear rate is an important parameter in a surface aerator, but it is not easy to characterize. Knowledge of the shear rate is essential for the design and operation of surface aerators. The specific energy dissipation rate in a stirred tank well known to depend on the shear rate γ and the shear stress τ, as follows:

$$\frac{P}{V} = \tau\gamma \tag{1}$$

where P is the power input and V is the volume of fluid in the tank. Furthermore, for Newtonian fluids, the viscosity μ is the ratio of shear stress and shear rate, i.e.

$$\mu = \frac{\tau}{\gamma} \tag{2}$$

Therefore, the Equation (1) can be written as follows:

$$\frac{P}{V} = \mu\gamma^2 \tag{3}$$

or

$$\gamma = \sqrt{\frac{1}{\mu}\frac{P}{V}} \tag{4}$$

The Equation (4) applies to laminar, turbulent and transitional flow. For agitation under laminar flow, the power number (N_p) and the agitator Reynolds number (R_r) are related as follows:

$$N_p = \frac{C}{R_r} \tag{5}$$

where the constant C depends on the geometry of the tank and impeller.

3. Finite element analysis for fluid dynamic in stirred tank

To analyze the shear rate with dealt to the fluid dynamic in the tank, it is considered the equation of Navier-Stokes is assumed to give the dynamic as follows.

$$\rho\frac{\partial u}{\partial t} + \rho(u\cdot\nabla)u = \nabla\cdot\left[\rho i + \mu(\nabla u + (\nabla u)^T) - \frac{2}{3}\mu(\nabla\cdot u)i\right] + F \tag{6}$$

$$\frac{\partial\rho}{\partial t} + \nabla\cdot(\rho u) = 0 \tag{7}$$

The wall is assumed in the model as in the Equation (8), and the rotating wall equation as in Equations (9) and (10). The rotating domain is using the Equation (11). In the model it is also assumed the symmetry for upper of domain as in the Equations (12) - (14).

$$u = 0 \tag{8}$$

$$u = v_{wall} \tag{9}$$

$$\left.\left(\frac{\partial x}{\partial t}\right)\right|_x = v_{wall} \tag{10}$$

$$dx = dx(r_{bp}, \omega, t) \tag{11}$$

$$u \cdot n = 0 \tag{12}$$

$$K - (K \cdot n)n = 0 \tag{13}$$

$$K = [\mu(\nabla u + (\nabla u)^T)]n \tag{14}$$

The geometric model is imported from the existing from the COMSOL document [7]. By using Finite Element Method, the mesh is generated using tetrahedron as depicted in Figure 1 . A careful selection of the mesh type and number is considered before the simulation. These have a significant influence for calculation accuracy and efficiency. In 90% of the overall volume of the CSTR hexahedron mesh type is used in the meshing scheme for its higher calculation accuracy in 3-dimension multi-phase simulation.

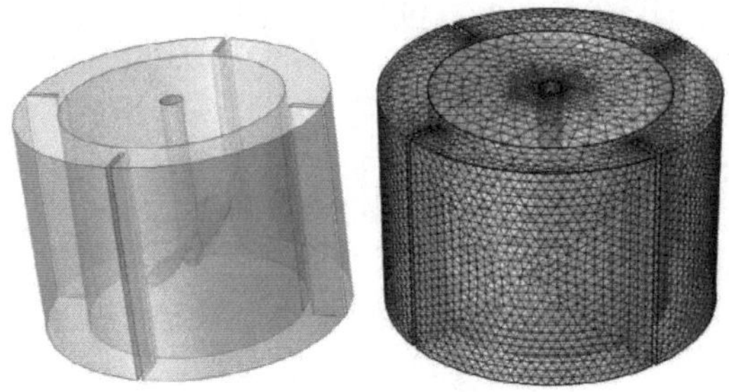

Figure 1. Model and mesh of the domain

4. Result and discussion

This section will show the distribution of shear rate in the slice of domain an the level of baffles. The distributions are considered in the speed of 10 rpm, 15 rpm, 20 rpm, and 25 rpm time by time. Figure 2 shows the distribution of shear rates in the case of speed 25 rpm.

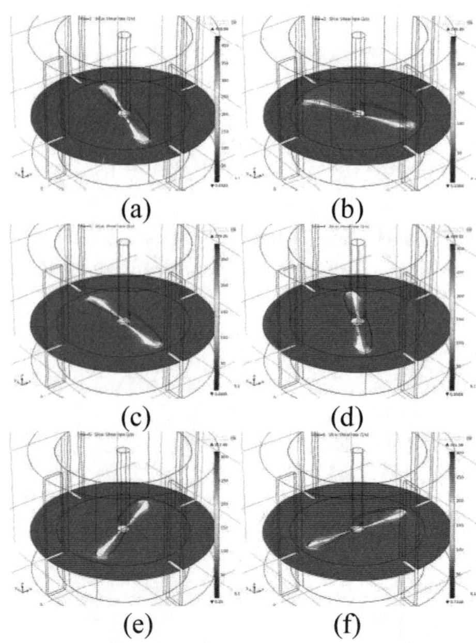

Figure 2. Shear rate distribution at the slice in the level of blade for (a) 1s, (b) 2s, (c) 3s, (d) 4s, (e) 5s, and (f) 6s in the case of speed 25 rpm.

Table 1 Maximum shear rates

Time (s)	Maximum shear rate (s^{-1})			
	25rpm	**20rpm**	**15rpm**	**10rpm**
0.25	23.904	19.163	14.397	9.5939
0.5	122.39	100.12	77.292	53.175
0.75	274.18	222.04	169.46	115.95
1	429.59	344.45	261.23	178.13
1.25	517.25	414.87	313.25	211.74
1.5	478.69	391.35	295.41	200.93
1.75	334.91	299.39	221.64	149.56
2	246.65	212.67	171.33	111.67
2.25	285.25	212.58	169.98	115.04
2.5	303.6	218.47	169.33	119.14
2.75	317.95	239.04	165.26	117.72
3	279.23	242.73	165.85	118.76
3.25	286.27	240.35	173.03	114.38
3.5	292.41	234.9	172.6	116.81
3.75	300.67	214.9	176.74	113.3
4	309.32	218.58	181.09	113.55
4,25	289.08	220.03	176.61	108.77
4.5	285.13	235.17	170.08	111.59
4.75	301.77	239.91	170.38	114.67
5	311.49	227.6	155.69	117.19
5.25	301.12	211.79	157.82	118.67
5.5	287.79	226.8	159.75	118.12
5.75	284.12	221.33	161.17	118.26
6	301.58	243.29	168.8	116.65

The distributions are depicted at the several time started by 0.25 second and followed by 0.5 second in sequence until 6 second. From the Figure 2, it can be seen that the highest shear rates are happen near the end of the baffle.

According to the maximum shear rate depend on the time (Table 1), it can be seen that the highest share rate in all type of speed are happen in the time of 1.25 second, i.e. 517.25 s^{-1}, 414.87 s^{-1}, 313.25 s^{-1}, and 211.74 s^{-1} for the speeds of 10rpm, 15rpm, 20rpm and 25rpm, respectively. The shear rate comes to a stable condition after the time 2 second.

Figure 3 shows the graphic of the maximum shear rates vs time. From the figure it can be seen that there is a relationship between the speeds and the maximum shear rates. It can also be seen that the shear rates tend to stable condition after the time of 2 second.

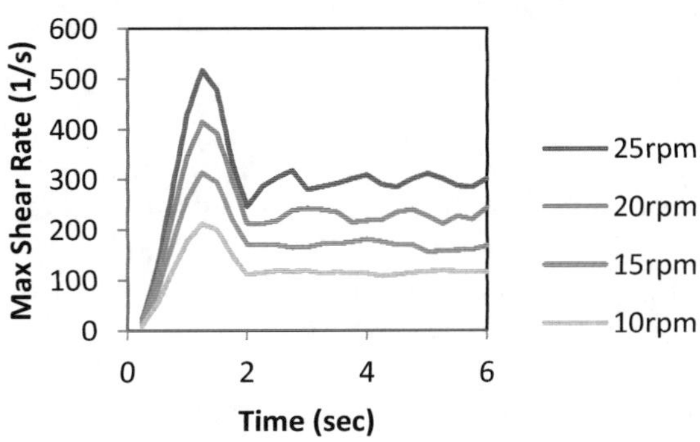

Figure 3. Maximum shear rate vs time

IOP Conf. Series: Materials Science and Engineering **308** (2018) 012048 doi:10.1088/1757-899X/308/1/012048

Table 2 The average maximum shear rate after the time of 2 second

Speed (rpm)	Maximum shear rate (s^{-1})
10	115.546
15	168.559
20	227.067
25	293.143

Table 2 shows the average maximum shear rates after the time of 2 second. From the table it can be found that the average shear rates have a relationship depend on the speed. According to the average shear rates. Figure 4 shows that the relationship form a linear relation as follows.

$$y = 11.826x + 5.875 \tag{15}$$

with $R^2 = 0.9976$. It can be concluded that the relation is linear.

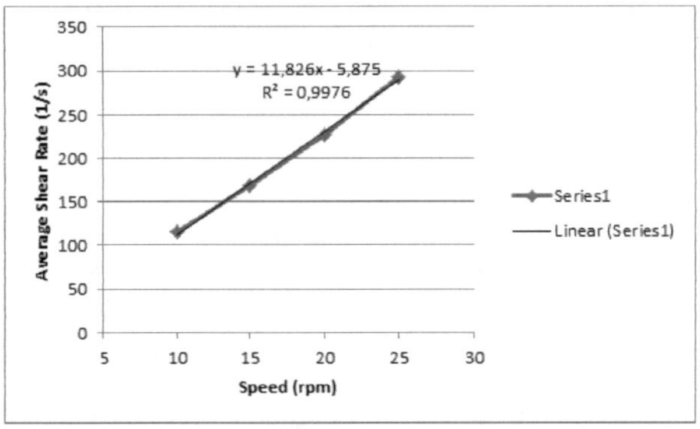

Figure 4. Graph of average shear rate and speed

5. Conclusion

The computational reasoning presented here leads to equations for correlating the average shear rate in continuous stirred tank operated with various types speed in laminar flow regimes. In Newtonian fluids, the average shear rate in laminar flow is confirmed to depend on the impeller rotational speed N.

Acknowledgements

This research was supported by the Minister of Research, Technology and Higher Education, Republic of Indonesia, by the scheme of Eexcellent Research of Universities (PUPT), 2015, by the research project 120/SP2H/PL/Dit.Litabmas/II/2015.

References

[1] Zhang S, Müller D, Arellano-Garcia H, Wozny G 2013 CFD Simulation of the Fluid Hydrodynamics in a Continuous Stirred-Tank *Chemical Engineering Transactions* **32** pp 1441-1446

[2] Kumar B 2010 Energy Dissipation and Shear Rate with Geometry of Baffled Surface Aerator *Chemical Engineering Research Bulletin* **14** pp 92-96

[3] Pérez J A S, Porcel E M R, López J L C, Sevilla J M F, Chisti Y 2006 Shear rate in stirred tank and bubble column bioreactors *Chemical Engineering Journal* **124** pp 1–5

[4] Safrit B T, Schlager G E, Li Z 2015 Modeling and Simulation of Polymerization of Lactide to Polylactic Acid and Co-Polymers of Polylactic Acid using High Viscosity Kneader Reactors

[http://www.listdryprocessing.com/fileadmin/user_upload/download/publications/modeling-and-simulation-of- polymerization-of-lactide-to-polylactic.pdf.] Accessed: June 29, 2015

[5] Johnsen M S 2014 *Particle Transport and Hole Cleaning in Wells During Drilling* Master Thesis (University of Stavanger)

[6] Vlaev S, Georgiev D, Nikov I, Elqotbi M 2007 The CFD Approach for Shear Analysis of Mixing Reactor: Verification and Examples of Use *Journal of Engineering Science and Technology* **2** pp 177–187

[7] COMSOL 2014 Laminar Flow in a Baffled Stirred Mixer [https://www.comsol.it/shared/downloads/ models/cfd_baffled_mixer.pdf] Accessed: June 29 2015

IOP Conf. Series: Materials Science and Engineering **308** (2018) 012049 doi:10.1088/1757-899X/308/1/012049

Development of high velocity gas gun with a new trigger system-numerical analysis

Z Husin[*1], **H Homma**[2]

[1]Graduate School of Department of Mechanical Engineering, University of Sumatera Utara, Medan, Indonesia
[2]Visiting Professor, Department of Mechanical Engineering, Faculty of Engineering, University of Sumatera Utara, Medan, Indonesia

[*]E-mail: zakirngn@yahoo.co.id

Abstract. In development of high performance armor vests, we need to carry out well controlled experiments using bullet speed of more than 900 m/sec. After reviewing trigger systems used for high velocity gas guns, this research intends to develop a new trigger system, which can realize precise and reproducible impact tests at impact velocity of more than 900 m/sec. A new trigger system developed here is called a projectile trap. A projectile trap is placed between a reservoir and a barrel. A projectile trap has two functions of a sealing disk and triggering. Polyamidimide is selected for the trap material and dimensions of the projectile trap are determined by numerical analysis for several levels of launching pressure to change the projectile velocity. Numerical analysis results show that projectile trap designed here can operate reasonably and stresses caused during launching operation are less than material strength. It means a projectile trap can be reused for the next shooting.

1. Introduction

According to National Institute of Justice, USA, more than 3000 police officers have saved their lives by body armor in a period of 43 years since 1970s [1]. US National Institute of Justice (NIJ) set standard of body armor. In the standard, performance of body armor and test methods to evaluate the performance are elaborated. According to the NIJ standard 0101.06 issued in 2009 [2], body armor is classified into type IIA, II, IIIA, III, and IV. Minimum performance of type IV, which is most tough in five types, is set forth as "Type IV flexible armor shall be tested in both the "as new" state and the conditioned state with .30 caliber AP bullets (U.S. Military designation M2 AP) with a specified mass of 10.8 g and a velocity of 878 m/s ± 9.1 m/s."

On the other hand, improvement in fire arms or development of new fire arms is always promoted. As result, powers of fire arms are increased further and further. As mentioned above, in order to save human lives from bullet shooting, armor vests have been constantly developed. Unfortunately, performance of armor vest does not defeat newly developed fire weapon. Therefore, US Defense Advanced Research Agency started a project of armor vest development widely calling for proposal from related inventors in 2009 [3], it is called "Armor Challenge Phase 2". This project aims to develop an armor vest satisfying the following conditions:

(1) Projectile size: 7.62mm in Bullet diameter, and 51mm in Case length, weight: 9.7g, or 11.3g
(2) Projectile speed: 915-930 m/sec
(3) Weight of vest: less than 30 kg/m^2

(4) Cost: competitive with current body armors

For impact studies, gas guns are used to accelerate a projectile through a barrel and to generate the required impact onto a target under controlled conditions. Many light gas guns have been developed and installed for impact studies in laboratories over the world. Gas gun facilities can be applied to simulate different situations, varying with the characteristics of the gun itself. To obtain highly accurate data, precise and reproducible impact tests must be carried out.

A key device for precise and reproducible impact tests is a trigger system of a light gas gun. Gas guns developed until now utilize one of the following three trigger systems;

- Mechanical system [4]
- Solenoid valve [5]
- Rapture disk [6]

A mechanical trigger system in principle uses a mechanical sliding system to open valves between a gas reservoir and a barrel. A mechanical trigger system is often utilized by air rifles and air hand guns.

In studies of impact phenomena, a gas gun installed in a laboratory also utilizes a mechanical trigger system [4]. Mechanical trigger system is typically used in low-speed gas guns. During the operation of a mechanical trigger system takes some milliseconds to fully open the valve. Therefore, a certain time lag for a projectile to be accelerated fully by the compressed gas is always necessary. This time lag is sometime fatal disadvantage to get high muzzle velocity.

A solenoid valve trigger mechanism is similar to a mechanical trigger system. The difference is a driving force to open a valve. In a mechanical trigger system, the driving force to open a valve is given by a mechanical mechanism while in a solenoid valve trigger system, a driving force is generated by an electro-magnetic mechanism. Flow volume of a solenoid valve is constantly improved, and one example of flow volume in the latest developed solenoid valve is around 1.67×10^{-2} m^3/sec (normal) for back pressure of 10 MPa. If we use this solenoid valve for a trigger system of a high speed gas gun of which a barrel diameter is 6.0 mm, fluid flow in the barrel through the solenoid valve is around 590.9 m/sec when the gas pressure is 10 MPa. It is obvious that we cannot obtain high muzzle speed of more than 900 m/sec.

A rapture disk trigger system is usually used for a highspeed gas gun. Then, a rapture disk is placed between a gas reservoir and a barrel. When pressure of a reservoir gas is gradually increased and the rapture disk is broken by the pressure, the rapture disk opens up very quickly, and large compressed gas volume flows in the barrel to accelerate a projectile. As compared with a solenoid valve, a rapture disk can pour a larger volume of gas into the barrel. As a result, a gas gun using a rapture disk can produce high muzzle velocity. Thus, a rapture disk is successfully used for high velocity gas gun, but it is based on fracture of thin plate of a metallic material. Fracture process of materials is substantially a random process or a probabilistic process. This means that rapture disk fracture involves randomnesses in several senses. The first one is fracture pattern. The second is breaking pressure. The third is gas flow resulted from the former factors. The reproducibility of impact test is therefore, of poor quality.

As mentioned above, usual trigger systems have substantial time lag for a propellant gas to reach a projectile after the trigger system is activated. This may cause disturbance in flow and loss in acceleration of projectile

Consequently, this research aims to develop a new trigger system for a high velocity gas gun, which can realize precise and reproducible impact tests at impact velocity of higher than 900 m/sec, and eventually, to develop an armor vest preventing a bullet of more than 900 m/sec from penetration. A new trigger system developed here is called a projectile trap. This new trigger system is based on a novel concept and simultaneously possesses two functions of sealing between a reservoir and a barrel and launching without time lag.

IOP Conf. Series: Materials Science and Engineering **308** (2018) 012049 doi:10.1088/1757-899X/308/1/012049

2. Developent of New Trigger System

2.1. Basic Concept of Gas Gun System

This research aims to develop a new trigger system for a single stage gas gun using helium gas as a propellant medium. A single stage gas gun has the following specifications:

(1) Spherical projectile: 6.0 mm in diameter, 0.36 g in mass, and made of silicon nitride Si_3N_4
(2) Maximum projectile speed: more than 900 m/sec
(3) Barrel: 6.05 mm in inner diameter, and 650.0 mm or more in length
(4) Pressure media: Nitrogen gas for low velocity, and Helium gas for high velocity

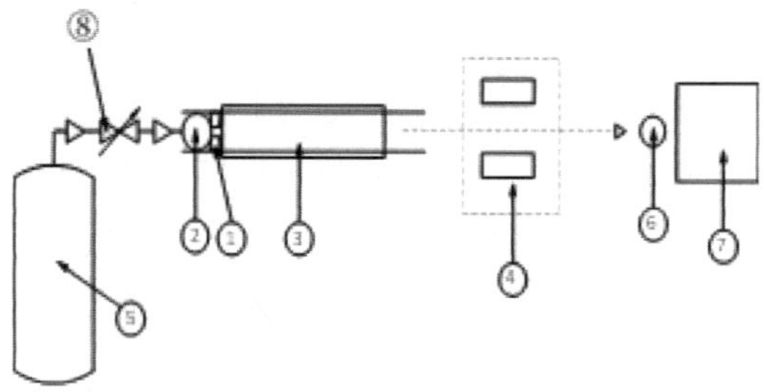

Figure 1. Scematic view of a high velocity gas gun with a new trigger system

In the usual trigger systems, a sealing or trigger position is a little far from a projectile. To eliminate this disadvantage, a new trigger system has simultaneously two functions of sealing and propelling. In Figure 1, basic concept of a high velocity gas gun with a new trigger system is schematically illustrated.

Items ①and ② are a projectile trap and a projectile. A new trigger system consists of these two components. Item ③ is a barrel, item④ is a projectile-velocity measurement device, item ⑤ is a helium gas reservoir, item ⑥ is a flying projectile and item ⑦ is a projectile catcher and item ⑧is a solenoid valve. A projectile and a projectile trap, a new trigger system seals a boundary between a room behind of a projectile and a barrel.

A compressed helium gas from the reservoir flows into a room behind a projectile. However, the new trigger system prevents gas leak into the barrel. When a compressed helium gas flows in further, the pressure of the room behind the projectile is elevated and the projectile trap is pressed and bent. The projectile can pass through the projectile trap at a certain pressure.

The projectile trap is changed in an inner hole diameter and thickness so that the pressure for the projectile to pass through the projectile trap can be adjusted for a required level. The required pressure level is determined by a projectile muzzle velocity.

The projectile triggered by this system is accelerated in the barrel and launched from an end of the barrel at a certain velocity. To measure the projectile velocity, a measurement device is constructed by use of a photodiode and two laser beams.

2.2. Materials of Projectile and Projectile Trap

In development of a new trigger system for a high velocity gas gun, material selection for projectiles and projectile trap is one of importance factors. The Torlon is a polymer of polyamidimide and high wear resistance without lubrication. Its fracture elongation is more than 10 %. This material is quite suitable for the projectile trap. Torlon can be treated as a non-linear material with a hyperelastic property. In Torlon products, Torlon 4203L is selected because of high strength and large fracture strain. For the projectile material, silicon nitride Si3N4 is selected. The mechanical properties of two materials are shown in table 1.

Table 1. Material properties of Projectile and Projectile Trap [7][8]

Material Properties	units	Projectile	Projectile Trap
Young's Modulus	GPa	290	-
Density (ρ)	Kg/m^3	3200	1420
Poison Ratio (υ)	-	0.28	0.45
Tensile Strength	MPa	525	192
Compressive Strength	MPa	5500	220
Tensile Strain at fracture	%	-	15
Compressive Strain at fracture	%	-	80

2.3. Alignment of a projectile trap and a projectile

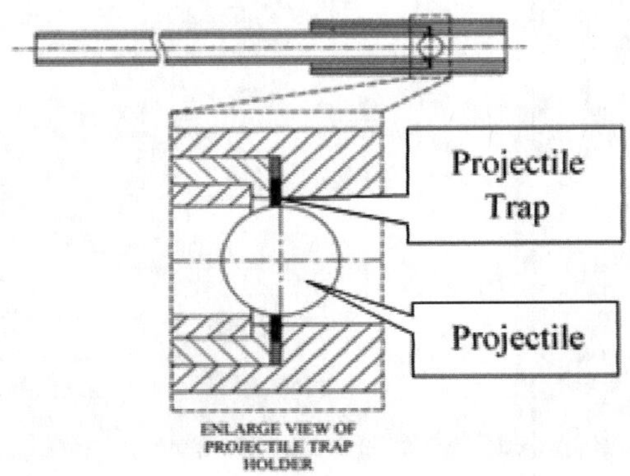

Figure 2. Schematic illustration of projectile trap

A projectile and a projectile trap must firmly contact with each other for sealing function. To assure the firm contact, a projectile trap is machined for a center hole that fits to a spherical projectile surface. In figure 2, schematic view of cross section of a projectile trap is illustrated. A diameter of the hole at the trap upper surface is the same as a spherical projectile diameter. In this research, a trap hole is automatically decided when thickness of a projectile trap is identified. The geometries and dimensions of projectile trap is determined by simulation using ANSYS Multiphysics v13.0. The projectile trap trigger mechanism is simulated by creating a model of projectile and projectile trap with contact. Then the launching pressure is applied to the model. In addition, we must know stresses generated in a projectile trap during a launching process in order to design a practical dimensions of a projectile trap and make sure that a new launching system is feasible, realistic, practical, and reliable.

IOP Conf. Series: Materials Science and Engineering **308** (2018) 012049 doi:10.1088/1757-899X/308/1/012049

3. Numerical Analysis of Projectile Trap

3.1. Model Geometry

In simulation, a situation of a spherical projectile contacting with a projectile trap is considered as shown in figure 3. The problem is axisymmetric and the model is a solid of revolution in respect of z axis. In the figure, a half of cross section is shown. A half circle of the radius R, 3.0 mm represents a projectile and a rectangle represents a projectile trap. L1 denotes trap length and is fixed at 0.5 mm, L2 denotes fixed support area, and t denotes thickness of the trap. The projectile trap geometries will be changed only in thickness to obtain appropriate launching pressure.

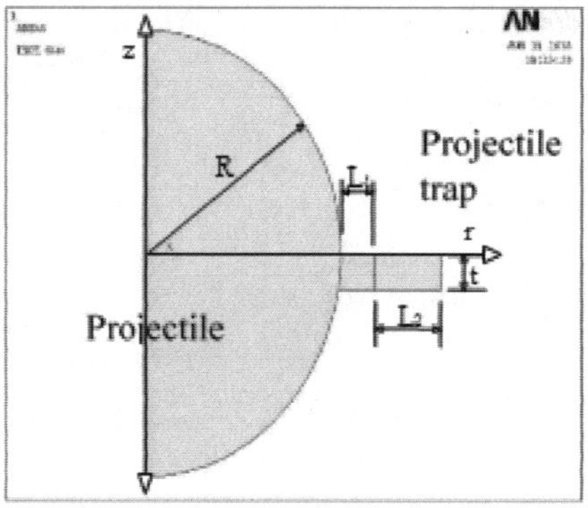

Figure 3. Projectile trap trigger system model

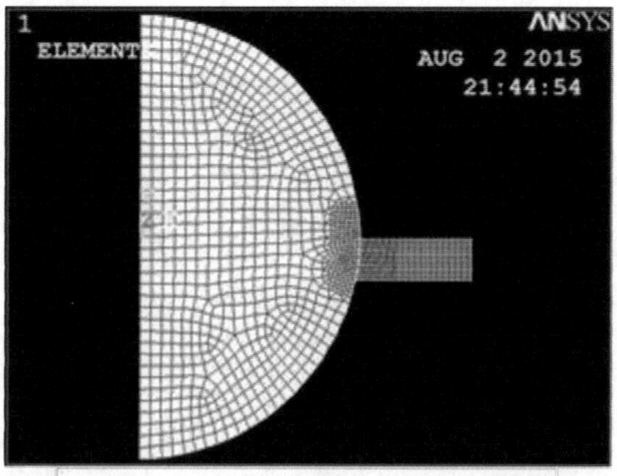

Figure 4. Mesh model of projectile and projectile trap

In figure 4, mesh model is shown. Total mesh number is 1354 and total node number is 4105. At the contact boundary, contact elements are generated.

3.2. Material Model

Projectile material is silicon nitride Si3N4. Its mechanical properties are shown in table 1. This material is regarded as linear elastic under this research conditions. On the other hand, projectile trap material is polyamidimide, Torlon, in which large non-linear property can be seen in the stress and strain relation as shown in figure 8. In this numerical analysis, this material is represented as isotropic hyper-elastic one. Mooney-Rivlin five-parameter model is utilized to express material behaviours. In figure 5, Mooney-Rivlin five-parameter model is well compared with the experimental data.

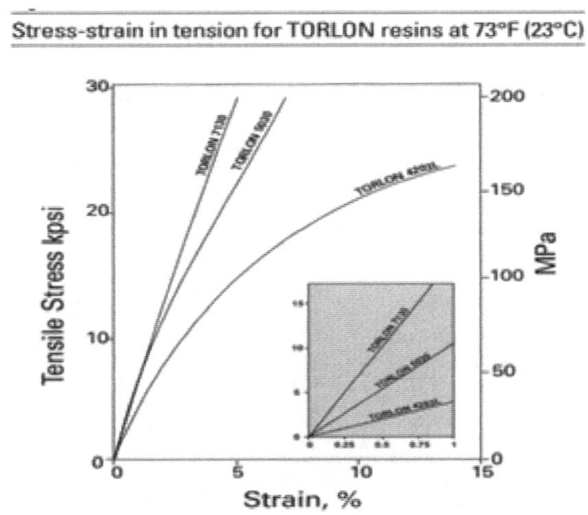

Figure 5. Stress strain graph of Torlon materials [8]

3.3 Numerical Analysis Results

3.2.1 Flat projectile trap

In the numerical analysis, pressure load is applied on the projectile surface above the projectile trap. The pressure is gradually increased until the projectile is launched from the trap. First, stress analysis is carried out for a flat projectile trap. One numerical analysis result is shown in Figure 6 for the trap thickness of 0.5 mm and the trap length L1 of 0.5 mm. For applied pressure of 13.0 MPa, von Mises equivalent stress contours are plotted in the figure.

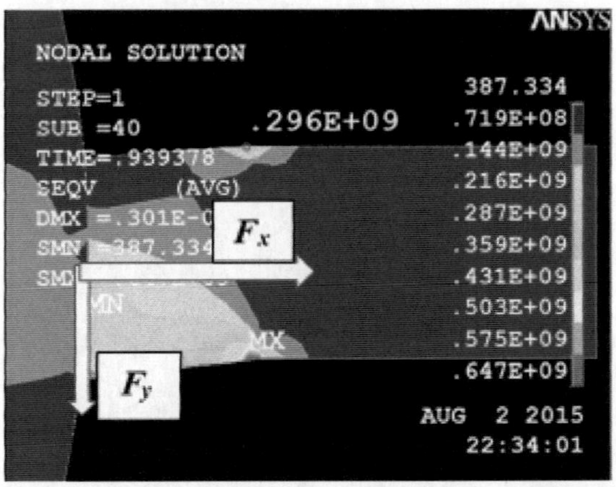

Figure 6. Contours of von Mises equivalent stress in a flat trap

The figure indicates the stress states just before launching of the projectile. The stress concentration takes place at the end of the support, right hand side of trap, where the contour color is blue. Main principal stress is compressive. However, von Mises equivalent stress is more than 647 MPa at the compression side and 296 MPa at the tension side, which is much larger than Torlon compressive and tensile strength, 172 MPa and 192 MPa, respectively.

This means that the flat projectile trap will be failed before launching of the projectile. Then, we consider applied forces to the projectile trap. One is a friction force generated by contact between the projectile and the projectile trap, denoted by Fy in the figure. Another force is compressive force generated by press of the projectile, denoted by Fx. Because Fy generates bending moment in the projectile trap, both the forces cause compression stresses at the edge of the support.

IOP Conf. Series: Materials Science and Engineering **308** (2018) 012049 doi:10.1088/1757-899X/308/1/012049

3.2.2 Slant projectile trap

To reduce the stress concentration, a new trap configuration is considered. In the flat trap, compressive stresses due to the friction force and the press force are superimposed at the supporting edge. If the trap is inclined as shown in Figure 7, bending moment due to the press force becomes clockwise direction while the friction force generates counterclockwise bending moment. Eventually, resultant moment at the supporting edge can be reduced significantly.

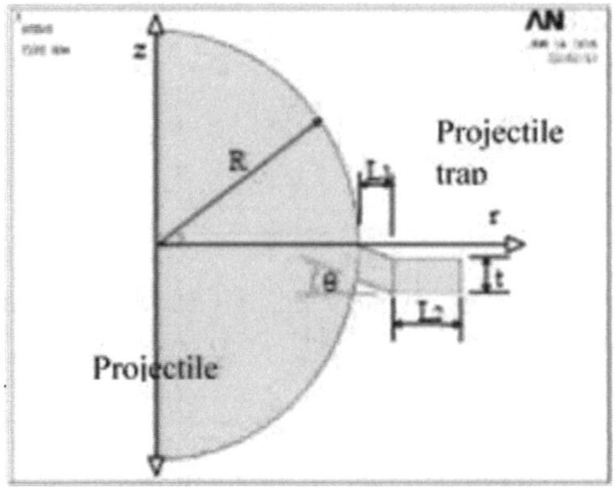

Figure 7. Slant projectile trap

In figure 8, von Mises equivalent stress is plotted as contours for applied pressure of around 13.0 MPa and in case of slant angle θ of 20 degrees. At the compression side, von Mises equivalent stress is more than 390 MPa, while at the tension side, von Mises equivalent stress is tremendously reduced to 68.8 MPa, which is much less than the tensile strength of Torlon.

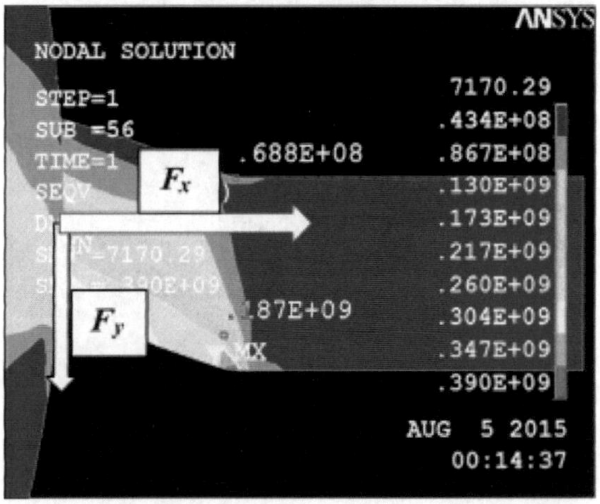

Figure 8. Von Mises equivalent stress distribution in the slant projectile trap

4. Final shape and dimension of projectile trap

Calculation is done for von Mises equivalent stress changing slant angles. Then, it is found that when the slant angle θ is 20 degrees, the offset effect is largest. When the slant angle θ is small, clockwise moment generated by the press force is small and then offset effect is small. On the other hand, when the slant angle is large, the clockwise moment becomes large and the large bending stress is caused at the supporting edge.

As shown in figure 8, von Mises equivalent stress due to the compressive stress reaches the maximum at the lower side supporting edge. The stress at the area exceeds compressive strength of Torlon. The von Mises equivalent stress is about two times of Torlon compressive strength. However, when we look at strain in the projectile trap, it should be noted that von Mises equivalent strain is not so large like the equivalent stress as shown in figure 9, and it is less than 9.0 % when the launching pressure is around 14.0 MPa.

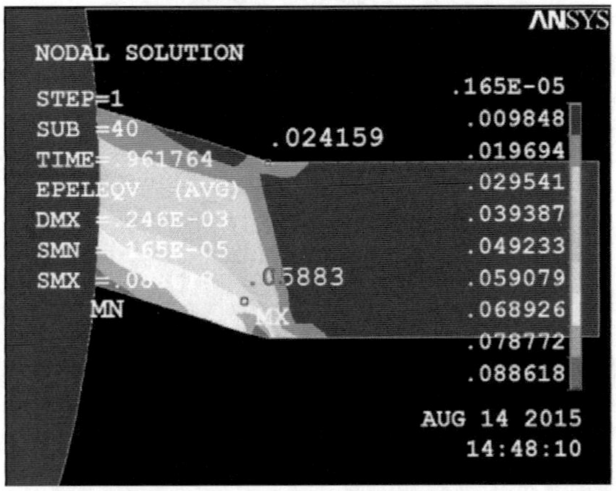

Figure 9. Von Mises equivalent strain distribution in the slant projectile trap

Compressive stress-strain curves of Torlon are not available in any brochures. Only one data of Torlon 4203 is published [9]. According to this data, Torlon 4203 has 150 MPa compressive yield strength and it can be compressed up to strain of more than 80 % without failure. Therefore, it can be speculated that because even if the von Mises equivalent stress is higher than compressive strength of the material, von Mises equivalent strain is rather small, catastrophic failure of projectile trap does not arise in this situation. Figure 10 indicates the maximum von Mises equivalent strain against the launching pressure for flat projectile trap and slant projectile trap at angle 20 degrees.

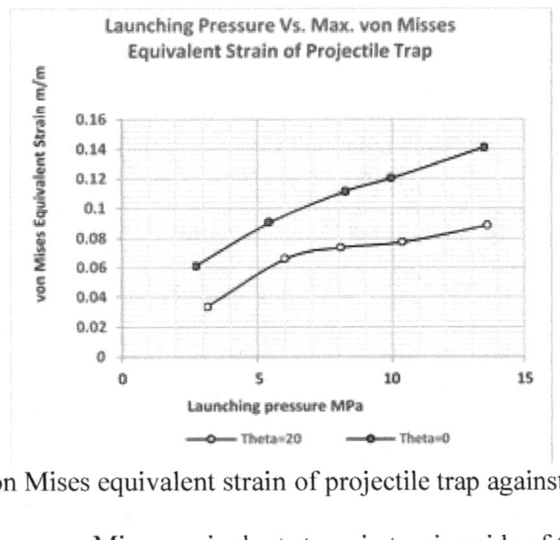

Figure 10. Max. von Mises equivalent strain of projectile trap against launching pressure

In figure 11, the maximum von Mises equivalent stress in tension side of the projectile trap is plotted against the launching pressure. According to simple calculation, the launching pressure of 13.0 MPa can accelerate a projectile more than 900 m/sec in 650 mm long barrel. In the figure, the results for slant angle of zero and 20 degrees are compared. It can be seen that the slant projectile trap is subjected too much smaller stress as compared with the flat trap.

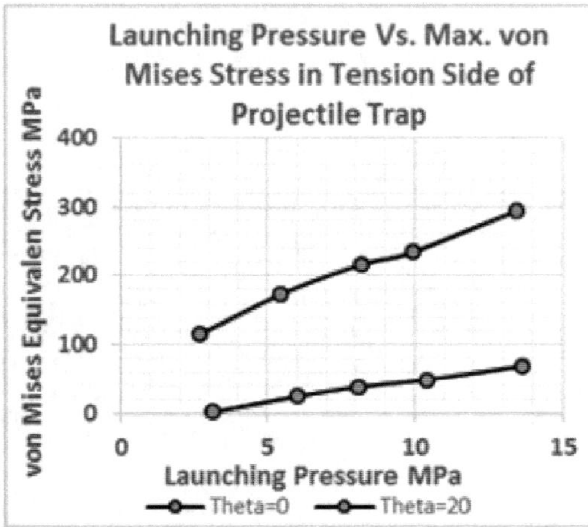

Figure 11. Max. von Mises equivalent stress at the upper supporting edge of projectile trap against launching

Finally, numerical analysis results can suggest that the shape and dimension of projectile trap shown in table 2 and figure 10 can be adopted for experiments.

Table 2. Projectile trap thickness and launching pressure associate with it

Trap Thickness t mm	Launching Pressure P MPa
0.5	13.5
0.46	10.4
0.42	8.0
0.38	6.0
0.34	3.2

Then the figure 12 displays the geometry and dimension of projectile trap for machined

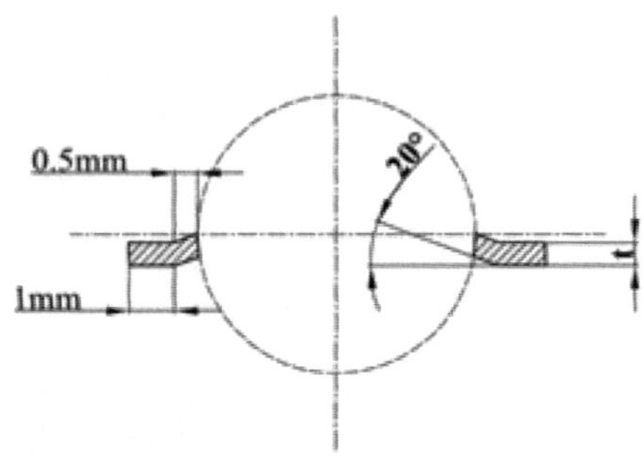

Figure 12. Geometry and dimension of projectile trap for machined

5. Conclusion

Aiming to develop a new trigger system for light gas gun, this research carried out numerical analysis for a trigger system developed by a new concept. The following conclusions are obtained:

(1) Projectile trap is a new idea for trigger system of light gas gun. This has two functions of sealing and launching.

(2) When the projectile is launched, the gas immediately accelerates a projectile. There is not time lag like a rupture disk trigger system.

(3) Numerical analysis results show that if slant projectile traps are adopted, stress in the trap is significantly reduced.

We need to carry out experiment to assure the conclusions obtained above.

References

[1] Thecnology Body Armore, Retrieved From http://www.nij.gov/topics/technology/body-armor/pages/welcome.aspx

[2] Haggy D, Morgan J, Caplan M, Stoke D 2008 *Ballistic resistance of body armor NIJ Standard-0101.06* (US Department of Justice, National Institute of Justice)

[2] DARPA_Armor Challenge, Retrieved From http://www.directedtechnologies.com/DARPA_ArmorChallenge.html

[3] Homma H, Shockey D A, Murayama Y 1983 *Response of Cracks in Structural Materials to Short Pulse Loads* (SRI International Medrol Park USA)

[4] Rohrbach Z J, Buresh T R, Madsen M J 2011 *Modeling the exit velocity of a compressed air cannon* (Department of Physics, Wabash College, Crawfordsville, Indiana USA)

[5] Solenoid valve basics, Retrieved From http://www.solenoid-valve-info.com/solenoid-valve-basics.html 16

[6] Retrieved From https://www.ceramicindustry.com/

[7] Torlon Design Manual, Solvay Advanced Polymers, L.L.C

[8] Richeton J, Ahzi S, Vecchiro K S, F.C. Jiang F C, Adharapurapu R R 2006 Influence of temperature and strain rate on the mechanical behavior of three amorphous polymers: Characterization and modeling of the compressive yield stress *International Journal of Solids and Structures* **43** pp 2318-2335

IOP Conf. Series: Materials Science and Engineering **308** (2018) 012050 doi:10.1088/1757-899X/308/1/012050

Design and fabrication hazard stakes golf course polymeric foam material empty bunch (EFB) fiber reinforced

Zulfahmi*, B Syam, B Wirjosentono
Impact and Fracture Research Center, Department of Mechanical Engineering University of Sumatera Utara (USU) Jl. Almamater, Kampus USU Medan 20155 INDONESIA, Phone/Fax.: +62-813-236899

*E-mail: fahmigam@gmail.com

Abstract. A golf course with obstacles in the forms of water obstacle and lateral water obstacle marked with the stakes which are called golf course obstacle stake in this study. This study focused on the design and fabrication of the golf course obstacle stake with a solid cylindrical geometry using EFB fiber-reinforced polimeric foam composite materials. To obtain the EFB fiber which is free from fat content and other elements, EFB is soaked in the water with 1% (of the watre total volume) NaOH. The model of the mould designed is permanent mould that can be used for the further refabrication process. The mould was designed based on resin-compound paste materials with talc powder plus E-glass fiber to make the mould strong. The composition of polimeric foam materials comprised unsaturated resin Bqtn-Ex 157 (70%), blowing agent (10%), fiber (10%), and catalyst (10%). The process of casting the polimeric foam composit materials into the mould cavity should be at vertical casting position, accurate interval time of material stirring, and periodical casting. To find out the strength value of the golf course obstacle stake product, a model was made and simulated by using the software of Ansys workbench 14.0, an impact loading was given at the height of 400 mm and 460 mm with the variation of golf ball speed (USGA standard) v = 18 m/s, v = 35 m/s, v = 66.2 m/s, v = 70 m/s, and v = 78.2 m/s. The clarification showed that the biggest dynamic explicit loading impact of Fmax = 142.5 N at the height of 460 mm with the maximum golf ball speed of 78.2 m/s did not experience the hysteresis effect and inertia effect. The largest deformation area occurred at the golf ball speed v = 66.2 mm/s, that is 18.029 mm (time: 2.5514e-004) was only concentrated around the sectional area of contact point of impact, meaning that the golf course obstacle stakes made of EFB fiber-reinforced polymeric foam materials have the geometric functional strength that are able to absorb the energy of golf ball impact.

1. Introduction

1.1 Background

In general stakes golf course is used (figure 1) is made of wood, bamboo, metal pipe or using materials pseudoplastis (survey: Arun Golf Club - Lhokseumawe). Technically the election stakes conventional materials do not meet the standards, as well as on other aspects of the design have not provided better visibility.

Figure 1. Hazard stakes golf course

To rectify this deficiency, and the rift Impact Research Center (Impact and Fracture Research Center) USU Mechanical Engineering Department, in collaboration with PTPN-III golf course designed stakes new model uses a polymeric foam material reinforced Empty Fruit Bunch(EFB).

Model design tailored to function on its placement on the golf course, in order to obtain the product look elegant, economical, aesthetic and visual appeal. To determine the stress distribution due to impact of golf balls, golf stakes made modeling using software ANSYS workbench 14.0, then simulated.

1. Formulation of the problem Limitations of this research:
2. Design and fabrication stakes golf course.
3. Model design and mold-making molds election materials.
4. Treatment EFB into fibers.
5. Composites process materials.
6. Time integration with a time of fabrication processes.
7. ANSYS workbench 14.0 simulation.

1.2 Research Objectives

The general objective of this research is the design and fabrication hazard stakes golf course new models are more robust against the possible impact and have high economic value, resistant to tropical situations, and easily moved or removed.

1.3 Special Purpose:

1. Design and fabrication hazard stakes golf course polymeric foam material.
2. Utilization of waste fibers EFB as polymeric foam composite material.
3. Analysis of stress distribution due to impact loading (ANSYS workbench 14.0).

1.4 Research benefits:

1. Scientifically development of new technologies in the model memfabrikasi stakes golf courses of polymeric foammaterial.
2. EFB commercial waste at palm oil mills can produce new products (sustainable).
3. Reducing deforestation and environmental impact.
4. Allows for the birth of new patent products.

2. Literature

2.1 Conceptual design and fabrication.

This phase is the most creative. Broadly speaking, the products are designed to fulfill its function (DeGarmo,1998). Typically, golf stick swing speed of 100 mph (using a 11 ° lofted driver, golf clubs head weight of 200 grams, 0.825 COR, and punches perfectly on center hit), then the ball speed reached 148 mph (Bishoy Emmanuel,2006).

IOP Conf. Series: Materials Science and Engineering **308** (2018) 012050 doi:10.1088/1757-899X/308/1/012050

Parameters of tensile properties, tensile strength of polyester 55-175 MPa. Izod Impact 10-30 J/25 elongation 3%, while the tensile strength of polyurethane is 1 MPa - 70 MPa, 30 J/25 Izod impact has not been broken and the elongation 100-1000%. Differences in the characteristics of the formulation ingredients in polymerization reactions must generate its own advantages (Purwanto and Sparringa, 1998).

Table 1. Blowing agent vs Resin variation,

B.Agent	20%	30%	40%	10%	0
Resin	60%	50%	40%	70%	80%
EFB	10%	10%	10%	10%	10%
Mekpo	10%	10%	10%	10%	10%
ρ kg/m^3	***1097***	***945***	***936***	***1022***	***1124***

Sources: Data Research Zulfikar, 2010.

Resin material variation 70%, 10% Blowing Agent, 10% fiber, 10% catalyst, ρ = 1022 (kg/m3), Mechanical properties with the following specifications:

a. Tensile strength = 7.427 MPa.
b. Modulus elasticity = 288.8759 MPa.
c. Elongasi = 2.571%.
d. Syt (MPa) = 7.427.
e. Regangan = 0.02571 mm / mm.

Results of experimental testing of high-speed impact of the specimen material fiber reinforced *polymeric foam* EFB fracture stress occurs at 13.41 MPa (Zulfikar, 2010). Mechanical strength of the fiber diameter EFB an average of 0.4 mm have a price elasticity modulus of 11.88 GPa with tensile stress max. 156.3 MPa (Gunawan, F.E., et al). Aim of the composite are:

1. Improve the mechanical properties and / or certain specific properties.
2. Simplify the design that are difficult to manufacture.
3. Flexibility in design, save cost.
4. Make lighter materials.

3. Methodology

3.1 Place and Time Research

3.1.1 Place

1. Survey in Arun Golf Club.
2. Design and fabrication in Impact dan Fracture Research Center of Mechanical Engineering USU Medan.
3. Simulated of software ANSYS 14.0 in (IC-STAR) USU Medan.

3.1.2 Time

Started since the approval of the proposal until the completion of the study.

3.2 Criteria design process

1. Loading of force capacity.
2. Mechanical analysis of the specific strength (strength and density, modulus of elasticity).
3. Optimizing the design and consider the cost implications.

3.3 Construction Planning Stake Stakes Golf Courses.

Design models is a factor of attraction for application use products stakes golf course.

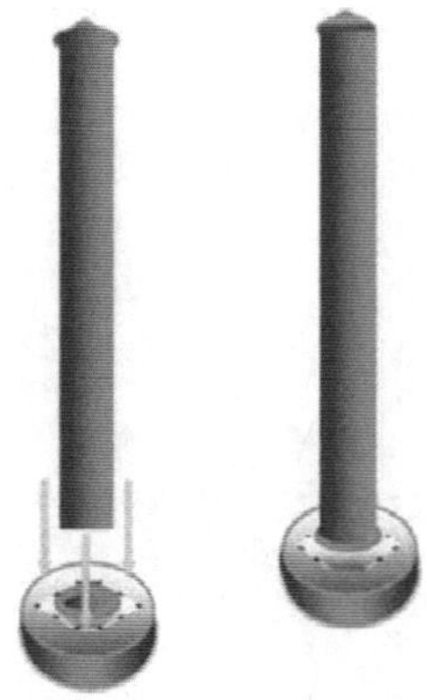

Figure 2. Design model

Fabrication stakes golf course is formed from compound materials:

a. Resin unsaturated poliester (orthophtalic 157 BQTN-EX)
b. Blowing Agent (polyol and isocyanate)
c. EFB fiber.
d. MEKPO (Metyl Ethyl Kethon Peroxide).

3.3.1 Tools and Methods

Table 2 List of Test Equipment and Tools.

Instrument	Variable
ANSYS 14.0	Distribution of stress due to impact loads.
PRODUCTION EQUIPMENT	
Mould - PVC pipe - Talc powder, resin, E-glass	Making molds stakes golf course.
Chemical equipment - Scales - Measuring cup	Chemical equipment - Scales - Measuring cup - Composities materials

3.4 Research Design

3.4.1. Design of mold model.

Mold base polymer material can be made of wood, which has been reinforced plastic, plates, gypsum materials. Factors in planning the pattern:

1. Shrinkage
2. Excess material
3. Draft angle
4. Distortion
5. Pating line

This research used permanent mold type for fabrication, it's superiority also can be used for refabrication.

IOP Conf. Series: Materials Science and Engineering **308** (2018) 012050 doi:10.1088/1757-899X/308/1/012050

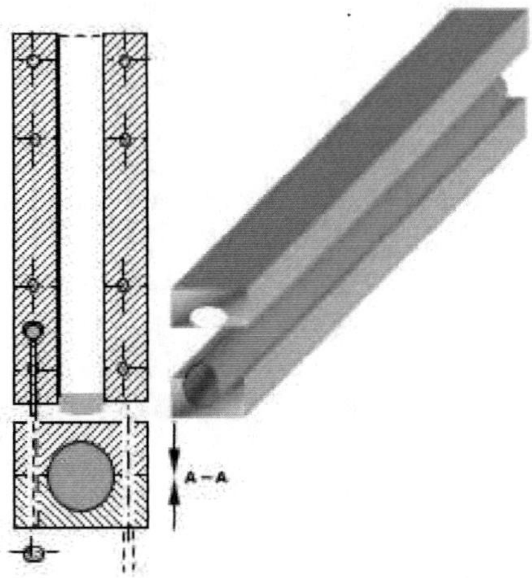

Figure 3. Planning model of mold.

Materials used: Talc powder: 2 kg, Resin BTQN-Ex 157: 2 litre, Mekpo: 2%, E-glass fiber: 3 piece, Materials mold: PVC pipe d=2½ inc, length 90 mm.

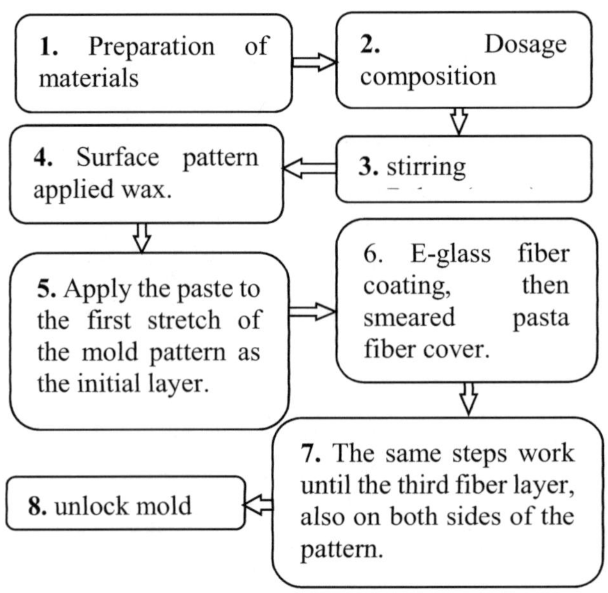

Figure 4. Flowchart of making mold

3.4.2. Reconstruction of mold fabrication

To make the mold into two symmetrical parts then, in the middle of the slab given separator pattern of paper materials.

Figure 5. Mold making

Giving mirror glaze (Figure 6) aims to facilitate the separation of the pattern material with cast product.

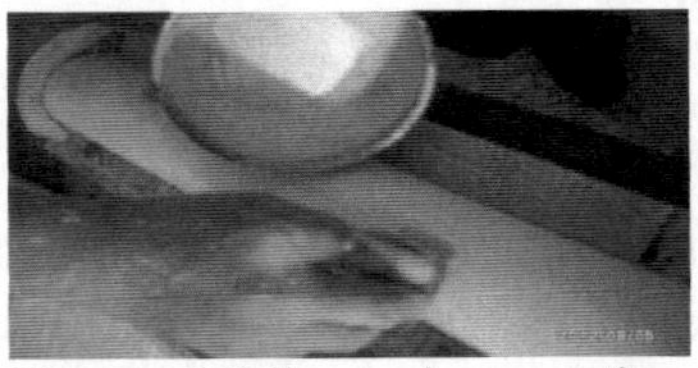

Figure 6. Mirror glaze coating.

3.4.3 Workmanship molding composition.
Talc powder and resin and catalyst in the ratio 1: 1: 1% mixed evenly in the container until it becomes a paste can be seen in Figure 7.

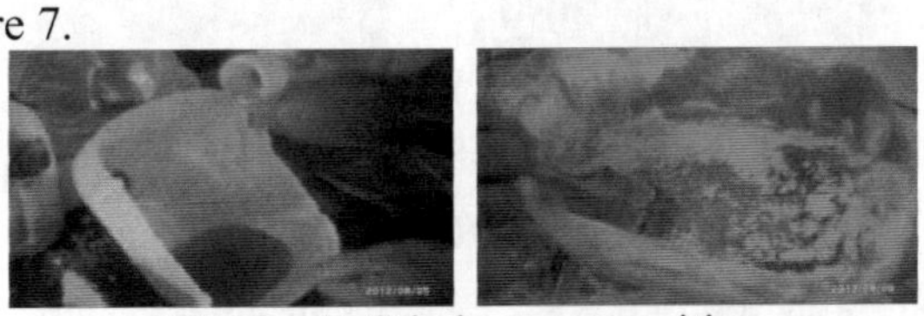

Figure 7. Stirring composition

Surface pattern that has been smeared with pasta included E-glass fibers to the third layer (Figure 8) and dry for 24 hours.

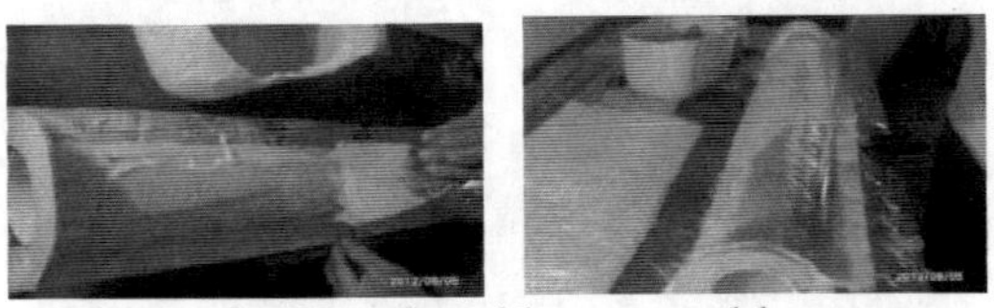

Figure 8. *Hand and lay-up* mold pattern.

Figure 2. Hazard stakes golf coursea disease can be tested through a blood test, one of which can be seen from the form of red blood cells. The normal and abnormal morphology of the red blood cells of a patient is very helpful to doctors in detecting a disease [2]

IOP Conf. Series: Materials Science and Engineering **308** (2018) 012050 doi:10.1088/1757-899X/308/1/012050

3.4.4 Construction mold.

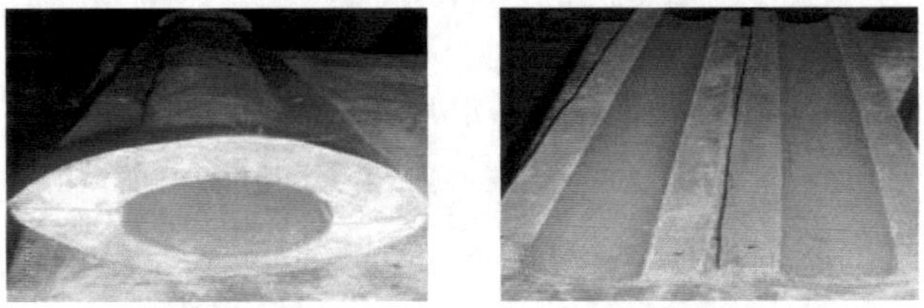

Figure 9. Mold when closed and opened.

Variabel observed in the fabrication process:

1. The time interval stirring composite materials to the casting time.
1. Distribution of composite materials in the field of mold cavity (fluidity).
2. Geometry and structural products.
3. Toughness products.

3.4.5 Fabrication for base stakes.

The proportion of portland cement mortar materials: cement, sand and water, 1: 2: 0.5. specification SNI 03-6882-2002. Central point graphic is void d = 25 mm, a depth of 4 mm hole.

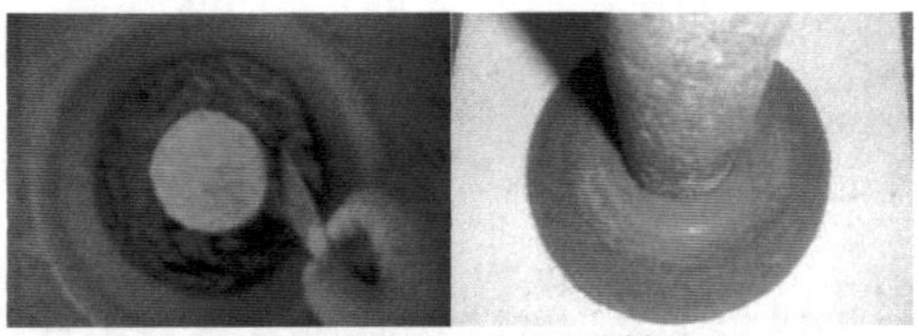

Figure 10. Produce of base

3.5 Computer Simulation

The height of the impact point was 460 mm, the trajectory of the golf ball elevation angle of 30 ° using a 11 ° driver golf ball is lofted speed, v: 66.2 m / s, 70 m / s, 78.2 m / s (velocity USGA research, speed of a golf ball).

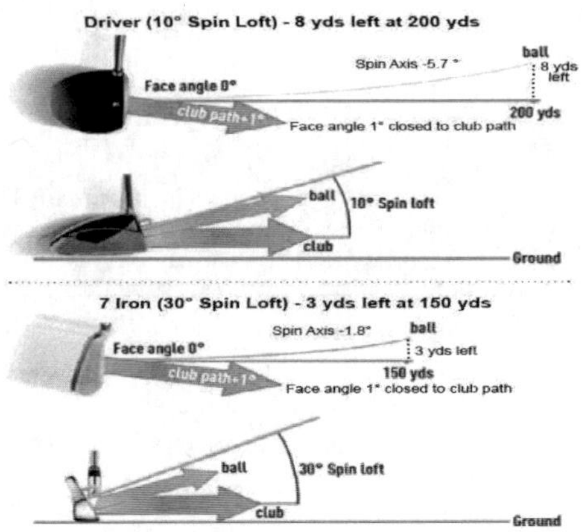

Figure 11. Driver golf ball

IOP Conf. Series: Materials Science and Engineering **308** (2018) 012050 doi:10.1088/1757-899X/308/1/012050

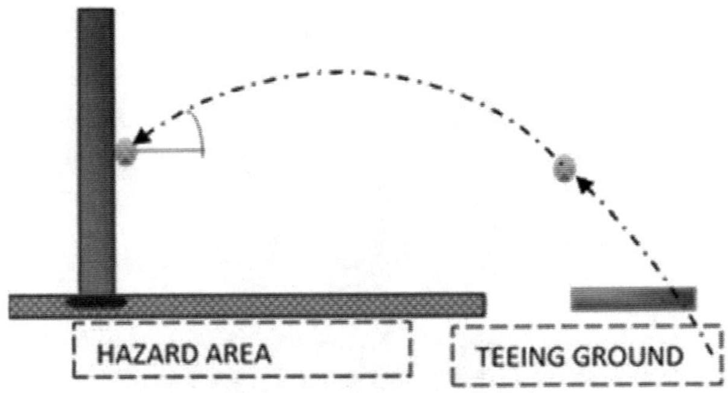

Figure 12. Impact Simulation scheme.

Golf ball with an initial velocity v_o and 30° elevation angle α_o the initial velocity component in the X-axis, vox, and the Y-axis, $v_o y$ then when v = 66.2 m / s:

$$v_0x = 66{,}2 \cos 30 = 57{,}33 \text{ m/s} \quad (1)$$

$$v_0y = 66{,}2 \sin 30 = 33{,}1 \text{ m/s} \quad (2)$$

Moment of inertia which occur on the hazard stakes golf course is:

$$\sigma = \frac{Mc}{I} = \frac{F.l.\frac{1}{2}d}{\frac{\pi}{64}d^4} = \frac{32F.l}{\pi d^3} \quad (3)$$

Creating 3D objects to stakes golf courses, as follows:

1. Open ANSYS Workbench 14.0 as shown in (Figure 13, select the Geometry tab Component System.

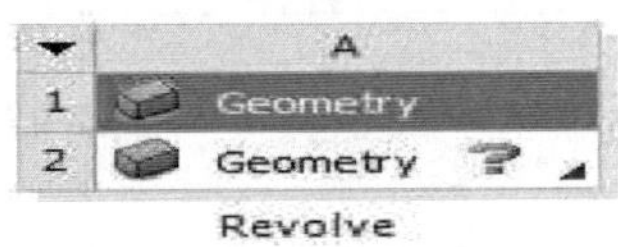

Figure 13. *Start* ANSYS.

2. The exchange into mode sketching and select line.
3. Click on Data Engineering (Figure 14).

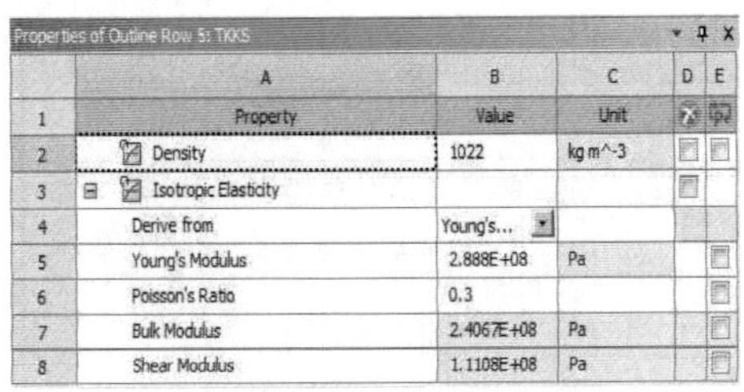

Properties of Outline Row 5: TKKS

	A	B	C	D	E
1	Property	Value	Unit		
2	Density	1022	kg m^-3		
3	Isotropic Elasticity				
4	Derive from	Young's...			
5	Young's Modulus	2.888E+08	Pa		
6	Poisson's Ratio	0.3			
7	Bulk Modulus	2.4067E+08	Pa		
8	Shear Modulus	1.1108E+08	Pa		

Figure 14. Engineering Data.

1. *Meshing* (Figure 15),→ *Generate Mesh.*

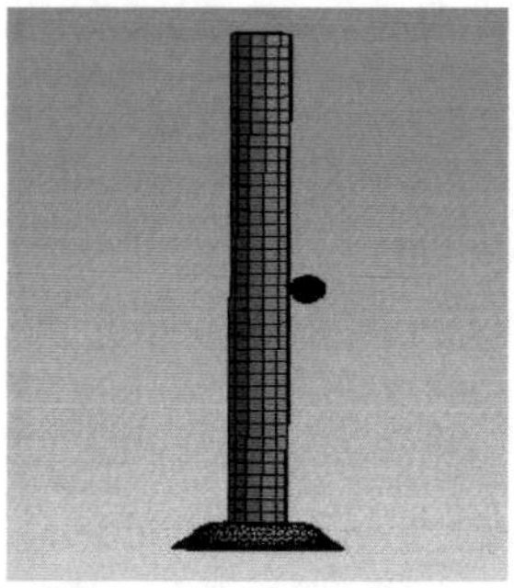

Figure 15. *Meshing.*

2. Right click on *solution* (Figure 16) → *Deformation* → *Total Deformation.*
3. *Solve.*

Figure 16. Simulation result.

3.6 Flowchart of Research

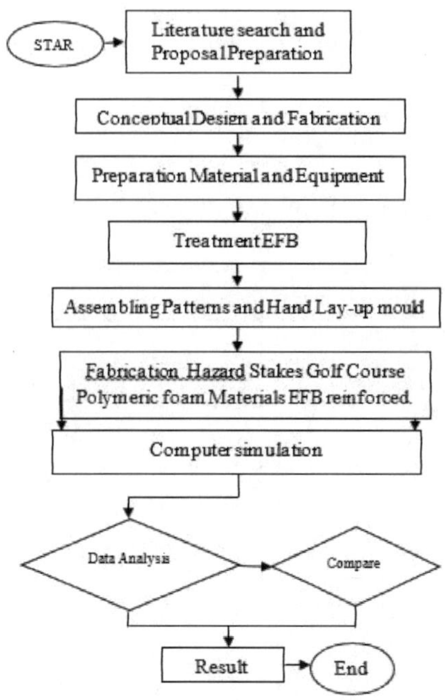

Figure 17. Flowchart of Research

IOP Conf. Series: Materials Science and Engineering **308** (2018) 012050 doi:10.1088/1757-899X/308/1/012050

3.7 Variable Research

Variables observed in this research:

1. Design of product models.
2. Treatment of EFB into fibers.
3. Selection of materials and process engineering hand lay-up mold.
4. Dimensional of geometric mold.
5. Kompositas engineering polymeric materials EFB fiber reinforced foam.
6. Kompositas the time interval until pouring into molds.
7. Stress distribution / computer simulation.

4. Results and Discussion

4.1. Treatment of Oil Palm Empty Fruit Bunch.

The addition of 1% NaOH while soaking can reduce the fat content and acid, eliminates odors resulting yarn is soft fibers. Fiber size in dry conditions after crushed by crusher machine with a mesh size of 0.8 mm diameter fiber length obtained 2 mm - 5 mm.

4.2. Component Fabrication procedure Stakes.

1. Container I: Compound Resin and EFB fiber stirred evenly over time 0:00:24.
2. Pour catalyst into the firs tcontainer and stirred for a time 0:01:07.
3. Stirring polyol and isocyanate in containers II during a time 0:00:15.
4. Compounding the container material is poured into the container I to container II, during the time of 0:02:30.
5. Pouring into molds at a time 0:03:20

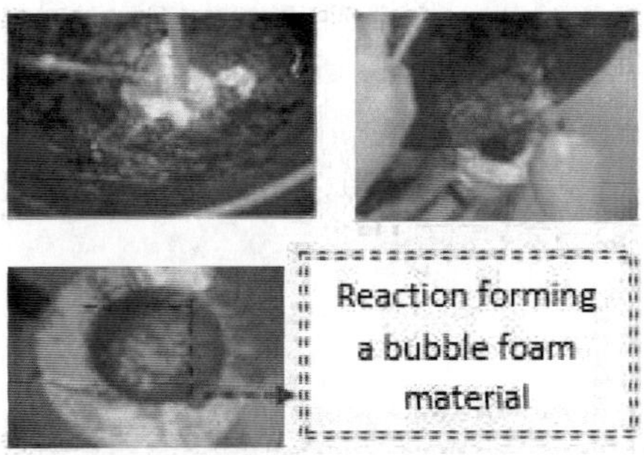

Figure 18. Fabrication circle.

Stirring time blowing agent effective: 10-15 seconds, the time on the verge of blowing agent to foam the granules begin to harden, these events may lead to material failure (Figure 19).

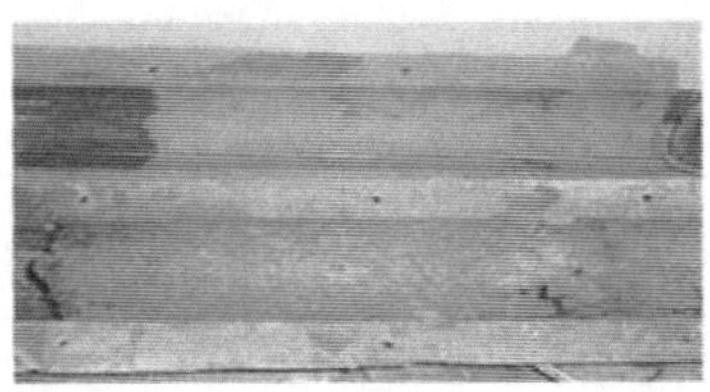

Figure 19. failure of the mechanical properties.

4.3 Fabrication Process observation.

Form in proportion to the conceptual design is obtained, if:

1. Engineering choosing the right materials.
2. The accuracy of dosing material.

IOP Conf. Series: Materials Science and Engineering **308** (2018) 012050 doi:10.1088/1757-899X/308/1/012050

3. The time interval composition is right.
4. Understand that the integration process with interval compotition rapid pouring in a record short time.

Base Type Model:
Tipe I=2407 gr, Tipe II=1942 gr, **Tipe III=1549 gr.** Proporsi Mortar 1:2:0,5 (SNI 03-6882-2002).

Figure 20. Base Type Model.

Fabrication yield components assembled as shown in Figure 21

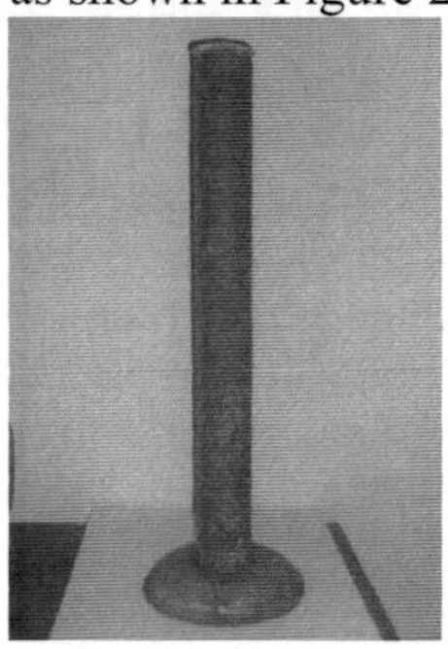

Figure. 21 Fabrication results (weight 1550 g).

4.4. Simulation result

4.4.1. Impact point height 460 mm.

Simulations show that the stress distribution caused by impact force with the lowest values occur at v = 18 m / s as shown in Figure 22. Total deformation of 8.7461 mm, maximum strain of 0.08099 mm / mm, maximum stress (von-Mises) at 48.863 MPa, and the normal stress (X - axis) of 14.006 MPa (time: 4.6952 E-004).

Table 3. Stress response of the impact point 460

V VELOCITY (m/s)	Force (KN)	Total Deformasi (mm)	Eq.Elastic Strain (mm/mm)	Eq.von-mises Stress (MPa)	Normal Stress (x Axis) (Mpa)
18	14,842	8,7461	0,08099	48,863	14,006
35	31,982	10,186	0,13073	101,63	53,625
66,2	42,881	18,029	0,2617	262,24	41,219
70	41,225	16,786	0,33499	336,61	46,276
78,2	46,450	9,0478	0,96072	727,39	119,04

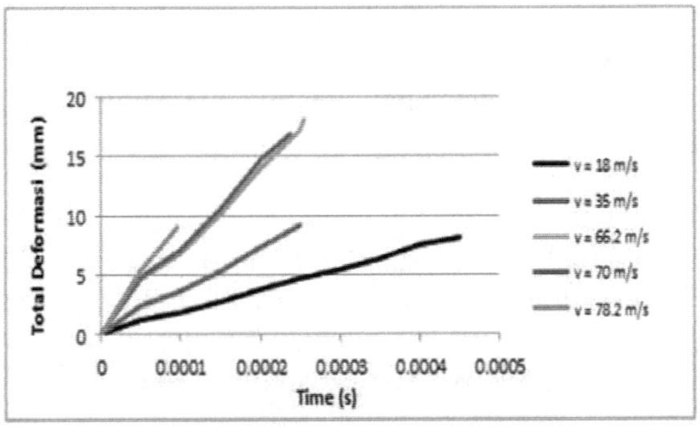

Figure 22. Time vs. Deformasi

Figure 23 shows the elastic strain when $v = 78.2$ m / s and impact loading, F = 46.450 KN increased strain, ϵmaks = 0.96072 mm / mm, at the time (time: 2.3922 E-004). Strain$_{min}$ occurs at, $v = 18$ m / s, ϵmin = 0.8099 mm / mm over time (time: 4.2635 E-004).

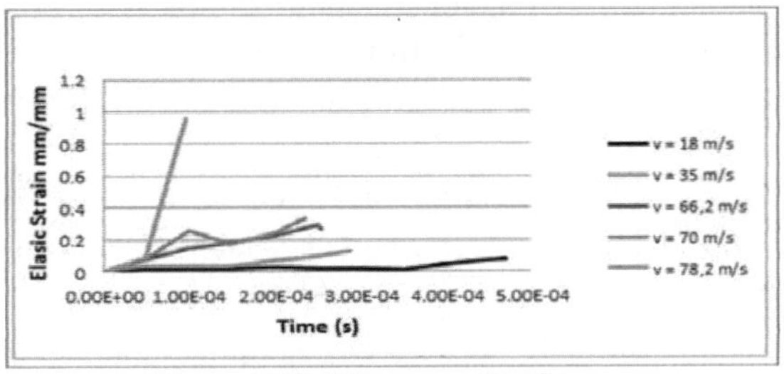

Figure 23. Time vs. Eq.Stress.

Figure 24 shows Normal stress min. = -57.947 MPa and a maximum of 58.609 MPa (time: 2.8474 e-004) occurs at a constant velocity, $v = 66.2$ m/s.

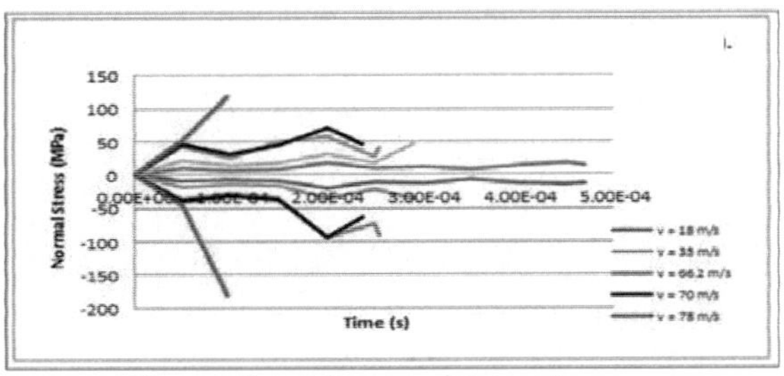

Figure 24. Time vs. Normal stress.

5. Conclusion

1. Design product model has been developed using solid work 2010.
2. Permanent mold type (permanent mold) using paste material: resin, talc powder and E-glass fiber reinforced.
3. Soaking TKKS using NaOH 1% of the volume of water for 48 hours can reduce the fat content of the resulting fiber TKKS approximately 9% - 10% to 1.5% - 2%.
4. Thoroughness and accuracy intervals stirring up the process of casting material into the mold cavity (cavity mold) on a periodic basis (mixing resin, blowing agent, catalyst and fiber is 00:02:30, pouring into molds 00:03:20).
5. Simulation results: area deformation and strain is concentrated at a point about cross-sectional area of impact loading. Impact simulation on height 460 mm, v = 78.2 m / s, an explicit dynamic

IOP Conf. Series: Materials Science and Engineering **308** (2018) 012050 doi:10.1088/1757-899X/308/1/012050

loading style, Fmaks = 119.34 KN yield stress (von-Mises) maximum, σ_{maks} = 727.39 MPa, ϵ = 0.960 mm / mm , the normal stress σn = 119.04 MPa, stress distribution is concentrated around the cross point impact, do not experience the effects of hysteresis and inertia, meaning that stakes golf course has the toughness of a geometric functional which is able to absorb impact energy.

Suggestion

1. Research needs to be supported by equipment golf game simulator.
2. EFB should still fresh pressure is applied using a hydraulic press before soaking the fiber treatment.
3. Need further research to design creative models and maximum formative.

References

[1] Bergstrom J S 2006 Advanced Finite Element Modeling of Polymer Foam Component *ABAQUS Conference*

[2] Syam B, Weriono, Rahmawati 2009 Analysis of Concrete and Rubber Alas of Traffic Cone Subjected to Impact Loading (accepted to be presented in the 6th Conf. on Numerical Analisys in Engineering, Lombok Island, Indonesia)

[3] Hermiati E 2007 Efforts to Reduce Dirt and Extracts on Oil Palm Bunches by Washing (UPT Balai Litbang, LIPI)

[4] Stewart J 1998 Calculus, Fourth Edition (By: A Division of International Thomson Publishing Inc)

[5] Syam B, Weriono, Rahmawati Analysis of Concrete and Rubber of Traffic Cone Subjected to Impact Loading

IOP Conf. Series: Materials Science and Engineering **308** (2018) 012051 doi:10.1088/1757-899X/308/1/012051

Electromagnetic Modelling of MMIC CPWs for High Frequency Applications

E P Sinulingga[*1)], P B K Kyabaggu[2)], A A Rezazadeh [3)]

[1]Electrical Engineering Department, Universitas Sumatera Utara, Indonesia
[2]Bukoola General Enterprises, Kampala, Uganda
[3]Elecrical and Electronic Engineering, The University of Manchester, United Kingdom

[*]E-mail: emerson.sinulingga@usu.ac.id

Abstract. Realising the theoretical electrical characteristics of components through modelling can be carried out using computer-aided design (CAD) simulation tools. If the simulation model provides the expected characteristics, the fabrication process of Monolithic Microwave Integrated Circuit (MMIC) can be performed for experimental verification purposes. Therefore improvements can be suggested before mass fabrication takes place. This research concentrates on development of MMIC technology by providing accurate predictions of the characteristics of MMIC components using an improved Electromagnetic (EM) modelling technique. The knowledge acquired from the modelling and characterisation process in this work can be adopted by circuit designers for various high frequency applications.

1. Introduction

The multilayer technique in MMIC technology provides the flexibility needed for microwave designers and engineers to be able to realise various low loss characteristic impedance CPW transmission lines [1]. Several multilayer transmission lines have been fabricated using photolithography to realise 3D CPWs which yield wide range characteristic impedances. This is achieved by utilising either overlap metal designs (denoted by OV) or elevation signal conductor designs (denoted by ELEV). CPW lines with an impedance between 65 and 70 ohms and dissipation loss ranging between 0.2 and 0.5 dB/mm, as well as CPW lines with characteristic impedance as low as 13 ohms with dissipation loss of 0.4 dB/mm, have been reported [1]. However the frequency observed was limited to between 16 GHz and 21 GHz. Currently there is no available data and analysis indicating that fabricated multilayer CPW configurations with relatively low attenuation characteristics (below 1 dB/mm) can be used at frequencies beyond 21 GHz. Furthermore, data presented in [1] shows a tendency towards higher dissipation loss as frequency increases beyond 16 GHz for all the fabricated structures.

In many applications (such as wireless applications up to 70 GHz) or in the integration of MMIC with nano scale devices, it is essential to utilise low-loss multilayer CPW transmission lines as interconnects or a feed lines. These narrow bandwidth observations make the wideband characteristic investigations of various multilayer CPW transmission lines presented in [1] become essential. In addition, MMIC components which have been modelled and fabricated need to be tested through on-wafer measurements in order to validate the desired components characteristics. As in some designs, CPW MMIC transmission lines may have a very narrow signal conductor track (S) or gap (W) between their grounds and their signal conductor track. Therefore, these CPW transmission lines require bonding pads and feeders (also known as probe pads) to accommodate the size of the probes used in the on-wafer measurement. Often the probe pads provide unwanted parasitic effects which influence the characteristics of the intrinsic components.

2. Probe Pads Design

Several investigations have been reported to de-embed the effect of these pads [2, 3]. Although these indicate some improvements, the utilised methods do not completely remove the distortions caused by the probe pads. In addition, in some cases the de-embedding technique may alter the intrinsic characteristic of the components. One therefore needs to analyse the actual effect of the probe pads in multilayer CPW structures based on their designs. Furthermore, one may suspect there to be a relationship between probe positions and the magnitude of the unwanted distortions (either in the simulation or on-wafer measurement). This has never before been reported in any literature. It is thus essential to begin investigations by analysing the effect of the shape of the probe pads, followed by the probe positions in multilayer CPW MMIC transmission lines.

In this work, two designs of probe pads (each consists of bonding pad and feeder) have been considered. The cross-sectional view of the bonding pads and the top views of both probe pads are shown in Figures 1 and 2 respectively. Each design in Figure 2 has two different lengths of feeder, i.e. 75 μm and 100 μm. These are investigated to establish the effect of the feeder to the magnitude of the unwanted distortion. Both probe pads are embedded into a multilayer transmission line with a signal conductor width of 20 μm, a gap width of 15 μm and an intrinsic length of 2000 μm. The thickness of semi-insulating GaAs (h_1) is 600 μm. The thickness of polyimide (h_2) and metal (t) are 5 μm and 0.7 μm respectively. The top view of the transmission line with probe pad 1 is shown in Figure 3. The simulated S-parameters for both probe pad designs are shown in Figures 4 and 5.

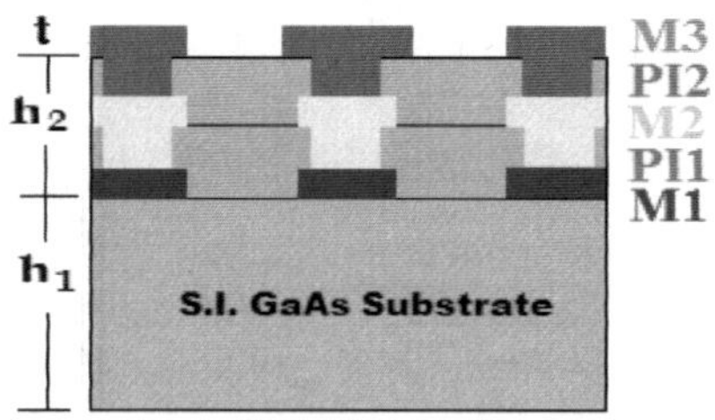

Figure 1. Cross-sectional view of a bonding pad for a multilayer CPW transmission line.

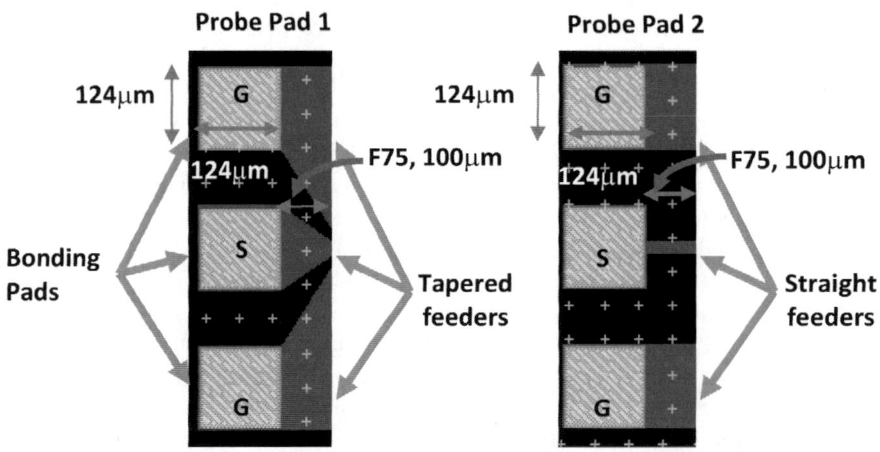

Figure 2. Top views of two different shapes of probe pads, each consisting of bonding pads and feeders with different shape.

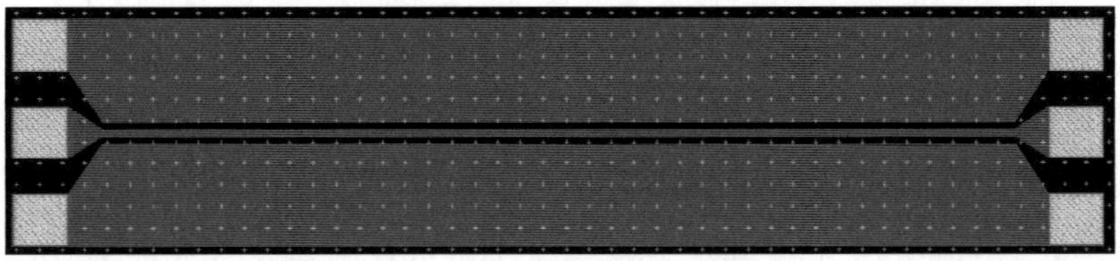

Figure 3. Top view of the layout of a multilayer CPW transmission line embedded with probe pad 1.

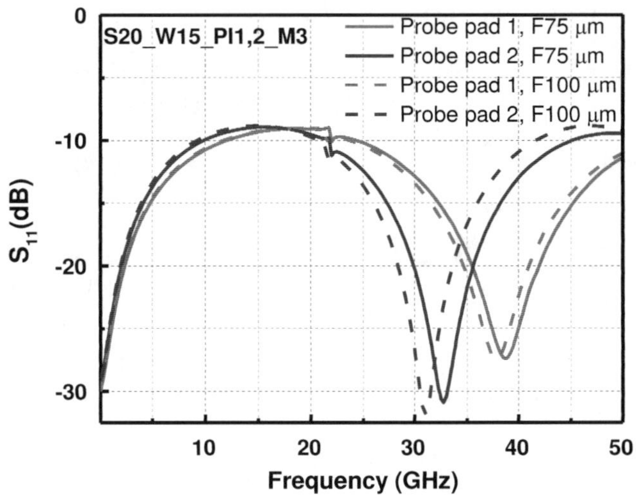

Figure 4. Simulated S_{11} parameter of multilayer CPW transmission lines with various probe pads.

The data simulation shows that probe pad 1 with a tapered feeder have higher resonance frequency than probe pad 2 with a straight feeder. This is due to the shape of the feeders which allows the signal to have a smooth transition between probes and the intrinsic transmission lines. There is no significant effect of having a longer feeder length, except that the resonance frequency shifts slightly to a lower frequency. Based on this finding, probe pad 1 with feeder length 75 μm is chosen as the default probe pads in this work.

3. Optimum Electromagnetic Modelling

Observations of data presented in figures 4 and 5 also demonstrate a probe pad effect which may disturb the intrinsic characteristics of the transmission lines. These are shown in both S_{11} and S_{21} parameters at 22 GHz. Two reference probe positions (A and B, in the on-wafer measurement and layout simulation) as illustrated in figure 6 have therefore been investigated. The horizontal distances of position A and B from the end of the pads are 30 μm and 62 μm respectively. The fabricated CPW transmission line in figure 3 has been measured with these two probe positions. The results are presented in figure 7. One can demonstrate using the on-wafer measurement that by shifting the probe from position A to B, the distortions which occur at 22 GHz and 44 GHz can be slightly reduced. This is due to the optimum distance available for the CPW mode to settle as the signal incident to the probe pads.

In order to gain more understanding of this distortion, a similar approach has been employed in the simulation. This investigation involves EM modelling using port reference offset positions A and B, with some setup modifications. The simulation with reference position A utilises the edge mesh, while the simulation with reference position B utilises the edge and transmission line meshes. The motivation driving these modifications is to understand the effect of complex current computation within the transmission line, including the multilayer bonding pads which may yield some uncertain characteristics. This can provide a clear understanding of the source of distortions at 22 GHz and 44

GHz. The results are presented in figure 8. One may observe the improvement achieved by using port reference B and utilising both edge and transmission line meshes. The simulation data show good agreement at a lower frequency before the unwanted distortions occur at 22 GHz and 44 GHz. The utilisation of edge and transmission line meshes together with the reference offset position B reduces these distortions. This finding demonstrates that simulator and measurement software faces complex transverse current distribution [4,5] at the edge of the structure as well as at its port discontinuity, i.e. the multilayer bonding pads.

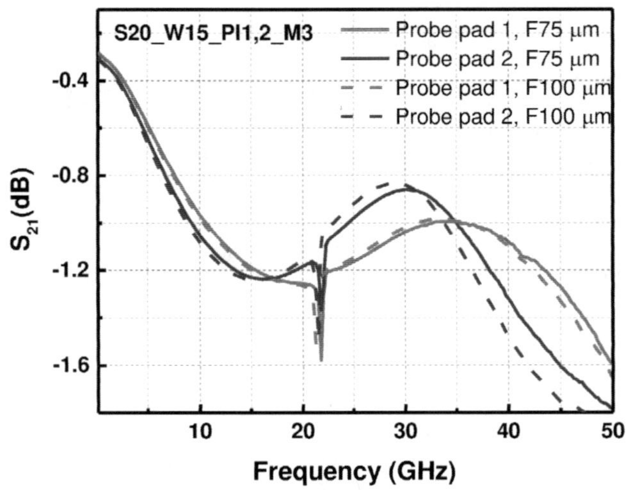

Figure 5. Simulated S_{21} parameter of multilayer CPW transmission lines with various probe pads.

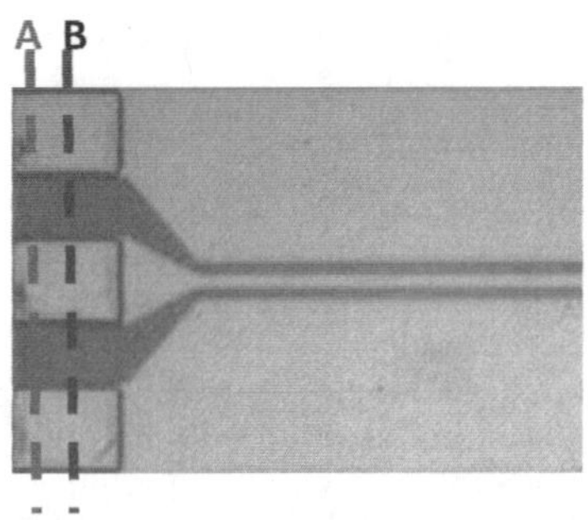

Figure 6. Micrograph of a multilayer CPW transmission line with illustration of different probe positions.

Furthermore, the Agilent ADS Momentum 2011 simulator is powered with an adaptive frequency sampling (AFS) algorithm. This feature compares sampled S-parameters data points to a rational fitting model in order to accurately represent the spectral response of the modelled components or circuits [4,5]. The simulation therefore yields less uncertainty characteristics for large frequency ranges than the measurement.

4. Conclusions

The results confirmed that the S-parameters characterisation is valid for use with the multilayer CPW structure. The results presented also confirm that the new simulation setup with additional edge and transmission line mesh (Ref B, EDGE) yields very good agreement with the measured data. It reduces distortions at 22 GHz and 44 GHz which resulted from uncertainties arising as a consequence of complex transverse current distribution on the edge of the multilayer CPW transmission line as well as at its discontinuity, i.e. the multilayer bonding pads.

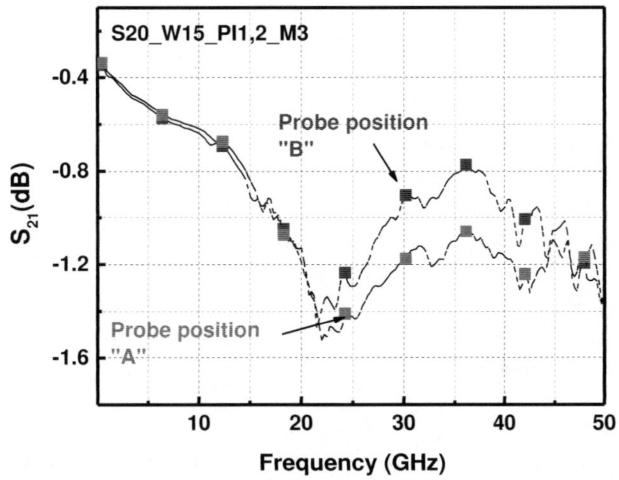

Figure 7. Measured S_{21} parameter of a multilayer CPW transmission line with different probe positions.

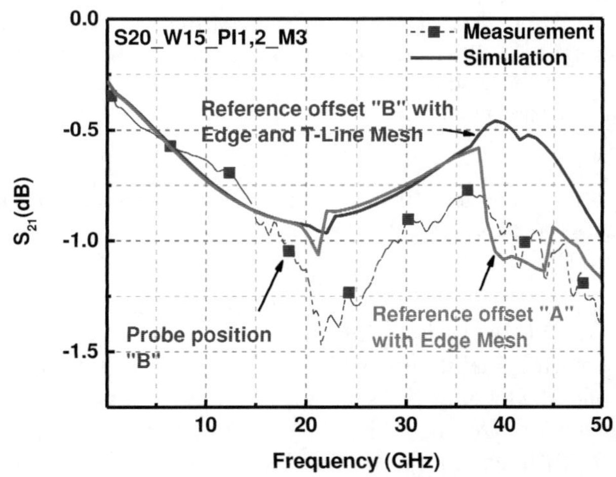

Figure 8. Simulated and measured S_{21} parameter of a multilayer CPW transmission line with different reference offset ports or probe position.

References

[1] Vo T V, L. Krishnamurthy, Sun Q and Rezazadeh A A 2006 3-D Low Loss Coplanar Waveguide Transmission Lines in Multilayer MMICs *Microwave Theory and Techniques, IEEE Transactions* **54** pp 2864 – 2871

[2] Chen C H and Deen M J 2001 A General Noise and S-parameter Deembedding Procedure for On-Wafer High Frequency Noise Measurements of MOSFETs *Microwave Theory and Techniques IEEE Transactions* **49** pp 1004 – 1005

[3] Heymann P, Prinzler H and Scnieder F 1994 De-Embedding Of MMIC Transmission-Line Measurements *International Microwave Symposium Digest, IEEE MTT-S* pp 1045 – 1048

[4] Agilent Technologies 2011 *Advanced Design System 2011 User's Guide*

[5] Agilent Technologies 2011 *Agilent AN154 -S-Parameter Design; Agilent Application Note*

IOP Conf. Series: Materials Science and Engineering **308** (2018) 012052 doi:10.1088/1757-899X/308/1/012052

The edge detection enhancement on satellite image using bilateral filter

I Fawwaz[1], M Zarlis[2], Suherman[3], R F Rahmat[4]

[1,2,3,4]Department of Information Technology, Faculty of Computer Science and Information Technology, Universitas Sumatera Utara, Medan, Indonesia.

Email: insi.dini@gmail.com, m.zarlis@usu.ac.id, suherman.usu@gmail.com, romi.fadillah@usu.ac.id

Abstract.
Satellite imagery is the image taken from satellites from outer space. It captures the appearance of the earth's surface through remote sensing. Usually, satellite image's object is hardly detected because many objects are covered with cloud shadows in the sky. Edge detection has an important role in image analysis. Edge detection aims to extract the boundary of an object contained in the image. Gaussian Filter is usually used on edge detection to smooth the image and reduce noise. However, in satellite image, we can improve it by using another filter. In this research, we compare gaussian filter with bilateral filter and analyze those two methods. We also using several operators to see the optimized process. After the experiment, by using the comparison parameter such as largest PSNR value, and the smallest MSE value, we can conclude that Bilateral Filter with Canny Operator is the most optimized edge detection for Satellite Imagery.

Keywords: Gaussian Filter, Bilateral Filter, Edge Detection, Satellite Imagery, Roberts Operator, Canny Operator, Frei Chen Operator

1. Introduction

Satellite imagery is very useful in various sectors such as geography, mining, and energy, meteorology, weather, forestry, etc. Satellite imagery is difficult to detect because many of the satellite images are covered with cloud shadows in the sky [1]. There are many methods for edge detection. Some methods for edge detection are using Roberts Operator, Canny Operator, and Frei Chen Operator. These methods have different characteristics. Edge detection becomes the main approach in image recognition and analysis can be based on the edge of the image containing useful information [2]. Image enhancement using the Gaussian Filter is commonly used for edge detection to reduce image noise. Gaussian filters are applied to smooth the image and reduce the noise, but will also smooth the edges [3]. Bilateral filter is a technique to smooth images while preserving the edges [4]. This technique is very similar to Gaussian filtering [5]. Bilateral filter is a non-iterative, local and simple nonlinear technique by combining gray degrees based on geometric proximity and photometric similarities [6]. Therefore, this paper proposes an optimized algorithm for satellite image edge detection based on image enhancement using Bilateral Filter by preserving edges. The purpose of this research is to analyze traditional edge detection of Gaussian Filter on satellite images and improve edge detection with Bilateral Filters, then the comparative analysis is conducted in this research to see the optimized edge detection.

In image processing, filtering is one of the bases used in various tasks such as interpolation, re-sampling, and noise removal. The filter used depends on the nature of the task performed by the filter and the type of data to be used for the operation [7]. Filtering is one tool used in many fields for various applications, especially in this section for image enhancement.

IOP Conf. Series: Materials Science and Engineering **308** (2018) 012052 doi:10.1088/1757-899X/308/1/012052

2. Gaussian Filter

Gaussian Filter is used in the field of image analysis especially for smoothing process, blurring, eliminating detail, and noise removal [8]. In Gaussian Filter, the intensity value of each pixel is replaced by the average of the weighted value for each of its neighboring pixels and pixels itself. The number of neighbors involved depends on the filter being designed. For digital image processing, the zero mean Gaussian of two variables is expressed by the equation:

$$f(x,y) = \frac{1}{2\pi\sigma^2} \exp\left(-\frac{x^2 + y^2)}{2\sigma^2}\right) \tag{1}$$

Discrete estimates of this continuous function are determined using two parameters [9]
(1) the desired size of the kernel (as N x N *filter mask*);
(2) the value of σ, the standard deviation of Gaussian function.

3. Bilateral Filter

Bilateral filters are one of the noise reduction methods that have two kernel filters: spatial kernel and range kernel [5]. The spatial kernel aims to smooth the image based on the Gaussian function [10]. Spatial Kernel is the distance between the pixels of the image (Euclidean distance), and range kernel is the similarity of the intensity between the two pixels in the image. Spatial kernel (W_{σ_s}) is the realization of spatial proximity measurement is by the equation

$$W_{\sigma_s} = \exp\left(\frac{-(\|p-q\|^2)}{2\sigma_s^2}\right) \tag{2}$$

While range kernel (W_{σ_r}) means the weighting of pixels in accordance with the size of the difference in intensity of the pixel with the intensity at the pixel which is the center of analysis on the image. Calculation of range kernel in each pixel is shown in the equation.

$$W_{\sigma_r} = \exp\left(\frac{-(|I(p)-I(q)|^2)}{2\sigma_s^2}\right) \tag{3}$$

Bilateral Filtering ($W^B(p)$) computation as follows:

$$W^B(p) = \frac{1}{W}\sum_{p\, q\, \in N\,(p)} G_{\sigma_s}((\|p-q\|)\, G_{\sigma_r}(|Ip - Iq|)\, Iq \tag{4}$$

Where I is a grey level image, $N(p)$ is a neighborhood of p, $Ip$is the intensity value at pixel location p, Wp is the normalization factor which ensures pixel weights sum to 1

$$W_p = \sum_{q\, \in N\,(p)} G_{\sigma_s}((\|p-q\|)\, G_{\sigma_r}(|Ip - Iq|) \tag{5}$$

4. Edge Detection

Edge is the essential characteristic of an image [4]. The edges of an image contain crucial information from the image Edge detection is a significant step in image processing. Image edges can represent the objects contained in the image, shape, size, and sometimes also information about the texture. Edge detection is used to extract the essential features [14].

4.1 Roberts Operator

The first operator used for edge detection is the Roberts Operator [11]. It was given by Robert Cross in 1963. Roberts operator is a first-order operator, which uses a partial differential operator to find the edge [2]. Roberts operator is a simple edge detection technique and has a fast computing rate. The image to be detected by the edge is combined with the mask of the operator and the gradient magnitude to be calculated directly. The Roberts operator uses the following 2x2 masks:

$$G_x = \begin{bmatrix} 1 & 0 \\ 0 & -1 \end{bmatrix} \quad G_y = \begin{bmatrix} 0 & 1 \\ -1 & 0 \end{bmatrix}$$

4.2. Canny Operator

Canny is one of the modern edge detection algorithms. In 1986 John Canny proposed three criteria to be the basis of filter mining to optimize edge detection in noise-sensitive images. The first step in implementing Canny is to smooth the image [2]. Then find the image gradient in both vertical and horizontal directions [12]. Canny uses a Gaussian filter before applying the mask. Gaussian filters can reduce the noise as much as possible [11]. Canny edge detectors are not sensitive to noise causing Canny can detect weak edges [2].

4.3. Frei Chen Operator

Frei Chen is a unique operator [12]. Frei Chen mask contains the base vector. This is implemented in the 3x3 image area represented by the number of nine Frei Chen masks [13]. Here we described the frei chen operator used in this research :

$$G_1 = \frac{1}{2\sqrt{2}}\begin{bmatrix} 1 & \sqrt{2} & 1 \\ 0 & 0 & 0 \\ -1 & -\sqrt{2} & -1 \end{bmatrix} G_2 = \frac{1}{2\sqrt{2}}\begin{bmatrix} 1 & 0 & -1 \\ \sqrt{2} & 0 & -\sqrt{2} \\ 1 & 0 & -1 \end{bmatrix} G_3 = \frac{1}{2\sqrt{2}}\begin{bmatrix} 0 & -1 & \sqrt{2} \\ 1 & 0 & -1 \\ -\sqrt{2} & 1 & 0 \end{bmatrix}$$

$$G_4 = \frac{1}{2\sqrt{2}}\begin{bmatrix} \sqrt{2} & -1 & 0 \\ -1 & 0 & 1 \\ 0 & 1 & -\sqrt{2} \end{bmatrix} G_5 = \frac{1}{2}\begin{bmatrix} 0 & 1 & 0 \\ -1 & 0 & -1 \\ 0 & 1 & 0 \end{bmatrix} G_6 = \frac{1}{2}\begin{bmatrix} -1 & 0 & 1 \\ 0 & 0 & 0 \\ 1 & 0 & -1 \end{bmatrix}$$

$$G_7 = \frac{1}{6}\begin{bmatrix} 1 & -2 & 1 \\ -2 & 4 & -2 \\ 1 & -2 & 1 \end{bmatrix} G_8 = \frac{1}{6}\begin{bmatrix} -2 & 1 & -2 \\ 1 & 4 & 1 \\ -2 & 1 & -2 \end{bmatrix} G_9 = \frac{1}{3}\begin{bmatrix} 1 & 1 & 1 \\ 1 & 1 & 1 \\ 1 & 1 & 1 \end{bmatrix}$$

4.4. Slope Magnitude Method

Slope magnitude can use the following equation where Gx and Gy is the image gradient in the horizontal and vertical direction [12]. Calculation of Gradient Magnitude (significant slope of the line) is denoted by Equation 6 below:

$$G = \sqrt{{G_x}^2 + {G_y}^2} \cong |G_x| + |G_y| \tag{6}$$

Where: G= gradient magnitude
G_x= gradient to point x
G_y = gradient to point y

5. Mean Square Error (MSE)

MSE is the mean of the square of the error value between the original image and the image of the image processing. The greater the value of MSE, the worse the quality of the resulting image.

Conversely, the smaller the value of MSE, the better the quality of the resulting image. Mean Square Error (MSE) can be formulated as follows [11]:

$$MSE = \frac{1}{MN}\sum_{x=0}^{M-1}\sum_{y=0}^{N-1}\left(f(x,y) - g(x,y)\right)^2 \quad (7)$$

Where: M = the length of the image in pixels
N = image width in pixels
$f(x,y)$ = pixel value after edge detection
$g(x,y)$ = pixel value after edge detection

6. Peak Signal to Noise Ratio (PSNR)

Peak Signal to Noise Ratio (PSNR) is a calculation that determines the value of an image produced. The value of PSNR is determined by the size or magnitude of the MSE value that occurs in the image. The greater the value of PSNR, the better the resulting image. Conversely, the smaller the value of PSNR, the resulting image gets worse. PSNR can be calculated by the formula [14]:

$$PSNR = 10 \text{ Log } 10\left(\frac{255^2}{MSE}\right) \quad (8)$$

7. Results and Discussion

The dataset used in this research are satellite images which has been resized into 300 x 300 pixels resolution. We retrieve the data from USGS (United States Geological Survey). In this research we use lakes and seas object from the image. Figure 1 shows the comparative result between gaussian filter and bilateral filter

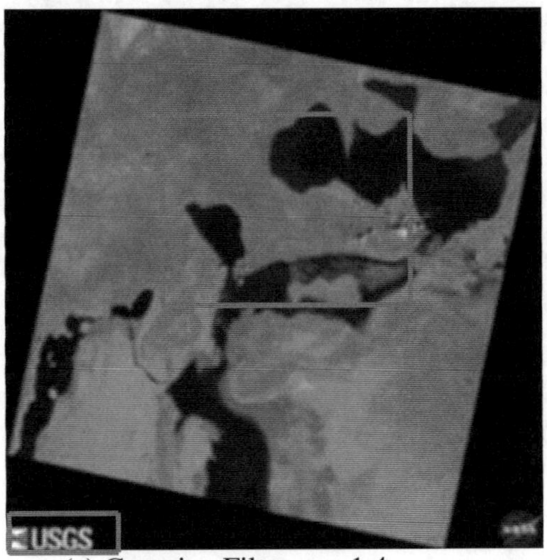

(a) Gaussian Filter, σ= 1.4
MSE = 10437.2
PSNR= 7.94495

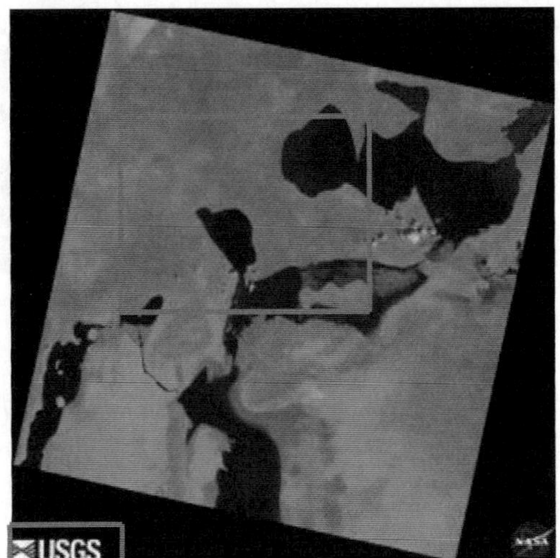

(b). Bilateral Filter, σ_s= 1.4, σ_r= 0,05
MSE = 10436.3
PSNR= 7.94543

Fig 1. Experiments for satellite imagery show the results of the image using Gaussian Filter looks more blur than the image of Bilateral Filter which preserves the edge

IOP Conf. Series: Materials Science and Engineering **308** (2018) 012052 doi:10.1088/1757-899X/308/1/012052

Table 1. Edge detection results on satellite images with Gaussian Filter

Satelite Imagery	Roberts	Canny	Frei Chen
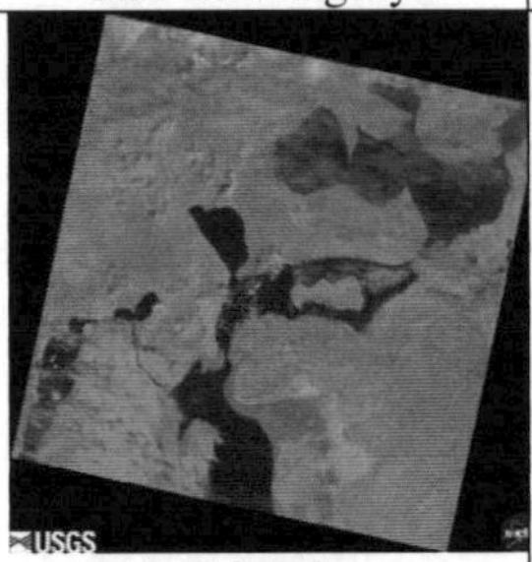 Aral Sea.jpg σ_S= 1.5	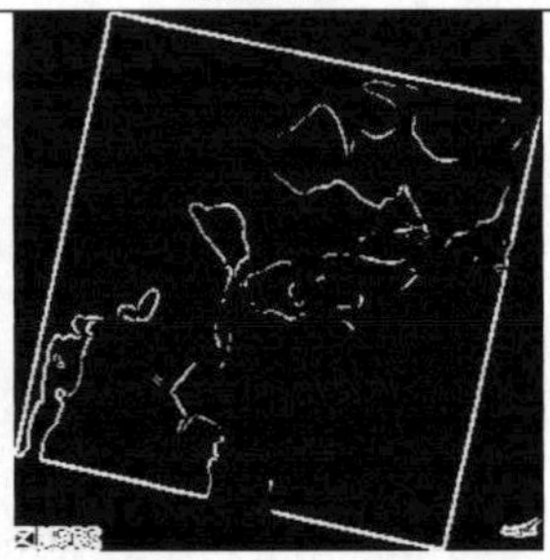	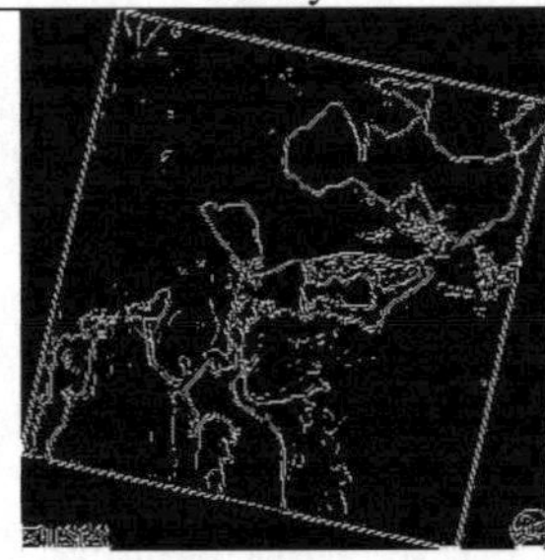	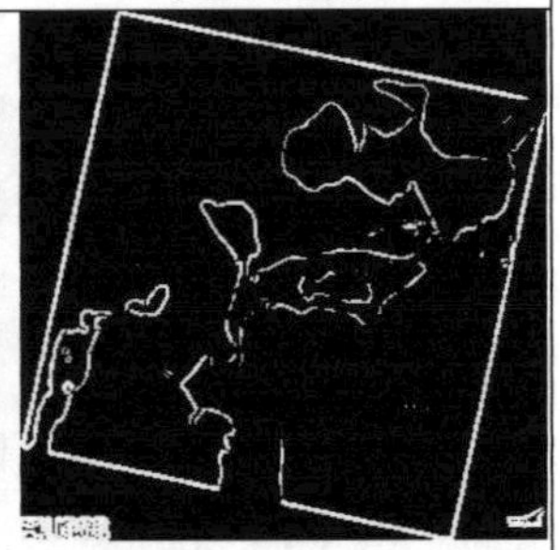
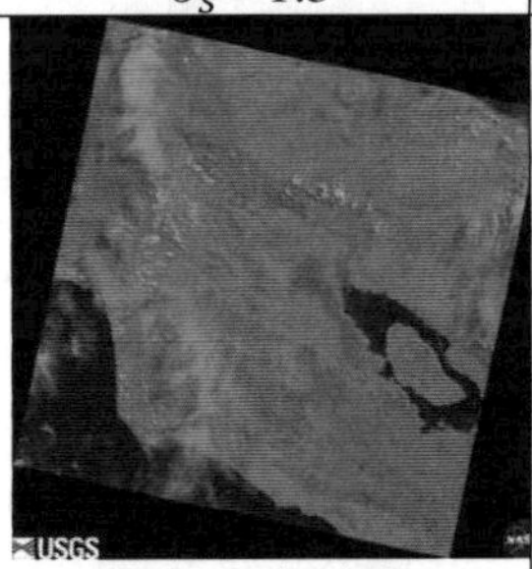 Toba Lake.jpg σ_S= 2	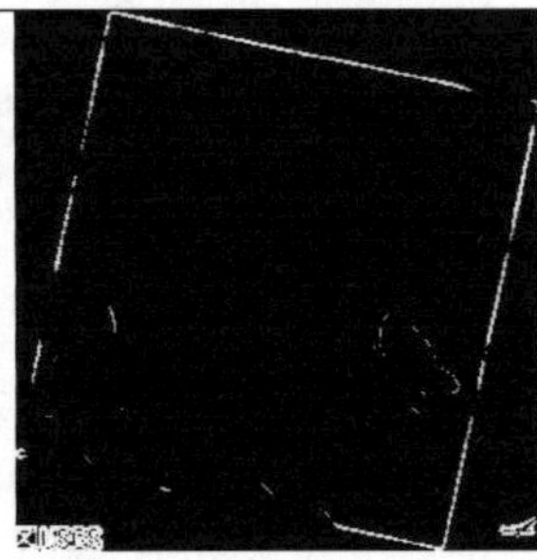	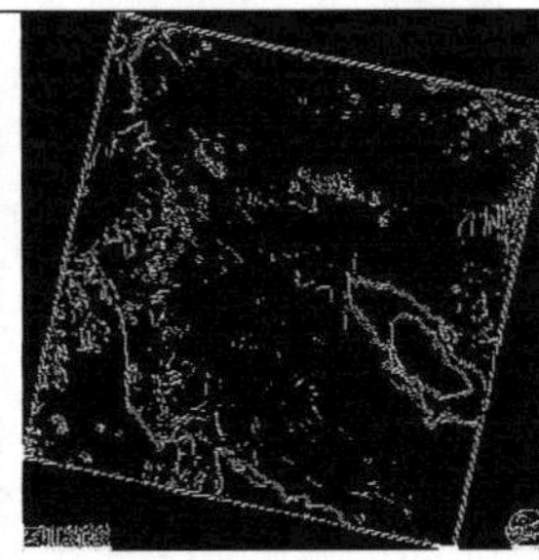	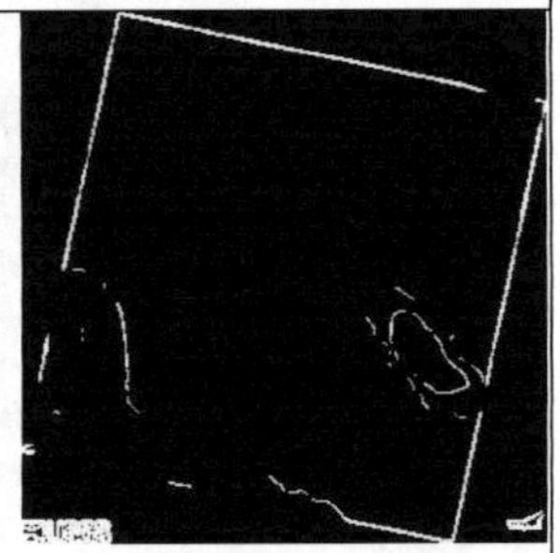
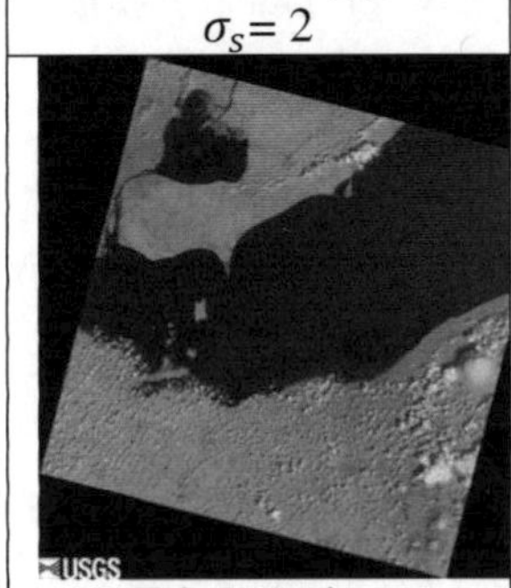 Erie Lake.jpg σ_S= 2.5	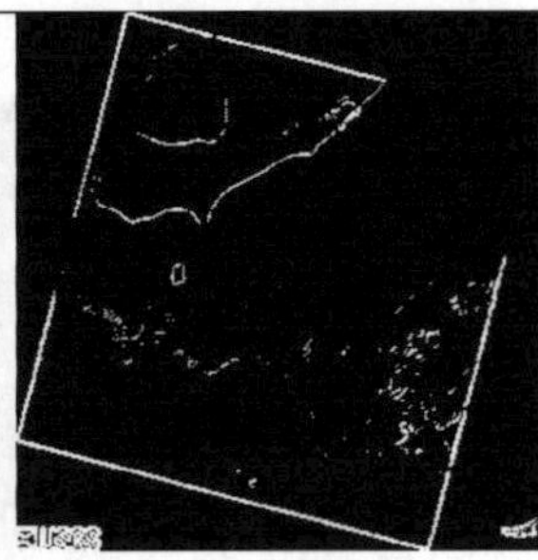	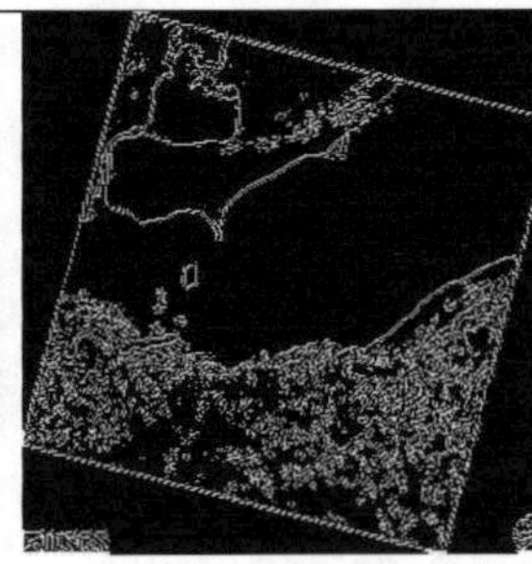	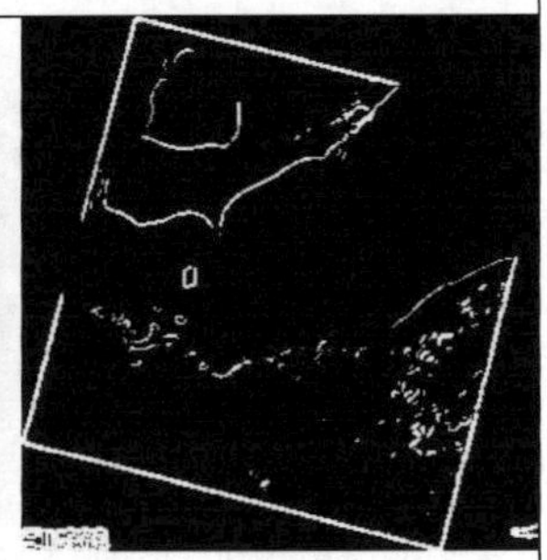

In table 1 we use different σ_S values for each image. The result shows edge detection on Roberts, and Frei Chen produces a smooth edge of the line and seen several lines cut off. While Canny produces more edge. Clouds on satellite imagery are also considered as edges.The higher the value of $\sigma_S$given to the Gaussian Filter will remove the edge of the image.

In table 2 we only use different $\sigma_S$values for each image and keep the σ_R value. The image on the edge detection of Roberts and Frei Chen has increased even the edge of the line on Frei Chen is thicker than using Gaussian Filter. Edge detection with Bilateral Filter can retain the edges although on Canny is not very visible.

We also experiment with increasing value of σ_R on the Bilateral Filter. The result shows that if the value of σ_R higher than before, the noise on the image becomes faded because σ_R controls the power of the noise filter on the image. As a result, the detected image loses many edges. While at σ_S, the higher the value of σ_S, then the object in the image faded. But it can still keep the edge so edge information can still be seen.

IOP Conf. Series: Materials Science and Engineering **308** (2018) 012052 doi:10.1088/1757-899X/308/1/012052

Table 2. Edge detection results on satellite images with Bilateral Filter

Satelite Imagery	Roberts	Canny	Frei Chen
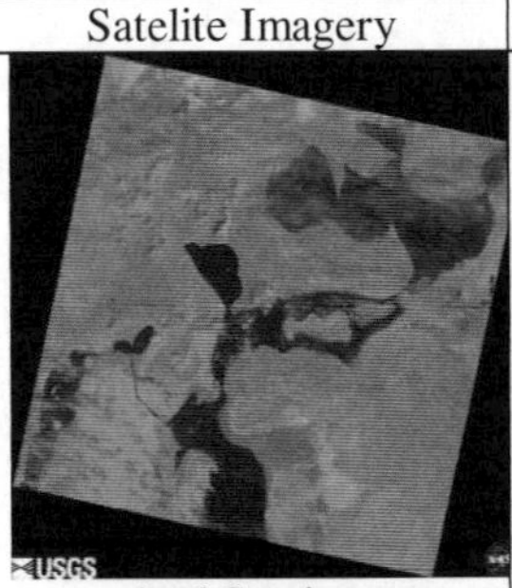 Aral Sea.jpg σ_S=1.5 σ_R= 0.05	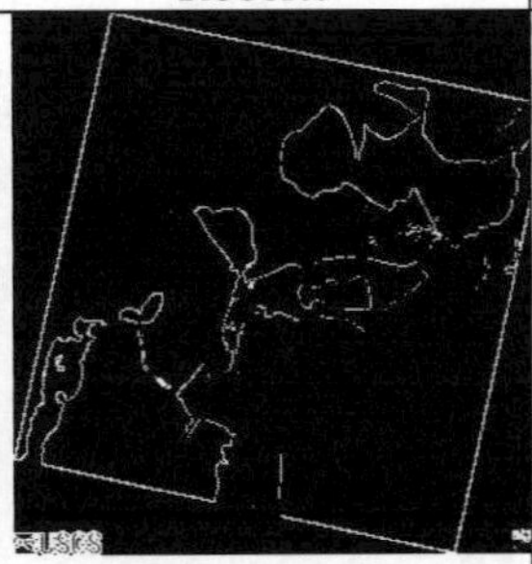	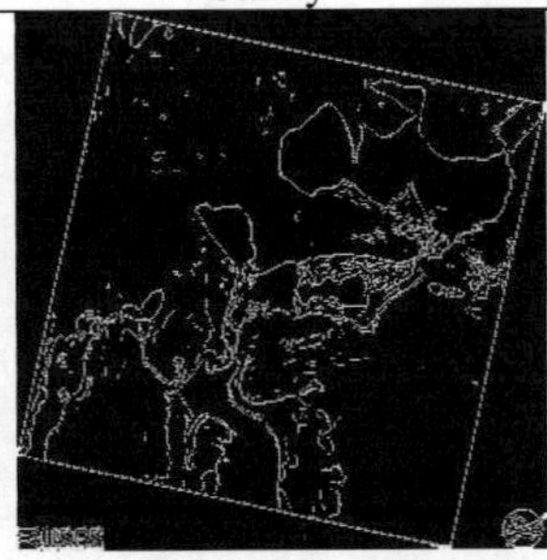	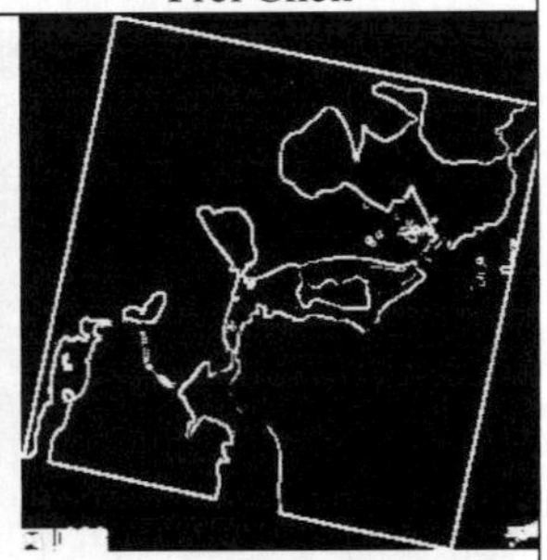
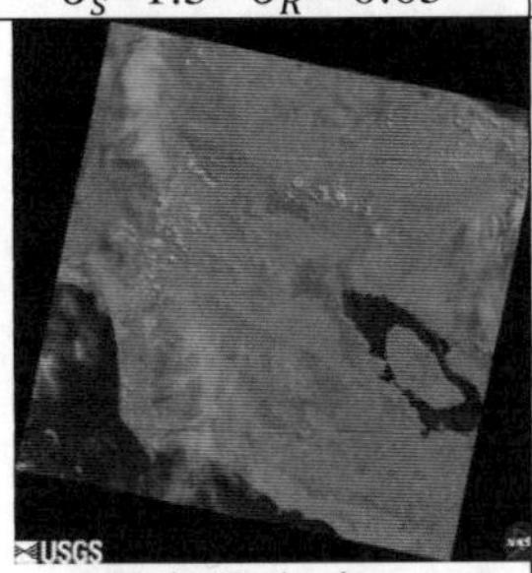 Toba Lake.jpg σ_S=2 σ_R= 0.05	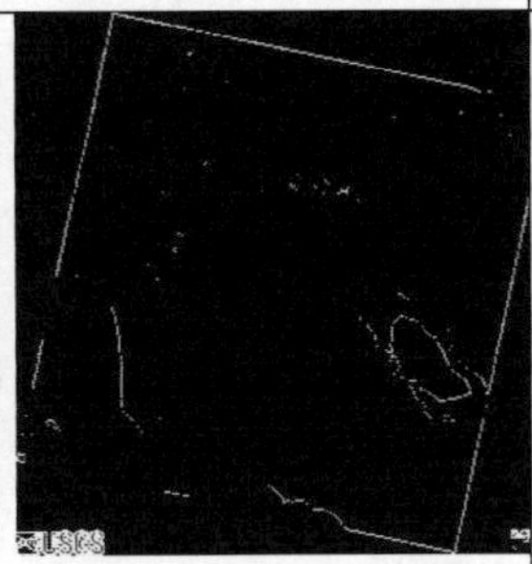	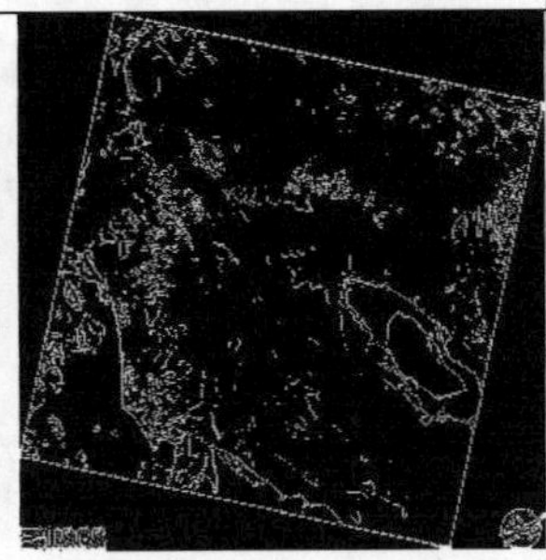	
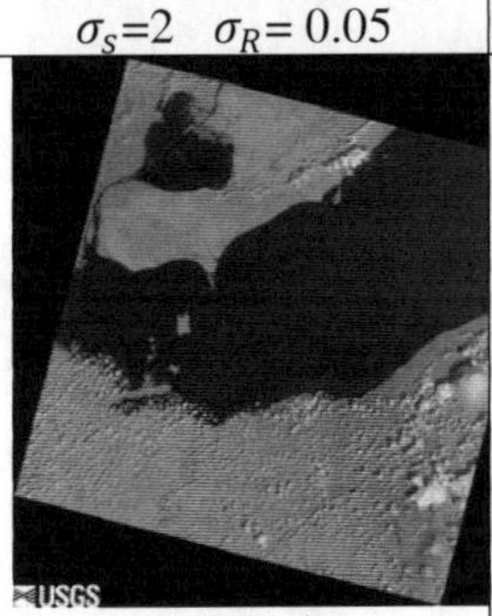 Erie Lake.jpg σ_S=2.5 σ_R= 0.05		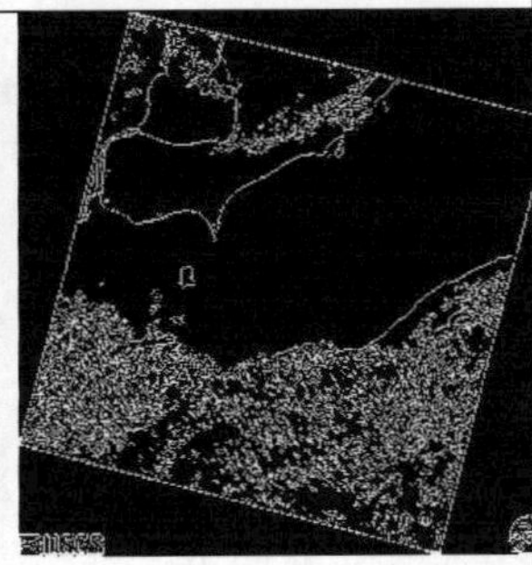	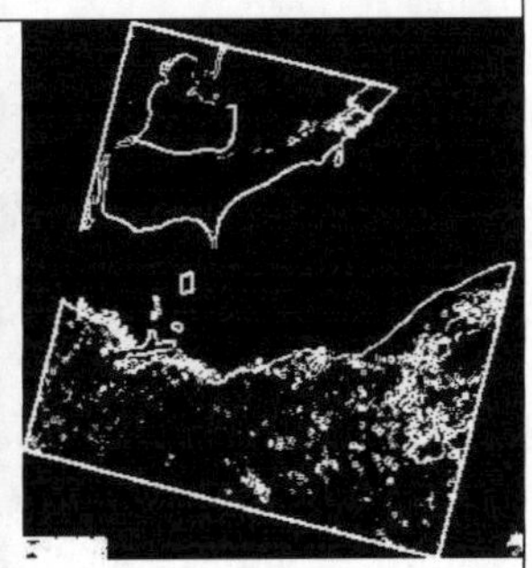

Table 3. MSE and PSNR result for 300*300 pixels satellite image with Gaussian Filter

Methods	**Aral.jpg** σ= 1.5	**Toba.jpg** σ= 2	**Erie.jpg** σ= 2.5	**Nyasa.jpg** σ= 3	**Malawi.jpg** σ= 4.5	**Tanganyika. jpg**σ= 6
Roberts	**MSE :** 10510.5 **PSNR:** 7.91457	**MSE :** 8108.37 **PSNR:** 9.04147	**MSE :** 6922.77 **PSNR:** 9.726	**MSE :** 5902.17 **PSNR:** 10.4207	**MSE :** 6406.89 **PSNR:** 10.0643	**MSE :** 6716.1 **PSNR:** 9.85963
Canny	**MSE :** 10196.8 **PSNR:** 8.04618	**MSE :** 7865.46 **PSNR:** 9.17356	**MSE :** 6444.9 **PSNR:** 10.0386	**MSE :** 5541.44 **PSNR:** 10.6946	**MSE :** 6101.34 **PSNR:** 10.2766	**MSE :** 6316.61 **PSNR:** 10.126
Frei Chen	**MSE :** 10508.3 **PSNR:** 7.91546	**MSE :** 8107 **PSNR:** 9.0422	**MSE :** 6920.95 **PSNR:** 9.72915	**MSE :** 5891.82 **PSNR:** 10.4283	**MSE :** 6400.17 **PSNR:** 10.0689	**MSE :** 6704.71 **PSNR:** 9.867

Table 4. MSE and PSNR result for 300*300 pixels satellite image with Bilateral Filter

Methods	Aral.jpg σ_S== 1.5 σ_R= 0.05	Toba.jpg σ_S== 2 σ_R= 0.05	Erie.jpg σ_S== 2.5 σ_R= 0.05	Nyasa.jpg σ_S== 3 σ_R= 0.05	Malawi.jpg σ_S== 4.5 σ_R= 0.05	Tanganyika. jpgσ_S== 6 σ_R= 0.05
Roberts	**MSE :** 10512.3 **PSNR:** 7.91384	**MSE :** 8108.6 **PSNR:** 9.04135	**MSE:** 6918.15 **PSNR:** 9.7309	**MSE:** 5701.52 **PSNR:** 10.5709	**MSE:** 6203.37 **PSNR:** 10.2045	**MSE:** 6502.57 **PSNR:** 9.99995
Canny	**MSE :** 10255.4 **PSNR:** 8.02128	**MSE:** 7850.84 **PSNR:** 9.18164	**MSE:** 6146.8 **PSNR:** 10.2447	**MSE:** 5462.4 **PSNR:** 10.757	**MSE:** 5994.87 **PSNR:** 10.353	**MSE:** 6250.33 **PSNR:** 10.1718
Frei Chen	**MSE :** 10505.6 **PSNR:** 7.91658	**MSE:** 8104.82 **PSNR:** 9.04337	**MSE:** 6910.57 **PSNR:** 9.73566	**MSE:** 5699.88 **PSNR:** 10.5721	**MSE:** 6202.02 **PSNR:** 10.2055	**MSE:** 6501.34 **PSNR:** 10.0008

Bilateral Filter is applied to improve the limitation of Gaussian Filter. The evaluation results of the proposed technique show that the lowest MSE value and the highest PSNR are owned by Canny Detector after using Bilateral Filter. Based on experimental results, Bilateral Filters can efficiently detect the edge of the image.

8. Conclusion and Future Work
In this research, the performance comparison between Gaussian and Bilateral Filters on image edge detection indicates that edge detection using Bilateral Filter is better at retaining the edge information of the image. Improved edge detection on the image can also be seen from the lowest value of MSE and the highest PSNR value with Bilateral Filter.

9. References
[1] Lalwani S and Ansari U 2012 *Enhancing the resolution of Satellite Image by Denoising Deblurring and Contrast Setting* **10** 809
[2] Bin L and Yeganeh M S 2012 *Comparison for Image Edge Detection Algorithms* **2** pp 01-04
[3] Tao Y and Yue-hong Q 2015 *Improvement and Implementation for Canny Edge Detection Algorithm* **9631**
[4] Srisuk S 2014 *Bilateral Filtering as a Tool for Image Smoothing with Edge Preserving Properties*
[5] Xiangguang Leng *et al* 2016 *Hybrid Bilateral Filtering Algorithm based on Edge Detection* **10** pp 809-816
[6] Tomasi C and Manduchi R 1998 *Bilateral filtering for gray and color images* pp 839-846.
[7] Sharma A and Khunteta A 2016 *Satelite Image Enhancement using Discreate Wavelet Transform Singular Value Decomposition and its Noise Performance Analysis* p 4639
[8] Haritha Ch *et al* 2017 *Combined Gaussian Filtering, DWT and Morphological Approach for Satelite Image Enhancement* **06**
[9] Solomon C and Breckon T 2011 *Fundamentals of Digital Image Processing A Practical Approach with Examples in Matlab*
[10] Jose A and Sekhar C 2013 *Bilateral Edge Detector*
[11] Singh S and Singh R 2015 *Comparison of Various Edge Detection Techniques*
[12] Thepade S D and Shinde Y D 2015 *Improvisation of Content Based Image retrieval using Color Edge Detection with various Gradient Filters and Slope Magnitude Method*
[13] Mishra A, Mahapatra S, and Banerjee S 2017 *Modified Frei-Chen Operator based Infrared and Visible Sensor Image Fusion for Real-Time Applications* **17** p 4639

[14] Jadhav B D and Patil P M 2015 *An Effective method for Satelite Image Enhancement* pp 1171-1175
[15] USGS landsat8 dataset downloaded [online].https://earthexplorer.usgs.gov/

IOP Conf. Series: Materials Science and Engineering **308** (2018) 012053 doi:10.1088/1757-899X/308/1/012053

Method of calculation overall equipment effectiveness in fertilizer factory

I Siregar[1*], M A Muchtar[2], R F Rahmat[2], U Andayani [2], T H Nasution[3] and R M Sari[1]

[1]Department of Industrial Engineering, Universitas Sumatera Utara, Almamater Rd., University of Sumatera Utara, Medan 20155
[2]Department of Information Technology, Universitas Sumatera Utara, Almamater Rd., University of Sumatera Utara, Medan 20155
[3]Department of Electrical Engineering, Universitas Sumatera Utara, Almamater Rd., University of Sumatera Utara, Medan 20155

[*]Email: ikhsan.siregar@usu.ac.id

Abstract. This research was conducted at a fertilizer company in Sumatra, where companies that produce fertilizers in large quantities to meet the needs of consumers. This company cannot be separated from issues related to the performance/effectiveness of the machinery and equipment. It can be seen from the engine that runs every day without a break resulted in not all of the quality of products in accordance with the quality standards set by the company. Therefore, to measure and improve the performance of the machine in the unit Plant Urea-1 as a whole then used method of Overall Equipment Effectiveness (OEE), which is one important element in the Total Productive Maintenance (TPM) to measure the effectiveness of the machine so that it can take measures to maintain that level. In July, August and September OEE values above the standard set at 85%. Meanwhile, in October, November and December have not reached the standard OEE values. The low value of OEE due to lack of time availability of machines for the production shut down due to the occurrence of the engine long enough so that the availability of reduced production time..

1. Introduction

A process on the production floor can be stopped due to a problem in the production machine / equipment, for example the engine stops unexpectedly, the length of setup time, the decline in the engine production speed, the engine produces defective products and damage to the engine. This causes losses to the company because in addition to lowering the efficiency and effectiveness of machinery / equipment and productivity companies also resulted in the cost to be incurred due to such damage.
Preventive maintenance is very important, as was done in previous research, especially for machines that have been used in the past 10 years [1]. Other studies have suggested that using preventive maintenance can reduce maintenance costs [2]. Even in previous research states that by doing preventive mintenance can improve availability and reability [3].
In this study, will be measured the value of OEE, where previous research states that OEE can be useful for the industrial world [4]. Even OEE can also be used for use in software and hardware [5].
The fertilizer plant where this research is conducted is a company that produces fertilizer in large quantities to meet the needs of consumers. The company is inseparable from problems related to the

IOP Conf. Series: Materials Science and Engineering **308** (2018) 012053 doi:10.1088/1757-899X/308/1/012053

performance / effectiveness of machinery and equipment. This can be seen from the machine that runs every day without a break resulting in not all product quality in accordance with the quality standards set by the company. Therefore, to measure and improve the performance level of the engine in the Urea-1 Plant unit as a whole system, the Overall Equipment Effectiveness (OEE) method is used which is one of the important elements in the TPM to measure the effectiveness of the machine so that it can take steps to maintain That level.

The problem to be solved in this research is how to increase the value of overall equipment effectiveness (OEE) in Plant Urea-1. While the goal to be achieved from this research is to calculate the value of availability, performance, quality and OEE on Urea-1 Plant Plant as a whole system.

2. Research method

The study was conducted in the Department of Reliability in Urea plant area.

Calculation of OEE

According to Nakajima (1988), OEE calculations are performed with several steps, i.e.:

1. The availability calculation, i.e. the ratio indicating the utilization of the time available for the operation of the machine or equipment. The required data is downtime and loading time, using the following calculation formula:

$$Availability = \frac{Operation Time}{Loading\ Time} \times 100\%$$

$$= \frac{Loading\ Time - Total Downtime}{Loading\ Time} \times 100\%$$

2. Performance calculation, which is the ratio that indicates the ability of the equipment to produce the goods. The required data are total production, cycle time, and operation time, using the calculation formula is as follows:

$$Performance\ Efficiency = \frac{processed amount \times Ideal Cycle Time}{Operation Time} \times 100\%$$

3. Calculation of quality, namely the ratio that shows the ability of the equipment to produce goods in accordance with the standards specified. The required data is the total production and the number of defects, using the calculation formula is as follows:

$$Rate\ of\ Quality\ Product = \frac{Processed\ Amount - Defect\ Amount}{Processed\ Amount} x100\%$$

4. Calculation of overall equipment effectiveness (OEE), obtained from the multiplication of the three categories. So that, the formula used for calculation is as follows:

OEE = *Availability* x *Performance x Quality*

3. Result and Analysis

Calculation of OEE Value

The calculation is done on Plant Urea-1 where the Plant is a Plant that produce urea fertilizer and always operate so often experience constraints during the production process. Prior to performing OEE calculations, the calculations were made on its constituent categories, namely availability, performance and quality. The data used are data in 2016 from July to December obtained based on historical data from the Reliability Department.

Table 1. Availability Ratio Calculation from July 2016 until December 2016

Month	Loading Time (Hours)	Downtime (Hours)	Operation Time (Hours)	Availability Ratio (%)
July 2016	661,10	22,75	638,35	96,56
August 2016	645,50	9	636,5	98,61
September 2016	715,90	13,25	702,65	98,15
October 2016	626,20	20,7	605,5	96,69

IOP Conf. Series: Materials Science and Engineering **308** (2018) 012053 doi:10.1088/1757-899X/308/1/012053

November 2016	720	-	720	100
December 2016	594,17	22,25	571,92	96,26

Table 2. Calculation of Performance Efficiency from July 2016 until December 2016

Month	Processed Amount (Ton)	Ideal Cycle Time (Hours/Ton)	Operation Time (Hours)	Performance Efficiency (%)
July 2016	69.359,24	0,8151	638,35	88,56
August 2016	80.766,00	0,6844	636,5	86,84
September 2016	88.490,10	0,7905	702,65	99,55
October 2016	96.180,60	0,5290	605,5	84,03
November 2016	71.988,86	0,6995	720	69,93
December 2016	64.768,94	0,7072	571,92	80,10

Table 3. OEE Calculation Result from July 2016 until December 2016

Bulan	*Availability Ratio* (%)	*Performance Efficiency* (%)	*Rate of Quality Product* (%)	OEE (%)
July 2016	96,56	88,56	100,00	85,52
August 2016	98,61	86,84	100,00	85,63
September 2016	98,15	99,55	100,00	97,71
October 2016	96,69	84,03	100,00	81,25
November 2016	100	69,93	100,00	69,93
December 2016	96,26	80,10	100,00	77,10

Table 1 shows the recapitulation of the calculation results for the availability category of Urea-1 Plant. The data in Table 1 shows the achievement of availability ratio on machines in Urea-1 Plant unit in July 2016 until December 2016 is above the predetermined standard of 90%. This indicates the availability of machines for production has been used effectively.

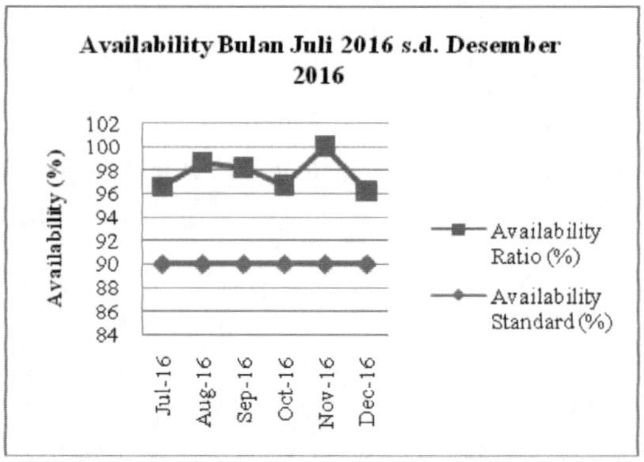

Figure 1. Availability Ratio Achievement from July 2016 until December 2016

Table 2 shows the recapitulation of performance performance calculation results. The data in Table 2 shows only in September that has reached an average of 95%, whereas in July, August, October, November and December have not reached the standard. This shows that the performance of the engine in the Urea-1 Plant unit has not been effective.

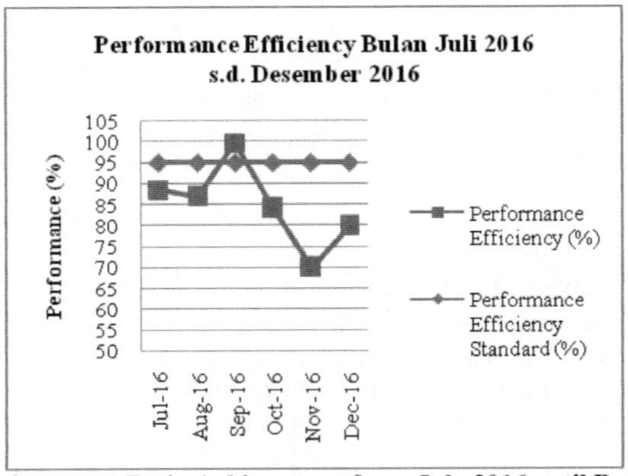

Figure 2. Performance Ratio Achievement from July 2016 until December 2016

Rate of Quality machine in Urea-1 Plant unit during July 2016 until December 2016 has reached the standard set that is 99%. This is because there are no defective urea fertilizer production (waste amount).

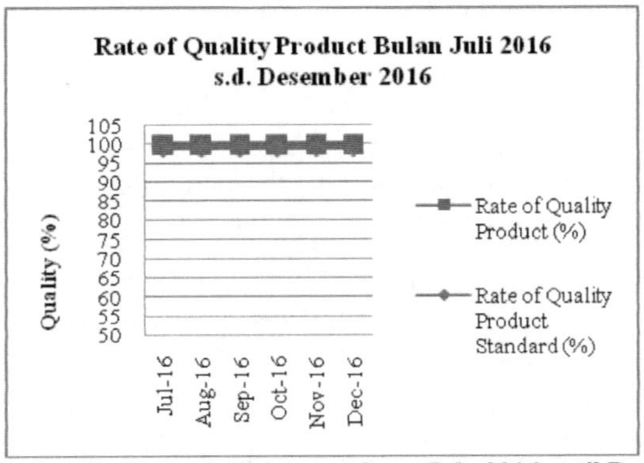

Figure 3. Quality Ratio Achievement from July 2016 until December 2016

Table 3 shows the recapitulation of OEE values and shows that in July, August and September the OEE score is above the prescribed standard of 85%. While in October, November and December the value of OEE has not reached the standard. The low value of OEE is due to the low Performance Efficiency value

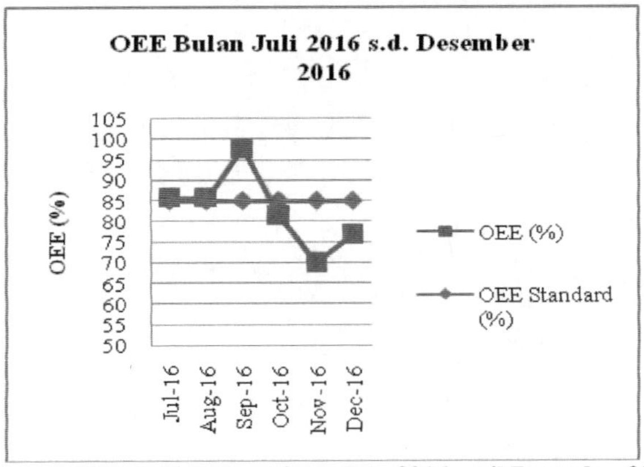

Figure 4. OEE Achievement from July 2016 until December 2016

The low value of OEE is due to the lack of machine availability time for the production due to the shut down on the machine long enough so that the availability of production time is reduced.

The company has indeed set preventive maintenance but only based on the economic life of the constituent components of the machine so that the engine maintenance is not regular and the machine damaged fluently. The proposed increase in the effectiveness of the machine in the Urea-1 Plant Unit can be done by increasing Preventive Maintenance. Efforts made Preventive Maintenance is to inspect the machine periodically and perform timely maintenance and use of quality materials so that the components of the engine in the Plant Urea-1 unit is not easily damaged. This can prevent breakdowns and delays on machines in the Urea-1 Plant unit, thus increasing the engine's effectiveness.

4. Conclusion and Discussion

The conclusions that can be obtained from the analysis and evaluation stage are as follows:

1. The value of OEE obtained during July, August and September 2016 has reached the expected standard, indicating that the machine is being used effectively.
2. Proposed in improving the effectiveness of the machine on the Urea-1 Plant Unit can be done with the increase of Preventive Maintenance that is to inspect the machine periodically and perform timely maintenance.

References

[1] Fujishima, M., Mori, M., Nishimura, K., Takayama, M., & Kato, Y. (2017). Development of Sensing Interface for Preventive Maintenance of Machine Tools. Procedia CIRP, (61), 796-799.

[2] Daneshkhah, A., Stocks, N. G., & Jeffrey, P. (2017). Probabilistic sensitivity analysis of optimised preventive maintenance strategies for deteriorating infrastructure assets. Reliability Engineering & System Safety, 163, 33-45.

[3] Mwanza, B. G., & Mbohwa, C. (2015). An assessment of the effectiveness of equipment maintenance practices in public hospitals. Procedia Manufacturing, 4, 307-314.

[4] Relkar, A. S., & Nandurkar, K. N. (2012). Optimizing & Analysing Overall Equipment Effectiveness (OEE) Through Design of Experiments (DOE). Procedia Engineering, 38, 2973-2980.

[5] Singh, R., Shah, D. B., Gohil, A. M., & Shah, M. H. (2013). Overall Equipment Effectiveness (OEE) calculation-Automation through hardware & software development. Procedia Engineering, 51, 579-584.